MOLECULAR SPECTROSCOPY

MOLECULAR SPECTROSCOPY

IRA N. LEVINE

Brooklyn College
City University of New York

A WILEY-INTERSCIENCE PUBLICATION

JOHN WILEY & SONS
New York • Chichester • Brisbane • Toronto • Singapore

Library of Congress Cataloging in Publication Data

Levine, Ira N date
Molecular spectroscopy.

"A Wiley-Interscience publication."

Rev. version of the book previously published as v. 2 of the author's Quantum chemistry.
Bibliography: p.
Includes index.
1. Molecular spectra. I. Title.

QC454.M6L48 541'.28 74-30477
ISBN 0-471-53128-6

Printed in the United States of America

20 19 18 17 16 15 14

Printed and bound by Quinn - Woodbine, Inc.

PREFACE

This book is intended for graduate and advanced undergraduate courses in molecular spectroscopy. It presupposes some knowledge of quantum mechanics, such as is gained in a one-semester introductory quantum-chemistry course.

This text is a much-revised version of the book previously published as Volume II of my book *Quantum Chemistry*. The first chapter of the present book reviews quantum mechanics and atomic and molecular electronic structure, so that this book can be used completely independently of any other book.

My aim in writing this book has been to give students a solid understanding of the theoretical basis of spectroscopic methods and an appreciation of how these methods provide information on molecular structure. Brief descriptions of experimental spectroscopic procedures are also included.

I have tried to present material as clearly and logically as possible, giving sufficient detail in the derivations to make them easy to follow. The text takes into account the limited mathematics and physics background of the average chemistry student. However, in no sense is the treatment superficial or "watered down."

The major changes in the new edition are as follows: There are three new chapters. Chapter 1 is a review and summary of aspects of quantum mechanics and electronic structure relevant to molecular spectroscopy. This chapter replaces the chapter on electronic structure of polyatomic molecules that was repeated from Volume I of *Quantum Chemistry*. Chapter 2 is a substantially expanded presentation of matrices. Previously, matrices were covered in the last chapter. The placement of matrices early in the book allows their use throughout the book; in particular, the very tedious and involved treatment of normal vibrations has been replaced by a simpler and clearer treatment using matrices. Chapter 7 covers molecular electronic spectroscopy, and contains two new sections, one on electronic spectra of polyatomic molecules, and one on photoelectron spectroscopy, together with the section on electronic spectra of diatomic molecules from the previous edition. In addition to the new material on matrices, electronic spectra of polyatomic molecules, and photoelectron spectroscopy, the following topics have been added: line widths and shapes (Section 3.5), lasers (Section 3.6), spin relaxation (Section 8.8), Einstein coefficients, oscillator strengths, Fourier-transform NMR, calculation of excited-state wave functions, Koopmans' theorem, *ab initio* and semiempirical calculations of ionization potentials, force constants, and magnetic-resonance parameters.

I have left group theory as the last chapter. If it is desired to cover group theory earlier in the course, Sections 9.1–9.8 of the group-theory chapter can be studied prior to Chapter 6 on molecular vibrations, and Sections 9.9–9.12 read in conjunction with Chapter 6.

Professors James Bolton and Harry King reviewed the manuscript of the previous edition of this book; I have belatedly incorporated many of their excellent suggestions in this edition. I am also grateful to Professors Frank Meeks and Kenneth Sando who reviewed an earlier version of the present edition. Professor Vicki Chuckrow provided helpful advice on some mathematical points. The help of Dr. Theodore Hoffman of Wiley-Interscience is gratefully acknowledged.

Any suggestions that readers may have for improvement of the book will be welcomed.

IRA N. LEVINE

Brooklyn, New York
September 1974

CONTENTS

CHAPTER 1 QUANTUM MECHANICS AND ELECTRONIC STRUCTURE **1**

1.1 Quantum Mechanics **1**
1.2 Mathematics Review **11**
1.3 Units **23**
1.4 The Particle in a Box **23**
1.5 The Harmonic Oscillator **25**
1.6 Orbital Angular Momentum **27**
1.7 Angular-Momentum Ladder Operators **30**
1.8 Parity **32**
1.9 The Variation Method **33**
1.10 Perturbation Theory **35**
1.11 The Central-Force Problem **38**
1.12 The Two-Particle Problem **39**
1.13 The Two-Particle Rigid Rotor **40**
1.14 The Hydrogen Atom **41**
1.15 Spin **42**
1.16 The Pauli Principle **45**
1.17 Many-Electron Atoms **47**
1.18 Symmetry Point Groups **53**
1.19 Molecular Electronic Structure **56**
1.20 The Hartree–Fock Method **63**
1.21 Semiempirical Methods **70**

CHAPTER 2 MATRICES 78

2.1 Matrices 78
2.2 Eigenvalues and Eigenvectors 88
2.3 Matrices and Quantum Mechanics 93

CHAPTER 3 TIME-DEPENDENT STATES AND SPECTROSCOPY 110

3.1 Time-Dependent Perturbation Theory 110
3.2 Absorption and Emission of Radiation 114
3.3 Selection Rules 123
3.4 Spectroscopy 129
3.5 Line Shapes and Widths 133
3.6 Lasers 135

CHAPTER 4 VIBRATION AND ROTATION OF DIATOMIC MOLECULES 142

4.1 Nuclear Motion in Diatomic Molecules 142
4.2 Anharmonicity, Vibration–Rotation Interaction, and Centrifugal Distortion 149
4.3 Potential-Energy Functions for Diatomic Molecules 160
4.4 Selection Rules for Rotational and Vibrational Transitions 162
4.5 Rotational Spectra of Diatomic Molecules 165
4.6 Vibration–Rotation Spectra of Diatomic Molecules 168
4.7 Parity of Diatomic-Molecule Wave Functions 175
4.8 Nuclear Spin and the Pauli Principle 178
4.9 Ortho and Para Modifications of Homonuclear Diatomic Molecules 183
4.10 The Raman Effect 186
4.11 Rotational Energies of Non-$^1\Sigma$ Diatomic Molecules 188

CHAPTER 5 ROTATION OF POLYATOMIC MOLECULES 195

5.1 Nuclear Motion of Polyatomic Molecules 195
5.2 Classical Mechanics of a Rigid Rotor 197
5.3 The Rotational Hamiltonian Operator 203
5.4 The Spherical Top 209
5.5 The Symmetric Top 210
5.6 The Asymmetric Top 213
5.7 Pure-Rotation Spectra of Polyatomic Molecules 216
5.8 Microwave Spectroscopy 219

CHAPTER 6 VIBRATION OF POLYATOMIC MOLECULES 235

6.1 Introduction 235
6.2 Classical Mechanics of Vibration 235
6.3 Symmetry and Normal Vibrations 246
6.4 Quantum Mechanics of Vibration 249
6.5 Selection Rules for Vibration–Rotation and Pure-Rotation Transitions 255
6.6 Infrared Spectroscopy 260
6.7 Raman Spectroscopy 269
6.8 Degenerate and Near-Degenerate Vibrational Levels 271
6.9 Parity of Polyatomic-Molecule Wave Functions 279
6.10 Nuclear Spin and the Pauli Principle 283

CHAPTER 7 ELECTRONIC SPECTROSCOPY 296

7.1 Electronic Spectroscopy 296
7.2 Electronic Spectra of Diatomic Molecules 298
7.3 Electronic Spectra of Polyatomic Molecules 306
7.4 Photoelectron Spectroscopy 314

CHAPTER 8 MAGNETIC RESONANCE SPECTROSCOPY 323

8.1 Magnetic Resonance 323
8.2 Nuclear Magnetic Moments 323
8.3 Nuclear Magnetic Resonance 325
8.4 NMR Chemical Shifts 331
8.5 NMR Spin–Spin Splittings 337
8.6 NMR Spectra of Nuclei Other Than Protons 356
8.7 Origin of NMR Spin–Spin Coupling 357
8.8 Spin Relaxation 361
8.9 NMR and Rate Processes 364
8.10 Nuclear Quadrupole Moments 365
8.11 Electron Spin Resonance 366

CHAPTER 9 GROUP THEORY 387

9.1 Groups 387
9.2 Multiplication Tables 389
9.3 Coordinate Transformations and Matrices 394
9.4 Representations 399
9.5 Character Tables 402
9.6 Basis Functions 409
9.7 Group Theory and Quantum Mechanics 412
9.8 Group Theory and Electronic Structure 417
9.9 Group Theory and Molecular Vibrations 427
9.10 Group Theory and Molecular Rotation 439
9.11 The Direct-Product Representation 440
9.12 Character Tables of Some Point Groups 458

APPENDIX 467

BIBLIOGRAPHY 471

ANSWERS TO SELECTED PROBLEMS 475

INDEX 477

MOLECULAR SPECTROSCOPY

CHAPTER 1

QUANTUM MECHANICS AND ELECTRONIC STRUCTURE

1.1 QUANTUM MECHANICS

Before beginning our study of molecular spectroscopy, we shall in this chapter review and summarize those aspects of quantum mechanics and atomic and molecular electronic structure that are relevant to molecular spectroscopy. Since it is assumed that the reader has had a previous course in quantum chemistry, we shall not give detailed derivations in this chapter. If the reader is puzzled by any of the discussions in Chapter 1, he can consult *Levine, Gatz, Pilar*, or *Pauling and Wilson* for fuller treatments. (References with the author's name italicized are to books listed in the Bibliography.) Rather than going through Chapter 1 in detail, it might be a good idea to simply skim it, and refer back to it as needed when reading later chapters

The state of an n-particle quantum-mechanical system is described by its *state function* $\Psi(q,t)$, where q symbolizes the $3n$ spatial and n spin coordinates of the particles, and t is the time. The function Ψ must be single-valued and continuous; in addition, for bound states Ψ must be quadratically integrable—by this we mean that $\int \Psi^* \Psi \, d\tau$ is finite, where $\int d\tau$ denotes integration over all space, including a summation over spin variables. (Although $\int \Psi^* \Psi \, d\tau$ looks like an indefinite integral, it is understood to be a definite integral.) A function that satisfies the preceding conditions is said to be *well behaved*.

To each physical property of a quantum-mechanical system, there corresponds a quantum-mechanical *operator*. The operator is formed by taking the classical-mechanical expression for the property and replacing each Cartesian coordinate x_i by the operator multiplication by x_i and each linear momentum component p_{x_i} by the operator $(\hbar/i)(\partial/\partial x_i)$. Using a circumflex to denote an operator, we have

$$\hat{x}_i = x_i \cdot \tag{1.1}$$

$$\hat{p}_{x_i} = \frac{\hbar}{i}\frac{\partial}{\partial x_i} \tag{1.2}$$

where

$$\hbar \equiv h/2\pi, \qquad i = \sqrt{-1}$$

h being Planck's constant.

If $\hat{A}$ is the operator corresponding to the physical property A, then any measurement of the property A in any quantum-mechanical system must give a result that is one of the eigenvalues of the operator $\hat{A}$. The *eigenvalues* a_i and the *eigenfunctions* φ_i of $\hat{A}$ are defined as those numbers and functions that satisfy

$$\hat{A}\varphi_i = a_i\varphi_i \tag{1.3}$$

If we take a large number of identical systems, each with the same state function Ψ, and measure A in each system, we shall in general get different results in different systems. (This is one of the striking differences between classical and quantum mechanics.) The *average value* of A, denoted by $\langle A \rangle$, is postulated to be given by

$$\langle A \rangle = \int \Psi^* \hat{A} \Psi \, d\tau \tag{1.4}$$

where the integration is over all space and Ψ is normalized. [To find $\langle A \rangle$ experimentally, we average the results of separate observations of A in a very large number N of identical systems, each in the same state Ψ: $\langle A \rangle = \Sigma_i a_i / N$, where the sum is over the observed values $a_1, a_2, \ldots, a_N$. This last equation is equivalent to

$$\langle A \rangle = \sum_a P(a) a$$

where the sum is over all possible values of A, and where $P(a)$ is the probability of observing the value a for the property A.]

The way in which the state function Ψ changes with time is postulated to be given by the *time-dependent Schrödinger equation*:

$$-\frac{\hbar}{i}\frac{\partial\Psi}{\partial t}=\hat{H}\Psi \tag{1.5}$$

where $\hat{H}$ is the *Hamiltonian* (or energy) *operator* for the system; $\hat{H}$ is obtained by converting the classical-mechanical expression for the system's energy in terms of Cartesian coordinates and linear momenta to an operator by the procedure described above. Thus for a one-particle, three-dimensional system, the classical-mechanical Hamiltonian function is the sum of the kinetic energy T and the potential energy V:

$$H=T+V=\left(p_x^2+p_y^2+p_z^2\right)/2m+V(x,y,z,t) \tag{1.6}$$

where p_x,p_y,p_z are the linear-momentum components [Equation (1.144)] and the potential energy is defined by

$$-\partial V/\partial x=F_x, \qquad -\partial V/\partial y=F_y, \qquad -\partial V/\partial z=F_z \tag{1.7}$$

where F_x, F_y, F_z are the components of the force on the particle. Replacement of classical quantities by operators gives as the Hamiltonian operator

$$\hat{H}=\hat{T}+\hat{V}=-\frac{\hbar^2}{2m}\left(\frac{\partial^2}{\partial x^2}+\frac{\partial^2}{\partial y^2}+\frac{\partial^2}{\partial z^2}\right)+V(x,y,z,t)$$

$$\hat{H}=-\frac{\hbar^2}{2m}\nabla^2+V(x,y,z,t) \tag{1.8}$$

where $\nabla^2\equiv\partial^2/\partial x^2+\partial^2/\partial y^2+\partial^2/\partial z^2$ is the *Laplacian* operator. For an n-particle system, the Hamiltonian is

$$\hat{H}=-\sum_{i=1}^{n}\frac{\hbar^2}{2m_i}\nabla_i^2+V(x_1,y_1,z_1,\ldots,x_n,y_n,z_n,t) \tag{1.9}$$

$$\nabla_i^2\equiv\partial^2/\partial x_i^2+\partial^2/\partial y_i^2+\partial^2/\partial z_i^2 \tag{1.10}$$

Integration of (1.5) with respect to time introduces one arbitrary constant. If we know Ψ at the present time t_0, we can evaluate this constant and thereby determine Ψ at any future time.

Quantum-mechanical operators that correspond to physical properties must be linear and Hermitian. A *linear* operator $\hat{A}$ satisfies

$$\hat{A}\,(f+g)=\hat{A}f+\hat{A}g \tag{1.11}$$

$$\hat{A}\,(cf)=c\hat{A}f \tag{1.12}$$

where f and g are any functions and c is any constant. A *Hermitian* operator is a linear operator that satisfies

$$\int f^* \hat{A} g \, d\tau = \int g (\hat{A} f)^* \, d\tau \tag{1.13}$$

for all well-behaved functions f and g, where the integration is over all space.

The eigenfunctions φ_i of a Hermitian operator form a *complete set*. By this we mean that any well-behaved function f that satisfies the same boundary conditions as the φ_i's can be expanded as

$$f = \sum_i c_i \varphi_i \tag{1.14}$$

where the expansion coefficients c_i are constants.

The preceding statements together with the postulates of spin and the Pauli principle (Sections 1.15 and 1.16) can be taken as the postulates of quantum mechanics.

If the potential energy V in the Hamiltonian is independent of time, then there exist solutions of (1.5) that have the form

$$\Psi(q,t) = f(t)\psi(q) \tag{1.15}$$

where f is a function of time only and ψ is a function of the coordinates only. To verify (1.15), we substitute it into (1.5) to get $-(\hbar/i)f'(t)\psi(q) = f(t)\hat{H}\psi(q)$, since $\hat{H}$ does not affect $f(t)$. Division by $f\psi$ gives

$$-\frac{\hbar}{i}\frac{f'(t)}{f(t)} = \frac{\hat{H}\psi(q)}{\psi(q)} \tag{1.16}$$

Since $\hat{H}$ is independent of t, the right side of (1.16) is a function of the coordinates only, while the left side is a function of t only. By the usual separation-of-variables argument (Section 1.2), each side of (1.16) must equal a constant, which we call E. Therefore, $-(\hbar/i)(df/dt)/f = E$, which integrates to $f(t) = A\exp(-iEt/\hbar)$, where A is an arbitrary constant. Since A can be absorbed into the function $\psi(q)$ in (1.15), we drop it to give

$$f = e^{-iEt/\hbar} \tag{1.17}$$

Also we have $\hat{H}\psi/\psi = E$ or

$$\hat{H}\psi = E\psi \tag{1.18}$$

which is the *time-independent Schrödinger equation* or, simply, the *Schrödinger equation*. Since the Hamiltonian operator $\hat{H}$ is formed from the classical-mechanical expression for the energy, we have $\hat{E}=\hat{H}$. The numbers E in (1.18) are the eigenvalues of the energy operator, and by the postulate preceding (1.3), the numbers E that satisfy (1.18) are thus the possible energies of the system. States of the form (1.15) are called *stationary states*; for a stationary state

$$\Psi(q,t)=e^{-iEt/\hbar}\psi(q) \tag{1.19}$$

where E is the energy of the state and ψ is the *wave function*.

Common notations for the integrals that occur in quantum mechanics are

$$\int \varphi_i^* \hat{A}\varphi_j\,d\tau \equiv \langle\varphi_i|\hat{A}|\varphi_j\rangle \equiv \langle i|\hat{A}|j\rangle \equiv A_{ij} \tag{1.20}$$

where φ_i and φ_j are functions, $\hat{A}$ is an operator, and the integration is over all space. The integral in (1.20) is often called a *matrix element* of $\hat{A}$. A notation used when no operator occurs in the integral is

$$\int \varphi_i^* \varphi_j\,d\tau \equiv \langle\varphi_i|\varphi_j\rangle \equiv \langle i|j\rangle \tag{1.21}$$

Note that

$$\langle i|j\rangle=\langle j|i\rangle^* \tag{1.22}$$

The Hermitian property (1.13) expressed in bracket notation and in matrix-element notation is

$$\langle\varphi_i|\hat{A}|\varphi_j\rangle=\langle\varphi_j|\hat{A}|\varphi_i\rangle^* \tag{1.23}$$

$$A_{ij}=(A_{ji})^* \tag{1.24}$$

Ket notation is sometimes used for functions in quantum mechanics. In this notation, the function f is denoted by the symbol $|f\rangle$: $f=|f\rangle$. Ket notation is convenient for denoting eigenfunctions by listing their eigenvalues. Thus $|nlm\rangle$ denotes the hydrogen-atom stationary-state wave function with quantum numbers n, l, and m.

The following properties of Hermitian operators follow from the definition (1.23). The eigenvalues of a Hermitian operator are real. Two eigenfunctions of a Hermitian operator that correspond to different eigenvalues are orthogonal; that is, if $\hat{A}$ is Hermitian and $\hat{A}\varphi_1=a_1\varphi_1$, $\hat{A}\varphi_2=a_2\varphi_2$, with $a_1\neq a_2$, then

$$\langle\varphi_1|\varphi_2\rangle=\int \varphi_1^*\varphi_2\,d\tau=0 \tag{1.25}$$

If n linearly independent eigenfunctions of an operator $\hat{A}$ correspond to the same eigenvalue, we say that eigenvalue is *n-fold degenerate*. (Linear independence is defined in Section 1.2.) For example, for the particle in a cubic box (Section 1.4), the three states $|n_x n_y n_z\rangle = |112\rangle$, $|121\rangle$, and $|211\rangle$ have the same energy, and this energy level is threefold degenerate. Because $\hat{A}$ is linear, it follows that any linear combination of eigenfunctions of a degenerate level is an eigenfunction of $\hat{A}$ with the same eigenvalue as each member of the linear combination.

Although (1.25) need not hold when the eigenfunctions φ_1 and φ_2 correspond to the same eigenvalue of $\hat{A}$ (i.e., when $a_1 = a_2$), we can form linear combinations of φ_1 and φ_2 that will still be eigenfunctions of $\hat{A}$ with the same eigenvalue, but that will be orthogonal. One way (Schmidt orthogonalization) to take the linear combinations is

$$f_1 = \varphi_1, \qquad f_2 = \varphi_2 - \frac{\langle \varphi_1 | \varphi_2 \rangle \varphi_1}{\langle \varphi_1 | \varphi_1 \rangle} \tag{1.26}$$

The reader can verify that f_1 and f_2 are orthogonal. We shall always assume (unless stated otherwise) that the eigenfunctions of degenerate levels have been chosen to be orthogonal.

Except for continuum states (these are states that correspond to eigenvalues that take on a continuous range of values, rather than discrete values), we can always normalize the eigenfunctions by multiplying them by normalization constants, and we shall generally assume this has been done:

$$\langle \varphi_i | \varphi_i \rangle = 1 \tag{1.27}$$

Thus the eigenfunctions of a Hermitian operator form an *orthonormal* (orthogonal and normalized) set:

$$\langle \varphi_i | \varphi_j \rangle = \delta_{ij} \tag{1.28}$$

where the *Kronecker delta* is defined by

$$\delta_{ij} = 1 \quad \text{if } i = j; \qquad \delta_{ij} = 0 \quad \text{if } i \neq j \tag{1.29}$$

Let φ_i symbolize a complete, orthonormal set of functions (e.g., they may be the eigenfunctions of some Hermitian operator). Expansion of an arbitrary well-behaved function f in terms of the φ_i's gives

$$f = \sum_i c_i \varphi_i \tag{1.30}$$

Multiplication by φ_j^* followed by integration over all space gives

$$\int \varphi_j^* f \, d\tau = \sum_i c_i \delta_{ij} = c_j \tag{1.31}$$

where (1.28) was used, and where the Kronecker delta causes all but one term of the sum to vanish. Hence

$$c_j = \langle \varphi_j | f \rangle \tag{1.32}$$

which is the desired relation for the expansion coefficients. Equation (1.30) becomes

$$f = \sum_i \langle \varphi_i | f \rangle \varphi_i \tag{1.33}$$

Using ket notation, we write f as $|f\rangle$ and φ_i as $|\varphi_i\rangle$:

$$|f\rangle = \sum_i |\varphi_i\rangle\langle \varphi_i | f \rangle \tag{1.34}$$

If the complete set of functions φ_i are eigenfunctions that correspond to a continuous range of eigenvalues, the summation in (1.30) becomes an integration. An operator with continuous eigenvalues is $\hat{p}_x$ for the x component of momentum. The eigenvalue equation is $(\hbar/i)(\partial\varphi_k/\partial x) = k\varphi$, where the eigenvalues are k. Integration gives

$$\varphi_k = Ae^{ikx/\hbar} \tag{1.35}$$

where A is an arbitrary constant. To keep φ from becoming infinite as $|x|$ goes to infinity, the eigenvalues k must be real numbers; there are no other restrictions on them:

$$-\infty < k < \infty \tag{1.36}$$

The expansion (1.30) becomes

$$f(x) = \int_{-\infty}^{\infty} c(k) e^{ikx/\hbar} \, dk \tag{1.37}$$

For an operator with both discrete and continuous eigenvalues, f is given by a summation over the discrete eigenfunctions plus an integration over the continuum eigenfunctions.

Suppose the complete set φ_i in (1.30) is the set of eigenfunctions of the Hermitian operator $\hat{B}$: $\hat{B}\varphi_i = b_i\varphi_i$. Further suppose that the function f being expanded in (1.30) and (1.33) is also an eigenfunction of $\hat{B}$: $\hat{B}f = b_f f$. Since $\hat{B}$ is Hermitian, it follows from (1.25) that the expansion coefficient

$c_i = \langle \varphi_i | f \rangle$ equals zero unless φ_i and f have the same eigenvalue ($b_i = b_f$). Thus when f and the φ_i's are eigenfunctions of the same operator, we only include functions φ_i in the expansion (1.30) that have the same $\hat{B}$ eigenvalue as f. There may be several such functions, due to degeneracy.

The *commutator* $[\hat{A}, \hat{B}]$ of two operators is defined as

$$[\hat{A}, \hat{B}] \equiv \hat{A}\hat{B} - \hat{B}\hat{A} \tag{1.38}$$

If

$$[\hat{A}, \hat{B}] = 0 \tag{1.39}$$

we say that $\hat{A}$ and $\hat{B}$ *commute*. If two Hermitian operators $\hat{A}$ and $\hat{B}$ commute, then it can be proven that there exists a common complete set of eigenfunctions for them; that is, if (1.39) holds, where $\hat{A}$ and $\hat{B}$ are Hermitian, then we can find some complete set of functions φ_i such that $\hat{A}\varphi_i = a_i\varphi_i$ and $\hat{B}\varphi_i = b_i\varphi_i$. If one or more of the eigenvalues of $\hat{A}$ are degenerate, we have an element of choice in that we can take many different linear combinations of the eigenfunctions of a degenerate eigenvalue. Not all these linear combinations need be eigenfunctions of $\hat{B}$; however, the theorem does show that we can choose linear combinations that will be eigenfunctions of $\hat{B}$. The converse of the preceding theorem is also true: if $\hat{A}$ and $\hat{B}$ are linear operators that possess a common complete set of eigenfunctions, then $\hat{A}$ and $\hat{B}$ commute. For three or more operators, *all* pairs of operators must commute for there to exist a complete set of eigenfunctions common to all the operators. Thus $\hat{A}$, $\hat{B}$, and $\hat{C}$ have a common complete set of eigenfunctions if and only if $[\hat{A}, \hat{B}] = 0$, $[\hat{A}, \hat{C}] = 0$, and $[\hat{B}, \hat{C}] = 0$.

Let a system be in the state Ψ at some instant of time. Let φ_i be the complete set of orthonormal eigenfunctions of the Hermitian operator $\hat{G}$ that corresponds to the physical property G:

$$\hat{G}\varphi_i = g_i\varphi_i$$

Expanding Ψ, we have from (1.30) and (1.33)

$$\Psi = \sum_i c_i\varphi_i = \sum_i \langle \varphi_i | \Psi \rangle \varphi_i \tag{1.40}$$

Substitution of (1.40) into $\langle G \rangle = \langle \Psi | \hat{G} | \Psi \rangle$ and use of orthonormality gives

$$\langle G \rangle = \sum_i |c_i|^2 g_i \tag{1.41}$$

However, as noted earlier, we have

$$\langle G\rangle = \sum_i P(g_i)g_i \tag{1.42}$$

where $P(g_i)$ is the probability of observing the value g_i for G. Hence we conclude that

$$P(g_i) = |c_i|^2 = |\langle\varphi_i|\Psi\rangle|^2 \tag{1.43}$$

Thus the squares of the absolute values of the coefficients in the expansion (1.40) give the probabilities for observing the various possible values g_i of the physical property G. For the special case where the state function is an eigenfunction of $\hat{G}$, $\Psi = \varphi_k$, we have $c_k = 1$ in (1.40), and we are certain to observe g_k in a measurement of G. Thus, for a stationary state, we have $\hat{H}\Psi = E\Psi$, and an energy measurement is certain to give E.

For a property with a continuous range of allowed values (e.g., x, p_x), (1.42) becomes

$$\langle G\rangle = \int P(g)g\,dg \tag{1.44}$$

where $P(g)$ is the *probability density* for property G; that is, $P(g)\,dg$ is the probability of observing a value of G between g and $g + dg$. Consider a one-particle, one-dimensional system. From (1.4) with $A = x$, we have

$$\langle x\rangle = \int \Psi^* x\Psi\,d\tau = \int_{-\infty}^{\infty} |\Psi|^2 x\,dx \tag{1.45}$$

Comparison of (1.44) and (1.45) shows that $|\Psi|^2$ is the probability density for the x coordinate; that is, the probability of finding the particle between x and $x + dx$ equals $|\Psi|^2\,dx$.

Suppose that state function Ψ at time t_1 is not an eigenfunction of $\hat{G}$. Let the eigenfunctions and eigenvalues of $\hat{G}$ be φ_i and g_i: $\hat{G}\varphi_i = g_i\varphi_i$. A measurement of G at time t_1 will give one of the eigenvalues of $\hat{G}$. Suppose we find the value g_k. For a measurement to have any meaning, it ought to be reproducible; that is, if we immediately remeasure G at a time $t_1 + dt$ that differs from t_1 only by an infinitesimal time interval, we ought to get g_k again. But if the second measurement is certain to give g_k, then the state function at time $t_1 + dt$ must be φ_k, the eigenfunction of $\hat{G}$ with eigenvalue g_k. We conclude that the measurement of G at time t_1 changed the state function from Ψ (which is not an eigenfunction of $\hat{G}$) to φ_k, the eigenfunction of $\hat{G}$ that corresponds to the observed value of G. This sudden change in state function caused by a measurement is called the *reduction of the state function*. Note that reduction is a probabilistic change, since we cannot predict with certainty which eigenvalue of $\hat{G}$ will be found by the measurement at t_1.

When Ψ is not an eigenfunction of $\hat{A}$, measurement of A in identical systems each in the same state Ψ gives various values. As a measure of the spread in the observed values, we define the *uncertainty* (or *standard deviation*) ΔA by

$$(\Delta A)^2 \equiv \langle (A - \langle A \rangle)^2 \rangle = \langle \Psi | (\hat{A} - \langle A \rangle)^2 | \Psi \rangle \tag{1.46}$$

For any two Hermitian operators, one can prove that

$$\Delta A \Delta B \geqslant \tfrac{1}{2} \left| \langle \Psi | [\hat{A}, \hat{B}] | \Psi \rangle \right| \tag{1.47}$$

which is the general statement of the *uncertainty principle*. For the case $A = x$, $B = p_x$, Eqs. (1.1), (1.2), and (1.47) give

$$[\hat{x}, \hat{p}_x] \equiv \hat{x}\hat{p}_x - \hat{p}_x\hat{x} = i\hbar \tag{1.48}$$

$$\Delta x \Delta p_x \geqslant \tfrac{1}{2}\hbar \tag{1.49}$$

There is also an uncertainty relation involving energy and time:

$$\Delta E \tau \geqslant \tfrac{1}{2}\hbar \tag{1.50}$$

In this relation, ΔE is the uncertainty in the energy of a state whose lifetime is τ. Equation (1.50) is not a special case of (1.47); rather, (1.50) must be derived by a special treatment. [See Y. Aharonov and D. Bohm, *Phys. Rev.*, **122**, 1649 (1961).]

Consider the matrix element $\langle \varphi_i | \hat{B} | \varphi_j \rangle$, where φ_i and φ_j are eigenfunctions of a Hermitian operator $\hat{A}$ that commutes with $\hat{B}$: $\hat{A}\varphi_i = a_i\varphi_i$, $\hat{A}\varphi_j = a_j\varphi_j$, and $[\hat{A}, \hat{B}] = 0$. Then, provided the eigenvalues a_i and a_j are not equal, it follows (Problem 1.3) that $\langle \varphi_i | \hat{B} | \varphi_j \rangle$ vanishes:

$$\langle \varphi_i | \hat{B} | \varphi_j \rangle = 0 \quad \text{if } a_i \neq a_j \tag{1.51}$$

From (1.33), we can derive a useful sum rule. Let R and S be any two well-behaved functions, let $\hat{A}$ and $\hat{B}$ be linear operators, and let φ_i be a complete orthonormal set of functions. If we set $f = \hat{B}S$ in (1.33), and then operate on both sides with $\hat{A}$, we get

$$\hat{A}\hat{B}S = \sum_i \langle \varphi_i | \hat{B}S \rangle \hat{A}\varphi_i$$

Multiplication by R^* followed by integration over all space gives the desired sum rule:

$$\sum_i \langle R|\hat{A}|\varphi_i\rangle\langle\varphi_i|\hat{B}|S\rangle = \langle R|\hat{A}\hat{B}|S\rangle \tag{1.52}$$

1.2 MATHEMATICS REVIEW

In this section we review some mathematics useful for quantum chemistry and spectroscopy.

Vectors. A physical quantity that has both magnitude and direction can be represented by a directed line segment or *vector* in three-dimensional space. Let **A** be a vector quantity. (We use boldface type for vectors.) We set up a Cartesian coordinate system xyz and denote vectors of unit length along the x, y, and z axes by **i**, **j**, and **k**, respectively. If A_x, A_y, and A_z are the projections of **A** on the x, y, and z axes, then **A** is given by

$$\mathbf{A} = A_x\mathbf{i} + A_y\mathbf{j} + A_z\mathbf{k} \tag{1.53}$$

The magnitude of **A** is

$$A \equiv |\mathbf{A}| = \left(A_x^2 + A_y^2 + A_z^2\right)^{1/2} \tag{1.54}$$

Since **A** is completely specified by the three numbers A_x, A_y, A_z, we can regard a vector in three-dimensional space as an ordered set of three numbers; thus the vector $(2,0,3)$ has an x component of 2, a y component of 0, and a z component of 3. [More generally, we can regard an ordered set of n numbers $(A_1, A_2, \ldots, A_n)$ as a vector in an abstract n-dimensional "space"; an n-dimensional vector space is a mathematical concept rather than a physical one (unless $n \leqslant 3$).] Equality of vectors means equality of their corresponding components. If $\mathbf{A} = \mathbf{B}$, then $A_x = B_x$, $A_y = B_y$, $A_z = B_z$.

The *dot product* $\mathbf{A} \cdot \mathbf{B}$ of two vectors is given by

$$\mathbf{A} \cdot \mathbf{B} = A_x B_x + A_y B_y + A_z B_z = \mathbf{B} \cdot \mathbf{A} \tag{1.55}$$

In particular

$$\mathbf{A} \cdot \mathbf{A} = A_x^2 + A_y^2 + A_z^2 = |\mathbf{A}|^2$$

The dot product can be shown to be a scalar equal to

$$\mathbf{A} \cdot \mathbf{B} = |\mathbf{A}|\,|\mathbf{B}| \cos\theta \tag{1.56}$$

where θ is the angle between **A** and **B**. The *cross product* $\mathbf{A} \times \mathbf{B}$ is given by

$$\mathbf{A} \times \mathbf{B} = (A_y B_z - A_z B_y)\mathbf{i} + (A_z B_x - A_x B_z)\mathbf{j} + (A_x B_y - A_y B_x)\mathbf{k} \quad (1.57)$$

which is most easily remembered in the form

$$\mathbf{A} \times \mathbf{B} = \begin{vmatrix} \mathbf{i} & \mathbf{j} & \mathbf{k} \\ A_x & A_y & A_z \\ B_x & B_y & B_z \end{vmatrix} = \mathbf{i}\begin{vmatrix} A_y & A_z \\ B_y & B_z \end{vmatrix} - \mathbf{j}\begin{vmatrix} A_x & A_z \\ B_x & B_z \end{vmatrix} + \mathbf{k}\begin{vmatrix} A_x & A_y \\ B_x & B_y \end{vmatrix} \quad (1.58)$$

(Determinants are discussed below.) The cross product can be shown to be a vector of magnitude $|\mathbf{A}||\mathbf{B}|\sin\theta$ (where θ is the angle $\leqslant 180°$ between **A** and **B**) and direction perpendicular to the plane defined by **A** and **B**, such that **A**, **B**, and $\mathbf{A} \times \mathbf{B}$ form a right-handed system.

The vector operator ∇ (*del*) is defined as

$$\nabla \equiv \mathbf{i}\partial/\partial x + \mathbf{j}\partial/\partial y + \mathbf{k}\partial/\partial z \quad (1.59)$$

Operators. An *operator* is a rule for transforming any given function to some other function. If the operator $\hat{A}$ transforms the function f to the function g, we write $g = \hat{A}f$ (where the circumflex denotes an operator). The *sum* and the *product* of any two operators are defined by $(\hat{A} + \hat{B})f \equiv \hat{A}f + \hat{B}f$ and $(\hat{A}\hat{B})f \equiv \hat{A}(\hat{B}f)$. Note that $\hat{B}f$ is some function, and applying $\hat{A}$ to $\hat{B}f$, we get the function defined as $(\hat{A}\hat{B})f$. Two operators $\hat{A}$ and $\hat{B}$ are *equal* if $\hat{A}f = \hat{B}f$ for all functions f. The *null operator* $\hat{0}$ and the *unit operator* $\hat{1}$ are defined by $\hat{0}f = 0$ and $\hat{1}f = f$. [Note that (1.48) and (1.39) are more precisely written as

$$[\hat{x}, \hat{p}_x] = i\hbar\hat{1} \quad (1.60)$$

$$[\hat{A}, \hat{B}] = \hat{0} \quad (1.61)$$

However, we shall usually omit circumflexes over operators that are simply multiplication by a constant.] Repeated application of the definition of operator multiplication shows that the associative law holds for all operators:

$$\hat{A}(\hat{B}\hat{C}) = (\hat{A}\hat{B})\hat{C} \quad (1.62)$$

The *commutator* $[\hat{A},\hat{B}]$ of two operators is defined by $[\hat{A},\hat{B}] \equiv \hat{A}\hat{B} - \hat{B}\hat{A}$. The *square* of $\hat{A}$ is defined by $\hat{A}^2 \equiv \hat{A}\hat{A}$, with similar definitions for higher powers.

A *linear* operator is one that obeys (1.11) and (1.12) for all functions f and g and all constants c. A *Hermitian* operator is a linear operator that obeys (1.13) for all well-behaved functions f and g.

The *eigenvalues* and *eigenfunctions* of an operator are defined by (1.3).

Some identities valid for linear operators are

$$(\hat{A} + \hat{B})\hat{C} = \hat{A}\hat{C} + \hat{B}\hat{C} \tag{1.63}$$

$$\hat{A}(\hat{B} + \hat{C}) = \hat{A}\hat{B} + \hat{A}\hat{C} \tag{1.64}$$

$$[\hat{A},\hat{B}] = -[\hat{B},\hat{A}] \tag{1.65}$$

$$[k\hat{A},\hat{B}] = [\hat{A},k\hat{B}] = k[\hat{A},\hat{B}] \qquad (k \text{ a constant}) \tag{1.66}$$

$$[\hat{A},\hat{B}+\hat{C}] = [\hat{A},\hat{B}] + [\hat{A},\hat{C}] \tag{1.67}$$

$$[\hat{A},\hat{B}\hat{C}] = [\hat{A},\hat{B}]\hat{C} + \hat{B}[\hat{A},\hat{C}] \tag{1.68}$$

$$[\hat{A}\hat{B},\hat{C}] = [\hat{A},\hat{C}]\hat{B} + \hat{A}[\hat{B},\hat{C}] \tag{1.69}$$

Complex numbers. A complex number z has the form $z = x + iy$, where $i = \sqrt{-1}$ and x and y are real numbers. We can represent z by a point in the complex xy plane by associating the complex number $z = x + iy$ with the point whose Cartesian coordinates are (x,y). The distance r of the point z from the origin is the *absolute value* $|z|$ of z; the angle θ that the radius vector from the origin to z makes with the positive x axis is the *phase* of z. We have

$$|z| = r = (x^2 + y^2)^{1/2}, \qquad \tan\theta = y/x \tag{1.70}$$

$$x = r\cos\theta, \qquad y = r\sin\theta \tag{1.71}$$

$$z = x + iy = r(\cos\theta + i\sin\theta) = re^{i\theta} \tag{1.72}$$

where we used

$$e^{i\theta} = \cos\theta + i\sin\theta \tag{1.73}$$

From (1.70) and (1.72), it follows that $|z_1 z_2| = |z_1||z_2|$.

The *complex conjugate* z^* of z is defined by

$$z^* = (x + iy)^* \equiv x - iy \tag{1.74}$$

We have $z=z^*$ if and only if z is a real number (with $y=0$). Some identities are

$$zz^* = x^2 + y^2 = r^2 = |z|^2 \tag{1.75}$$

$$(z_1z_2)^* = z_1^* z_2^*, \qquad (z_1+z_2)^* = z_1^* + z_2^*, \qquad (z_1/z_2)^* = z_1^*/z_2^*$$

Let $\omega^n = 1$, where n is a positive integer. The roots of this equation are the nth roots of unity. The number 1 has absolute value 1 and its phase can be taken as 0 or 2π or 4π or 6π or .... Hence $1 = e^{i2\pi k}$, where $k = 0, 1, 2, \ldots$. If we take the number $e^{i2\pi k/n}$ and raise it to the nth power, we get $(e^{i2\pi k/n})^n = e^{i2\pi k} = 1$. Hence the nth roots of unity are

$$\omega = e^{i2\pi k/n}, \qquad k = 0, 1, 2, \ldots, n-1, \qquad i = \sqrt{-1} \tag{1.76}$$

(With $k = n$, $n+1, \ldots$ we get the same roots as with $k = 0, 1, \ldots$.)

Determinants. A determinant of order n is a square array of n^2 elements; the determinant has a value calculated by the procedure described below. Let a_{ij} be the element in the ith row and jth column of the nth-order determinant A. The *minor* of element a_{ij} is the determinant of order $n-1$ obtained by striking out the ith row and jth column of A. The *cofactor* C_{ij} of a_{ij} is its minor multiplied by $(-1)^{i+j}$. When $n > 1$, the value of the determinant A is obtained by taking the n elements of any one row (or column), multiplying each such element by its cofactor, and adding the n products. Thus, if we use the elements of the kth row, we have

$$A = a_{k1}C_{k1} + a_{k2}C_{k2} + \cdots + a_{kn}C_{kn} = \sum_{m=1}^{n} a_{km}C_{km} \tag{1.77}$$

where k can be 1 or 2 or $\cdots$ or n. We have thus defined a determinant of order n in terms of determinants of order $n-1$. A determinant of order 1 has one element, and the determinant's value is defined as the value of that element; thus $|-5| = -5$, where the vertical lines denote a determinant (and not absolute value). For a determinant of order 2, use of (1.77) with $k=1$ gives

$$\begin{vmatrix} a_{11} & a_{12} \\ a_{21} & a_{22} \end{vmatrix} = a_{11}|a_{22}| - a_{12}|a_{21}| = a_{11}a_{22} - a_{12}a_{21} \tag{1.78}$$

where the vertical lines indicate determinants. For a determinant of order

3, use of (1.77) with $k=1$ and (1.78) gives

$$\begin{vmatrix} a_{11} & a_{12} & a_{13} \\ a_{21} & a_{22} & a_{23} \\ a_{31} & a_{32} & a_{33} \end{vmatrix} = a_{11}\begin{vmatrix} a_{22} & a_{23} \\ a_{32} & a_{33} \end{vmatrix} - a_{12}\begin{vmatrix} a_{21} & a_{23} \\ a_{31} & a_{33} \end{vmatrix} + a_{13}\begin{vmatrix} a_{21} & a_{22} \\ a_{31} & a_{32} \end{vmatrix}$$

$$= a_{11}a_{22}a_{33} - a_{11}a_{32}a_{23} - a_{12}a_{21}a_{33} + a_{12}a_{31}a_{23} + a_{13}a_{21}a_{32} - a_{13}a_{31}a_{22}$$

The *principal diagonal* of a determinant is the diagonal that runs from the upper left to the lower right. A determinant is in *diagonal* form when its only nonzero elements lie on the principal diagonal. A determinant is in *block-diagonal* form when its only nonzero elements lie in square blocks centered along the principal diagonal.

Some theorems on determinants are the following: (1) If all the elements in any one row (or any one column) are zero, the determinant equals zero. (2) Interchange of any two rows (or any two columns) multiplies a determinant's value by -1. (3) If any two rows (or any two columns) are identical, the determinant equals zero. (4) Multiplication of every element of any one row (or any one column) by the constant k multiplies the determinant's value by k. (5) Addition of the same constant multiple of the elements of one row to the corresponding elements of another row leaves the determinant unchanged in value. (The corresponding statement holds for columns.) (6) A determinant in diagonal form equals the product of the diagonal elements. (7) A determinant in block-diagonal form equals the product of the block determinants. For example,

$$\begin{vmatrix} u & v & 0 & 0 & 0 \\ i & r & 0 & 0 & 0 \\ 0 & 0 & p & e & s \\ 0 & 0 & n & m & r \\ 0 & 0 & e & s & r \end{vmatrix} = \begin{vmatrix} u & v \\ i & r \end{vmatrix} \begin{vmatrix} p & e & s \\ n & m & r \\ e & s & r \end{vmatrix} \tag{1.79}$$

Simultaneous linear equations. A set of n simultaneous linear equations in the n unknowns $x_1, x_2, \ldots, x_n$ has the form

$$\begin{aligned} a_{11}x_1 + a_{12}x_2 + \cdots + a_{1n}x_n &= b_1 \\ a_{21}x_1 + a_{22}x_2 + \cdots + a_{2n}x_n &= b_2 \\ \cdot \quad \cdot \quad \cdot \quad \cdot \quad \cdots \quad \cdot \quad &\cdot \quad \cdot \\ a_{n1}x_1 + a_{n2}x_2 + \cdots + a_{nn}x_n &= b_n \end{aligned} \tag{1.80}$$

where the a's and b's are constants. If $b_1 = b_2 = \cdots = b_n = 0$, the set of equations is *homogeneous*. If at least one of the b's is nonzero, the set is *inhomogeneous*. Let $\det(a_{ij})$ be the determinant of the coefficients in (1.80):

$$\det(a_{ij}) = \begin{vmatrix} a_{11} & a_{12} & \cdots & a_{1n} \\ a_{21} & a_{22} & \cdots & a_{2n} \\ \cdot & \cdot & \cdots & \cdot \\ a_{n1} & a_{n2} & \cdots & a_{nn} \end{vmatrix} \tag{1.81}$$

Let R_k be the determinant obtained by replacement of the elements of column k in $\det(a_{ij})$ with $b_1, b_2, \ldots, b_n$. *Cramer's rule* states that for inhomogeneous equations, the unknown x_k is given by

$$x_k = \frac{R_k}{\det(a_{ij})} \tag{1.82}$$

A set of n linear homogeneous equations in n unknowns always has the solution $x_1 = x_2 = \cdots = x_n = 0$, which is the *trivial* solution. Suppose the coefficient determinant $\det(a_{ij})$ is not equal to zero for a set of linear homogeneous equations; we can then use Cramer's rule (1.82). Since the equations are homogeneous, the determinant R_k will have a column of zeros, and will equal zero; hence $x_1 = x_2 = \cdots = x_n = 0$, and we have only the trivial solution. Thus for a nontrivial solution of the homogeneous equations to exist, we must have $\det(a_{ij}) = 0$. This condition can also be shown to be sufficient to insure the existence of a nontrivial solution. A system of n simultaneous, linear, homogeneous equations in n unknowns has a nontrivial solution if and only if the determinant of the coefficients equals zero.

If $x_1 = d_1, x_2 = d_2, \ldots, x_n = d_n$ is a solution of a set of n linear homogeneous equations, then $x_1 = cd_1, x_2 = cd_2, \ldots, x_n = cd_n$ (where c is any constant) is readily seen to be a solution also. Since we have an arbitrary constant in the solution, to solve a set of linear homogeneous equations, we set one of the unknowns (e.g., x_1) equal to an arbitrary constant ($x_1 = c$), thereby converting the equations to n equations in $n-1$ unknowns; we discard one equation and proceed to solve the set of $n-1$ inhomogeneous equations in $n-1$ unknowns. (The most efficient way to do this is by successive elimination of unknowns, rather than by Cramer's rule.) In quantum mechanics, the arbitrary constant c is usually evaluated by normalization.

Linear independence. The set of functions $f_1, f_2, \ldots, f_n$ is said to be *linearly independent* if the only way the equation

$$c_1 f_1 + c_2 f_2 + \cdots + c_n f_n = 0 \tag{1.83}$$

can be satisfied is with $c_1 = c_2 = \cdots = c_n = 0$. If the functions $f_1, \ldots, f_n$ are linearly dependent, then (1.83) holds with some of the c's being nonzero. This means we can solve (1.83) for one of the f's and express it as a linear combination of the other members of the set. For a linearly independent set, no member of the set can be expressed as a linear combination of the other members.

Taylor series. Let $f(x)$ be a real function all of whose derivatives exist at the point $x = a$ and within some neighborhood of a. Then $f(x)$ can usually be expanded in an infinite series (called a *Taylor series*) as follows:

$$f(x) = f(a) + \frac{f'(a)}{1!}(x-a) + \frac{f''(a)}{2!}(x-a)^2 + \cdots = \sum_{n=0}^{\infty} \frac{f^{(n)}(a)}{n!}(x-a)^n \tag{1.84}$$

where $f^{(n)}(a)$ is the nth derivative of $f(x)$ evaluated at $x = a$:

$$f^{(n)}(x) = \left.\frac{df(x)}{dx}\right|_{x=a} \tag{1.85}$$

[There are a few exceptional cases where all the derivatives exist, but the Taylor series does not converge to $f(x)$.] The series (1.84) will converge within some interval

$$a - c < x < a + c \tag{1.86}$$

centered about a, and diverge outside this interval. To find the interval of convergence requires that we consider f as a function of the complex variable z; the constant c in (1.86) equals the distance from a to the nearest singularity of $f(z)$ in the complex plane. A *singularity* is a point where $f(z)$ or its derivatives do not exist. For example, consider the Taylor-series expansion of $1/(1+x^2)$ about $a = 0$. The function $1/(1+z^2)$ has singularities at $z = +i$ and $-i$ (where f becomes infinite), and the distance from a to the nearest singularity is thus 1; hence $c = 1$, and the Taylor series converges in the interval $-1 < x < 1$. As another example, let $f(x) = e^x$, and let us choose $a = 0$. We find $f(0) = 1$, $f'(0) = 1$, and so on. Hence (1.84) gives

$$e^x = 1 + x + \frac{x^2}{2!} + \frac{x^3}{3!} + \cdots = \sum_{n=0}^{\infty} \frac{x^n}{n!} \tag{1.87}$$

It can be shown that the complex function e^z and all its derivatives are defined for all values of z; hence the series (1.87) converges for all x. The interval of convergence is

$$-\infty < x < \infty \tag{1.88}$$

Dirac delta function. The Dirac delta function $\delta(x)$ is a function defined to have the following properties:

$$\delta(x) = 0 \quad \text{for } x \neq 0 \tag{1.89}$$

$$\delta(x) = \infty \quad \text{for } x = 0 \tag{1.90}$$

$$\int_{-\infty}^{\infty} \delta(x)\, dx = 1 \tag{1.91}$$

In reality, no mathematical function has these three properties; however, we can regard the delta function as the limiting case of a function that becomes successively more peaked at the origin, while the area under the function remains equal to 1.

Differential equations. A linear, homogeneous, second-order differential equation has the form

$$y''(x) + F(x)y'(x) + G(x)y(x) = 0 \tag{1.92}$$

[If the coefficient of y'' is not 1, we can divide by this coefficient to obtain the form (1.92).] The one-particle, one-dimensional Schrödinger equation has the form (1.92) with $F=0$ and $G=(2m/\hbar^2)[E-V(x)]$. If $y_1(x)$ and $y_2(x)$ are linearly independent solutions of (1.92), then the general solution is

$$y = c_1 y_1 + c_2 y_2 \tag{1.93}$$

where c_1 and c_2 are arbitrary constants.

In the special case that F and G in (1.92) are constants, we have

$$y'' + ay' + by = 0 \tag{1.94}$$

where a and b are constants. If we try the solution $y = e^{sx}$, substitution into (1.94) gives (after division by e^{sx})

$$s^2 + as + b = 0 \tag{1.95}$$

This quadratic equation in s is called the auxiliary equation. If its roots are s_1 and s_2 ($\neq s_1$), then the general solution of (1.94) is

$$y = c_1 e^{s_1 x} + c_2 e^{s_2 x} \tag{1.96}$$

For example, suppose we want to solve

$$y'' + k^2 y = 0 \tag{1.97}$$

where k is a real constant. The auxiliary equation $s^2 + k^2 = 0$ has the roots $s = \pm ik$, and the general solution of (1.97) is

$$y = c_1 e^{ikx} + c_2 e^{-ikx} = (c_1 + c_2)\cos kx + i(c_1 - c_2)\sin kx \tag{1.98}$$

$$y = A\cos kx + B\sin kx \tag{1.99}$$

where (1.73) was used, and where we defined two new arbitrary constants A and B. The solution (1.99) can be written in the alternative form

$$y = c\sin(kx + d) \tag{1.100}$$

where c and d are arbitrary constants. The equivalence of (1.99) and (1.100) follows from the formula for $\sin(w+z)$.

When F and G in (1.92) are not constants, a useful approach is to try a power-series solution, a topic treated in most quantum-chemistry texts.

Separation of variables. Let the Hamiltonian of an n-particle system have the form

$$\hat{H} = \hat{H}_1 + \hat{H}_2 + \cdots + \hat{H}_n \tag{1.101}$$

where $\hat{H}_i$ contains coordinates and derivatives that refer to particle i only. The Hamiltonian (1.101) corresponds to a system with no interactions between the particles. The Schrödinger equation is

$$\hat{H}\psi(q_1,\ldots,q_n) = E\psi(q_1,\ldots,q_n) \tag{1.102}$$

where q_i symbolizes the coordinates of particle i. If we assume that

$$\psi(q_1,\ldots,q_n) = \psi_1(q_1)\psi_2(q_2)\cdots\psi_n(q_n) \tag{1.103}$$

then substitution of (1.103) into (1.102) and use of (1.101) gives (after division by ψ)

$$\frac{\hat{H}_1\psi_1(q_1)}{\psi_1(q_1)} + \frac{\hat{H}_2\psi_2(q_2)}{\psi_2(q_2)} + \cdots + \frac{\hat{H}_n\psi_n(q_n)}{\psi_n(q_n)} = E \tag{1.104}$$

$$(\hat{H}_1\psi_1)/\psi_1 = E - (\hat{H}_2\psi_2)/\psi_2 - \cdots - (\hat{H}_n\psi_n)/\psi_n \equiv E_1 \quad (1.105)$$

where E_1 is defined as $(\hat{H}_1\psi_1)/\psi_1$. Since E_1 equals $[\hat{H}_1\psi_1(q_1)]/\psi_1(q_1)$, it is independent of $q_2,\ldots,q_n$. Since E_1 equals $E-(\hat{H}_2\psi_2)/\psi_2 - \cdots - (\hat{H}_n\psi_n)/\psi_n$, it is independent of q_1. Hence E_1 is independent of all the coordinates and is a constant. We have

$$\hat{H}_1\psi_1(q_1) = E_1\psi_1(q_1) \quad (1.106)$$

which is a Schrödinger equation for particle 1. Similarly,

$$\hat{H}_2\psi_2(q_2) = E_2\psi_2(q_2), \quad \ldots, \quad \hat{H}_n\psi_n(q_n) = E_n\psi_n(q_n) \quad (1.107)$$

From (1.106), (1.107), and (1.104), we have

$$E = E_1 + E_2 + \cdots + E_n \quad (1.108)$$

Thus the wave function of a system of noninteracting particles is the product of wave functions for each particle, and the energy is the sum of energies for each particle.

The same reasoning applies to a one-particle Hamiltonian that is the sum of terms that each involve only one coordinate: $\hat{H} = \hat{H}_x + \hat{H}_y + \hat{H}_z$. Here

$$\psi = \psi_x(x)\psi_y(y)\psi_z(z) \quad (1.109)$$

$$E = E_x + E_y + E_z \quad (1.110)$$

$$\hat{H}_x\psi_x = E_x\psi_x, \qquad \hat{H}_y\psi_y = E_y\psi_y, \qquad \hat{H}_z\psi_z = E_z\psi_z \quad (1.111)$$

The chain rule. Let f be a function of x, y, and z: $f=f(x,y,z)$. Suppose we carry out the following change of variables:

$$x = x(r,s,t), \qquad y = y(r,s,t), \qquad z = z(r,s,t)$$

Substitution into f converts it to a function of the new independent variables r, s, and t:

$$f(x,y,z) = f[x(r,s,t), y(r,s,t), z(r,s,t)] = g(r,s,t)$$

The *chain rule* relates the partial derivatives of g to those of f:

$$\left(\frac{\partial g}{\partial r}\right)_{s,t} = \left(\frac{\partial f}{\partial x}\right)_{y,z}\left(\frac{\partial x}{\partial r}\right)_{s,t} + \left(\frac{\partial f}{\partial y}\right)_{x,z}\left(\frac{\partial y}{\partial r}\right)_{s,t} + \left(\frac{\partial f}{\partial z}\right)_{x,y}\left(\frac{\partial z}{\partial r}\right)_{s,t} \quad (1.112)$$

with similar equations for $(\partial g/\partial s)_{r,t}$ and $(\partial g/\partial t)_{r,s}$.

Spherical polar coordinates. The spherical polar coordinates r,θ,φ are defined by drawing a radius vector **r** from the origin to point (x,y,z): θ is the angle between **r** and the positive z axis; r is the distance of (x,y,z) from the origin; φ is the angle the projection of **r** in the xy plane makes with the positive x axis. The ranges of the variables are $0 \leqslant r \leqslant \infty$, $0 \leqslant \theta \leqslant \pi$, $0 \leqslant \varphi \leqslant 2\pi$. The relation between spherical polar and Cartesian coordinates is

$$\begin{aligned} z &= r\cos\theta, \qquad x = r\sin\theta\cos\varphi, \qquad y = r\sin\theta\sin\varphi \\ x^2+y^2+z^2 &= r^2, \qquad \tan\varphi = y/x, \qquad \cos\theta = z/r \end{aligned} \tag{1.113}$$

In Cartesian coordinates, the infinitesimal volume element is $dv = dx\,dy\,dz$. In spherical polar coordinates, the volume element is

$$dv = r^2 \sin\theta\, dr\, d\theta\, d\varphi \tag{1.114}$$

If one uses the chain rule to transform the Laplacian $\nabla^2 = \partial^2/\partial x^2 + \partial^2/\partial y^2 + \partial^2/\partial z^2$ into spherical polar coordinates, the result is

$$\nabla^2 = \frac{\partial^2}{\partial r^2} + \frac{2}{r}\frac{\partial}{\partial r} + \frac{1}{r^2}\frac{\partial^2}{\partial \theta^2} + \frac{1}{r^2}\cot\theta\frac{\partial}{\partial\theta} + \frac{1}{r^2\sin^2\theta}\frac{\partial^2}{\partial\varphi^2} \tag{1.115}$$

Direction cosines and rotation of coordinates. Suppose we have a set of Cartesian coordinates x,y,z. Let line L pass through the origin and make angles α, β, and γ with the positive x, y, and z axes, respectively. The *direction cosines* l, m, and n of line L are defined as

$$l \equiv \cos\alpha, \qquad m \equiv \cos\beta, \qquad n \equiv \cos\gamma \tag{1.116}$$

Let (x_L,y_L,z_L) be a point on line L, and let r be the distance from the origin to this point. From (1.113) with θ replaced by γ, we have $z_L = r\cos\gamma$, with two similar equations for x_L and y_L:

$$x_L = r\cos\alpha, \qquad y_L = r\cos\beta, \qquad z_L = r\cos\gamma \tag{1.117}$$

Equation (1.113) gives $x_L^2 + y_L^2 + z_L^2 = r^2$, and use of (1.117) then gives $\cos^2\alpha + \cos^2\beta + \cos^2\gamma = 1$, or

$$l^2 + m^2 + n^2 = 1 \tag{1.118}$$

Let $\mathbf{r}_1$ and $\mathbf{r}_2$ be vectors from the origin to the points (x_1,y_1,z_1) and (x_2,y_2,z_2). Equations (1.55) and (1.56) give $\mathbf{r}_1 \cdot \mathbf{r}_2 = x_1x_2 + y_1y_2 + z_1z_2 = r_1r_2\cos\theta_{12}$, where θ_{12} is the angle between the vectors. Hence $\cos\theta_{12}$

$=(x_1/r_1)(x_2/r_2)+(y_1/r_1)(y_2/r_2)+(z_1/r_1)(z_2/r_2)$; using (1.117) and (1.116), we have

$$\cos\theta_{12}=l_1l_2+m_1m_2+n_1n_2 \tag{1.119}$$

where l_1,m_1,n_1 and l_2,m_2,n_2 are the direction cosines of $\mathbf{r}_1$ and $\mathbf{r}_2$.

Let the Cartesian coordinate axes $x'y'z'$ have the same origin as the xyz axes. The $x'y'z'$ set is obtainable from the xyz set by rotation, reflection, or inversion, or some combination of these operations. (If the $x'y'z'$ set is left handed while the xyz set is right handed, we must perform a reflection or inversion as well as a rotation to generate the $x'y'z'$ axes from the xyz axes.) Let the vector $\mathbf{r}$ have coordinates (x,y,z) and (x',y',z') in the two coordinate systems. If $\mathbf{i}'$ is a vector of unit length along the x' axis, then (1.55) gives $\mathbf{r}\cdot\mathbf{i}'=x'$. Let l_1,m_1,n_1 be the direction cosines of the x' axis with respect to the xyz system; let l_2,m_2,n_2 and l_3,m_3,n_3 be the direction cosines of the y' and z' axes (respectively) with respect to the xyz system. Since the vector $\mathbf{i}'$ has unit length, it follows from (1.117) and (1.116) that the coordinates of $\mathbf{i}'$ in the xyz system are l_1, m_1, and n_1. Hence the equation $x'=\mathbf{r}\cdot\mathbf{i}'$ becomes

$$x'=(\mathbf{i}x+\mathbf{j}y+\mathbf{k}z)\cdot(l_1\mathbf{i}+m_1\mathbf{j}+n_1\mathbf{k})=l_1x+m_1y+n_1z$$

We can derive similar equations for y' and z' from $y'=\mathbf{r}\cdot\mathbf{j}'$ and $z'=\mathbf{r}\cdot\mathbf{k}'$. Therefore, the coordinates of a point in the rotated coordinate system $x'y'z'$ are related to its coordinates in the unrotated xyz system by

$$\begin{aligned} x'&=l_1x+m_1y+n_1z\\ y'&=l_2x+m_2y+n_2z\\ z'&=l_3x+m_3y+n_3z \end{aligned} \tag{1.120}$$

where the coefficients are the direction cosines of the rotated axes with respect to the unrotated axes. One final point. The angle between any two of the axes x',y',z' is 90°; hence $\mathbf{i}'\cdot\mathbf{j}'=\mathbf{i}'\cdot\mathbf{k}'=\mathbf{j}'\cdot\mathbf{k}'=0$, so that

$$\begin{aligned} l_1l_2+m_1m_2+n_1n_2&=0\\ l_1l_3+m_1m_3+n_1n_3&=0\\ l_2l_3+m_2m_3+n_2n_3&=0 \end{aligned} \tag{1.121}$$

1.3 UNITS

We shall use mainly the cgs Gaussian system of units. This is a mixed system with electrical quantities measured in cgs electrostatic units (esu) and magnetic quantities measured in cgs electromagnetic units (emu).

In 1960, the eleventh General Conference on Weights and Measures recommended the International System of Units (Système International d'Unités), abbreviated as SI units, for use in science; SI units are essentially the rationalized mks system of units. Relations between SI units and Gaussian units are given in Table A.4 of the Appendix. Table A.5 allows one to convert equations from SI to Gaussian units.

Quantum chemists often use atomic units, a system of units in which $\hbar$, the proton charge e, and the electron mass m each have a numerical value of 1. In this system, the unit of length is the bohr, defined by 1 bohr $\equiv a_0 \equiv \hbar^2/me^2 = 0.52918$ Å, where a_0 is the Bohr radius and the angstrom (Å) is defined by 1 Å $\equiv 10^{-8}$ cm $= 10^{-10}$ m; the unit of energy is the hartree, defined by 1 hartree $\equiv e^2/a_0 = 27.212$ eV.

1.4 THE PARTICLE IN A BOX

The one-dimensional particle-in-a-box problem is that of a single particle subject to the following potential-energy function:

$$V(x) = 0 \quad \text{for } 0 \leqslant x \leqslant l$$

$$V(x) = \infty \quad \text{for } x < 0 \text{ and for } x > l$$

where l is the length of the box. The stationary-state wave functions and energies are found from the time-independent Schrödinger equation. Outside the box, the wave function must be zero, because V is infinite. Inside the box, where $V = 0$, the Hamiltonian is $-(\hbar^2/2m)d^2/dx^2$, and the Schrödinger equation is

$$-(\hbar^2/2m)d^2\psi/dx^2 = E\psi \tag{1.122}$$

Equation (1.122) can be put in the form (1.97), and (1.99) gives as the general solution

$$\psi = A\cos\left[(2mE/\hbar^2)^{1/2}x\right] + B\sin\left[(2mE/\hbar^2)^{1/2}x\right] \tag{1.123}$$

Since ψ is zero outside the box, for ψ to be continuous, the function (1.123) must vanish at the ends of the box, $x=0$ and l. The condition $\psi(0)=0$ gives $A=0$, leaving only the sine term in (1.123). For the condition $\psi(l)=0$ to hold, the argument of the sine function must equal $n\pi$ at $x=l$, where $n=1,2,3,\ldots$. (Negative values of n do not give linearly independent solutions; $n=0$ makes $\psi=0$ everywhere.) Hence $E=n^2h^2/8ml^2$. The constant B is determined by normalization. The allowed wave functions and energies are then

$$\psi_n=(2/l)^{1/2}\sin(n\pi x/l), \qquad 0\leqslant x\leqslant l \tag{1.124}$$

$$E_n=n^2h^2/8ml^2, \qquad n=1,2,3,\ldots \tag{1.125}$$

A three-dimensional analog of the one-dimensional particle in a box is a particle in a rectangular parallelepiped with sides a, b, and c; we have

$$V=0 \quad \text{for the region } 0\leqslant x\leqslant a, 0\leqslant y\leqslant b, 0\leqslant z\leqslant c$$

$$V=\infty \quad \text{elsewhere}$$

Outside the box, $\psi=0$. Within the box, the Schrödinger equation is

$$(\hat{H}_x+\hat{H}_y+\hat{H}_z)\psi=E\psi$$

$$\hat{H}_x=-(\hbar^2/2m)(\partial^2/\partial x^2), \quad \hat{H}_y=-(\hbar^2/2m)(\partial^2/\partial y^2),$$

$$\hat{H}_z=-(\hbar^2/2m)(\partial^2/\partial z^2)$$

$\hat{H}_x$, $\hat{H}_y$, and $\hat{H}_z$ are Hamiltonians for a particle in a one-dimensional box. Using the separation-of-variables theorem of Section 1.2, we conclude that $\psi(x,y,z)$ is the product of one-dimensional particle-in-a-box wave functions for the x, y, and z directions, and E is the sum of three one-dimensional particle-in-a-box energies. Calling the quantum numbers associated with the x, y, and z directions n_x, n_y, and n_z, we have from (1.124), (1.125), (1.109), and (1.110)

$$\psi=(8/abc)^{1/2}\sin(n_x\pi x/a)\sin(n_y\pi y/b)\sin(n_z\pi z/c) \tag{1.126}$$

$$E=(h^2/8m)(n_x^2/a^2+n_y^2/b^2+n_z^2/c^2) \tag{1.127}$$

$$n_x=1,2,3,\ldots \qquad n_y=1,2,3,\ldots \qquad n_z=1,2,3,\ldots$$

where the three quantum numbers vary independently.

For a cubic box ($a=b=c$), we have

$$E=(h^2/8ma^2)(n_x^2+n_y^2+n_z^2) \tag{1.128}$$

Some of the energy levels are degenerate. For example, the states $|112\rangle$, $|121\rangle$, and $|211\rangle$ have the same energy. [We are using ket notation (Section 1.1), denoting each state by listing the quantum numbers n_x, n_y, n_z.]

A *free particle* is a particle subject to no forces, so that $V=0$ everywhere. For a free particle moving in one dimension, the Schrödinger equation is (1.122) and its solution is [Eq. (1.98)]

$$\psi=c_1 \exp\left[i(2mE)^{1/2}x/\hbar\right]+c_2 \exp\left[-i(2mE)^{1/2}x/\hbar\right] \tag{1.129}$$

To keep ψ from going to infinity as $|x|$ goes to infinity, the energy E must be nonnegative: $E \geqslant 0$.

1.5 THE HARMONIC OSCILLATOR

The one-dimensional harmonic oscillator has the potential energy

$$V=\tfrac{1}{2}kx^2=2\pi^2\nu^2 mx^2 \tag{1.130}$$

where k is the *force constant* (the force on the particle is $F=-dV/dx=-kx$), m is the particle's mass, and

$$\nu=(1/2\pi)(k/m)^{1/2} \tag{1.131}$$

is the classical vibration frequency. The Schrödinger equation is

$$\left(-\frac{\hbar^2}{2m}\frac{d^2}{dx^2}+\tfrac{1}{2}kx^2\right)\psi=E\psi \tag{1.132}$$

Carrying out a power-series solution, one finds that the quadratically integrable normalized solutions of (1.132) are

$$\psi_v(x)=(\alpha/\pi)^{1/4}(2^v v!)^{-1/2}e^{-\alpha x^2/2}H_v(\alpha^{1/2}x) \tag{1.133}$$

$$\alpha \equiv 2\pi\nu m/\hbar \tag{1.134}$$

$$v=0,1,2,\ldots \tag{1.135}$$

where the functions H_v are certain polynomials (the *Hermite polynomials*) defined by

$$H_n(z) \equiv (-1)^n e^{z^2}(d^n e^{-z^2}/dz^n), \qquad n=0,1,2,\ldots \tag{1.136}$$

The reader can verify that

$$H_0=1, \qquad H_1=2z, \qquad H_2=4z^2-2 \tag{1.137}$$

The Hermite polynomials obey the recursion relation

$$zH_n(z)=nH_{n-1}(z)+\tfrac{1}{2}H_{n+1}(z) \tag{1.138}$$

The explicit form of the ground-state ($v=0$) wave function is

$$\psi_0=(\alpha/\pi)^{1/4}e^{-\alpha x^2/2} \tag{1.139}$$

The $v=0$ and $v=1$ wave functions are sketched in Fig. 1.1. For any stationary state of the oscillator, the number of nodes in ψ equals the quantum number v. For states with v even, the Hermite polynomial H_v contains only even powers of x, and ψ_v is an even function; for states with v odd, H_v contains only odd powers, and ψ_v is an odd function.

The energy levels are found to be

$$E_v=(v+\tfrac{1}{2})h\nu \tag{1.140}$$

Note that ψ is nonzero everywhere (except at isolated nodal points). Classically, the oscillator is confined to the region where its total energy is

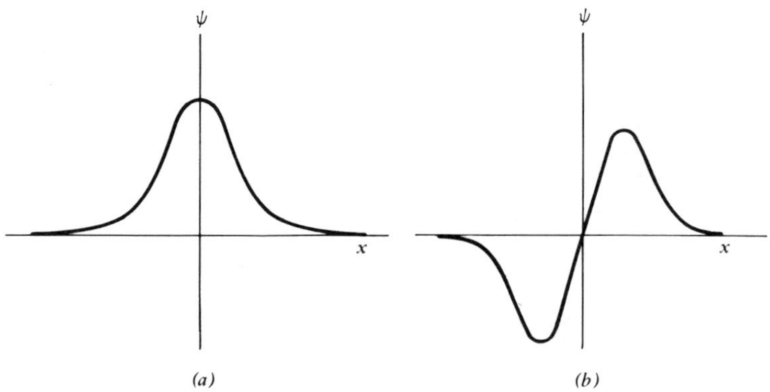

Fig. 1.1 Harmonic-oscillator wave functions for (*a*) $v=0$; (*b*) $v=1$.

greater than or equal to the potential energy: $E \geqslant V$. Quantum mechanically, there is some probability for finding the particle in classically forbidden regions where $E < V$.

For the three-dimensional harmonic oscillator, $V = \frac{1}{2}k_x x^2 + \frac{1}{2}k_y y^2 + \frac{1}{2}k_z z^2$. Separation of variables gives the wave function as the product of three one-dimensional harmonic oscillator wave functions, and gives the energy as

$$E = (v_x + \tfrac{1}{2})h\nu_x + (v_y + \tfrac{1}{2})h\nu_y + (v_z + \tfrac{1}{2})h\nu_z \tag{1.141}$$

$$v_x = 0, 1, 2, \ldots \qquad v_y = 0, 1, 2, \ldots \qquad v_z = 0, 1, 2, \ldots$$

The isotropic three-dimensional harmonic oscillator has $k_x = k_y = k_z$ and $\nu_x = \nu_y = \nu_z$.

1.6 ORBITAL ANGULAR MOMENTUM

Let **r** be the position vector from the origin to a particle in motion; let **v** be the particle's velocity and **p** its linear momentum:

$$\mathbf{r} = x\mathbf{i} + y\mathbf{j} + z\mathbf{k} \tag{1.142}$$

$$\mathbf{v} \equiv d\mathbf{r}/dt = (dx/dt)\mathbf{i} + (dy/dt)\mathbf{j} + (dz/dt)\mathbf{k} \tag{1.143}$$

$$\mathbf{p} \equiv m\mathbf{v} = mv_x\mathbf{i} + mv_y\mathbf{j} + mv_z\mathbf{k} \tag{1.144}$$

where m is the particle's mass, x, y, and z are its coordinates, and **i**, **j**, and **k** are the unit vectors along the axes of a Cartesian coordinate system. (The particle's acceleration **a** is $\mathbf{a} \equiv d\mathbf{v}/dt = d^2\mathbf{r}/dt^2$.) The *orbital angular momentum* **L** of the particle is defined as

$$\mathbf{L} \equiv \mathbf{r} \times \mathbf{p} = \begin{vmatrix} \mathbf{i} & \mathbf{j} & \mathbf{k} \\ x & y & z \\ p_x & p_y & p_z \end{vmatrix} \tag{1.145}$$

$$L_x = yp_z - zp_y, \qquad L_y = zp_x - xp_z, \qquad L_z = xp_y - yp_x \tag{1.146}$$

The square of the magnitude of the orbital angular momentum is

$$L^2 = \mathbf{L} \cdot \mathbf{L} = L_x^2 + L_y^2 + L_z^2 \tag{1.147}$$

Replacement of classical quantities by operators gives

$$\hat{L}_x = -i\hbar(y\partial/\partial z - z\partial/\partial y) \tag{1.148}$$

$$\hat{L}_y = -i\hbar(z\partial/\partial x - x\partial/\partial z) \tag{1.149}$$

$$\hat{L}_z = -i\hbar(x\partial/\partial y - y\partial/\partial x) \tag{1.150}$$

$$\hat{L}^2 = \hat{L}_x^2 + \hat{L}_y^2 + \hat{L}_z^2 \tag{1.151}$$

One finds the following commutation relations to hold:

$$\left[\hat{L}_x, \hat{L}_y\right] = i\hbar\hat{L}_z, \qquad \left[\hat{L}_y, \hat{L}_z\right] = i\hbar\hat{L}_x, \qquad \left[\hat{L}_z, \hat{L}_x\right] = i\hbar\hat{L}_y \tag{1.152}$$

$$\left[\hat{L}^2, \hat{L}_x\right] = \left[\hat{L}^2, \hat{L}_y\right] = \left[\hat{L}^2, \hat{L}_z\right] = 0 \tag{1.153}$$

Hence there exists a complete set of common eigenfunctions for $\hat{L}^2$ and any one of its components. The eigenvalue equations for $\hat{L}^2$ and $\hat{L}_z$ are found to be separable in spherical polar coordinates (but not in Cartesian coordinates). Using the chain rule to transform the derivatives, we can find $\hat{L}_x$, $\hat{L}_y$, and $\hat{L}_z$ in spherical polar coordinates, and then use (1.151) to find $\hat{L}^2$. One finds

$$\hat{L}_z = -i\hbar\partial/\partial\varphi \tag{1.154}$$

$$\hat{L}^2 = -\hbar^2\left(\frac{\partial^2}{\partial\theta^2} + \cot\theta\frac{\partial}{\partial\theta} + \frac{1}{\sin^2\theta}\frac{\partial^2}{\partial\varphi^2}\right) \tag{1.155}$$

$\hat{L}_x$ and $\hat{L}_y$ have more complicated forms than $\hat{L}_z$, but also involve only θ and φ, and not r.

Use of (1.8), (1.115), and (1.155) shows that the one-particle kinetic-energy operator can be expressed as

$$\hat{T} = -\frac{\hbar^2}{2m}\left(\frac{\partial^2}{\partial r^2} + \frac{2}{r}\frac{\partial}{\partial r}\right) + \frac{1}{2mr^2}\hat{L}^2 \tag{1.156}$$

Let $Y(\theta,\varphi)$ be the common eigenfunctions of $\hat{L}^2$ and $\hat{L}_z$. To solve $\hat{L}^2Y = aY$, $\hat{L}_zY = bY$ (where a and b are the eigenvalues), we try a separation of variables, writing $Y(\theta,\varphi) = S(\theta)T(\varphi)$. Solving the differential equations, one finds that the well-behaved, normalized solutions are

$$T_m(\varphi)=(2\pi)^{-1/2}e^{im\varphi} \tag{1.157}$$

$$S_{l,m}(\theta)=\left[\frac{(2l+1)}{2}\frac{(l-|m|)!}{(l+|m|)!}\right]^{1/2}P_l^{|m|}(\cos\theta) \tag{1.158}$$

$$l=0,1,2,\ldots$$

$$m=-l,-l+1,\ldots,0,\ldots,l-1,l$$

where the associated Legendre functions $P_l^{|m|}(w)$ are defined by

$$P_l^{|m|}(w)=\frac{1}{2^l l!}(1-w^2)^{|m|/2}\frac{d^{l+|m|}}{dw^{l+|m|}}(w^2-1)^l, \qquad l=0,1,2,\ldots \tag{1.159}$$

The eigenvalues of $\hat{L}^2$ and $\hat{L}_z$ are found to be $l(l+1)\hbar^2$ and $m\hbar$, respectively. The angular-momentum eigenfunctions $Y_l^m(\theta,\varphi)=S_{l,m}(\theta)T_m(\varphi)$ are called *spherical harmonics*:

$$Y_l^m(\theta,\varphi)=\left[\frac{2l+1}{4\pi}\frac{(l-|m|)!}{(l+|m|)!}\right]^{1/2}P_l^{|m|}(\cos\theta)e^{im\varphi} \tag{1.160}$$

$$\hat{L}^2Y_l^m(\theta,\varphi)=l(l+1)\hbar^2Y_l^m(\theta,\varphi) \tag{1.161}$$

$$\hat{L}_zY_l^m(\theta,\varphi)=m\hbar Y_l^m(\theta,\varphi) \tag{1.162}$$

The eigenfunctions Y_l^m have been normalized; moreover, being eigenfunctions of the Hermitian operators $\hat{L}^2$ and $\hat{L}_z$, they are orthogonal; hence

$$\int_0^{2\pi}\int_0^{\pi}[Y_l^m(\theta,\varphi)]^*Y_{l'}^{m'}(\theta,\varphi)\sin\theta\,d\theta\,d\varphi=\delta_{l,l'}\delta_{m,m'} \tag{1.163}$$

where the $\sin\theta$ factor comes from the volume element (1.114).

A moving electric charge generates a magnetic field. For a particle with charge q and mass m, its orbital angular momentum $\mathbf{L}$ gives rise to a magnetic dipole moment $\boldsymbol{\mu}_L$ given (in Gaussian units) by

$$\boldsymbol{\mu}_L=(q/2mc)\mathbf{L} \tag{1.164}$$

where c is the speed of light.

1.7 ANGULAR-MOMENTUM LADDER OPERATORS

The last section discussed angular momentum from the viewpoint of solving the differential eigenvalue equations that result from expressing $\hat{L}^2$ and $\hat{L}_z$ as differential operators. The ladder-operator approach uses just the angular-momentum commutation relations, and is also applicable to spin angular momentum (Section 1.15). Rather than using L, we shall use the symbol M to indicate that we are dealing with any kind of angular momentum.

We are given three linear operators $\hat{M}_x, \hat{M}_y, \hat{M}_z$ that by hypothesis obey the commutation relations [*cf.* (1.152)]:

$$\left[\hat{M}_x, \hat{M}_y\right] = i\hbar\hat{M}_z, \qquad \left[\hat{M}_y, \hat{M}_z\right] = i\hbar\hat{M}_x, \qquad \left[\hat{M}_z, \hat{M}_x\right] = i\hbar\hat{M}_y \tag{1.165}$$

The $\hat{M}^2$ operator is defined as

$$\hat{M}^2 \equiv \hat{M}_x^2 + \hat{M}_y^2 + \hat{M}_z^2 \tag{1.166}$$

It follows from (1.165) and (1.166) that

$$\left[\hat{M}^2, \hat{M}_x\right] = \left[\hat{M}^2, \hat{M}_y\right] = \left[\hat{M}^2, \hat{M}_z\right] = 0 \tag{1.167}$$

so that we can have simultaneous eigenfunctions of $\hat{M}^2$ and $\hat{M}_z$:

$$\hat{M}^2 Y_{jm} = a_j Y_{jm}, \qquad \hat{M}_z Y_{jm} = b_m Y_{jm} \tag{1.168}$$

where a_j and b_m are the eigenvalues, and Y_{jm} (not necessarily the spherical harmonics) are the common eigenfunctions.

The raising operator $\hat{M}_+$ and the lowering operator $\hat{M}_-$ are defined by

$$\hat{M}_+ \equiv \hat{M}_x + i\hat{M}_y, \qquad \hat{M}_- \equiv \hat{M}_x - i\hat{M}_y \tag{1.169}$$

From (1.169) and (1.165), we find that

$$\hat{M}_z\left(\hat{M}_+ Y_{jm}\right) = (b_m + \hbar)\left(\hat{M}_+ Y_{jm}\right) \tag{1.170}$$

$$\hat{M}_z\left(\hat{M}_- Y_{jm}\right) = (b_m - \hbar)\left(\hat{M}_- Y_{jm}\right) \tag{1.171}$$

so that application of $\hat{M}_+$ to a given eigenfunction Y_{jm} converts it to an eigenfunction of $\hat{M}_z$ with eigenvalue $\hbar$ higher than that of Y_{jm}; $\hat{M}_-$ lowers the $\hat{M}_z$ eigenvalue by $\hbar$. From (1.169) and (1.167), we find that

$$\hat{M}^2(\hat{M}_+ Y_{jm}) = a_j(\hat{M}_+ Y_{jm}), \qquad \hat{M}^2(\hat{M}_- Y_{jm}) = a_j(\hat{M}_- Y_{jm}) \quad (1.172)$$

so that $\hat{M}_+ Y_{jm}$ and $\hat{M}_- Y_{jm}$ are still eigenfunctions of $\hat{M}^2$ with the same eigenvalue a_j as Y_{jm}. From (1.170) and (1.171), the eigenvalues of $\hat{M}_z$ for a given a_j are spaced by $\hbar$.

Using

$$\hat{M}_x^2 + \hat{M}_y^2 = \hat{M}^2 - \hat{M}_z^2 \quad (1.173)$$

and (1.168), we get

$$(\hat{M}_x^2 + \hat{M}_y^2) Y_{jm} = (a_j - b_m^2) Y_{jm} \quad (1.174)$$

Since $M_x^2 + M_y^2$ is a nonnegative physical quantity, its eigenvalues $a_j - b_m^2$ must be nonnegative; therefore $-a_j^{1/2} \leqslant b_m \leqslant a_j^{1/2}$. Thus for a given a_j, the eigenvalue b_m is bounded above and below. Let $b_{\max}$ be the highest $\hat{M}_z$ eigenvalue for a given a_j, and let $Y_{j,\max}$ be the corresponding eigenfunction. In general, $\hat{M}_+$ acting on any Y_{jm} converts it to an eigenfunction with $\hat{M}_z$ eigenvalue increased by $\hbar$ [(1.170)]. However, since $Y_{j,\max}$ is the eigenfunction with the highest $\hat{M}_z$ eigenvalue, application of $\hat{M}_+$ to $Y_{j,\max}$ cannot give a new eigenfunction, but must give zero: $\hat{M}_+ Y_{j,\max} = 0$. Applying $\hat{M}_-$, we get $\hat{M}_- \hat{M}_+ Y_{j,\max} = 0$. We now use (1.169), (1.173), the first equation of (1.165), and (1.168) to get

$$0 = \hat{M}_- \hat{M}_+ Y_{j,\max} = (\hat{M}^2 - \hat{M}_z^2 - \hbar \hat{M}_z) Y_{j,\max} = (a_j - b_{\max}^2 - \hbar b_{\max}) Y_{j,\max}$$

$$a_j = b_{\max}^2 + \hbar b_{\max} \quad (1.175)$$

Similar reasoning starting from $\hat{M}_- Y_{j,\min} = 0$ leads to $a_j = b_{\min}^2 - \hbar b_{\min}$, which, when substituted into (1.175), gives $b_{\max}^2 + \hbar b_{\max} + \hbar b_{\min} - b_{\min}^2 = 0$. Regarding this as a quadratic equation in the unknown $b_{\max}$, we find the roots: $b_{\max} = -b_{\min}$ and $b_{\max} = b_{\min} - \hbar$. The second root is rejected, since $b_{\max}$ cannot be less than $b_{\min}$. We found previously that the $\hat{M}_z$ eigenvalues for a fixed $\hat{M}^2$ eigenvalue are spaced by $\hbar$; hence $b_{\max} = b_{\min} + n\hbar$, where n is a nonnegative integer $(0, 1, \ldots)$. Since $b_{\min} = -b_{\max}$, we have $b_{\max} = -b_{\max} + n\hbar$, or

$$b_{\max} = \tfrac{1}{2} n\hbar, \qquad b_{\min} = -\tfrac{1}{2} n\hbar, \qquad n = 0, 1, \ldots$$

From (1.175), we have $a_j = (\frac{1}{4}n^2 + \frac{1}{2}n)\hbar^2$. Setting $j \equiv \frac{1}{2}n$, we have $b_{\max} = j\hbar$,

$b_{\min} = -j\hbar$, $a_j = (j^2 + j)\hbar^2$. Thus the eigenvalues of $\hat{M}^2$ are

$$a_j = j(j+1)\hbar^2, \qquad j = 0, \tfrac{1}{2}, 1, \tfrac{3}{2}, 2, \ldots \tag{1.176}$$

Since the eigenvalues of $\hat{M}_z$ are spaced by $\hbar$, start at $j\hbar$, and end at $-j\hbar$, the $\hat{M}_z$ eigenvalues are

$$b_m = m\hbar, \qquad m = -j, -j+1, \ldots, j-1, j \tag{1.177}$$

Note the possibility of half-integral (as well as integral) values for the quantum number j. For orbital angular momentum, only integral quantum numbers occur, but for spin angular momentum, we can have half-integral values.

1.8 PARITY

An *even* function of the $3n$ Cartesian coordinates is defined as one that obeys

$$f(-x_1, -y_1, -z_1, \ldots, -x_n, -y_n, -z_n) = f(x_1, y_1, z_1, \ldots, x_n, y_n, z_n) \tag{1.178}$$

An *odd* function obeys

$$f(-x_1, -y_1, -z_1, \ldots, -x_n, -y_n, -z_n) = -f(x_1, y_1, z_1, \ldots, x_n, y_n, z_n) \tag{1.179}$$

The parity operator $\hat{\Pi}$ is defined by

$$\hat{\Pi} g(x_1, y_1, z_1, \ldots, x_n, y_n, z_n) = g(-x_1, -y_1, -z_1, \ldots, -x_n, -y_n, -z_n) \tag{1.180}$$

$\hat{\Pi}$ is a linear operator. Note that $\hat{\Pi}(fg) = (\hat{\Pi}f)(\hat{\Pi}g)$, where f and g are any two functions. Clearly $\hat{\Pi}^2 g = g$, so that $\hat{\Pi}^2 = \hat{1}$. The eigenvalue equation for $\hat{\Pi}$ is $\hat{\Pi}\varphi_i = k_i\varphi_i$. Application of $\hat{\Pi}$ gives $\hat{\Pi}^2\varphi_i = k_i\hat{\Pi}\varphi_i = k_i^2\varphi_i = \varphi_i$, since $\hat{\Pi}^2 = \hat{1}$. Hence $k_i^2 = 1$, and the eigenvalues of the parity operator are $k_i = \pm 1$. For $k_i = +1$, the eigenvalue equation reads $\hat{\Pi}\varphi_i(x_1, \ldots, z_n) = \varphi_i(x_1, \ldots, z_n)$; using (1.180) we have

$$\varphi_i(-x_1, \ldots, -z_n) = \varphi_i(x_1, \ldots, z_n) \tag{1.181}$$

Hence the eigenfunctions of $\hat{\Pi}$ that correspond to a $+1$ eigenvalue are all well-behaved even functions. Similarly, the $\hat{\Pi}$ eigenfunctions with -1 eigenvalue are all well-behaved odd functions.

For the commutator of $\hat{\Pi}$ with the Hamiltonian, we have $[\hat{\Pi}, \hat{H}] = [\hat{\Pi}, \hat{T}] + [\hat{\Pi}, \hat{V}]$. Since $\hat{T}$ involves the second derivatives of the Cartesian coordinates, we have $\hat{\Pi}(\hat{T}g) = \hat{T}(\hat{\Pi}g)$, so that $\hat{\Pi}$ and $\hat{T}$ commute. Hence $[\hat{\Pi}, \hat{H}] = [\hat{\Pi}, \hat{V}]$. We have $\hat{\Pi}(\hat{V}g) = (\hat{\Pi}V)(\hat{\Pi}g)$; if V is an even function, then $\hat{\Pi}V = V$, and $\hat{\Pi}\hat{V}g = \hat{V}\hat{\Pi}g$, so that $\hat{\Pi}$ and $\hat{V}$ commute. Hence when the potential energy V is an even function, the parity operator commutes with the Hamiltonian, and the stationary-state eigenfunctions of $\hat{H}$ can be chosen to be eigenfunctions of $\hat{\Pi}$. But the eigenfunctions of $\hat{\Pi}$ are even functions and odd functions. Hence when V is even, each of the wave functions ψ_i can be chosen as either even or odd. (Recall, e.g., the harmonic oscillator.) A state whose wave function is even is said to have *even parity*; similarly for odd wave functions.

The following equations hold for even and odd functions:

$$\int_{-\infty}^{\infty} f(x)\,dx = 0 \quad \text{for } f \text{ odd} \tag{1.182}$$

$$\int_{-\infty}^{\infty} f(x)\,dx = 2\int_{0}^{\infty} f(x)\,dx \quad \text{for } f \text{ even} \tag{1.183}$$

Similarly, if $g(x_1, \ldots, z_n)$ is an odd function of some or all of the $3n$ Cartesian coordinates, its integral over all space vanishes.

1.9 THE VARIATION METHOD

Because of interelectronic repulsions, the Schrödinger equation for many-electron atoms and molecules cannot be solved exactly. The two main approximation methods used are the variation method and perturbation theory. The variation method is based on the following theorem. Given a system with time-independent Hamiltonian $\hat{H}$, then if φ is any normalized, well-behaved function that satisfies the boundary conditions of the problem, one can show (by expanding φ in terms of the eigenfunctions of $\hat{H}$) that

$$W \equiv \int \varphi^* \hat{H} \varphi \, d\tau \geqslant E_0 \tag{1.184}$$

where E_0 is the true ground-state energy. If φ is not normalized, then (1.184) is replaced by

$$W \equiv \langle \varphi | \hat{H} | \varphi \rangle / \langle \varphi | \varphi \rangle \geqslant E_0 \tag{1.185}$$

A trial variation function φ thus provides an upper bound for the ground-state energy. One usually includes variational parameters in φ and looks for those values of the parameters that minimize the variational integral W; the function φ then serves as an approximation to the ground-state wave function.

A special type of variational function is the *linear variation function*. Here, φ has the form

$$\varphi = c_1 f_1 + c_2 f_2 + \cdots + c_n f_n = \sum_{i=1}^{n} c_i f_i \tag{1.186}$$

where the functions $f_1, \ldots, f_n$ are n linearly independent well-behaved functions, and the parameters $c_1, \ldots, c_n$ are numbers chosen to minimize the variational integral W. Substituting (1.186) into (1.185) and setting $\partial W/\partial c_j = 0$ $(j = 1, \ldots, n)$ so as to minimize W, we are led to the following set of n linear, homogeneous equations in the n unknowns $c_1, \ldots, c_n$:

$$\sum_{j=1}^{n} [(H_{ij} - S_{ij} W)c_j] = 0, \qquad i = 1, 2, \ldots, n \tag{1.187}$$

$$H_{ij} \equiv \int f_i^* \hat{H} f_j \, d\tau, \qquad S_{ij} \equiv \int f_i^* f_j \, d\tau \tag{1.188}$$

For a nontrivial solution of the equations (1.187), the coefficient determinant must vanish (Section 1.2). Hence

$$\det(H_{ij} - S_{ij} W) = 0 \tag{1.189}$$

$$\begin{vmatrix} H_{11} - S_{11} W & H_{12} - S_{12} W & \cdots & H_{1n} - S_{1n} W \\ H_{21} - S_{21} W & H_{22} - S_{22} W & \cdots & H_{2n} - S_{2n} W \\ \vdots & \vdots & \vdots\;\vdots\;\vdots & \vdots \\ H_{n1} - S_{n1} W & H_{n2} - S_{n2} W & \cdots & H_{nn} - S_{nn} W \end{vmatrix} = 0 \tag{1.190}$$

The *secular equation* (1.189) is an algebraic equation of nth degree, which we solve for the n roots:

$$W_0 \leqslant W_1 \leqslant W_2 \leqslant \cdots \leqslant W_{n-1} \tag{1.191}$$

It can be proven that these n roots are upper bounds for the true energies $E_0, \ldots, E_{n-1}$ of the n lowest states:

$$E_0 \leqslant W_0, \quad E_1 \leqslant W_1, \quad \ldots, \quad E_{n-1} \leqslant W_{n-1} \tag{1.192}$$

Having found the roots (1.191), we substitute them one at a time into the set of equations (1.187), which we then solve for the c's; for each root W_k, we find a set of coefficients $c_1^{(k)}, \ldots, c_n^{(k)}$, where the superscript indicates that these coefficients go with the root W_k. Use of these n sets of coefficients in (1.186) provides approximate wave functions for the n lowest states.

1.10 PERTURBATION THEORY

Perturbation theory is an approximation method useful when the Hamiltonian $\hat{H}$ of the system is "close" to the Hamiltonian $\hat{H}^0$ of a system which we can solve exactly. We call the system with Hamiltonian $\hat{H}$ the *perturbed* system, and the system with Hamiltonian $\hat{H}^0$ the *unperturbed* system; we define the perturbation $\hat{H}'$ as

$$\hat{H}' \equiv \hat{H} - \hat{H}^0 \tag{1.193}$$

The Schrödinger equations for the unperturbed and perturbed systems are

$$\hat{H}^0\psi_n^{(0)} = E_n^{(0)}\psi_n^{(0)} \tag{1.194}$$

$$\hat{H}\psi_n = E_n\psi_n \tag{1.195}$$

where the subscripts indicate state n of the unperturbed and perturbed systems.

To aid in relating the solutions of (1.195) to those of (1.194), we imagine that the perturbation is applied in a continuous manner; this corresponds to the introduction of a parameter λ in (1.193), so that

$$\hat{H} = \hat{H}^0 + \lambda\hat{H}' \tag{1.196}$$

As λ increases from 0 to 1, the unperturbed system is transformed into the perturbed system. The presence of λ makes the eigenfunctions and eigenvalues of $\hat{H}$ depend on the parameter λ: $\psi_n = \psi_n(q;\lambda)$ and $E_n = E_n(\lambda)$, where q symbolizes the coordinates. We expand ψ_n and E_n as Taylor series in λ, denoting the coefficients of successive powers of λ by $\psi_n^{(0)}, \psi_n^{(1)}, \ldots$ and $E_n^{(0)}, E_n^{(1)}, \ldots$:

$$\psi_n = \psi_n^{(0)} + \lambda\psi_n^{(1)} + \lambda^2\psi_n^{(2)} + \cdots \tag{1.197}$$

$$E_n = E_n^{(0)} + \lambda E_n^{(1)} + \lambda^2 E_n^{(2)} + \cdots \tag{1.198}$$

Note that in (1.197) we have assumed that $\lim_{\lambda\to 0}\psi_n = \psi_n^{(0)}$. However, if $\psi_n^{(0)}$ belongs to a degenerate energy level, any linear combination of the unperturbed eigenfunctions of the level is an eigenfunction of $\hat{H}^0$, and we must set $\lim_{\lambda\to 0}\psi_n$ equal to a linear combination of the unperturbed eigenfunctions of the degenerate level; degenerate perturbation theory is outlined later in this section. For now, we consider only nondegenerate levels.

The parameter λ is just a mathematical device, and we ultimately set $\lambda = 1$ in (1.197) and (1.198).

Substitution of the expansions (1.197) and (1.198) into (1.195) followed by expansion of the first-order wave-function correction $\psi_n^{(1)}$ according to

$$\psi_n^{(1)} = \sum_m b_m \psi_m^{(0)} \tag{1.199}$$

leads to the following expressions for $\psi_n^{(1)}$ and the first- and second-order energy corrections:

$$\psi_n^{(1)} = \sum_{m\neq n} \frac{\langle \psi_m^{(0)}|\hat{H}'|\psi_n^{(0)}\rangle}{E_n^{(0)} - E_m^{(0)}} \psi_m^{(0)} \tag{1.200}$$

$$E_n^{(1)} = \langle \psi_n^{(0)}|\hat{H}'|\psi_n^{(0)}\rangle \tag{1.201}$$

$$E_n^{(2)} = \sum_{k\neq n} \frac{|\langle \psi_k^{(0)}|\hat{H}'|\psi_n^{(0)}\rangle|^2}{E_n^{(0)} - E_k^{(0)}} \tag{1.202}$$

where the sums are over all states except state n.

Because of the difficulty in evaluating the infinite sums needed to find $E^{(2)}, E^{(3)}, \ldots$, one sometimes uses a *variation–perturbation* approach, in which $E^{(2)}, E^{(3)}, \ldots$ are evaluated by looking for functions that minimize certain integrals involving $\hat{H}'$; see *Hameka*, p. 223, for details.

Now consider perturbation theory for an unperturbed energy level that is n-fold degenerate, with the states

$$\psi_i^{(0)}, \qquad i = 1, \ldots, n \tag{1.203}$$

all having energy $E_n^{(0)}$. Any linear combination of the n degenerate functions (1.203) is an eigenfunction of $\hat{H}^0$; as $\lambda \to 0$, each of the n perturbed wave functions ψ_j ($j = 1, \ldots, n$) that correspond to the n-fold-degenerate unperturbed level will go over into some particular linear combination of the functions (1.203):

$$\lim_{\lambda \to 0} \psi_j = \sum_{i=1}^{n} c_{ij} \psi_i^{(0)} \equiv \varphi_j^{(0)}, \qquad j = 1, \ldots, n \tag{1.204}$$

where we used $\varphi_j^{(0)}$ to denote these n *correct* zeroth-order wave functions; which are the correct zeroth-order functions depends on $\hat{H}'$. (The j subscript in c_{ij} indicates the set of coefficients for $\varphi_j^{(0)}$.) Substitution of (1.198) and (1.197) with $\psi_j^{(0)}$ replaced by $\varphi_j^{(0)}$ into (1.195) leads to the following set of linear, homogeneous equations for the coefficients c_{ij} in $\varphi_j^{(0)}$:

$$\sum_{i=1}^{n} \left[\left(\hat{H}'_{mi} - E_j^{(1)} \delta_{mi} \right) c_{ij} \right] = 0, \qquad m = 1, \ldots, n \tag{1.205}$$

$$H'_{mi} \equiv \langle \psi_m^{(0)} | \hat{H}' | \psi_i^{(0)} \rangle \tag{1.206}$$

[The Kronecker delta occurs in (1.205) because we assumed that the $\psi_i^{(0)}$'s were chosen to be orthogonal.] For a nontrivial solution of (1.205), the coefficient determinant vanishes:

$$\det \left(H'_{mi} - E_j^{(1)} \delta_{mi} \right) = 0 \tag{1.207}$$

This *secular equation* is an algebraic equation of degree n with n roots $E_1^{(1)}, \ldots, E_n^{(1)}$ that give the n first-order energy corrections. Substitution of each $E_j^{(1)}$ in turn into (1.205) allows one to solve for the set of coefficients c_{ij} $(i = 1, \ldots, n)$ that go with the root $E_j^{(1)}$. Having found the n correct zeroth-order wave functions $\varphi_j^{(0)}$ $(j = 1, \ldots, n)$, we can then proceed to find $E_j^{(2)}$, $\psi_j^{(1)}$, and so on; the formulas for these corrections turn out to be essentially the same as for the nondegenerate case, provided that $\varphi_j^{(0)}$ is used in place of $\psi_n^{(0)}$.

When the secular determinant in (1.207) is in diagonal form (all off-diagonal elements equal to zero), it follows that the initially chosen unperturbed wave functions $\psi_j^{(0)}$ of the degenerate level are the correct zeroth-order wave functions $\varphi_j^{(0)}$. Conversely, if the initially chosen unperturbed wave functions are the correct zeroth-order functions, then the secular determinant will be in diagonal form.

When the secular determinant is in block-diagonal form, the secular equation (1.207) splits up into several equations. For example, if we have a five-fold degenerate unperturbed level and the secular determinant happens to consist of a 2×2 block and a 3×3 block [as in (1.79)], then the secular equation splits into two separate equations; moreover, two of the correct zeroth-order wave functions are linear combinations of $\psi_1^{(0)}$ and $\psi_2^{(0)}$ only, and the other three correct zeroth-order functions are linear combinations of $\psi_3^{(0)}$, $\psi_4^{(0)}$, and $\psi_5^{(0)}$ only.

One method to put the secular determinant in block-diagonal form (and thereby simplify the calculation) is to choose the initial zeroth-order functions $\psi_i^{(0)}$ to be eigenfunctions of some operator $\hat{A}$ that commutes with both $\hat{H}^0$ and $\hat{H}'$. Since $[\hat{H}',\hat{A}]=0$, it follows from (1.51) that the integrals H'_{ij} vanish when $\psi_i^{(0)}$ and $\psi_j^{(0)}$ correspond to different eigenvalues of $\hat{A}$. Therefore, the secular determinant will be in block-diagonal form, with each block corresponding to a different eigenvalue of $\hat{A}$.

1.11 THE CENTRAL-FORCE PROBLEM

For a one-particle central-force problem, the potential energy is a function only of the particle's distance from the origin: $V=V(r)$; from (1.156), the Hamiltonian is

$$\hat{H}=(-\hbar^2/2m)(\partial^2/\partial r^2+2r^{-1}\partial/\partial r)+(1/2mr^2)\hat{L}^2+V(r) \tag{1.208}$$

Because V depends only on r, one finds that this Hamiltonian commutes with the orbital angular-momentum operators $\hat{L}^2$ and $\hat{L}_z$. Hence the eigenfunctions ψ of $\hat{H}$ can also be chosen as eigenfunctions of $\hat{L}^2$ and $\hat{L}_z$:

$$\hat{H}\psi=E\psi \tag{1.209}$$

$$\hat{L}^2\psi=l(l+1)\hbar^2\psi, \qquad l=0,1,2,\ldots \tag{1.210}$$

$$\hat{L}_z\psi=m\hbar\psi, \qquad m=-l,\ldots,+l \tag{1.211}$$

The eigenfunctions of $\hat{L}^2$ and $\hat{L}_z$ are the spherical harmonics $Y_l^m(\theta,\varphi)$; these eigenfunctions can be multiplied by an arbitrary function of r and still be eigenfunctions of $\hat{L}^2$ and $\hat{L}_z$ (which do not involve r). Hence for a central-force problem, ψ has the form

$$\psi=R(r)Y_l^m(\theta,\varphi) \tag{1.212}$$

Use of (1.212), (1.208), and (1.210) in (1.209) gives as the differential equation satisfied by the radial function $R(r)$

$$(-\hbar^2/2m)[R''(r)+(2/r)R'(r)]+[l(l+1)\hbar^2/2mr^2+V(r)]R(r)=ER(r) \tag{1.213}$$

1.12 THE TWO-PARTICLE PROBLEM

Consider a two-particle system with the particles having masses m_1 and m_2 and coordinates (x_1,y_1,z_1) and (x_2,y_2,z_2). We define the *relative* (or *internal*) coordinates x,y,z of the system by

$$x \equiv x_2 - x_1, \qquad y \equiv y_2 - y_1, \qquad z \equiv z_2 - z_1 \tag{1.214}$$

The coordinates X,Y,Z of the center of mass of the system are given by

$$X = \frac{m_1 x_1 + m_2 x_2}{m_1 + m_2}, \qquad Y = \frac{m_1 y_1 + m_2 y_2}{m_1 + m_2}, \qquad Z = \frac{m_1 z_1 + m_2 z_2}{m_1 + m_2} \tag{1.215}$$

We shall impose the restriction that the potential energy V depends only on the relative coordinates of the particles: $V = V(x,y,z)$. Substitution of (1.214) and (1.215) into the kinetic-energy expression leads to the following expression for the classical-mechanical Hamiltonian:

$$H = p_M^2/2M + \left[p_\mu^2/2\mu + V(x,y,z)\right] \tag{1.216}$$

where M is the total mass ($M \equiv m_1 + m_2$), the *reduced mass* μ is defined as

$$\mu \equiv m_1 m_2/(m_1 + m_2) \tag{1.217}$$

and the linear momenta $\mathbf{p}_M$ and $\mathbf{p}_\mu$ have components defined by

$$p_X = M(dX/dt), \qquad p_Y = M(dY/dt), \qquad p_Z = M(dZ/dt)$$

$$p_x = \mu(dx/dt), \qquad p_y = \mu(dy/dt), \qquad p_z = \mu(dz/dt)$$

Since the Hamiltonian (1.216) is the sum of a part dependent only on the center-of-mass coordinates X,Y,Z and a part (the bracketed terms) dependent only on the internal coordinates x,y,z, separation of variables allows us to conclude that the (complete) wave function is the product of a function $\psi_{\text{trans}}(X,Y,Z)$ and a function $\psi_{\text{int}}(x,y,z)$. Since $p_M^2/2M$ is a free-particle Hamiltonian, ψ_{trans} is a free-particle wave function. The function ψ_{int} is an eigenfunction of the operator corresponding to the bracketed terms in (1.216). The energy is the sum of an energy of translational motion of the entire system of mass M and an energy of internal (relative) motion of the two particles. We solve for the internal wave functions and energies by setting up a one-particle Schrödinger equation using the reduced mass μ and the relative coordinates x,y,z.

1.13 THE TWO-PARTICLE RIGID ROTOR

The two-particle rigid rotor consists of masses m_1 and m_2 attached to the ends of a rigid, massless rod of length d. Since the particles are held at a fixed distance from each other, the potential energy does not change as the rotor moves through space, and V is a constant, which we may take as zero: $V=0$. Thus the energy is entirely kinetic. $V=0$ is a special case of V depending only on the relative coordinates of the particles, so we may use the results of Section 1.12 to separate the translational and internal motions; the energy is the sum of translational energy of the system and energy of internal motion. (Ordinarily, the energy of internal motion of a two-particle system is of two types—vibrational and rotational. In this case, the interparticle distance is held fixed, so there is only rotational internal motion.) To find the internal wave functions and energies, we must find the eigenfunctions of the Hamiltonian corresponding to the bracketed terms in (1.216). The potential energy $V=0$ is a special case of a central force (Section 1.11), so we use spherical polar coordinates to deal with the internal motion. However, since the interparticle distance is constant, the relative spherical polar coordinate r is not a variable; the only variables are the internal angular coordinates θ and φ (which define the orientation of the rotor interparticle axis with respect to a coordinate system that translates with the rotor but that does not rotate). The internal wave function ψ is a function of θ and φ; but, since this is a central-force problem, $\psi(\theta,\varphi)$ must be a spherical harmonic $Y_l^m(\theta,\varphi)$. The Hamiltonian is (1.208) with the r derivatives omitted (since r is not a variable), with r in $\hat{L}^2/2mr^2$ set equal to d, with m replaced by the reduced mass $\mu = m_1m_2/(m_1+m_2)$, and with $V(r)=0$. Thus the Schrödinger equation $\hat{H}\psi = E\psi$ for the rotor's internal motion is

$$(1/2\mu d^2)\hat{L}^2 Y_l^m(\theta,\varphi) = E Y_l^m(\theta,\varphi)$$

Use of (1.161) gives

$$E = l(l+1)\hbar^2/2\mu d^2, \qquad l=0,1,2,\ldots \tag{1.218}$$

It is conventional to use J and M, rather than l and m, for the rotational angular-momentum quantum numbers. Also, the moment of inertia I of the rotor about an axis passing through the center of mass and perpendicular to the interparticle line is found to equal μd^2. (The moment of inertia of a system of particles about an axis is defined by $I=\Sigma_i m_i r_i^2$, where r_i is the perpendicular distance from particle i to the axis.) Hence we write

$$E = J(J+1)\hbar^2/2I, \qquad J=0,1,2,\ldots \tag{1.219}$$

The two-particle rotor levels are $(2J+1)$-fold degenerate, since for each J there are $2J+1$ possible values of the quantum number M (ranging from $-J$ to $+J$), and E is independent of M.

1.14 THE HYDROGEN ATOM

The hydrogenlike atom consists of a nucleus of charge Ze and one electron. ($Z=1$ for H, $Z=2$ for He^+, etc.) The potential energy is given by Coulomb's law as

$$V=-Ze^2/r \tag{1.220}$$

where r is the distance of the electron from the nucleus. Since V depends on only the relative coordinates of the particles, the energy is the sum of translational energy of the atom as a whole and energy of internal motion (Section 1.12). The Hamiltonian for internal motion is

$$\hat{H}=-(\hbar^2/2\mu)\nabla^2-Ze^2/r \tag{1.221}$$

where μ is the reduced mass [Equation (1.217)] and ∇^2 refers to relative coordinates. The Hamiltonian (1.221) is a central-force problem; hence

$$\psi=R(r)Y_l^m(\theta,\varphi) \tag{1.222}$$

where the radial function R is found by solving (1.213) with $V=-Ze^2/r$. Using a power-series solution, one finds that the bound-state ($E\leqslant 0$) radial functions are

$$R_{nl}(r)=Nr^le^{-Zr/na}L_{n+l}^{2l+1}(2Zr/na) \tag{1.223}$$

where N is the normalization constant, a is defined as

$$a\equiv\hbar^2/\mu e^2 \tag{1.224}$$

n is a quantum number such that

$$n=1,2,3,\ldots,\qquad l\leqslant n-1 \tag{1.225}$$

and L_{n+l}^{2l+1} is an associated Laguerre polynomial, whose highest power of r is $n-l-1$, and whose explicit form we mercifully omit.

The bound-state energy levels are found to be

$$E=-(Z^2/n^2)(e^2/2a) \tag{1.226}$$

Since E depends on only n, states with the same n but different l or m have the same energy.

If internal nuclear motion is neglected, then the reduced mass μ is approximated by the electron mass m, and a in these formulas is replaced by a_0 (the Bohr radius), where

$$a_0 \equiv \hbar^2/me^2 = 0.52918\ \text{Å} \tag{1.227}$$

In addition to bound-state wave functions and energies, there are continuum wave functions and energies; all positive energies are allowed.

A letter code is used to specify the value of the quantum number l: the letters $s,p,d,f,g,h,i,k,l,\ldots$ indicate l values of $0,1,2,3,4,5,6,7,8,\ldots$, respectively. The notations $2p_{-1} \equiv |21-1\rangle$ indicate the state with $n=2$, $l=1$, $m=-1$. The explicit forms of the $n=1$ and $n=2$ hydrogenlike-atom wave functions are

$$\begin{aligned}
1s &= \pi^{-1/2}(Z/a)^{3/2} e^{-Zr/a} \\
2s &= \pi^{-1/2}(Z/2a)^{3/2}(1 - Zr/2a)e^{-Zr/2a} \\
2p_1 &= (1/8\pi^{1/2})(Z/a)^{5/2}\, re^{-Zr/2a} \sin\theta\, e^{i\phi} \\
2p_0 &= \pi^{-1/2}(Z/2a)^{5/2} re^{-Zr/2a} \cos\theta \\
2p_{-1} &= (1/8\pi^{1/2})(Z/a)^{5/2} re^{-Zr/2a} \sin\theta\, e^{-i\phi}
\end{aligned} \tag{1.228}$$

The $e^{im\phi}/(2\pi)^{1/2}$ factor makes the spherical harmonics complex for $m \neq 0$. We can get real functions that are still eigenfunctions of $\hat{H}$ and $\hat{L}^2$ by taking linear combinations of $|nlm\rangle$ and $|nl{-}m\rangle$. For example,

$$2p_x \equiv 2^{-1/2}(2p_1 + 2p_{-1}) = \tfrac{1}{4}(2\pi)^{-1/2}(Z/a)^{5/2} xe^{-Zr/2a} \tag{1.229}$$

$$2p_y \equiv -i2^{-1/2}(2p_1 - 2p_{-1}) = \tfrac{1}{4}(2\pi)^{-1/2}(Z/a)^{5/2} ye^{-Zr/2a} \tag{1.230}$$

where (1.73) and (1.113) were used. The notation $2p_z$ is used for $2p_0$, since $2p_0$ has the z coordinate as a factor: $2p_z = 2p_0$.

1.15 SPIN

In addition to orbital angular momentum $\mathbf{L}$ (which arises from motion through space), elementary particles possess an intrinsic (or *spin*) angular momentum **S**. The spin **S** has no classical analog and is a relativistic effect. The Schrödinger equation is a nonrelativistic equation, and in nonrelativistic quantum mechanics, the existence of spin must be introduced as an additional postulate. Analogous to the orbital angular-momentum operators $\hat{L}_x$, $\hat{L}_y$, $\hat{L}_z$, we have spin angular-momentum operators $\hat{S}_x$, $\hat{S}_y$, $\hat{S}_z$, which are hypothesized to obey the same commutation relations as for orbital angular momentum:

$$\left[\hat{S}_x, \hat{S}_y\right] = i\hbar\hat{S}_z, \qquad \left[\hat{S}_y, \hat{S}_z\right] = i\hbar\hat{S}_x, \qquad \left[\hat{S}_z, \hat{S}_x\right] = i\hbar\hat{S}_y \tag{1.231}$$

The operator for the square of the magnitude of the spin angular momentum is $\hat{S}^2 = \hat{S}_x^2 + \hat{S}_y^2 + \hat{S}_z^2$. We find

$$\left[\hat{S}^2, \hat{S}_x\right] = \left[\hat{S}^2, \hat{S}_y\right] = \left[\hat{S}^2, \hat{S}_z\right] = 0 \tag{1.232}$$

The square of the magnitude of the spin $|\mathbf{S}|^2$ of a particle equals

$$s(s+1)\hbar^2 \tag{1.233}$$

where the spin quantum number s can be integral or half-integral. The value of s is characteristic of the particle. Some values of s are the following: electron, $\frac{1}{2}$; proton, $\frac{1}{2}$; neutron, $\frac{1}{2}$; pi-meson, 0. The component of **S** along the z axis has the possible values

$$m_s\hbar, \qquad m_s = -s, -s+1, \ldots, s-1, s \tag{1.234}$$

The quantum numbers s and m_s are the analogs of the orbital angular-momentum quantum numbers l and m.

For an electron, $s = \frac{1}{2}$, and m_s can be $-\frac{1}{2}$ or $+\frac{1}{2}$. We call the electron-spin eigenfunctions corresponding to these two eigenvalues α and β:

$$\hat{S}_z\alpha = \tfrac{1}{2}\hbar\alpha, \qquad \hat{S}_z\beta = -\tfrac{1}{2}\hbar\beta \tag{1.235}$$

Also, since $[\hat{S}^2, \hat{S}_z] = 0$, we have from (1.233):

$$\hat{S}^2\alpha = \tfrac{3}{4}\hbar^2\alpha, \qquad \hat{S}^2\beta = \tfrac{3}{4}\hbar^2\beta \tag{1.236}$$

It is convenient to consider the quantum number m_s as the variable in the spin functions α and β: $\alpha = \alpha(m_s)$ and $\beta = \beta(m_s)$. Since m_s takes on only two values, rather than a continuous range of values, we use sums rather than integrals to express the orthonormality of α and β, and the Hermitian property of $\hat{S}_x$ and other spin operators:

$$\sum_{m_s=-1/2}^{1/2} |\alpha(m_s)|^2 = 1 = \sum_{m_s=-1/2}^{1/2} |\beta(m_s)|^2 \tag{1.237}$$

$$\sum_{m_s=-1/2}^{1/2} \alpha^*(m_s)\beta(m_s) = 0 \tag{1.238}$$

$$\sum_{m_s} f^*(m_s)\hat{S}_x g(m_s) = \sum_{m_s} g(m_s)\left[\hat{S}_x f(m_s)\right]^* \tag{1.239}$$

The symbol $\int d\tau$ will denote a definite integral over the full range of the coordinates of the system, including a summation over the spin variables. Thus for a one-electron, three-dimensional system, $\int d\tau$ stands for

$$\sum_{m_s=-1/2}^{1/2} \int_{-\infty}^{\infty}\int_{-\infty}^{\infty}\int_{-\infty}^{\infty} dx\, dy\, dz$$

The symbol $\int dv$ will denote integration over the full range of the spatial coordinates only, without summation over spin variables.

Analogous to (1.169), the electron-spin raising and lowering operators are

$$\hat{S}_+ \equiv \hat{S}_x + i\hat{S}_y, \qquad \hat{S}_- \equiv \hat{S}_x - i\hat{S}_y \tag{1.240}$$

Analogous to the equations $\hat{M}_+ Y_{j,\max} = 0 = \hat{M}_- Y_{j,\min}$ of Section 1.7, we have

$$\hat{S}_+\alpha = 0, \qquad \hat{S}_-\beta = 0 \tag{1.241}$$

Also, since $\hat{S}_+$ is a raising operator, we must have $\hat{S}_+\beta = c\alpha$, where c is some constant. We can evaluate c by use of the normalization of the spin functions and the Hermitian property of $\hat{S}_x$ and $\hat{S}_y$; one finds (Problem 1.8) $|c| = \hbar$. Choosing the phase of c as zero, we have

$$\hat{S}_+\beta = \hbar\alpha \tag{1.242}$$

One finds similarly that

$$\hat{S}_-\alpha = \hbar\beta \tag{1.243}$$

From (1.240)–(1.243), we get

$$\hat{S}_x\alpha = \tfrac{1}{2}\hbar\beta, \qquad \hat{S}_x\beta = \tfrac{1}{2}\hbar\alpha \tag{1.244}$$

$$\hat{S}_y\alpha = \tfrac{1}{2}i\hbar\beta, \qquad \hat{S}_y\beta = -\tfrac{1}{2}i\hbar\alpha \tag{1.245}$$

Just as orbital angular momentum **L** gives rise to a magnetic dipole moment $\boldsymbol{\mu}_L$, spin angular momentum **S** gives rise to a spin magnetic dipole moment $\boldsymbol{\mu}_S$. Dirac's relativistic theory of the electron showed that

$$\boldsymbol{\mu}_S = -(e/mc)\mathbf{S} \equiv -g_e(e/2mc)\mathbf{S} \tag{1.246}$$

where g_e is the electron g factor. Although Dirac's treatment gave $g_e = 2$, subsequent more refined theoretical work and experimental data show that g_e is not exactly 2; rather

$$g_e = 2(1 + \alpha/2\pi + \cdots) = 2.0023 \tag{1.247}$$

where the *fine-structure constant* α is defined as

$$\alpha \equiv e^2/\hbar c = 1/137.036 \tag{1.248}$$

The dots in (1.247) indicate smaller terms involving α^2, α^3, and so on.

1.16 THE PAULI PRINCIPLE

Consider a system of n particles with state function $\Psi(q_1, \ldots, q_i, \ldots, q_k, \ldots, q_n)$, where q_i stands for the four coordinates (three spatial and one spin) of particle i. Suppose we interchange the coordinates of particles i and k in Ψ. If the n particles of the system are identical, then experimental evidence and relativistic quantum field theory show that the state function Ψ must be restricted in form so as to be symmetric with respect to exchange if the particles have integral spin ($s = 0$ or 1 or 2 or...) or antisymmetric with respect to exchange if the particles have half-integral spin ($s = \frac{1}{2}$ or $\frac{3}{2}$ or...). This is the *Pauli principle*. In the nonrelativistic quantum mechanics we are using, the Pauli principle is an additional postulate. By "symmetric with respect to exchange," we mean that Ψ is unchanged on interchange of the coordinates of identical particles:

$$\Psi(q_1, \ldots, q_k, \ldots, q_i, \ldots, q_n) = \Psi(q_1, \ldots, q_i, \ldots, q_k, \ldots, q_n)$$

By "antisymmetric with respect to exchange," we mean that interchange of the coordinates of identical particles multiplies Ψ by -1:

$$\Psi(q_1, \ldots, q_k, \ldots, q_i, \ldots, q_n) = -\Psi(q_1, \ldots, q_i, \ldots, q_k, \ldots, q_n)$$

Particles having half-integral spin and requiring antisymmetric wave functions are called *fermions*; particles having integral spin and requiring symmetric wave functions are called *bosons*.

Note that if we put $q_k = q_i$ in this last equation, we get $\Psi = -\Psi$, so that $\Psi = 0$ when $q_i = q_k$, that is, when $x_i = x_k$, $y_i = y_k$, $z_i = z_k$, and $m_{si} = m_{sk}$. Hence, there is zero probability for two fermions with the same spin (i.e., the same m_s value) to be at the same point in space. The Pauli principle causes electrons of like spin to keep apart from each other.

As an example, consider two identical, noninteracting particles, 1 and 2. Since the particles are noninteracting, solution of the Schrödinger equation using separation of variables gives for the spatial part of the system's wave function: $f(x_1, y_1, z_1)g(x_2, y_2, z_2)$, which we abbreviate to $f(1)g(2)$. The function $f(2)g(1)$ is degenerate with $f(1)g(2)$, since particles 1 and 2 are identical. The functions $f(1)g(2)$ and $f(2)g(1)$ are each neither symmetric nor antisymmetric with respect to exchange. To satisfy the Pauli principle, instead of $f(1)g(2)$ and $f(2)g(1)$, we must use as the spatial part of the wave function the following normalized linear combinations:

$$2^{-1/2}[f(1)g(2) + g(1)f(2)] \tag{1.249}$$

$$2^{-1/2}[f(1)g(2) - g(1)f(2)] \tag{1.250}$$

[Of course, if f and g are the same functions, then (1.250) vanishes, and (1.249) is replaced by $f(1)f(2)$.]

The spatial functions (1.249) and (1.250) must be multiplied by a spin function. For electrons, we have the following four possible two-electron spin functions, three symmetric and one antisymmetric:

		S	M_S	
symmetric:	$\alpha(1)\alpha(2)$	1	1	(1.251)
	$2^{-1/2}[\alpha(1)\beta(2) + \beta(1)\alpha(2)]$	1	0	(1.252)
	$\beta(1)\beta(2)$	1	−1	(1.253)
antisymmetric:	$2^{-1/2}[\alpha(1)\beta(2) - \beta(1)\alpha(2)]$	0	0	(1.254)

The quantum numbers listed are for the eigenvalues of the total-spin operators $\hat{S}^2$ and $\hat{S}_z$, where the total spin $\mathbf{S}$ is defined as $\mathbf{S} \equiv \mathbf{S}_1 + \mathbf{S}_2$. Since electrons are fermions, the symmetric two-electron spatial function (1.249) must be multiplied by the antisymmetric spin function (1.254) to give an overall wave function that is antisymmetric; the antisymmetric spatial function (1.250) must be multiplied by one of the symmetric spin functions (1.251)–(1.253).

An example is the perturbation treatment of the He atom in which we neglect the interelectronic repulsion e^2/r_{12}. The correct zeroth-order wave function for the ground state is

$$1s(1)1s(2)[\alpha(1)\beta(2)-\beta(1)\alpha(2)]/2^{1/2} \tag{1.255}$$

where $1s$ denotes a hydrogenlike orbital. The correct zeroth-order functions for the states arising from the $1s2s$ configuration are

$$\begin{aligned}
&{}^1S\text{: } 2^{-1/2}[1s(1)2s(2)+2s(1)1s(2)]2^{-1/2}[\alpha(1)\beta(2)-\beta(1)\alpha(2)]\\
&{}^3S\text{: } 2^{-1/2}[1s(1)2s(2)-2s(1)1s(2)]\alpha(1)\alpha(2)\\
&{}^3S\text{: } 2^{-1/2}[1s(1)2s(2)-2s(1)1s(2)]2^{-1/2}[\alpha(1)\beta(2)+\beta(1)\alpha(2)]\\
&{}^3S\text{: } 2^{-1/2}[1s(1)2s(2)-2s(1)1s(2)]\,\beta(1)\beta(2)
\end{aligned} \tag{1.256}$$

The term symbols ${}^1S$ and ${}^3S$ are defined in the next section. For the states of the $1s2p$ configuration, we get zeroth-order wave functions similar to those of (1.256), except that there are 12 ($=4\times3$) zeroth-order functions instead of 4; the factor 3 arises from the spatially degenerate $2p_x$, $2p_y$, and $2p_z$ orbitals.

For three or more electrons, the wave function cannot be factored into a simple product of a space part and a spin part; see the next section.

1.17 MANY-ELECTRON ATOMS

With internal nuclear motion neglected, the Hamiltonian for a many-electron atom with atomic number Z is

$$\hat{H} = -\frac{\hbar^2}{2m}\sum_i \nabla_i^2 - Ze^2\sum_i \frac{1}{r_i} + \sum_i \sum_{j>i} \frac{e^2}{r_{ij}} + \hat{H}_{\text{s.o.}} \tag{1.257}$$

The first term on the right is the operator for the electrons' kinetic energy; the second term is the operator for the potential energy of attraction between the electrons and the nucleus (r_i being the distance between electron i and the nucleus); the third term is potential energy of repulsion between all pairs of electrons (r_{ij} being the distance between electrons i and j); the last term is the spin–orbit interaction (discussed below). In addition, there are other relativistic terms besides spin–orbit interaction, which we neglect.

If we neglect interelectronic repulsions and spin–orbit interaction, then the atomic Hamiltonian is approximated by $\hat{H}^0$, where

$$\hat{H}^0 = \sum_i \left(-\frac{\hbar^2}{2m}\nabla_i^2 - \frac{Ze^2}{r_i} \right) \tag{1.258}$$

Since $\hat{H}^0$ is the sum of hydrogenlike Hamiltonians, the zeroth-order wave function is the product of hydrogenlike functions, one for each electron. We call any one-electron spatial wave function an *orbital*. To allow for electron spin, each spatial orbital is multiplied by a spin function (either α or β) to give a *spin-orbital*. To introduce the required antisymmetry into the wave function, we take the zeroth-order wave function as a *Slater determinant* of spin-orbitals. For example, for the Li ground state, the normalized zeroth-order wave function is

$$\frac{1}{\sqrt{6}} \begin{vmatrix} 1s(1)\alpha(1) & 1s(1)\beta(1) & 2s(1)\alpha(1) \\ 1s(2)\alpha(2) & 1s(2)\beta(2) & 2s(2)\alpha(2) \\ 1s(3)\alpha(3) & 1s(3)\beta(3) & 2s(3)\alpha(3) \end{vmatrix} \tag{1.259}$$

Since interchange of the rows of a determinant multiplies it by -1, (1.259) is antisymmetric. To fully specify the Slater determinant, it suffices to simply list the spin-orbitals; hence we abbreviate (1.259) to $|1s\alpha 1s\beta 2s\alpha|$. Adopting the convention that a bar indicates spin function β and the absence of a bar indicates spin function α, we further abbreviate (1.259) to

$$|1s\,\overline{1s}\,2s| \tag{1.260}$$

Note that if the spin-orbital in column three of (1.259) were the same as that of either column one or column two, the determinant would vanish (since two columns would be identical). Hence antisymmetry of ψ means that no two electrons can occupy the same spin-orbital in an approximate wave function in which electrons are assigned to orbitals.

Use of $\hat{H}^0$ as the approximation to the Hamiltonian gives rise to electron *configurations*, in which the electrons are assigned to orbitals. Thus for ground-state Li, we have the configuration $1s^2 2s$, with two electrons in the $1s$ orbital and one in the $2s$ orbital. The ground-state C configuration is $1s^2 2s^2 2p^2$. Orbitals having the same n and the same l are said to belong to the same subshell; the electron configuration is specified by stating how many electrons are in each subshell.

In general, a given electron configuration has several wave functions, each corresponding to the same eigenvalue of the zeroth-order Hamiltonian $\hat{H}^0$. This degeneracy is partly removed by inclusion of the electron repulsion term $\hat{H}_{\text{rep}}$,

$$\hat{H}_{\text{rep}} \equiv \sum_i \sum_{j>i} \frac{e^2}{r_{ij}} \tag{1.261}$$

thereby splitting each configuration into several *terms* (see below). Because of $\hat{H}_{\text{rep}}$, the operators for orbital angular momenta of individual electrons do not commute with the atomic Hamiltonian. Defining the *total* electronic orbital angular momentum **L** by

$$\mathbf{L} \equiv \sum_i \mathbf{L}_i \tag{1.262}$$

one finds that the total orbital angular momentum operators $\hat{L}^2$ and $\hat{L}_z$ commute with the Hamiltonian $\hat{H}^0 + \hat{H}_{\text{rep}}$. Similar to (1.262), one defines the total electronic spin angular momentum **S** of the atom by $\mathbf{S} = \Sigma_i \mathbf{S}_i$.

Since we are interested in the total angular momentum, we need the rule for addition of angular momenta. Let $\mathbf{M}_1$ and $\mathbf{M}_2$ be two angular momenta of any type (both orbital, both spin, or one spin and one orbital). Let the eigenvalues of $\hat{M}_1^2$, $\hat{M}_{1z}$, $\hat{M}_2^2$, $\hat{M}_{2z}$ be $j_1(j_1+1)\hbar^2$, $m_1\hbar$, $j_2(j_2+1)\hbar^2$, $m_2\hbar$, respectively. Defining the total angular momentum **M** as

$$\mathbf{M} \equiv \mathbf{M}_1 + \mathbf{M}_2 \tag{1.263}$$

one finds that the components $\hat{M}_x$, $\hat{M}_y$, $\hat{M}_z$ of $\hat{\mathbf{M}}$ obey the usual angular-momentum commutation relations; hence by the arguments of Section 1.7, the eigenvalues of $\hat{M}^2$ and $\hat{M}_z$ are

$$J(J+1)\hbar^2, \quad J = 0, \tfrac{1}{2}, 1, \ldots \quad \text{and} \quad M_J\hbar, \quad M_J = -J, \ldots, J \tag{1.264}$$

One can show that the total angular-momentum quantum number J takes on the possible values

$$J = j_1 + j_2, j_1 + j_2 - 1, \ldots, |j_1 - j_2| \tag{1.265}$$

(For example, if $j_1 = 2$ and $j_2 = 3$, then J has the possible values 5, 4, 3, 2, 1.)

The eigenvalues of the total electronic angular-momentum operators $\hat{L}^2$ and $\hat{S}^2$ of the atom are

$$L(L+1)\hbar^2, \qquad L = 0, 1, 2, \ldots \tag{1.266}$$

$$S(S+1)\hbar^2, \qquad S = 0, \tfrac{1}{2}, 1, \ldots \tag{1.267}$$

One finds that states of a given electronic configuration that have the same values of L and S have the same energy eigenvalue for the Hamiltonian $\hat{H}^0 + \hat{H}_{\text{rep}}$; such states are said to belong to the same *term*. Different terms arising from the same electronic configuration have different energies. For example, consider the $1s2p$ He configuration. We have $s_1 = \frac{1}{2}$, $s_2 = \frac{1}{2}$ for the two electrons. Hence from (1.265), we have $S = 1, 0$ for the total electronic spin quantum number. We have $l_1 = 0$, $l_2 = 1$ for the electrons, and (1.265) gives $L = 1$ for the total electronic orbital angular-momentum quantum number. To designate each term, we specify the L value by a code letter ($S, P, D, F, G, \ldots$ indicate L values of $0, 1, 2, 3, 4, \ldots$), and we write the *multiplicity* $2S + 1$ as a left superscript on the L code letter. Thus the terms arising from the He $1s2p$ configuration are ^3P and ^1P, each having a different energy. Equation (1.256) gives the zeroth-order wave functions of the He $1s2s$ terms ^3S and ^1S. Electrons in filled subshells make no contribution to the total spin or orbital angular momentum. Thus a $1s^2 2s^2 2p3p$ configuration gives the terms $^3S, ^3P, ^3D, ^1S, ^1P, ^1D$.

The terms arising from a configuration with more than one electron in a given partly filled subshell are not so easily found, since the Pauli principle must be taken into account. We omit discussion.

Finally, consider $\hat{H}_{\text{s.o.}}$. From an electron's viewpoint, the nucleus is moving around the electron, thereby producing a magnetic field (proportional to the electronic orbital angular momentum **L**); this field interacts with the electronic magnetic moment (which is proportional to **S**), giving the *spin–orbit interaction*. When $\hat{H}_{\text{s.o.}}$ is included in the Hamiltonian, the total electronic orbital and spin angular momentum operators $\hat{L}^2$ and $\hat{S}^2$ no longer commute with the atomic Hamiltonian, but the operator for the *total* electronic angular momentum **J** does commute with $\hat{H}^0 + \hat{H}_{\text{s.o.}} + \hat{H}_{\text{rep}}$. The definition of **J** is $\mathbf{J} \equiv \mathbf{L} + \mathbf{S}$. The eigenvalues of $\hat{J}^2$ are $J(J+1)\hbar^2$. Inclusion of $\hat{H}_{\text{s.o.}}$ splits each term into several closely spaced *levels*, each level corresponding to a different J value. (Different values of J correspond to different relative orientations of **L** and **S**, and different orientations of these vectors give different amounts of spin–orbit interaction.) The possible values of J are found from the L and S values of the term using the rule (1.265). We designate a level by adding its J value as a right subscript to the term symbol. Thus for a ^3P term, $L = 1$ and $S = 1$, so J can be 2, 1, or 0; hence the levels of the term are $^3P_2$, $^3P_1$, $^3P_0$. Even though $\hat{L}^2$ and $\hat{S}^2$ don't commute with $\hat{H}$ when $\hat{H}_{\text{s.o.}}$ is included, provided $\hat{H}_{\text{s.o.}}$ is small, $\hat{L}^2$ and $\hat{S}^2$ "almost" commute with $\hat{H}$, and it is a good approximation to use the quantum numbers L and S to characterize atomic energy levels. Tables of atomic energy levels as derived from atomic spectra are available: C. E. Moore, *Atomic Energy Levels*, Nat. Bur. Stds. Circular 467, vols. I, II, and III, 1949, 1952, and 1958.

Each level is $(2J+1)$-fold degenerate, corresponding to the $2J+1$ values of M_J, where $M_J\hbar$ is the eigenvalue of $\hat{J}_z$. This degeneracy can be removed by application of an external magnetic field (the Zeeman effect), thereby splitting each level into its component *states*.

The preceding discussion, in which we first add the $\mathbf{L}_i$'s to get $\mathbf{L}$ and add the $\mathbf{S}_i$'s to get $\mathbf{S}$, and then add $\mathbf{L}$ and $\mathbf{S}$ to get $\mathbf{J}$, is called *Russell–Saunders coupling* (or *L–S* coupling). Russell–Saunders coupling is valid when $\hat{H}_{\text{s.o.}}$ is small compared to $\hat{H}_{\text{rep}}$. Spin–orbit interaction increases with increasing atomic number, so that for heavy atoms $\hat{H}_{\text{s.o.}}$ exceeds $\hat{H}_{\text{rep}}$ and Russell–Saunders coupling does not hold. For very heavy atoms, *j–j* coupling holds; here, one adds the spin and orbital angular momenta of electron i to get the total angular momentum $\mathbf{j}_i$ for electron i; the $\mathbf{j}_i$'s are then added to give the total electronic angular momentum $\mathbf{J}$.

The zeroth-order antisymmetric wave functions of closed-subshell states, states that have only one electron outside a closed-subshell configuration, and states that are one electron short of having a closed-subshell configuration can be expressed as a single Slater determinant [e.g. (1.259)]. However, for open-subshell states in general, one has to take an appropriate linear combination of a few Slater determinants to get a state that is an eigenfunction of $\hat{L}^2$ and $\hat{S}^2$.

As noted above, application of an external magnetic field $\mathbf{B}$ splits each atomic energy level into states, each state having a different M_J value. The electronic orbital angular momentum $\mathbf{L}$ gives rise [Equation (1.164)] to an orbital magnetic dipole moment $\boldsymbol{\mu}_L = -(e/2mc)\mathbf{L}$. The spin magnetic dipole moment is given by (1.246) as $\boldsymbol{\mu}_S = -(e/mc)\mathbf{S}$. The interaction energy E_B between an applied magnetic field $\mathbf{B}$ and a magnetic dipole moment $\boldsymbol{\mu}$ is [*Halliday and Resnick*, Equation (33-12)]

$$E_B = -\boldsymbol{\mu} \cdot \mathbf{B} \tag{1.268}$$

Hence the Hamiltonian for the Zeeman interaction is

$$\hat{H}_B = -(\hat{\boldsymbol{\mu}}_L + \hat{\boldsymbol{\mu}}_S) \cdot \mathbf{B} = \beta_e \hbar^{-1}(\hat{\mathbf{L}} + 2\hat{\mathbf{S}}) \cdot \mathbf{B} = \beta_e B \hbar^{-1}(\hat{J}_z + \hat{S}_z) \tag{1.269}$$

where the *Bohr magneton* β_e is defined as $\beta_e \equiv e\hbar/2mc$, where $\mathbf{J} = \mathbf{L} + \mathbf{S}$ was used, and where the z direction was taken to be the direction of $\mathbf{B}$. Since $\hat{H}_B$ depends on $\hat{J}_z$, states with different M_J values are split by the magnetic field.

The magnetic-field vector $\mathbf{B}$ is called the *magnetic induction* or the *magnetic flux density*. A more suitable name for $\mathbf{B}$ would be the "magnetic field strength"; however, historically, the magnetic-field vector $\mathbf{H}$ was given the name *magnetic field strength*. This is because it was formerly believed that $\mathbf{H}$ was the fundamental magnetic-field vector; it is now known that $\mathbf{B}$ is the fundamental magnetic-field vector (just as $\mathbf{E}$ is the fundamental electric-field vector). Many texts write **H** instead of **B**; in Gaussian units (which we are using), **H**=**B** in vacuum; since **B** is the fundamental vector, it is best to use **B** in equations like (1.268). The SI unit of B is the tesla (T), also called the weber/m^2; the Gaussian unit of B is the gauss (G); 1 T=10^4 G.

In classical mechanics, application of an external magnetic field **B** to a magnetic dipole moment produces a torque on the magnetic moment, causing the magnetic moment vector (and the corresponding angular momentum vector) to revolve about the field direction at a fixed angle to **B**; this motion is called *precession*. In quantum mechanics, one can show that the average values of the angular-momentum components precess about the field direction.

If the atomic nucleus has spin angular momentum $\mathbf{I}\neq 0$, then the nuclear magnetic moment interacts with the electronic orbital and spin magnetic moments to give atomic *hyperfine structure*; the total atomic angular momentum **F** is the sum of **I** and **J**: $\mathbf{F}=\mathbf{I}+\mathbf{J}$. In the H atom 1$s$ ground state, $J=\frac{1}{2}$; a proton has spin $\frac{1}{2}$, so $I=\frac{1}{2}$; hence F can be 0 or 1; the $F=1\rightarrow 0$ transition gives a line at 1420 megacycles per second.

The parity of atomic states is important in spectroscopy. A radial function is an even function [see (1.113)]; the spherical harmonic Y_l^m is found to be an even or odd function of the Cartesian coordinates according to whether l is an even or odd number. For a many-electron atom, it follows that states arising from a configuration for which the sum of the l values of all the electrons is an even number are even functions; when $\Sigma_i l_i$ is odd, the state has odd parity.

Let us briefly consider calculation of atomic wave functions. Consideration of just $\hat{H}^0$ gives a zeroth-order wave function that is an antisymmetrized product of hydrogenlike spin-orbitals. Such a wave function is very poor—for one thing, it uses the full nuclear charge for each orbital; in reality, inner electrons screen the outer electrons from the nucleus. We can take an antisymmetrized product of spin-orbitals and vary the forms of the orbitals until we get the best possible wave function of this restricted form, as determined by minimization of the variational integral. This best possible orbital wave function is the *Hartree–Fock* function, and the orbitals are the Hartree–Fock orbitals. The procedure for finding the Hartree–Fock wave function is discussed in Section 1.20. Because of its restricted form, the Hartree–Fock ψ must deviate significantly from the true ψ; the most important method of improving on Hartree–Fock wave functions is configuration interaction, discussed in Section 1.20.

1.18 SYMMETRY POINT GROUPS

A *symmetry operation* is a transformation of a body that sends it to a position physically indistinguishable from its original position, and that preserves the distances between all pairs of points in the body. A *symmetry element* is a point, line, or plane with respect to which a symmetry operation is carried out. In discussing the symmetry of molecules, we consider the symmetry of the nuclear framework.

We consider four kinds of symmetry elements. For an *n-fold proper rotation axis of symmetry* C_n, rotation by $2\pi/n$ radians about the axis is a symmetry operation. For a *plane of symmetry* σ, reflection through the plane is a symmetry operation. For a *center of symmetry* i, inversion through this center point is a symmetry operation. For an *n-fold improper rotation axis* S_n, rotation by $2\pi/n$ radians about the axis followed by reflection in a plane perpendicular to the axis is a symmetry operation. To denote symmetry operations, we add a circumflex to the symbol for the corresponding symmetry element. Thus $\hat{C}_n$ is a rotation by $2\pi/n$ radians. Note that since $\hat{S}_1=\hat{\sigma}$, a plane of symmetry is equivalent to an S_1 axis. It is easy to see that a 180° rotation about an axis followed by reflection in a plane perpendicular to the axis is equivalent to inversion; hence $\hat{S}_2=\hat{i}$, and a center of symmetry is equivalent to an S_2 axis.

We always choose the z axis of the coordinate system as coinciding with the highest-order (proper or improper) rotation axis of the molecule. A symmetry plane that contains this axis is called a σ_v plane; a symmetry plane perpendicular to this axis is called a σ_h plane (where v and h stand for *vertical* and *horizontal*).

The dipole moment of a molecule with a symmetry axis must lie along the axis; hence a molecule with two noncoincident axes of symmetry has no dipole moment.

The *product* of two symmetry operations is defined to mean successive performance of them, with the operation on the right of the product being performed first. We use the convention that the coordinate axes remain fixed in space when symmetry operations are performed on a body. For example, consider the product $\hat{\sigma}(xz)\hat{C}_2(z)$, where $\hat{C}_2(z)$ is a counterclockwise rotation about the z axis and $\hat{\sigma}(xz)$ is a reflection in the xz plane. The $\hat{C}_2(z)$ operation converts the x and y coordinates of a point to their negatives, and leaves the z coordinate unchanged. The $\hat{\sigma}(xz)$ operation converts the y coordinate of a point to its negative, and leaves the x and z coordinates unchanged. Hence the product $\hat{\sigma}(xz)\hat{C}_2(z)$ has the effect of bringing a point originally at (x,y,z) to $(-x,y,z)$. Since a $\hat{\sigma}(yz)$ reflection brings a point from (x,y,z) to $(-x,y,z)$, we conclude that $\hat{\sigma}(xz)\hat{C}_2(z) = \hat{\sigma}(yz)$. Two symmetry operations are equal when they have the same effect on every point in space.

A slight knowledge of trigonometry will enable the reader to verify that the $\hat{C}_4(z)$ operation brings a point from (x,y,z) to $(-y,x,z)$. Using this fact, one finds that the product $\hat{C}_2(x)\hat{C}_4(z)$ brings a point from (x,y,z) to $(-y,-x,-z)$, whereas the product $\hat{C}_4(z)\hat{C}_2(x)$ brings a point from (x,y,z) to $(y,x,-z)$. Hence $\hat{C}_4(z)\hat{C}_2(x) \neq \hat{C}_2(x)\hat{C}_4(z)$. Symmetry operations do not always commute.

From the definition of a symmetry operation, it follows that the product of any two symmetry operations of a body must be a symmetry operation. With a C_n symmetry axis, we thus have the following symmetry operations: $\hat{C}_n, \hat{C}_n^2, \hat{C}_n^3, \ldots, \hat{C}_n^{n-1}, \hat{C}_n^n$, where the exponent denotes the number of times the operation is performed. A $\hat{C}_n^n$ operation is a 360° rotation and therefore brings the body back to its original position; we call such an operation the *identity operation* $\hat{E}$. The set of all symmetry operations (including $\hat{E}$) forms a mathematical *group*. (The definition of a group is given in Chapter 9.) We now list the possible symmetry point groups of molecules. (The term "point group" is used because one point, the molecular center of mass, remains fixed for all symmetry operations.) We use a script letter to denote a symmetry group.

Groups with no C_n axis: $\mathcal{C}_1$, $\mathcal{C}_s$, $\mathcal{C}_i$. A molecule with no symmetry elements belongs to the group $\mathcal{C}_1$; this group consists of the symmetry operation $\hat{E}$. CHFBrI belongs to $\mathcal{C}_1$. The point group of a molecule with only a plane of symmetry is $\mathcal{C}_s$; the group consists of the symmetry operations $\hat{\sigma}$ and $\hat{E}(=\hat{\sigma}^2)$. HOCl belongs to $\mathcal{C}_s$. The point group of a molecule with only a center of symmetry is $\mathcal{C}_i$.

Groups with a single C_n axis: $\mathcal{C}_n$, $\mathcal{C}_{nh}$, $\mathcal{C}_{nv}$, $\mathcal{S}_{2n}$. A molecule with only a C_n axis belongs to the group $\mathcal{C}_n$ $(n=2, 3, 4, \ldots)$; the symmetry operations are $\hat{C}_n \hat{C}_n^2, \ldots, \hat{C}_n^{n-1}, \hat{E}$. The point group of H_2O_2 (dihedral angle 112°) is $\mathcal{C}_2$. A molecule whose symmetry elements are a C_n axis and a σ_h plane (perpendicular to the C_n axis) belongs to $\mathcal{C}_{nh}$ $(n=2, 3, 4, \ldots)$. The point group of *trans*-1,2-dibromoethylene is $\mathcal{C}_{2h}$. A molecule whose symmetry elements are a C_n axis and n σ_v planes (each containing the C_n axis) belongs to $\mathcal{C}_{nv}$. The point group of H_2O is $\mathcal{C}_{2v}$; that of NH_3 is $\mathcal{C}_{3v}$. The symmetry operations arising from an S_n axis constitute the group $\mathcal{S}_n$. One finds that for $n=3,5,7,\ldots$ the group $\mathcal{S}_n$ is the same as $\mathcal{C}_{nh}$; also, $\mathcal{S}_2$ is identical to $\mathcal{C}_i$ and $\mathcal{S}_1$ is identical to $\mathcal{C}_s$. Hence only for $n=4,6,8,\ldots$ is $\mathcal{S}_n$ a new group; we shall use the symbol $\mathcal{S}_{2n}$ $(n=2, 3, 4, \ldots)$ to emphasize the even order of the improper axis. One finds that the S_{2n} axis is also a C_n axis.

Groups with one C_n axis and n C_2 axes: $\mathcal{D}_n$, $\mathcal{D}_{nh}$, $\mathcal{D}_{nd}$. For group $\mathcal{D}_n$ ($n=2, 3, 4, \ldots$), the molecular symmetry elements are a C_n axis and n C_2 axes perpendicular to the C_n axis. A molecule with a C_n axis, n C_2 axes perpendicular to the C_n axis, and a σ_h symmetry plane (perpendicular to the C_n axis) belongs to group $\mathcal{D}_{nh}$ ($n=2, 3, 4, \ldots$). Because of the σ_h plane, the C_n axis is also an S_n axis. If n is even, the C_n axis is also an S_2 axis, and the molecule has a center of symmetry; there are also n σ_v symmetry planes, each containing the C_n axis and one C_2 axis. The trigonal planar molecule BF_3 belongs to $\mathcal{D}_{3h}$; benzene, C_6H_6, belongs to $\mathcal{D}_{6h}$. A molecule that has a C_n axis, n C_2 axes perpendicular to the C_n axis, and n vertical planes of symmetry each of which contains the C_n axis and bisects the angles between adjacent C_2 axes belongs to group $\mathcal{D}_{nd}$. The vertical symmetry planes in $\mathcal{D}_{nd}$ are called σ_d planes (for *diagonal*). The C_n axis is also an S_{2n} axis. In allene, $CH_2{=}C{=}CH_2$, one CH_2 group is rotated 90° with respect to the other, and the point group is $\mathcal{D}_{2d}$; the staggered conformation of ethane belongs to $\mathcal{D}_{3d}$.

Groups with more than one C_n axis, $n>2$: $\mathcal{T}_d$, $\mathcal{O}_h$, $\mathcal{I}_h$, $\mathcal{K}_h$, and so on. The group $\mathcal{T}_d$ consists of the symmetry operations of a regular tetrahedron. CH_4 belongs to $\mathcal{T}_d$; its symmetry elements are four C_3 axes (one along each C—H bond), three S_4 axes which are also C_2 axes (each bisecting two of the HCH bond angles), and six symmetry planes (each containing two C—H bonds). $\mathcal{O}_h$ is the group of symmetry operations of a regular octahedron or of a cube. (The cube and the regular octahedron have the same symmetry elements.) The octahedral molecule SF_6 belongs to $\mathcal{O}_h$. $\mathcal{I}_h$ is the group of symmetry operations of a regular pentagonal dodecahedron or icosahedron. $\mathcal{K}_h$ is the group of symmetry operations of a sphere. The groups $\mathcal{T}$, $\mathcal{T}_h$, $\mathcal{O}$, and $\mathcal{I}$ are related to $\mathcal{T}_d$, $\mathcal{O}_h$, and $\mathcal{I}_h$, but are chemically unimportant and will not be discussed.

A linear molecule belongs to $\mathcal{C}_{\infty v}$ if it has no center of symmetry, or to $\mathcal{D}_{\infty h}$ if it has a center of symmetry.

Let the symmetry operation $\hat{R}$ bring the point at (x,y,z) to (x',y',z'):

$$\hat{R}\,(x,y,z)\rightarrow(x',y',z') \tag{1.270}$$

Corresponding to the symmetry operation $\hat{R}$ (which acts on points in space), we define the operator $\hat{O}_R$, which acts on functions as follows:

$$\hat{O}_R f(x',y',z')=f(x,y,z) \tag{1.271}$$

The function $\hat{O}_R f$ has the same value at point (x',y',z') as f has at (x,y,z). For example, $\hat{C}_4(z)$ rotates points by 90° counterclockwise around the z axis; the operator $\hat{O}_{C_4(z)}$ applied to a $2p_x$ orbital changes it to a $2p_y$ orbital.

The operator $\hat{O}_i$ is the parity operator. To show this, we note that for $\hat{R}=\hat{i}$, Equations (1.270) and (1.271) become $\hat{i}(x,y,z)\to(-x,-y,-z)$ and $\hat{O}_i f(-x,-y,-z)=f(x,y,z)$. If we rename the variables according to $\bar{x}=-x$, $\bar{y}=-y$, $\bar{z}=-z$, then we have $\hat{O}_i f(\bar{x},\bar{y},\bar{z})=f(-\bar{x},-\bar{y},-\bar{z})$. Since $(\bar{x},\bar{y},\bar{z})$ is a general point in space, the bars can be dropped to give $\hat{O}_i f(x,y,z)=f(-x,-y,-z)$, which is the definition of $\hat{\Pi}$.

1.19 MOLECULAR ELECTRONIC STRUCTURE

The Born–Oppenheimer approximation. With spin–orbit and other relativistic interactions omitted, the Hamiltonian of a polyatomic molecule is

$$\hat{H}=-\frac{\hbar^2}{2}\sum_{\alpha}\frac{1}{m_\alpha}\nabla_\alpha^2-\frac{\hbar^2}{2m}\sum_i\nabla_i^2+\sum_{\alpha}\sum_{\beta>\alpha}\frac{Z_\alpha Z_\beta e^2}{r_{\alpha\beta}}$$

$$-\sum_{\alpha}\sum_{i}\frac{Z_\alpha e^2}{r_{i\alpha}}+\sum_{j}\sum_{i>j}\frac{e^2}{r_{ij}} \tag{1.272}$$

where nucleus α has mass m_α and atomic number Z_α, $r_{\alpha\beta}$ is the distance between nuclei α and β, $r_{i\alpha}$ is the distance between electron i and nucleus α, and r_{ij} is the distance between electrons i and j. The molecular Schrödinger equation is

$$\hat{H}\psi(q_i,q_\alpha)=E\psi(q_i,q_\alpha) \tag{1.273}$$

where q_i and q_α symbolize the electronic and nuclear coordinates, respectively. Because the nuclear masses are much greater than the electronic masses, it is an excellent approximation to treat the electronic motion separately from the nuclear motion, with the nuclei held in a fixed configuration while the electronic motion is solved. With the omission of nuclear kinetic-energy operators from (1.272), the electronic Schrödinger equation is

$$\left(\hat{H}_{\text{el}}+V_{NN}\right)\psi_{\text{el}}=U\psi_{\text{el}} \tag{1.274}$$

$$\hat{H}_{\text{el}}\equiv-\frac{\hbar^2}{2m}\sum_i\nabla_i^2-\sum_{\alpha}\sum_{i}\frac{Z_\alpha e^2}{r_{i\alpha}}+\sum_{j}\sum_{i>j}\frac{e^2}{r_{ij}} \tag{1.275}$$

$$V_{NN}\equiv\sum_{\alpha}\sum_{\beta>\alpha}\frac{Z_\alpha Z_\beta e^2}{r_{\alpha\beta}} \tag{1.276}$$

$\hat{H}_{\text{el}}$ is the purely electronic Hamiltonian; V_{NN} is the nuclear repulsion. Since V_{NN} is constant for a given nuclear configuration, it can be omitted from (1.274); this gives the same wave functions ψ_{el} and simply reduces the energy values by the constant V_{NN}. Thus we have

$$\hat{H}_{\text{el}}\psi_{\text{el}} = E_{\text{el}}\psi_{\text{el}} \tag{1.277}$$

$$U = E_{\text{el}} + V_{NN} \tag{1.278}$$

where E_{el} is the purely electronic energy, and U is the electronic energy including nuclear repulsion. We solve the electronic Schrödinger equation (1.277) or (1.274) to obtain electronic wave functions and energies that depend parametrically on the nuclear configuration:

$$\psi_{\text{el}} = \psi_{\text{el},n}(q_i; q_\alpha) \quad \text{and} \quad U = U_n(q_\alpha) \tag{1.279}$$

where n stands for the electronic quantum numbers. (For a given nuclear configuration, there are many possible electronic states.)

Having solved for the electronic wave functions and energies, we use the electronic energy including nuclear repulsion U as the potential energy in the Schrödinger equation for nuclear motion:

$$\hat{H}_N\psi_N = E\psi_N \tag{1.280}$$

$$\hat{H}_N = -\frac{\hbar^2}{2}\sum_\alpha \frac{1}{m_\alpha}\nabla_\alpha^2 + U(q_\alpha) \tag{1.281}$$

The constant E in (1.280) is the total energy of the molecule, since $\hat{H}_N$ includes operators for both nuclear and electronic energies. Each electronic state has a different electronic energy $U(q_\alpha)$; hence we must solve a different nuclear Schrödinger equation for each electronic state of the molecule.

The complete molecular wave function is

$$\psi(q_i, q_\alpha) = \psi_{\text{el}}(q_i; q_\alpha)\psi_N(q_\alpha) \tag{1.282}$$

This approximation of treating electronic and nuclear motions separately is the Born–Oppenheimer approximation. In the rest of this chapter, we consider the electronic Schrödinger equation.

Diatomic molecules. For a diatomic molecule, the nuclear configuration is a function of only one variable, the internuclear distance R. The curve $U(R)$ for a bound state resembles one of the heavy curves in Figs. 4.4, 7.1, and 7.2. At $R = 0$, the V_{NN} term makes U go to infinity. At $R = \infty$, U

equals the sum of the separated-atoms energies. The minimum in U occurs at the equilibrium internuclear distance R_e; the quantity $U(\infty) - U(R_e)$ is the equilibrium dissociation energy D_e (Fig. 4.4). For an unbound state, $U(R)$ has no minimum, but increases monotonically as R decreases from ∞ to 0.

The molecular electronic wave functions ψ_{el} are classified using the operators that commute with $\hat{H}_{\text{el}}$. For diatomic (and linear polyatomic) molecules, the operator $\hat{L}_z$ for the component of the total electronic orbital angular momentum along the internuclear axis commutes with $\hat{H}_{\text{el}}$ (although $\hat{L}^2$ does not commute with $\hat{H}_{\text{el}}$). The $\hat{L}_z$ eigenvalues are $M_L\hbar$, $M_L = 0, \pm 1, \pm 2, \ldots$. The quantum number Λ is defined as $\Lambda \equiv |M_L|$. The code letters Σ, Π, Δ, Φ, Γ, ... indicate Λ values of 0, 1, 2, 3, 4, ... For each nonzero Λ value, there are two M_L values: $+\Lambda$ and $-\Lambda$; it is found that the electronic energy depends on M_L^2, and there is a twofold degeneracy associated with Π, Δ, Φ, ... levels. Besides Λ, we have the quantum number S, where $S(S+1)\hbar^2$ is the square of the magnitude of the total electronic spin angular momentum.

If interelectronic repulsions are neglected (or taken into account in only an average way, as in the Hartree–Fock method), each electron is assigned to a molecular orbital (MO), and ψ_{el} is an antisymmetrized product of spin-orbitals. Each diatomic MO is characterized by the quantum number m, where $m\hbar$ is the axial (z) component of orbital angular momentum of an electron in that MO. The MO quantum number λ is defined as $\lambda \equiv |m|$, and the letter code σ, π, δ, ϕ, γ, ... indicates λ values of 0, 1, 2, 3, 4, ... (As in atoms, lowercase quantum numbers refer to individual electrons, whereas capital quantum numbers refer to the state of the atom or molecule.) The MO energy depends on m^2, so π MOs (and δ MOs, and ...) occur as degenerate pairs; a σ MO is nondegenerate. Each σ MO holds two electrons (one with spin α and one with spin β); each pair of degenerate π (or δ or ...) MOs holds four electrons. A set of degenerate MOs constitutes a molecular subshell; specification of the number of electrons in each subshell defines the electronic *configuration*.

Diatomic electronic states arising from the same electronic configuration and having the same values of Λ and of S belong to the same electronic *term*. As in atoms, states of the same term have the same energy if spin–orbit interaction is neglected. To specify a diatomic term, the *multiplicity* $2S+1$ is affixed as a left superscript to the Λ code letter. (The terminology *singlet, doublet, triplet, quartet*, ... is used for atomic and molecular terms according to whether $2S+1$ is 0, 1, 2, 3, 4, ...) For Σ terms (i.e., $\Lambda = 0$ terms), one adds a right superscript of plus or minus, according to whether the eigenvalue of ψ_{el} for reflection in a σ_v plane (containing the axis) is $+1$ or -1. [Σ^+ and Σ^- terms of the same configuration have different energies. As mentioned, non-Σ terms have a twofold spatial degeneracy, and we can form linear combinations of the two wave functions to give eigenfunctions of $\hat{O}_{\sigma_v}$. Since these linear combinations have the same energy, one does not ordinarily use the plus and minus superscripts for non-Σ terms. However, interaction between the electronic orbital angular momentum and the molecular rotational angular momentum causes a very slight splitting (Λ-type doubling) between the two spatially degenerate states of non-Σ terms, and here one uses the plus and minus superscripts.] If the diatomic (or linear polyatomic) molecule has a center of symmetry, a right subscript of g or u (for *gerade* or *ungerade*, German for even or odd) is added to the term symbol according to whether the inversion eigenvalue of ψ_{el} is $+1$ or -1. As an example, consider the terms arising from a $\pi\pi$ electronic configuration (two electrons in different π subshells). We have $s_1 = s_2 = \frac{1}{2}$, so $S = 1, 0$. Each electron has $m = \pm 1$, so the M_L values (found as the algebraic sum of the m's) are $2, -2, 0, 0$; the $+2$ and -2 values belong to the two spatially degenerate states of a Δ term; the two 0 values belong to two Σ terms, which are found (*Levine*, Section 13.7) to have different $\hat{O}_{\sigma_v}$ eigenvalues. Hence the terms are $^1\Sigma^+$, $^3\Sigma^+$, $^1\Sigma^-$, $^3\Sigma^-$, $^1\Delta$, and $^3\Delta$. A filled-subshell configuration gives the single term $^1\Sigma^+$ ($^1\Sigma_g^+$ for homonuclear diatomics).

The ground term of a molecule is prefixed by the letter X. Excited terms that have the same multiplicity as the ground term are usually prefixed by $A, B, C, \ldots$. The letters $a, b, c, \ldots$ usually denote excited terms of different multiplicity than the ground term.

Spin–orbit interaction splits a molecular term into closely spaced energy levels. For a diatomic molecule, the axial component of total electronic spin is $\Sigma\hbar$, where the quantum number Σ (which is the analog of M_S in atoms) ranges from $-S$ to $+S$. The axial components of electronic spin and orbital angular momenta add to give the total axial electronic angular momentum of a diatomic molecule as $(\Lambda + \Sigma)\hbar$, where $\Lambda + \Sigma$ ranges from $\Lambda + S$ to $\Lambda - S$. Diatomic energy levels are specified by adding the value of $\Lambda + \Sigma$ as a right subscript to the term symbol. For example, a $^4\Pi$ term has the energy levels $^4\Pi_{5/2}$, $^4\Pi_{3/2}$, $^4\Pi_{1/2}$, $^4\Pi_{-1/2}$. The quantum number Ω is defined as $\Omega \equiv |\Lambda + \Sigma|$.

Polyatomic molecules. The same term classifications hold for linear polyatomic molecules as for diatomic molecules. We now consider nonlinear polyatomics. With spin–orbit interaction neglected, the total electronic spin angular momentum operator $\hat{S}^2$ commutes with $\hat{H}_{\text{el}}$, and polyatomic-molecule terms are classified according to the multiplicity $2S + 1$. For nonlinear molecules, the electronic orbital angular momentum operators do not commute with $\hat{H}_{\text{el}}$. The symmetry operators $\hat{O}_R$, $\hat{O}_S$, ... (corresponding to the molecular symmetry operations $\hat{R}$, $\hat{S}$, ...) commute with $\hat{H}_{el}$, and electronic wave functions are classified according to their behavior on application of $\hat{O}_R$, $\hat{O}_S$, ...

Consider H_2O as an example (Fig. 1.2). The point group is $\mathcal{C}_{2v}$, and the symmetry operations are $\hat{E}$, $\hat{C}_2(z)$, $\hat{\sigma}_v(xz)$, $\hat{\sigma}_v(yz)$. All pairs of these operators are found to commute; hence ψ_{el} can be chosen to be an eigenfunction of all of them. $\hat{O}_E$ is the unit operator, so that $\hat{O}_E\psi_{el}=\psi_{el}$, and ψ_{el} has eigenvalue $+1$ for $\hat{O}_E$. For each of the other three symmetry operators, we have $\hat{O}_R^2=1$; the same reasoning used in Section 1.8 to find the eigenvalues of $\hat{\Pi}$ shows that any operator whose square is 1 has the eigenvalues ± 1. Hence each H_2O ψ_{el} has $\hat{O}_E$ eigenvalue $+1$ and eigenvalue $+1$ or -1 for each of the other three symmetry operators. One might thus think that there would be $1\times 2\times 2\times 2=8$ possible sets of symmetry eigenvalues. However, only certain combinations of symmetry eigenvalues are allowed: the symmetry operators $\hat{O}_R$, $\hat{O}_S$, ... multiply the same way the symmetry operations $\hat{R}$, $\hat{S}$, multiply; that is, if $\hat{R}\hat{S}=\hat{T}$, then $\hat{O}_R\hat{O}_S=\hat{O}_T$. Suppose that ψ_{el} is an eigenfunction of $\hat{O}_R$, $\hat{O}_S$, and $\hat{O}_T$ with eigenvalues r, s, and t, and let $\hat{R}\hat{S}=\hat{T}$. We then have $t\psi_{el}=\hat{O}_T\psi_{el}=\hat{O}_R(\hat{O}_S\psi_{el})=s\hat{O}_R\psi_{el}=rs\psi_{el}$; hence $t=rs$. Thus the symmetry eigenvalues must multiply the same way the corresponding symmetry operators multiply. For example, for H_2O we find $\hat{C}_2(z)\hat{\sigma}_v(xz)=\hat{\sigma}_v(yz)$. If a, b, c are the eigenvalues of ψ_{el} for $\hat{O}_{C_2(z)}$, $\hat{O}_{\sigma_v(xz)}$, $\hat{O}_{\sigma_v(yz)}$, then we must have $ab=c$. Thus if the a and b eigenvalues happen to be -1 and $+1$, respectively, then the c eigenvalue of $\hat{O}_{\sigma_v(yz)}$ must equal -1. One finds (Problem 1.11) that there are only four sets of symmetry eigenvalues that multiply properly, and these are

	$\hat{E}$	$\hat{C}_2(z)$	$\hat{\sigma}_v(xz)$	$\hat{\sigma}_v(yz)$
A_1	1	1	1	1
A_2	1	1	-1	-1
B_1	1	-1	1	-1
B_2	1	-1	-1	1

(1.283)

Each set has been labeled. The letter A indicates symmetry eigenvalue $+1$ for the highest-order $\hat{C}_n$ or $\hat{S}_n$ operation of the molecule, while B indicates eigenvalue -1 for this operation [which is $\hat{C}_2(z)$ in water]; the numerical subscripts distinguish different sets with the same letter label. The term *symmetry species* (or symmetry type or irreducible representation) denotes any given set of symmetry eigenvalues in (1.283). The *totally*

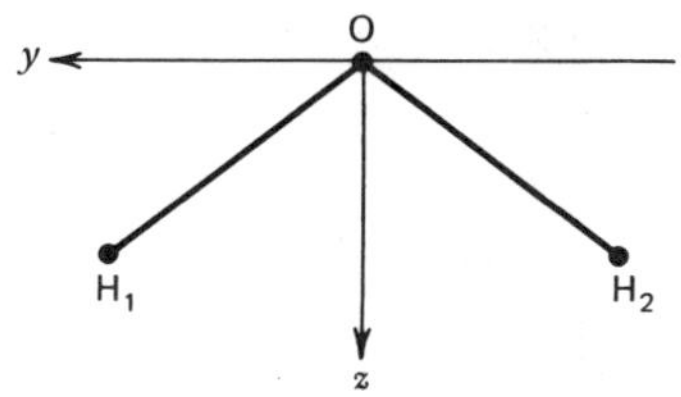

Fig. 1.2 Axes for the H_2O molecule; the x axis is perpendicular to the molecular plane.

symmetric symmetry species is the one with all symmetry eigenvalues equal to $+1$. (This is species A_1 for $\mathcal{C}_{2v}$.)

We denote an electronic term of H_2O by adding the multiplicity as a left superscript to the symmetry-species symbol. Thus the electronic states of a ${}^3A_1$ term of a $\mathcal{C}_{2v}$ molecule have $S=1$ and $\hat{O}_R\psi_{\text{el}}=\psi_{\text{el}}$ for each of the four symmetry operators.

The *orbital degeneracy* of a molecular electronic term is the degeneracy associated with the electronic spatial motion, and does not consider any additional spin degeneracy. (Thus Σ terms of diatomic molecules are orbitally nondegenerate, whereas Π, Δ, ... terms are orbitally doubly degenerate.) For $\mathcal{C}_{2v}$, all pairs of symmetry operators commute, and therefore ψ_{el} is a simultaneous eigenfunction of all symmetry operators; but for some point groups, the symmetry operators don't all commute, and ψ_{el} is not necessarily an eigenfunction of all the symmetry operators. To see what happens here, let $\hat{O}_R$ be a symmetry operator (which commutes with $\hat{H}_{\text{el}}$). Let $\psi_{\text{el},i}$ belong to an n-fold orbitally degenerate term. Application of $\hat{O}_R$ to the electronic Schrödinger equation gives $\hat{O}_R\hat{H}_{\text{el}}\psi_{\text{el},i} = \hat{O}_R E_{\text{el}}\psi_{\text{el},i}$; use of the commutation of $\hat{O}_R$ and $\hat{H}_{\text{el}}$ gives $\hat{H}_{\text{el}}(\hat{O}_R\psi_{\text{el},i}) = E_{\text{el}}(\hat{O}_R\psi_{\text{el},i})$. Hence $\hat{O}_R\psi_{\text{el},i}$ is an eigenfunction of $\hat{H}_{\text{el}}$ with the same energy eigenvalue as $\psi_{\text{el},i}$, and therefore must be a linear combination of the n orbitally degenerate electronic wave functions of the term that have the same $\hat{S}_z$ eigenvalue as $\psi_{\text{el},i}$: $\hat{O}_R\psi_{\text{el},i}=\sum_{j=1}^{n} r_{ij}\psi_{\text{el},j}$. The n^2 constants r_{ij} $(i=1,\ldots,n;\ j=1,\ldots,n)$ describe the symmetry behavior of the electronic wave functions of the n-fold orbitally degenerate term; these n^2 numbers arranged in a square array form a *matrix* (Chapter 2). For each symmetry operator $\hat{O}_R$ or $\hat{O}_S$ or ... there is a matrix with elements r_{ij} or s_{ij} or .... Let the number of symmetry operations in the molecular point group be h; the h matrices of coefficients that give the effects of the symmetry operations on the electronic wave functions of an n-fold orbitally degenerate term constitute the *symmetry species* (for that electronic term). Of course, if $n=1$, then the above equation becomes $\hat{O}_R\psi_{\text{el}}=r\psi_{\text{el}}$, and ψ_{el} is an eigenfunction of all the symmetry operators $\hat{O}_R$; each matrix of coefficients is then a single number (as discussed above for H_2O). Symmetry species are labeled by a letter, according to their orbital degeneracy n, as follows:

n	1	2	3	4	5
letter	A, B	E	T	G	H

(1.284)

Numerical subscripts are used when needed to distinguish different symmetry species with the same n. If there is a center of symmetry, a g or u subscript is added corresponding to $\hat{O}_i$ eigenvalue $+1$ or -1. The possible symmetry species for a molecule are found in a systematic way using group theory (Chapter 9); see Section 9.12 for tables listing the names of symmetry species of some point groups. Section 9.7 discusses the connection between symmetry and degeneracy in a systematic manner.

Dipole moments. Classically, the electric dipole moment $\mathbf{d}_{\text{cl}}$ of a system of point charges is $\mathbf{d}_{\text{cl}} = \Sigma_i q_i \mathbf{r}_i$, where $\mathbf{r}_i$ is the radius vector from the origin to charge q_i. Consider the application of a uniform external electric field $\mathbf{E}$ to a system of charges q_i. By definition, $\mathbf{E} = \mathbf{F}/q$, where $\mathbf{F}$ is the force the field exerts on charge q. Let $\mathbf{E}$ point in the z direction: $\mathbf{E} = \mathcal{E}_z \mathbf{k}$, where $\mathbf{k}$ is a unit vector. From (1.7) the potential energy V of a single charge in the field satisfies $dV/dz = -F_z = -q\mathcal{E}_z$, and $V = -q\mathcal{E}_z z$. Extension to a system of several charges in a field that points in an arbitrary direction gives

$$V = -\mathcal{E}_x \sum_i q_i x_i - \mathcal{E}_y \sum_i q_i y_i - \mathcal{E}_z \sum_i q_i z_i = -\mathbf{E} \cdot \mathbf{d}_{\text{cl}} \tag{1.285}$$

To find the quantum-mechanical expression for the molecular dipole moment, we consider (1.285) as a perturbation on the molecule; the perturbation operator $\hat{H}'$ is $\hat{H}' = -\mathbf{E} \cdot \hat{\mathbf{d}}$, where $\hat{\mathbf{d}}$ is the operator corresponding to $\mathbf{d}_{\text{cl}}$:

$$\hat{\mathbf{d}} = \sum_i q_i \hat{\mathbf{r}}_i = \mathbf{i}\hat{d}_x + \mathbf{j}\hat{d}_y + \mathbf{k}\hat{d}_z = \mathbf{i}\sum_i q_i x_i + \mathbf{j}\sum_i q_i y_i + \mathbf{k}\sum_i q_i z_i \tag{1.286}$$

The operator $\hat{\mathbf{d}}$ is the *electric-dipole-moment operator.* The sum is over electrons and nuclei:

$$\hat{\mathbf{d}} = \sum_i (-e)\mathbf{r}_i + \sum_\alpha Z_\alpha e \mathbf{r}_\alpha \tag{1.287}$$

The first-order energy correction $\langle \psi^{(0)} | \hat{H}' | \psi^{(0)} \rangle$ is

$$E^{(1)} = -\mathbf{E} \cdot \langle \psi_{\text{el}} | \hat{\mathbf{d}} | \psi_{\text{el}} \rangle \tag{1.288}$$

where ψ_{el} is the electronic wave function in the absence of the field. Comparison of (1.288) and (1.285) gives as the quantum-mechanical electric dipole moment **d** of the molecule

$$\mathbf{d} = \langle \psi_{el} | \hat{\mathbf{d}} | \psi_{el} \rangle \tag{1.289}$$

Since both ψ_{el} and $\hat{\mathbf{d}}$ depend on the nuclear configuration, the dipole moment **d** is a function of the nuclear configuration, and can be called the *dipole-moment function*. Ordinarily, one determines **d** averaged over the molecular zero-point vibrations.

1.20 THE HARTREE–FOCK METHOD

In the Hartree–Fock method, the molecular (or atomic) electronic wave function is approximated by an antisymmetrized product (Slater determinant) of spin-orbitals; each spin-orbital is the product of a spatial orbital and a spin function (α or β). Solution of the Hartree–Fock equations (given below) yields the orbitals that minimize the variational integral. Thus the Hartree–Fock wave function is the best possible electronic wave function in which each electron is assigned to a spatial orbital. For a closed-subshell state of an n-electron molecule, minimization of $\langle \Phi | \hat{H}_{el} | \Phi \rangle$, where $\hat{H}_{el}$ is the electronic Hamiltonian (1.275) and Φ is restricted to be an antisymmetrized product of n (one-electron) spin-orbitals, shows that the Hartree–Fock spatial orbitals ϕ_i satisfy the following Hartree–Fock differential equations:

$$\hat{F}(1)\phi_i(1) = \epsilon_i \phi_i(1) \tag{1.290}$$

$$\hat{F}(1) = \hat{H}_{core}(1) + \sum_j \left[2\hat{J}_j(1) - \hat{K}_j(1) \right] \tag{1.291}$$

$$\hat{H}_{core}(1) \equiv -(\hbar^2/2m)\nabla_1^2 - \sum_\alpha Z_\alpha e^2 / r_{1\alpha} \tag{1.292}$$

$$\hat{J}_j(1)\phi_i(1) \equiv \phi_i(1) \int |\phi_j(2)|^2 \frac{e^2}{r_{12}} \, dv_2 \tag{1.293}$$

$$\hat{K}_j(1)\phi_i(1) \equiv \phi_j(1) \int \phi_j^*(2)\phi_i(2) \frac{e^2}{r_{12}} \, dv_2 \tag{1.294}$$

The sum in (1.291) is over the occupied spatial orbitals of the molecule. $\hat{F}$ is the *Hartree–Fock operator*, and is a one-electron operator, as indicated by writing it as dependent on electron 1. $\hat{J}_j$ and $\hat{K}_j$, defined by (1.293) and (1.294), are the *Coulomb* and *exchange* operators, respectively. The first term in $\hat{H}_{\text{core}}$ is the kinetic-energy operator of electron 1; the second term in $\hat{H}_{\text{core}}$ is the potential energy of attraction between electron 1 and the nuclei (for an atom, there is only one nucleus). The terms involving $\hat{J}_j$ arise from the interactions between electron 1 and the other electrons; note that $\hat{J}_j(1)$ is the potential energy of interaction between the point charge of electron 1 and electron 2 considered to be smeared out into a hypothetical charge cloud of charge density (charge per unit volume) $-e|\phi_j(2)|^2$, where $\phi_j(2)$ is the orbital of electron 2. Thus the Hartree–Fock method considers average interelectronic interactions, rather than instantaneous interelectronic interactions (e^2/r_{ij}). The factor 2 in front of $\hat{J}_j$ arises from the two electrons in each MO. The $\hat{K}_j$ terms arise from writing the wave function as antisymmetric. Because the orbitals are all eigenfunctions of the same operator $\hat{F}$, they can be chosen as orthogonal. The operator $\hat{F}$ depends on its own eigenfunctions ϕ_i (through the dependence of $\hat{J}_j$ and $\hat{K}_j$ on ϕ_i and ϕ_j); hence we must solve (1.290) by an iterative process, guessing initial forms for the orbitals, computing $\hat{F}$ using the guessed orbitals, solving (1.290) with this $\hat{F}$ to get an improved set of orbitals, which we use to calculate an improved operator $\hat{F}$, which we use to solve (1.290) to get a further improved set of orbitals, which we use.... Eventually the solution converges to the true Hartree–Fock orbitals. The constant ϵ_i is the *orbital energy* of orbital ϕ_i. The electronic energy of the molecule (or atom) in the Hartree–Fock approximation is not simply the sum of orbital energies of the electrons; rather, evaluation of the variational integral shows the Hartree–Fock energy to also involve Coulomb and exchange integrals.

Rather than working with the Hartree–Fock differential equations (1.290) (with $i=1, 2, \ldots$) one usually uses the following procedure, due to Roothaan. Each orbital ϕ_i is expanded in terms of some chosen complete set of one-electron *basis functions* g_k:

$$\phi_i = \sum_k c_{ik} g_k \tag{1.295}$$

where the coefficients c_{ik} are unknown. Substitution of (1.295) into (1.290) followed by multiplication by g_j^* and integration over all space gives

$$\sum_k (F_{jk} - \epsilon_i S_{jk}) c_{ik} = 0, \qquad j = 1, 2, \ldots \tag{1.296}$$

$$F_{jk} \equiv \langle g_j | \hat{F} | g_k \rangle, \qquad S_{jk} \equiv \langle g_j | g_k \rangle \tag{1.297}$$

For a nontrivial solution of the homogeneous equations (1.296), the coefficient determinant must vanish:

$$\det(F_{jk} - \epsilon_i S_{jk}) = 0 \qquad (1.298)$$

To solve the Hartree–Fock–Roothaan equations, we proceed as follows. We start by guessing values for the coefficients c_{ik} in (1.295) to give a set of initial guessed orbitals. These initial orbitals are used to calculate an initial approximation to the $\hat{F}$ operator from (1.291)–(1.294). With this approximate $\hat{F}$, we evaluate the integrals F_{jk} and solve (1.298) for the initial approximations to the orbital energies ϵ_i; we substitute these orbital energies into (1.296) one at a time and solve for an improved set of coefficients c_{ik}. The improved coefficients give improved orbitals, from which we calculate an improved $\hat{F}$, from which we calculate improved F_{jk}'s, which are used in (1.298) to find improved ϵ_i's, which are used in (1.296) to get further improved orbitals, and so on. [The most efficient way to solve (1.296) and (1.298) is to use matrix algebra; see Chapter 2.] Of course, the calculation is done on a computer.

To find the true Hartree–Fock orbitals, one must use a complete set in (1.295), which means using an infinite number of g_k's. As a practical matter, one must use a finite number of basis functions, so that one gets approximations to the Hartree–Fock orbitals. However, with a well-chosen basis set, one can approach the true Hartree–Fock orbitals and energy quite closely with a not unreasonably large number of basis functions. Any MOs (or AOs) found by iterative solution of the Hartree–Fock–Roothaan equations are called self-consistent-field (SCF) orbitals, whether or not the basis set is large enough to give near-Hartree–Fock accuracy.

For atomic Hartree–Fock calculations, the basis functions used are the *Slater-type orbitals* (STOs), whose form is

$$(2\zeta/a_0)^{n+1/2}[(2n)!]^{-1/2} r^{n-1} e^{-\zeta r/a_0} Y_l^m(\theta,\phi)$$

The parameter ζ is the *orbital exponent*. The set of all STOs with integral values of n, l, and m and all possible positive values of ζ is a complete set.

For diatomic (and linear polyatomic) molecules, the basis set usually consists of several STOs centered on each atom. Thus the MOs are expressed as LCAOs—linear combinations of atomic orbitals. In a *minimal*-basis-set calculation, only inner-shell and valence-shell STOs are used. Thus a minimal basis calculation of HF would use as basis functions $1s_H$, $1s_F$, $2s_F$, $2p\sigma_F$, $2p\pi_F$, $2p\pi'_F$, where $2p\sigma_F$ is a fluorine $2p$ AO along the internuclear (z) axis (i.e., a $2p_z$ AO), and the $2p\pi$ and $2p\pi'$ AOs are perpendicular to the axis. The orbital exponents in the basis set are set at fixed values determined by experience to give good results. To improve the results, one can repeat the entire SCF calculation with several different sets of orbital exponents to determine the exponents that minimize the energy; this is called optimization of the orbital exponents. A minimal-basis-set calculation yields orbitals that differ significantly from the Hartree–Fock orbitals. To approach the Hartree–Fock solution, one uses an *extended* basis set, adding in orbitals higher than valence-shell orbitals on each atom.

For nonlinear polyatomic molecules, the MOs are expressed as linear combinations of functions centered on each atom. The functions used are either STOs or *Gaussian-type-orbitals* (GTOs). A GTO has the (unnormalized) radial factor $r^l \exp(-\zeta r^2/a_0^2)$, and a spherical harmonic as angular factor. (For nonlinear polyatomics, one uses the real, rather than the complex, forms of the spherical harmonics.) Use of STOs in polyatomic-molecule calculations makes the three- and four-center electron repulsion integrals time-consuming to evaluate. With Gaussian orbitals, integral evaluation is simplified, but at the expense of having to use more basis functions to get accurate results.

Each MO solution ϕ_i of the Hartree–Fock equations (1.290) can be classified as belonging to one of the possible molecular symmetry species. Thus for H_2O, the possible types of MOs are a_1, a_2, b_1, and b_2, where (1.283) gives the symmetry behavior. (We use lowercase letters for MO symmetry species.) The MOs of each symmetry species are numbered in order of increasing energy; thus $1a_1$, $2a_1$, $3a_1$ are the three lowest a_1 MOs of water. (For linear molecules, instead of letters, one uses σ, π, δ,... to indicate the $\hat{L}_z$ eigenvalue, and a g or u subscript to indicate the $\hat{O}_i$ eigenvalue if the molecule has a center of symmetry.) An MO holds two electrons of opposite spin. A set of MOs having the same energy constitutes a subshell. The electron configuration is defined by giving the number of electrons in each subshell. Since a and b MOs are nondegenerate, an a or b subshell holds two electrons. An e subshell consists of two degenerate e MOs (that are each transformed into linear combinations of the two e MOs by each molecular symmetry operator), and holds four electrons. A t subshell holds six electrons. As with atoms, a given molecular electronic configuration gives rise, in general, to several terms of different energy. However, a closed-subshell configuration always gives only a single nondegenerate term with multiplicity 1 and symmetry species the totally symmetric one. Molecular electronic ground states usually have a closed-subshell configuration. For a closed-subshell configuration, the MO wave function is a single Slater determinant; open-subshell wave functions are usually linear combinations of a few Slater determinants.

For a minimal-basis-set SCF MO calculation of H_2O (Fig. 1.2), the basis orbitals are the STOs H_11s, H_21s (on H atoms 1 and 2) and $O1s$, $O2s$, $O2p_x$, $O2p_y$, $O2p_z$ on oxygen. Each MO transforms according to one of the $\mathcal{C}_{2v}$ symmetry species (1.283). As a preliminary step in solving the Hartree–Fock–Roothaan equations, it is useful (but not essential) to form the basis AOs into linear combinations (called *symmetry orbitals* or *symmetry-adapted orbitals*) each of which transforms according to one of the molecular symmetry species. The use of basis functions that transform according to the molecular symmetry species puts the secular determinant (1.298) into block-diagonal form (each block corresponding to a different symmetry species), thereby simplifying the calculation. Proof of this statement follows from (1.51) and (1.25) for nondegenerate symmetry species; a general proof using group theory is given in Chapter 9.

Consider the symmetry behavior of the above seven minimal-basis H_2O AOs. A $\hat{C}_2(z)$ rotation sends $O2p_y$ into its negative, so $O2p_y$ has eigenvalue -1 for $\hat{O}_{C_2(z)}$; its eigenvalues for reflection in the yz and xz planes are 1 and -1, respectively. Hence $O2p_y$ has symmetry eigenvalues 1, -1, -1, 1 and belongs to symmetry species b_2. Each of the other oxygen basis AOs is found to transform according to one of the H_2O symmetry species. Thus each oxygen AO can serve as a symmetry orbital. However, a $\hat{C}_2(z)$ rotation sends H_11s into H_21s, so H_11s is not an eigenfunction of $\hat{O}_{C_2(z)}$ and does not transform according to a $\mathcal{C}_{2v}$ symmetry species. To form symmetry orbitals, we take the following linear combinations: H_11s+H_21s and H_11s-H_21s. We see that $\hat{O}_{C_2(z)}$ sends H_11s+H_21s into H_21s+H_11s, so that this linear combination has eigenvalue $+1$ for $\hat{O}_{C_2(z)}$. Considering the other symmetry operations, we find that H_11s+H_21s has symmetry species a_1; the linear combination H_11s-H_21s is found to belong to b_2. We thus have the following symmetry orbitals. Species a_1: $O1s$, $O2s$, $O2p_z$, H_11s+H_21s. Species b_2: $O2p_y$, H_11s-H_21s. Species b_1: $O2p_x$. With these basis functions, the secular determinant factors into 4×4, 2×2, and 1×1 blocks.

Thus four of the seven lowest H_2O MOs are linear combinations of the four a_1 symmetry orbitals listed above, and are a_1 MOs; similarly, the two lowest b_2 MOs are linear combinations of $O2p_y$ and H_11s-H_21s, and the lowest b_1 MO is (in this minimal-basis calculation) identical with $O2p_x$. The coefficients in the linear combinations and the orbital energies are found by iterative solution of the Hartree–Fock–Roothaan equations. One finds the ground-state electronic configuration of H_2O to be

$$(1a_1)^2(2a_1)^2(1b_2)^2(3a_1)^2(1b_1)^2 \tag{1.299}$$

The $1a_1$ MO is found to be mainly an inner-shell $O1s$ orbital. The $2a_1$ MO is a bonding combination of O2s, O2p_z, and H$_1$1s+H$_2$1s (plus a bit of O1s), with electron probability density spread over the whole molecule. The 1b_2 MO is a bonding combination of O2p_y and H$_1$1s−H$_2$1s. The 3a_1 MO is a combination of O2s, O2p_z, and H$_1$1s+H$_2$1s; its character is mainly lone pair (with some contribution to the bonding). The 1b_1 MO is a nonbonding, lone-pair O2p_x orbital perpendicular to the molecular plane.

Because seven basis functions were used, the minimal-basis SCF calculation yields approximations to the seven lowest H_2O MOs, five occupied and two unoccupied. The two unoccupied MOs are called *virtual* orbitals; the virtual orbitals are only approximations to MOs occupied in excited states, because the interelectronic interactions change when the electron configuration changes.

The MOs of a planar molecule are classified as *sigma* or *pi*, according to whether their eigenvalue for reflection in the molecular plane is +1 or −1, respectively. For $\mathcal{C}_{2v}$, the symmetry species a_1 and b_2 have +1 eigenvalue for $\hat{\sigma}_v(yz)$, whereas b_1 has −1 eigenvalue for $\hat{\sigma}_v(yz)$. Thus all the occupied H_2O ground-state MOs are sigma MOs, except 1b_1, which is pi.

For a minimal-basis-set calculation of H_2, the basis AOs are 1s AOs on H atoms a and b: $1s_a$ and $1s_b$. These are not symmetry orbitals, but the linear combinations $1s_a+1s_b$ and $1s_a-1s_b$ give symmetry orbitals. These two symmetry orbitals have different eigenvalues for a σ_h reflection; hence they belong to different symmetry species. Thus the minimal-basis secular determinant splits into two 1×1 blocks, and in this case the two lowest MOs are identical with the normalized symmetry orbitals. The MO with the plus sign builds up charge between the atoms and is clearly the lowest. Since they are composed of s AOs, these two MOs have $\hat{L}_z$ eigenvalue zero, and are σ MOs. Their designations are $1\sigma_g \approx N(1s_a+1s_b)$ and $1\sigma_u \approx N'(1s_a-1s_b)$, where N and N' are the normalization constants, the g and u subscripts give the $\hat{O}_i$ eigenvalue, and the prefix 1 indicates the lowest MO of each species. (Since these are only minimal-basis MOs, we use ≈, rather than =.) The often used separated-atoms notation is $\sigma_g 1s \approx N(1s_a+1s_b)$ and $\sigma_u^* 1s \approx N'(1s_a-1s_b)$, where the suffix 1$s$ indicates the MOs are linear combinations of the 1s separated-atoms AOs, and the absence or presence of a star indicates a +1 or −1 eigenvalue, respectively, for a $\hat{\sigma}_h$ reflection (in the plane perpendicular to the molecular axis). Starred homonuclear diatomic MOs are antibonding because they have a nodal plane midway between the nuclei.

The bonding MOs of water, discussed above, are spread out over the whole molecule. Since the SCF MO wave function is a Slater determinant, one can use Theorem 5 of the determinant discussion in Section 1.2 to change the form of the wave function. If $\phi_1,\ldots,\phi_5$ are the five occupied ground-state H_2O MOs, the ground-state SCF wave function is

$$|\phi_1 \overline{\phi_1} \phi_2 \overline{\phi_2} \phi_3 \overline{\phi_3} \phi_4 \overline{\phi_4} \phi_5 \overline{\phi_5}|$$

Addition of multiples of columns to one another leads to a description in which the occupied MOs are linear combinations of $\phi_1, \ldots, \phi_5$. Thus there is some degree of freedom in the MO description that we use. By forming appropriate linear combinations, we can get five occupied H_2O MOs that are not delocalized (as $\phi_1, \ldots, \phi_5$ are), but are largely localized, with one inner-shell MO, two equivalent localized lone-pair MOs, and two equivalent localized bond MOs. The delocalized MOs that transform according to the molecular symmetry species are called the *canonical* MOs to distinguish them from localized MOs. (For open-subshell excited states, the MO wave function is a linear combination of a few determinants, rather than one, so the localization procedure cannot be used.) Of the many ways to choose localized MOs, the most widely used is to choose them to minimize the interorbital repulsions, thereby giving the *energy-localized* MOs.

Near-Hartree–Fock wave functions are found to give accurate values of bond distances, bond angles, dipole moments, and rotational barriers. However, they give poor values for molecular dissociation energies.

A Hartree–Fock calculation takes interelectronic repulsions into account in only an average way. To take into account the instantaneous correlations in electronic motions, and thereby reach the true wave function, one must go beyond the Hartree–Fock approximation. (The difference between the true nonrelativistic molecular energy and the Hartree–Fock energy is called the *correlation energy*.) The most widely used procedure to do this is *configuration interaction* (CI). To do a CI calculation for, say, the ground state, one proceeds as follows. One starts with a large basis set of AOs and uses the Hartree–Fock–Roothaan equations to obtain an accurate SCF wave function. If N basis functions are used, the SCF calculation yields expressions for N molecular orbitals, many of them unoccupied. The ground state near-Hartree–Fock wave function is a Slater determinant with $n/2$ MOs doubly occupied (n being the number of electrons). One then uses the unoccupied (virtual) MOs to form Slater determinants with one or more electrons placed in virtual orbitals. The CI wave function ψ is then taken as a linear combination of the ground-state Slater determinant and Slater determinants with electrons in virtual MOs:

$$\psi = \sum_i c_i \Phi_i \tag{1.300}$$

where each Φ_i is an n-electron determinantal function. Of course, one includes only functions Φ_i that have the same spin eigenvalues and symmetry properties as the wave function that is being improved by CI. Since (1.300) is a linear variation function, minimization of the variational integral gives [Equation (1.189)]

$$\det(H_{ij} - ES_{ij}) = 0 \tag{1.301}$$

where the integrals H_{ij} and S_{ij} involve Φ_i and Φ_j. Solution of (1.301) gives the CI energy. If the original basis set consists of a complete set of one-electron functions (which requires that N be infinite), then the set of all possible determinantal functions Φ_i will be a complete set for expansion of an n-electron function; hence, by using larger and larger numbers of Φ_i's in (1.300), we can ultimately converge to the true wave function. As a practical matter, one is limited in the number of configurations that can be included.

The CI procedure just described uses a fixed set of orbitals in the functions Φ_i. An alternative approach is to vary the forms of the MOs in each determinantal function Φ_i in (1.300), in addition to varying the coefficients c_i. One uses an iterative process (which resembles the Hartree–Fock procedure) to find the optimum orbitals in the CI determinants. This form of CI is called the *multiconfiguration* SCF (MCSCF) method. Because the orbitals are optimized, the MCSCF method requires far fewer configurations than ordinary CI to get an accurate wave function. A particular form of the MCSCF approach developed for calculations on diatomic molecules is the optimized valence configuration (OVC) method.

An alternative to the SCF MO approach is the valence-bond (VB) method, which is discussed in most quantum-chemistry texts.

1.21 SEMIEMPIRICAL METHODS

An *ab initio* calculation uses the correct molecular electronic Hamiltonian (1.275) and does not introduce experimental data (other than the values of the fundamental physical constants) into the calculation. A *semiempirical* calculation uses a Hamiltonian simpler than the correct one, and takes some of the integrals as parameters whose values are determined using experimental data. The Hartree–Fock SCF MO method seeks the orbital wave function Φ that minimizes the variational integral $\langle\Phi|\hat{H}_{\text{el}}|\Phi\rangle$, where $\hat{H}_{\text{el}}$ is the true Hamiltonian, and is therefore an *ab initio* method. Semiempirical methods were developed because of the difficulties involved in *ab initio* calculation of medium-sized and large molecules. We can divide semiempirical methods into *one-electron* methods, which use a Hamiltonian that is the sum of one-electron terms, and *two-electron* methods, which use a Hamiltonian that contains electron repulsion terms e^2/r_{ij} in addition to one-electron terms. Some one-electron methods are the free-electron molecular-orbital (FE MO) method, the Hückel molecular-orbital (HMO) method, and the extended Hückel (EH) method. Some two-electron methods are the Pariser–Parr–Pople (PPP) method, the complete neglect of differential overlap (CNDO) method, the intermediate neglect of differential overlap (INDO) method, and the modified INDO (MINDO) method. All seven of these methods use the *valence-electron approximation*, in which it is assumed that the molecular valence electrons can be treated separately from the other electrons, with the core electrons and the nuclei providing some sort of effective potential for the valence electrons to move in.

The FE MO, HMO, and PPP methods are restricted to planar conjugated molecules (e.g., butadiene, benzene, pyridine), and make the *pi-electron approximation* of treating only those valence electrons that are in pi MOs (those that have eigenvalue -1 for reflection in the molecular plane); they assume the existence of a pi-electron Hamiltonian $\hat{H}_\pi$ of the form

$$\sum_i \hat{H}_{\text{core}}(i) + \sum_i \sum_{j>i} e^2/r_{ij} \tag{1.302}$$

$$\hat{H}_{\text{core}}(i) = -(\hbar^2/2m)\nabla_i^2 + V_i \tag{1.303}$$

where V_i is the potential energy of the ith pi electron in the field of the sigma electrons and the nuclei.

The FE MO method omits the internuclear repulsions e^2/r_{ij} from $\hat{H}_\pi$, leaving it as the sum of $\hat{H}_{\text{core}}(i)$'s, one for each pi electron. By separation of variables, the pi-electron wave function ψ_π is an antisymmetrized product of one-electron spin-orbitals, and the pi-electron energy E_π is the sum of energies of individual electrons. The FE MO method takes V_i in $\hat{H}_{\text{core}}(i)$ as a one-dimensional particle-in-a-box potential (Section 1.4). Hence E_π is the sum of particle-in-a-box energies e_i, with $e_i = n_i^2h^2/8ml^2$. We feed the pi electrons in pairs into the lowest few FE MOs. The method is crude, but it can be used to estimate the longest wavelength electronic absorption band of a conjugated molecule by taking the energy difference between the lowest vacant and highest occupied FE MOs. For the conjugated polyenes

$$CH_2{=}(CH{-}CH{=})_kCH_2 \tag{1.304}$$

the observed longest wavelength absorption λ_{max} increases monotonically from 1625 Å to 4470 Å as k goes from 0 to 9 (*Suzuki*, p. 204). For the polymethine ions

$$(CH_3)_2\overset{+}{N}{=}CH({-}CH{=}CH)_k{-}\ddot{N}(CH_3)_2 \qquad (1.305)$$

λ_{max} increases monotonically from 2240 Å to 8480 Å as k goes from 0 to 6 (*Suzuki*, p. 365). Taking l as the zigzag length along the chain, one finds the FE MO method works rather poorly for (1.304) (where the bonds alternate in length), but reasonably well for (1.305) (where the conjugated bond lengths are equal due to an equivalent resonance structure).

The HMO method also omits interelectronic repulsions from $\hat{H}_\pi$, taking $\hat{H}_\pi = \Sigma_i \hat{H}^{\text{eff}}(i)$, where $\hat{H}^{\text{eff}}(i)$ is $\hat{H}_{\text{core}}(i)$ plus some sort of average interaction with the other pi electrons; $\hat{H}^{\text{eff}}(i)$ is not explicitly specified, but its integrals are specified in a certain way (see below). By separation of variables, ψ_π is an antisymmetrized product of (one-electron) pi spin-orbitals, and E_π is the sum of pi-electron orbital energies. The pi MO ϕ_i is an eigenfunction of $\hat{H}^{\text{eff}}$: $\hat{H}^{\text{eff}}(i)\phi_i = e_i\phi_i$. For planar conjugated hydrocarbons, each pi MO is taken as a linear combination of $2p\pi$ AOs, one on each carbon (where $2p\pi$ denotes a $2p$ AO perpendicular to the molecular plane). Thus $\phi_i = \Sigma_r c_{ir} f_r$, where f_r is a $2p\pi$ AO on carbon r, and the sum goes over the n_C conjugated carbon atoms. This linear variation function leads to the secular equation $\det(H_{rs} - S_{rs}e_i) = 0$, where $H_{rs} \equiv \langle f_r|\hat{H}^{\text{eff}}|f_s\rangle$ and $S_{rs} \equiv \langle f_r|f_s\rangle$. The determinant is of order n_C. The following assumptions are used for the Hamiltonian integrals H_{rs} and the overlap integrals S_{rs}: $H_{rr} = \alpha$ for every carbon; $H_{rs} = \beta$ for every bonded pair of carbons; $H_{rs} = 0$ for every nonbonded pair; $S_{rs} = \delta_{rs}$ (the Kronecker delta). Thus overlap is neglected. The parameters α and β are chosen so that the property being calculated gives the best overall fit to experimental data for a series of compounds.

As an example, consider benzene, C_6H_6. The HMO secular determinant is of order six. The six principal-diagonal elements each equal $\alpha - e_i$; the off-diagonal elements are $H_{12} = H_{21} = H_{23} = H_{32} = H_{34} = H_{43} = H_{45} = H_{54} = H_{56} = H_{65} = H_{61} = H_{16} = \beta$, all others $= 0$. (The carbons are numbered 1 to 6 around the ring.) The secular equation gives as the HMO energies (Fig. 1.3) $\alpha + 2\beta$, $\alpha + \beta$, $\alpha + \beta$, $\alpha - \beta$, $\alpha - \beta$, $\alpha - 2\beta$. We omit giving the HMO coefficients. The $\mathcal{D}_{6h}$ symmetry species for the pi MOs are shown in Fig. 1.3; see Chapter 9 for more on the benzene symmetry species. Because each symmetry species occurs no more than once (this is not true for H_2O

discussed above), the minimal-basis SCF pi MOs of C_6H_6 are completely determined by symmetry (see Section 9.8 for more details); group theory shows that the pattern of pi MO energies in Fig. 1.3 follows from the symmetry of benzene. Of course, although the MO pattern of two pairs of doubly degenerate pi MOs lying between two nondegenerate pi MOs is correct, the quite crude Hückel theory does not give the correct spacings between pi MOs.

Benzene is an example of an *alternant* hydrocarbon, which is a planar conjugated hydrocarbon in which the carbons can be divided into two sets, "starred" and "unstarred," with each starred carbon bonded only to unstarred carbons, and vice versa. For alternant hydrocarbons, the HMO pi energies are symmetrically disposed above and below α, and the HMO coefficients in the MO with energy $\alpha - x\beta$ can be found simply by changing the signs of the unstarred-atom AO coefficients in the paired MO with energy $\alpha + x\beta$.

One can use the HMO method to accurately predict the longest wavelength electronic absorption of a benzenoid hydrocarbon, but the method is worthless for predicting the complete spectrum. For example, since interelectronic repulsions are ignored, HMO theory gives no separation between singlet and triplet terms that arise from the same configuration; experimentally, such terms differ significantly in energy.

The EH method (developed by Wolfsberg and Helmholz and by Hoffmann) is an extension of the Hückel method in which the pi-electron approximation is not made, but all valence electrons are treated. The method is thus applicable to nonplanar, as well as planar, molecules. The valence-electron Hamiltonian is taken as the sum of one-electron Hamiltonians: $\hat{H}_{\text{val}} = \sum_i \hat{H}^{\text{eff}}(i)$, where $\hat{H}^{\text{eff}}(i)$ is not explicitly defined. The valence-electron wave function is the antisymmetrized product of spin-

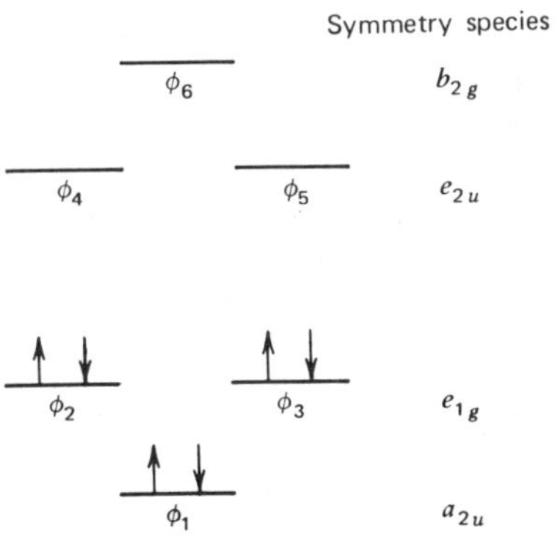

Fig. 1.3 Hückel pi MO's and ground-state pi-electron configuration of benzene.

orbitals, and E_{val} is the sum of valence-electron energies. Each MO ϕ_i is expanded as a linear combination of all the valence AOs: $\phi_i = \Sigma_j c_{ij} f_j$, where STOs are used for AOs. (For carbon, the valence AOs are one $2s$ and three $2p$ AOs.) This linear variation function leads to the secular equation: $\det(H_{jk}^{\text{eff}} - e_i S_{jk}) = 0$, where the integrals involve the AOs f_j and f_k. In EH theory, *all* overlap integrals S_{jk} are evaluated exactly. The integrals H_{jj}^{eff} are approximated as minus the atomic valence-state ionization potentials. (The valence state is the hypothetical state in which an atom exists in a molecule; the valence-state atomic wave function can be obtained from the molecular VB wave function.) Various formulas are used for H_{jk}^{eff}, $j \neq k$. One formula is $H_{jk}^{\text{eff}} = \frac{1}{2}K(H_{jj}^{\text{eff}} + H_{kk}^{\text{eff}})S_{jk}$, where K is a constant chosen at the whim of the person doing the calculation. Despite its crudity, EH theory is found to yield rather accurate bond angles for molecules not highly polar, and it gives a good estimate of the order of MO energies in such molecules. The method cannot be used to give accurate dipole moments, bond lengths, or rotational barriers.

We now consider the PPP, CNDO, INDO, and MINDO two-electron semiempirical methods. These are all SCF methods which iteratively solve the Hartree–Fock–Roothaan equations (1.296) and (1.298) until self-consistent MOs are obtained. However, instead of the true Hartree–Fock operator (1.291), they use a Hartree–Fock operator in which the sum in (1.291) goes over only the valence MOs. Thus, besides the terms in (1.292), $\hat{H}_{\text{core}}(1)$ in these methods also includes the potential energy of interaction of valence electron 1 with the field of the inner-shell electrons; rather than attempting a direct calculation of this interaction, the integrals of $\hat{H}_{\text{core}}(i)$ are given by various semiempirical schemes that make use of experimental data; furthermore, many of the electron repulsion integrals are neglected, so as to simplify the calculation.

The PPP method was developed first (in 1953), and is applicable only to planar conjugated molecules. The method uses the pi-electron approximation. The pi-electron Hamiltonian (1.302) gives an approximate Hartree–Fock operator like (1.291), but with only terms for the pi electrons, and with $\hat{H}_{\text{core}}(1)$ containing interaction terms between pi electron 1 and the sigma electrons [in addition to the terms in (1.292)]. The basis functions for the pi MOs are the $2p\pi$ AOs of the system: $\phi_i = \Sigma_r c_{ir} f_r$, where f_r is a $2p\pi$ AO on atom r. Overlap is neglected: $S_{rs} = \langle f_r | f_s \rangle = \delta_{rs}$. The integral $\beta_{rs}^{\text{core}} = \langle f_r | \hat{H}_{\text{core}} | f_s \rangle$, $r \neq s$, is taken as an empirical parameter if atoms r and s are bonded, and is set equal to 0 if r and s are not bonded. The integrals $\alpha_r^{\text{core}} = \langle f_r | \hat{H}_{\text{core}} | f_r \rangle$ are evaluated with a semiempirical formula that involves such things as valence-state ionization potentials and electron repulsion integrals. Consistent with the neglect of overlap, the PPP method uses the *zero-differential-overlap* (ZDO) *approximation* to eliminate many of the electron repulsion integrals; the ZDO approximation is $f_r(1)f_t(1)\,dv_1 = 0$ for $r \neq t$. As a consequence of the ZDO approximation, we have

$$(rt|su) \equiv \langle f_r(1)f_s(2)|e^2/r_{12}|f_t(1)f_u(2)\rangle = \delta_{rt}\delta_{su}\gamma_{rs}$$

$$\gamma_{rs} \equiv \langle f_r(1)f_s(2)|e^2/r_{12}|f_r(1)f_s(2)\rangle \equiv (rr|ss)$$

The notation $(rt|su)$ is widely used for electron repulsion integrals. The electron repulsion integrals γ_{rs} can all be evaluated theoretically from the known form of the STOs; however, better results are obtained if the integrals γ_{rr} are evaluated semiempirically as the difference of valence-state ionization potentials and electron affinities; also, the integrals γ_{rs} with $r \neq s$ are often evaluated semiempirically. Configuration interaction is included when doing an excited-state calculation. The PPP method gives quite a good account of the electronic spectra of many conjugated molecules.

The CNDO and INDO methods (developed by Pople and co-workers in 1965–1967) and the MINDO method (developed by Dewar and co-workers in 1969–1974) are extensions of the PPP method to nonplanar molecules. These methods include all the valence electrons in the Hartree–Fock operator $\hat{F}$, and iteratively solve the Hartree–Fock–Roothaan equations to find SCF valence-electron MOs.

The CNDO method uses as basis functions valence-shell Slater AOs with fixed orbital exponents. As in PPP theory, overlap integrals are neglected, and the ZDO approximation is used to eliminate most of the electron repulsion integrals, leaving only repulsion integrals of the form $(rr|ss)$. The PPP method has only one basis AO per atom, but the CNDO method has several basis AOs per atom (except for H); hence the ZDO approximation in CNDO causes neglect of electron repulsion integrals that contain the product $f_r(1)f_t(1)$, where f_r and f_t are different AOs on the *same* atom. In CNDO, some of the integrals of $\hat{H}_{\text{core}}$ are evaluated semiempirically with such data as atomic valence-state ionization potentials and electron affinities, and some of them are taken as parameters whose values are adjusted so that the theory will give a good fit with results of minimal-basis *ab initio* SCF calculations. The symbols CNDO/1 and CNDO/2 refer to two versions of CNDO, with CNDO/2 using an improved parametrization.

The INDO method differs from CNDO only in neglecting fewer electron repulsion integrals. In INDO, differential overlap between AOs on the same atom is not neglected in one-center electron repulsion integrals, but is neglected in two-center electron repulsion integrals. [An *n*-center integral ($n=1$, 2, 3, 4) contains basis functions centered on a total of n different nuclei.]

The CNDO and INDO methods give accurate bond angles and fairly accurate bond lengths and dipole moments, but poor dissociation energies.

The MINDO method (which exists in several versions: MINDO/1, MINDO/2, MINDO/2′, MINDO/3, MINDO-CI) was designed not to reproduce results of *ab initio* SCF calculations, but to give accurate molecular binding energies, which is an area where *ab initio* Hartree–Fock calculations fail. Even though MINDO involves approximations to the *ab initio* SCF method, by proper choice of parameters, MINDO can do better than *ab initio* SCF calculations in finding binding energies. The MINDO method is quite similar to the INDO method, except that the parameters are chosen so as to give accurate molecular heats of formation (rather than to give agreement with *ab initio* SCF calculations). MINDO gives heats of formation typically in error by about 3 kcal/mole, and gives generally accurate molecular geometries. It may be useful for calculating potential-energy surfaces for chemical reactions. [See M. J. S. Dewar, R. C. Haddon, and P. K. Weiner, *J. Am. Chem. Soc.*, **96**, 253 (1974) and references cited therein.]

Although semiempirical theories make some rather drastic approximations (e.g., the ZDO approximation), the use of empirical parameters partially compensates for these approximations, and allows the theories to give useful results for molecules too large to treat accurately by *ab initio* methods. For details of two-electron semiempirical theories, see *Murrell and Harget, Pople and Beveridge*.

Computer programs for the HMO, EH, PPP, CNDO, INDO, MINDO, and *ab initio* SCF MO methods can be obtained from the Quantum Chemistry Program Exchange, Chemistry Department, Indiana University, Bloomington, Indiana, 47401.

PROBLEMS

1.1 Prove that eigenfunctions of a Hermitian operator that correspond to different eigenvalues are orthogonal.

1.2 Calculate $\Delta x \Delta p_x$ for the harmonic-oscillator ground state (1.139).

1.3 Prove (1.51).

1.4 (a) Verify (1.63)–(1.69). (b) Which of these equations are valid for nonlinear operators?

1.5 (a) Find the Taylor series for $\ln x$ about $a=1$. (b) For $\ln x$, what is the nearest singularity to $x=1$ on the real axis? On the assumption (which happens to be true) that this is the nearest singularity in the complex plane, for what range of x does the series in (a) converge to $\ln x$?

1.6 Show that addition of a constant C to a time-independent Hamiltonian $\hat{H}$ leaves the stationary-state wave functions unchanged and adds C to each energy eigenvalue.

1.7 Suppose a system has the stationary-state wave function $\psi = N\exp(-bx^4)$, where N and b are constants. Find the potential-energy function $V(x)$ for the system. *Hint*: note that the zero level of energy is arbitrary, and can be chosen at our convenience.

1.8 Derive (1.242). *Hint*: Start by substituting $\hat{S}_+\beta/c$ for α in the first normalization equation of (1.237); then use (1.240) to replace the two $\hat{S}_+$ operators by $\hat{S}_x + i\hat{S}_y$; use the Hermitian property of $\hat{S}_x$ and $\hat{S}_y$, the first equation of (1.231), the relation $\hat{S}_x^2 + \hat{S}_y^2 = \hat{S}^2 - \hat{S}_z^2$, the equations (1.235) and (1.236), and the normalization equation for β to obtain $|c| = \hbar$.

1.9 A Hermitian operator $\hat{A}$ is often defined as one that satisfies $\langle\phi|\hat{A}|\phi\rangle = \langle\phi|\hat{A}|\phi\rangle^*$ for all well-behaved functions ϕ. Show that this definition leads to (1.13). *Hint*: Set $\phi = f + cg$ and simplify the resulting equation; then set $c = 1$ and $c = i$ to get two equations which when combined give (1.13).

1.10 Let Y_{jm} denote the *normalized* eigenfunctions of the generalized angular-momentum operators $\hat{M}^2$ and $\hat{M}_z$ of Section 1.7. Since $\hat{M}_+$ and $\hat{M}_-$ are ladder operators, we have $\hat{M}_+ Y_{jm} = CY_{j,m+1}$ and $\hat{M}_- Y_{jm} = DY_{j,m-1}$, where C and D are constants. Use the procedure of Problem 1.8 to show that

$$\hat{M}_\pm Y_{jm} = [j(j+1) - m(m\pm 1)]^{1/2}\hbar Y_{j,m\pm 1}$$

[For consistency with the phase choice of these equations, the factor $(-i)^{m+|m|}$ must be added to the right side of (1.160).]

1.11 Verify that (1.283) are the symmetry species for H_2O.

1.12 (a) Show that for $\mathcal{D}_{2h}$, all eight symmetry operations can be written as the product of one, two, or three of the reflections. (b) From (a) it follows that specification of the eigenvalues of the reflections fully defines the $\mathcal{D}_{2h}$ symmetry species. List the symmetry species of $\mathcal{D}_{2h}$. (Check with the table in Section 9.12.)

CHAPTER 2

MATRICES

2.1 MATRICES

We formulated quantum mechanics in Chapter 1 in terms of operators and their eigenfunctions. There exists an alternative formulation (Section 2.3) in terms of matrices and their eigenvectors, which is sometimes more convenient than the operator formulation. Matrix algebra was developed by Cayley in the mid-nineteenth century. Scientists remained unaware of matrices until 1925, when quantum mechanics was born. Today there is hardly a branch of science from physics to psychology that does not use matrices.

A matrix is a rectangular array of numbers; matrices obey certain rules of combination, which we shall shortly specify. Two examples of matrices are

$$\begin{pmatrix} 1 & 5 & 0 \\ -3 & 8 & 1.4 \end{pmatrix} \qquad \begin{pmatrix} a & b \\ c & d \end{pmatrix}$$

It is important to distinguish matrices from determinants. A determinant has a definite value attached to it; a matrix does not. A determinant is a square array; a matrix is rectangular and may or may not be square.

The curved lines around an array indicate it to be a matrix. The numbers that make up a matrix are called its *elements*. The general matrix has m rows and n columns. Denoting the matrix element in the ith row and jth column by a_{ij}, we write the general matrix as

$$\begin{pmatrix} a_{11} & a_{12} & \cdots & a_{1n} \\ a_{21} & a_{22} & \cdots & a_{2n} \\ \cdot & \cdot & \cdots & \cdot \\ a_{m1} & a_{m2} & \cdots & a_{mn} \end{pmatrix} \tag{2.1}$$

We shall use the symbol **A** to stand for the matrix (2.1).

Two matrices are *equal* if the corresponding elements of each matrix are equal. The matrix equation $\mathbf{A}=\mathbf{B}$ implies the *mn* scalar equations

$$a_{ij}=b_{ij}, \qquad i=1,2,\ldots,m, \qquad j=1,2,\ldots,n \tag{2.2}$$

Of course, equality of two matrices is possible only if they have the same number of rows and the same number of columns.

The *sum* of two matrices having the same dimensions is defined to be the matrix formed by adding corresponding elements of the two matrices. For example,

$$\begin{pmatrix} -1 & 7 \\ 0 & -9 \\ 1 & 8 \end{pmatrix} + \begin{pmatrix} 1 & 6 \\ 7 & 6 \\ 9 & 6 \end{pmatrix} = \begin{pmatrix} 0 & 13 \\ 7 & -3 \\ 10 & 14 \end{pmatrix}$$

If $\mathbf{C}=\mathbf{A}+\mathbf{B}$, then

$$c_{ij}=a_{ij}+b_{ij}, \qquad i=1,2,\ldots,m, \qquad j=1,2,\ldots,n \tag{2.3}$$

Since addition of numbers is commutative and associative, it follows that matrix addition is commutative and associative: $\mathbf{A}+\mathbf{B}=\mathbf{B}+\mathbf{A}$, $(\mathbf{A}+\mathbf{B})+\mathbf{C}=\mathbf{A}+(\mathbf{B}+\mathbf{C})$.

The product of a matrix and a scalar is defined to be the matrix formed by multiplying each element of the original matrix by the scalar. For example,

$$-0.2\begin{pmatrix} 3 & 9 & 9 \\ -1 & 8 & 6 \end{pmatrix} = \begin{pmatrix} -0.6 & -1.8 & -1.8 \\ 0.2 & -1.6 & -1.2 \end{pmatrix}$$

If $\mathbf{C}=k\mathbf{A}$, where k is a scalar, then $c_{ij}=ka_{ij}$. Matrix subtraction is defined by $\mathbf{A}-\mathbf{B}=\mathbf{A}+[(-1)\mathbf{B}]$.

A matrix containing only one row is called a *row matrix*; it has the form

$$(a_1 \quad a_2 \quad \cdots \quad a_n) \tag{2.4}$$

A *column matrix* has the form

$$\begin{pmatrix} b_1 \\ b_2 \\ \vdots \\ b_m \end{pmatrix} \tag{2.5}$$

In Section 1.2, we pointed out that an ordered set of n numbers can be regarded as defining a vector in an abstract n-dimensional space. In line with this idea, a row matrix is often called a *row vector*, and a column matrix is often called a *column vector*. The general m by n matrix (2.1) can be regarded as either a set of m ordered row vectors or a set of n ordered column vectors.

Next we discuss multiplication of one matrix by another. Before considering the general case, we define the matrix product of a row matrix and a column matrix. If **A** is a row matrix and **B** is a column matrix having the same number of elements as **A**, then the product **AB** is a 1 by 1 matrix, calculated according to

$$\mathbf{AB}=(a_1 \quad a_2 \quad \cdots \quad a_n)\begin{pmatrix} b_1 \\ b_2 \\ \vdots \\ b_n \end{pmatrix}=(a_1b_1+a_2b_2+\cdots+a_nb_n) \tag{2.6}$$

For example,

$$(6 \quad 2 \quad -2)\begin{pmatrix} 9 \\ 8 \\ -1 \end{pmatrix}=(54+16+2)=(72) \tag{2.7}$$

A 1 by 1 matrix is essentially a scalar, and the curved lines on the right of (2.6) can be omitted.

Now we consider the general rule for matrix multiplication. First of all, the matrix product **AB** is only defined when the number of columns in **A** equals the number of rows in **B**; when this condition is met, we may multiply **A** and **B** to form the matrix product **C**=**AB**. If **A** is an m by n matrix and **B** is an n by p matrix, then **C** is an m by p matrix, whose elements are found according to the following rule: The element c_{ij} is found by using (2.6) to form the product of the ith row vector of **A** and the jth column vector of **B**.

For example, let

$$\mathbf{A}=\begin{pmatrix} 1 & -1 & 3 \\ 6 & 2 & -2 \end{pmatrix} \qquad \mathbf{B}=\begin{pmatrix} 4 & 5 & 9 & 1 \\ 7 & 1 & 8 & 2 \\ 0 & -2 & -1 & \frac{1}{2} \end{pmatrix} \tag{2.8}$$

Since **A** is a 2 by 3 matrix and **B** is a 3 by 4 matrix, the product **C**=**AB** is defined and will be a 2 by 4 matrix:

$$\mathbf{C}=\begin{pmatrix} c_{11} & c_{12} & c_{13} & c_{14} \\ c_{21} & c_{22} & c_{23} & c_{24} \end{pmatrix}$$

To evaluate c_{23}, we form the product of the second row of **A** and the third column of **B**. This particular product was evaluated in (2.7); hence $c_{23}=72$. Evaluating the remaining elements, we get

$$\mathbf{C}=\begin{pmatrix} -3 & -2 & -2 & \frac{1}{2} \\ 38 & 36 & 72 & 9 \end{pmatrix}$$

By now, the reason for requiring the number of columns of **A** to equal the number of rows of **B** should be clear; if this requirement is not met, then a row vector of **A** would not have the same number of elements as a column vector of **B**, and we could not take their product.

We want the general formula for the elements of a matrix product. If **A** is an m by n matrix, its ith row vector is

$$(a_{i1} \quad a_{i2} \quad \cdots \quad a_{in}) \tag{2.9}$$

The jth column vector of the n by p matrix **B** is

$$\begin{pmatrix} b_{1j} \\ b_{2j} \\ \vdots \\ b_{nj} \end{pmatrix} \tag{2.10}$$

Forming the product of (2.9) and (2.10) according to (2.6), we have

$$c_{ij}=a_{i1}b_{1j}+a_{i2}b_{2j}+\cdots+a_{in}b_{nj}$$

The general matrix element of the product $\mathbf{C}=\mathbf{AB}$ is therefore

$$c_{ij}=\sum_{k=1}^{n} a_{ik}b_{kj} \tag{2.11}$$

Note that the summation index k is a dummy variable. (By this, we mean that the value of the sum is the same no matter what letter we use for the summation index.) Equation (2.11) is worth remembering.

Frequently we shall want to multiply an m by n matrix **A** by an n by 1 column vector **B** to get an m by 1 column vector $\mathbf{C}=\mathbf{AB}$. Equation (2.11) is of course applicable with $j=1$. However, we usually use only one index for the column-vector elements, as in (2.5). Hence we drop j altogether from (2.11) to get

$$c_i = \sum_{k=1}^{n} a_{ik} b_k, \qquad \mathbf{B} \text{ a column vector} \tag{2.12}$$

Is matrix multiplication commutative? In other words, is $\mathbf{AB}=\mathbf{BA}$? For the matrices of (2.8), although **AB** is defined, **BA** is not defined, since **B** has four columns while **A** has two rows. Let us take a case where both **AB** and **BA** are defined. Let

$$\mathbf{A}=\begin{pmatrix} 1 & 0 \\ -1 & 2 \end{pmatrix} \qquad \mathbf{B}=\begin{pmatrix} 1 & i \\ 0 & 0 \end{pmatrix}$$

where $i=\sqrt{-1}$. We have

$$\mathbf{AB}=\begin{pmatrix} 1 & i \\ -1 & -i \end{pmatrix} \qquad \mathbf{BA}=\begin{pmatrix} 1-i & 2i \\ 0 & 0 \end{pmatrix}$$

Thus matrix multiplication is not in general commutative. From the formula (2.11), we readily see this to be so. (Of course, there are particular pairs of matrices for which $\mathbf{AB}=\mathbf{BA}$.)

Matrix multiplication is associative, that is, $(\mathbf{AB})\mathbf{C}=\mathbf{A}(\mathbf{BC})$, assuming all the products are defined. The proof follows from (2.11) and is left as an exercise (Problem 2.2).

In many applications, we deal with *square* matrices:

$$\mathbf{A}=\begin{pmatrix} a_{11} & a_{12} & \cdots & a_{1n} \\ a_{21} & a_{22} & \cdots & a_{2n} \\ \cdot & \cdot & \cdots & \cdot \\ a_{n1} & a_{n2} & \cdots & a_{nn} \end{pmatrix} \tag{2.13}$$

The matrix (2.13) is a square matrix of *order n*. The elements $a_{11}, a_{22}, \ldots, a_{nn}$ lie on the *principal diagonal* of **A**; the sum of these elements is called the *trace* of the square matrix:

$$\mathrm{Tr}(\mathbf{A}) = \sum_{i=1}^{n} a_{ii} \tag{2.14}$$

Let **A** and **B** be any two square matrices of the same order. Although **AB** and **BA** are not, in general, equal, their traces are the same, as we now prove. The (i,j)th element of **AB** is

$$(\mathbf{AB})_{ij} = \sum_{k=1}^{n} a_{ik} b_{kj}$$

We have

$$\mathrm{Tr}(\mathbf{AB}) = \sum_{i=1}^{n} (\mathbf{AB})_{ii} = \sum_{i=1}^{n} \sum_{k=1}^{n} a_{ik} b_{ki}$$

$$\mathrm{Tr}(\mathbf{BA}) = \sum_{i=1}^{n} (\mathbf{BA})_{ii} = \sum_{i=1}^{n} \sum_{k=1}^{n} b_{ik} a_{ki}$$

Interchange of the dummy indices i and k in this last equation gives

$$\mathrm{Tr}(\mathbf{BA}) = \sum_{k=1}^{n} \sum_{i=1}^{n} b_{ki} a_{ik} = \sum_{i=1}^{n} \sum_{k=1}^{n} a_{ik} b_{ki} = \mathrm{Tr}(\mathbf{AB}) \qquad (2.15)$$

A matrix with every element equal to zero is called a *null matrix* and is symbolized by **0**.

A square matrix all of whose elements are zero, except those on the principal diagonal, is called a *diagonal matrix*.

A diagonal matrix with all diagonal elements equal to 1 is called a *unit matrix* **I**:

$$\mathbf{I} = \begin{pmatrix} 1 & 0 & 0 & \cdots & 0 \\ 0 & 1 & 0 & \cdots & 0 \\ \cdot & \cdot & \cdot & \cdots & \cdot \\ 0 & 0 & 0 & \cdots & 1 \end{pmatrix}$$

There is not one unit matrix, but rather there are unit matrices of order 1, order 2, and so on. The general element of a unit matrix is equal to the Kronecker delta:

$$(\mathbf{I})_{ij} = \delta_{ij} \qquad (2.16)$$

Multiplication of any square matrix **A** by a unit matrix of the same order leaves **A** unchanged. The proof is simple. We have for the (i,j)th element of **IA**

$$(\mathbf{IA})_{ij} = \sum_{k=1}^{n} (\mathbf{I})_{ik} a_{kj} = \sum_{k=1}^{n} \delta_{ik} a_{kj} = a_{ij} \qquad \text{Q.E.D.}$$

Similarly, $\mathbf{AI}=\mathbf{A}$.

The pth power of a square matrix $\mathbf{A}$ is defined as the product $\mathbf{AA}\cdots\mathbf{A}$ which contains p factors, where $p=1,2,3,\ldots$. The zeroth power of a square matrix $\mathbf{A}$ is defined to be the unit matrix of the same order as $\mathbf{A}$.

With any square matrix $\mathbf{A}$, we can associate a determinant, symbolized by $\det\mathbf{A}$. Obviously $\det\mathbf{I}=1$. A useful theorem is

$$\det(\mathbf{AB})=\det\mathbf{A}\det\mathbf{B} \tag{2.17}$$

where $\mathbf{A}$ and $\mathbf{B}$ are square matrices of the same order. (Proofs are given in *Pilar*, p. 203; *Hameka*, pp. 139–140.)

If $\mathbf{A}$ and $\mathbf{B}$ are square matrices of the same order, and if $\mathbf{AB}=\mathbf{BA}=\mathbf{I}$, where $\mathbf{I}$ is the unit matrix, then we say that $\mathbf{B}$ is the *inverse* of $\mathbf{A}$. The inverse of $\mathbf{A}$ is denoted by $\mathbf{A}^{-1}$, and we write $\mathbf{B}=\mathbf{A}^{-1}$. Of course, $\mathbf{A}=\mathbf{B}^{-1}$. We have

$$\mathbf{AA}^{-1}=\mathbf{A}^{-1}\mathbf{A}=\mathbf{I}$$

The unit matrix is its own inverse: $\mathbf{I}^{-1}=\mathbf{I}$, since $\mathbf{II}=\mathbf{I}$. Let $\mathbf{A}$ be a square matrix with elements a_{ij}, and let C_{ij} be the cofactor (Section 1.2) of a_{ij} in $\det\mathbf{A}$. The (i,j)th element of the matrix $\mathbf{A}^{-1}$ can be shown to be

$$(\mathbf{A}^{-1})_{ij}=\frac{C_{ji}}{\det\mathbf{A}} \tag{2.18}$$

(Note the reversal of the cofactor subscripts.) For the proof of (2.18), see Problem 2.4. If $\det\mathbf{A}=0$, then $\mathbf{A}$ has no inverse. When $\det\mathbf{A}\neq 0$, $\mathbf{A}$ is called *nonsingular*. Use of (2.18) to calculate the inverse is tedious and a variety of more efficient methods have been devised.

A matrix whose elements are all real numbers is called a *real* matrix.

A square matrix whose elements all satisfy the relation $a_{ij}=a_{ji}$ is called a *symmetric matrix*. The elements of a symmetric matrix are symmetric about the principal diagonal. An example is

$$\begin{pmatrix} 1 & 7i & 9 \\ 7i & 0 & -1 \\ 9 & -1 & 43 \end{pmatrix}$$

A square matrix whose elements all satisfy $a_{ij}=a_{ji}^*$ is called a *Hermitian matrix*. An example is

$$\begin{pmatrix} 1 & 7i & 9 \\ -7i & 0 & 2+3i \\ 9 & 2-3i & 34 \end{pmatrix}$$

The diagonal elements of a Hermitian matrix must be real. A real symmetric matrix is a special case of a Hermitian matrix. (The relation between Hermitian matrices and Hermitian operators will be shown in Section 2.3.)

If **A** is any matrix, its *complex conjugate* $\mathbf{A}^*$ is formed by taking the complex conjugate of every element. For a real matrix, $\mathbf{A}=\mathbf{A}^*$.

If **A** is any matrix, its transpose $\tilde{\mathbf{A}}$ is formed by interchanging the rows and columns of **A**. The elements $\tilde{a}_{ij}$ of $\tilde{\mathbf{A}}$ are

$$\tilde{a}_{ij}=a_{ji} \tag{2.19}$$

For a square matrix, taking the transpose amounts to reflecting the elements about the principal diagonal. A symmetric matrix is equal to its transpose: $\mathbf{A}=\tilde{\mathbf{A}}$. The transpose of a row matrix is a column matrix, and vice versa.

The matrix obtained by taking the complex conjugate of each element of **A** and then forming the transpose is called the *Hermitian conjugate* (or *conjugate transpose*) of **A** and is symbolized by $\mathbf{A}^\dagger$:

$$a_{ij}^\dagger=a_{ji}^* \tag{2.20}$$

(Physicists use the term *adjoint* to designate $\mathbf{A}^\dagger$; mathematicians use the term *associate* for $\mathbf{A}^\dagger$ and use the term *adjoint* with an entirely different meaning.) A Hermitian matrix is equal to its Hermitian conjugate: $\mathbf{A}=\mathbf{A}^\dagger$.

We illustrate the preceding definitions with an example:

$$\mathbf{A}=\begin{pmatrix} 1+i & 2-i \\ 3 & -1 \end{pmatrix} \qquad \tilde{\mathbf{A}}=\begin{pmatrix} 1+i & 3 \\ 2-i & -1 \end{pmatrix} \qquad \mathbf{A}^\dagger=\begin{pmatrix} 1-i & 3 \\ 2+i & -1 \end{pmatrix}$$

The following matrix identities are useful:

$$(\mathbf{AB})^*=\mathbf{A}^*\mathbf{B}^*, \qquad \widetilde{(\mathbf{AB})}=\tilde{\mathbf{B}}\tilde{\mathbf{A}}, \qquad (\mathbf{AB})^\dagger=\mathbf{B}^\dagger\mathbf{A}^\dagger \tag{2.21}$$

We shall prove the first and leave the other two as exercises. The (i,j)th element of **AB** is $\sum_k a_{ik}b_{kj}$. The (i,j)th element of $(\mathbf{AB})^*$ is $[\sum_k a_{ik}b_{kj}]^*$; using the equations following (1.75), we see that the (i,j)th element of $(\mathbf{AB})^*$ equals $\sum_k a_{ik}^* b_{kj}^*$. However, the (i,j)th element of $(\mathbf{A}^*)(\mathbf{B}^*)$ is $\sum_k a_{ik}^* b_{kj}^*$. Having proved corresponding matrix elements equal, we have shown that $(\mathbf{AB})^*=\mathbf{A}^*\mathbf{B}^*$.

An *orthogonal matrix* is one whose inverse is equal to its transpose: $\mathbf{A}^{-1}=\tilde{\mathbf{A}}$. A *unitary matrix* is one whose inverse is equal to its Hermitian conjugate: $\mathbf{A}^{-1}=\mathbf{A}^\dagger$. A real orthogonal matrix is a special case of a unitary matrix.

If $\mathbf{A}$ is orthogonal, then $\mathbf{AA}^{-1}=\mathbf{A\tilde{A}}=\mathbf{I}$. Equating corresponding matrix elements and using (2.11), we have

$$(\mathbf{A\tilde{A}})_{ij}=(\mathbf{I})_{ij}$$

$$\sum_k a_{ik}\tilde{a}_{kj}=\delta_{ij}$$

$$\sum_k a_{ik}a_{jk}=\delta_{ij} \tag{2.22}$$

where (2.16) and (2.19) were used. Similarly, from $\mathbf{\tilde{A}A}=\mathbf{I}$ we find that the elements of an orthogonal matrix also satisfy

$$\sum_k a_{ki}a_{kj}=\delta_{ij} \tag{2.23}$$

for all i and j. Reversal of the above procedure shows that the converse is valid; namely, a matrix whose elements satisfy (2.22) or (2.23) for all i and j is orthogonal.

If $\mathbf{U}$ is a unitary matrix, then $\mathbf{UU}^\dagger=\mathbf{U}^\dagger\mathbf{U}=\mathbf{I}$, and we find for all i and j that (Problem 2.5)

$$\sum_k u_{ki}^*u_{kj}=\delta_{ij}, \qquad \sum_k u_{ik}u_{jk}^*=\delta_{ij} \tag{2.24}$$

Consider two column vectors $\mathbf{b}$ and $\mathbf{c}$ with elements $b_1,\ldots,b_n$ and $c_1,\ldots,c_n$. We say that $\mathbf{b}$ and $\mathbf{c}$ are *orthogonal* when

$$\sum_{i=1}^{n} b_i^*c_i=b_1^*c_1+\cdots+b_n^*c_n=0 \tag{2.25}$$

The matrix $\tilde{\mathbf{b}}^*=\mathbf{b}^\dagger$ is a *row* vector with elements $b_1^*,\ldots,b_n^*$, and the orthogonality condition (2.25) can be written as [*cf.* (2.6)]

$$\mathbf{b}^\dagger\mathbf{c}=0$$

If we form $\mathbf{b}^\dagger\mathbf{b}$, we have

$$\mathbf{b}^\dagger\mathbf{b}=|b_1|^2+\cdots+|b_n|^2 \tag{2.26}$$

If $\mathbf{b}^\dagger\mathbf{b}=1$, then $\mathbf{b}$ is said to be *normalized*. If a set of column vectors has every member normalized and every member orthogonal to every other member, the set is called *orthonormal*. (Orthonormality was previously used to describe functions; we shall see in the next section that orthonormal functions can be represented by orthonormal column vectors.)

The first equation in (2.24) states that column vectors i and j of a unitary matrix are orthonormal; the second equation states that the row vectors of a unitary matrix form an orthonormal set. For a real orthogonal matrix, the row vectors are orthonormal, and so are the column vectors.

If we take a square matrix $\mathbf{S}$ and form the product $\mathbf{T}^{-1}\mathbf{S}\mathbf{T}$, where $\mathbf{T}$ is any nonsingular square matrix, we have carried out a *similarity transformation* on $\mathbf{S}$; if $\mathbf{R}=\mathbf{T}^{-1}\mathbf{S}\mathbf{T}$, then $\mathbf{R}$ and $\mathbf{S}$ are *similar* matrices. Similar matrices have equal traces. The proof uses (2.15):

$$\mathrm{Tr}(\mathbf{R})=\mathrm{Tr}(\mathbf{T}^{-1}\mathbf{S}\mathbf{T})=\mathrm{Tr}[(\mathbf{T}^{-1}\mathbf{S})\mathbf{T}]=\mathrm{Tr}[\mathbf{T}(\mathbf{T}^{-1}\mathbf{S})]=\mathrm{Tr}[(\mathbf{T}\mathbf{T}^{-1})\mathbf{S}]=\mathrm{Tr}(\mathbf{S})$$

since $(\mathbf{T}\mathbf{T}^{-1})\mathbf{S}=\mathbf{I}\mathbf{S}=\mathbf{S}$.

Matrices are useful in dealing with systems of linear equations. The set of n linear simultaneous equations (1.80) can be written as the following matrix equation:

$$\begin{pmatrix} a_{11} & a_{12} & \cdots & a_{1n} \\ a_{21} & a_{22} & \cdots & a_{2n} \\ \cdot & \cdot & \cdots & \cdot \\ a_{n1} & a_{n2} & \cdots & a_{nn} \end{pmatrix} \begin{pmatrix} x_1 \\ x_2 \\ \vdots \\ x_n \end{pmatrix} = \begin{pmatrix} b_1 \\ b_2 \\ \vdots \\ b_n \end{pmatrix}$$

This equation can be written more economically as

$$\mathbf{A}\mathbf{X}=\mathbf{B} \tag{2.27}$$

where the definitions of the column matrices $\mathbf{X}$ and $\mathbf{B}$ and the square matrix $\mathbf{A}$ are clear. If the determinant of the coefficient matrix $\mathbf{A}$ is nonzero, then $\mathbf{A}^{-1}$ exists, and multiplication by $\mathbf{A}^{-1}$ on the left of each side of (2.27) gives

$$\mathbf{X}=\mathbf{A}^{-1}\mathbf{B} \tag{2.28}$$

Solving a set of inhomogeneous linear equations amounts to finding the inverse of the coefficient matrix.

The equations (1.120) for a proper or improper rotation of coordinate axes can be written as

$$\mathbf{x}'=\mathbf{L}\mathbf{x} \tag{2.29}$$

where $\mathbf{x}'$ and $\mathbf{x}$ are column matrices with elements x',y',z' and x,y,z, respectively, and $\mathbf{L}$ is the direction-cosine matrix. Equations (1.118) and (1.121) show that $\mathbf{L}$ is an orthogonal matrix.

Matrices can be *partitioned* into *submatrices*, and the submatrices can be treated as single elements in finding a matrix product. For example, if **S** and **T** are third-order square matrices, we might partition them as follows:

$$\mathbf{S}=\left(\begin{array}{cc|c} s_{11} & s_{12} & s_{13} \\ s_{21} & s_{22} & s_{23} \\ \hline s_{31} & s_{32} & s_{33} \end{array}\right)=\begin{pmatrix} \mathbf{S}_{11} & \mathbf{S}_{12} \\ \mathbf{S}_{21} & \mathbf{S}_{22} \end{pmatrix}$$

$$\mathbf{T}=\left(\begin{array}{cc|c} t_{11} & t_{12} & t_{13} \\ t_{21} & t_{22} & t_{23} \\ \hline t_{31} & t_{32} & t_{33} \end{array}\right)=\begin{pmatrix} \mathbf{T}_{11} & \mathbf{T}_{12} \\ \mathbf{T}_{21} & \mathbf{T}_{22} \end{pmatrix} \tag{2.30}$$

where the submatrices are

$$\mathbf{S}_{11}=\begin{pmatrix} s_{11} & s_{12} \\ s_{21} & s_{22} \end{pmatrix} \qquad \mathbf{S}_{12}=\begin{pmatrix} s_{13} \\ s_{23} \end{pmatrix}$$

$$\mathbf{S}_{21}=(s_{31} \quad s_{32}) \qquad \mathbf{S}_{22}=(s_{33})$$

with similar definitions for the submatrices of **T**. (Of course, there are other ways of partitioning these matrices.) To form the product **ST**, we can write

$$\mathbf{ST}=\begin{pmatrix} \mathbf{S}_{11} & \mathbf{S}_{12} \\ \mathbf{S}_{21} & \mathbf{S}_{22} \end{pmatrix}\begin{pmatrix} \mathbf{T}_{11} & \mathbf{T}_{12} \\ \mathbf{T}_{21} & \mathbf{T}_{22} \end{pmatrix}$$

$$=\begin{pmatrix} \mathbf{S}_{11}\mathbf{T}_{11}+\mathbf{S}_{12}\mathbf{T}_{21} & \mathbf{S}_{11}\mathbf{T}_{12}+\mathbf{S}_{12}\mathbf{T}_{22} \\ \mathbf{S}_{21}\mathbf{T}_{11}+\mathbf{S}_{22}\mathbf{T}_{21} & \mathbf{S}_{21}\mathbf{T}_{12}+\mathbf{S}_{22}\mathbf{T}_{22} \end{pmatrix} \tag{2.31}$$

Equation (2.31) is readily verified by explicitly performing the submatrix multiplications and comparing with the result obtained by direct multiplication of **S** and **T**; this is left as an exercise for the reader.

2.2 EIGENVALUES AND EIGENVECTORS

Let **A** be a square matrix of some unspecified order. If

$$\mathbf{A}\mathbf{x}^{(n)}=\lambda_n\mathbf{x}^{(n)} \tag{2.32}$$

where $\mathbf{x}^{(n)}$ is a nonnull column vector with elements $x_1^{(n)}, x_2^{(n)}, \ldots$ and λ_n is a constant, we say that $\mathbf{x}^{(n)}$ is an *eigenvector* of the matrix $\mathbf{A}$, with *eigenvalue* λ_n. A matrix has a number of different eigenvectors, so we use the superscript (n) to distinguish them. (The parallelism in terminology with eigenfunctions and eigenvalues of operators is no accident; in the next section we will establish a correspondence between square matrices and operators and between column vectors and functions.)

Before proceeding, we note the following two simple theorems, which the reader should prove for himself. If $\mathbf{x}^{(n)}$ is an eigenvector of $\mathbf{A}$ with eigenvalue λ_n, then $c\mathbf{x}^{(n)}$ is also an eigenvector of $\mathbf{A}$ with eigenvalue λ_n, where c is any constant. If $\mathbf{x}^{(n)}$ and $\mathbf{x}^{(p)}$ are eigenvectors of $\mathbf{A}$ with the *same* eigenvalue $(\lambda_n = \lambda_p)$, then any linear combination $c_1\mathbf{x}^{(n)} + c_2\mathbf{x}^{(p)}$ is an eigenvector of $\mathbf{A}$ with eigenvalue λ_n, where c_1 and c_2 are constants. The first theorem allows us to multiply each element of an eigenvector by the same constant so as to normalize the eigenvector.

We are only interested in eigenvalues and eigenvectors of Hermitian matrices, so let $\mathbf{A}$ in (2.32) be Hermitian: $\mathbf{A} = \mathbf{A}^\dagger$. If we take the Hermitian conjugate of each side of (2.32) and use (2.21), we have

$$[\mathbf{A}\mathbf{x}^{(n)}]^\dagger = \mathbf{x}^{(n)\dagger}\mathbf{A}^\dagger = \mathbf{x}^{(n)\dagger}\mathbf{A}$$

$$[\lambda_n \mathbf{x}^{(n)}]^\dagger = \lambda_n^* \mathbf{x}^{(n)\dagger}$$

Hence for a Hermitian matrix

$$\mathbf{x}^{(n)\dagger}\mathbf{A} = \lambda_n^* \mathbf{x}^{(n)\dagger} \tag{2.33}$$

The eigenvector $\mathbf{x}^{(m)}$ of $\mathbf{A}$ satisfies

$$\mathbf{A}\mathbf{x}^{(m)} = \lambda_m \mathbf{x}^{(m)} \tag{2.34}$$

Now consider the matrix product $\mathbf{x}^{(n)\dagger}\mathbf{A}\mathbf{x}^{(m)}$. Using (2.33), we have

$$\mathbf{x}^{(n)\dagger}\mathbf{A}\mathbf{x}^{(m)} = \lambda_n^* \mathbf{x}^{(n)\dagger}\mathbf{x}^{(m)}$$

Using (2.34), we have

$$\mathbf{x}^{(n)\dagger}\mathbf{A}\mathbf{x}^{(m)} = \lambda_m \mathbf{x}^{(n)\dagger}\mathbf{x}^{(m)}$$

Equating the right-hand sides of these last two equations, we get

$$(\lambda_m - \lambda_n^*)\mathbf{x}^{(n)\dagger}\mathbf{x}^{(m)} = 0 \tag{2.35}$$

For $m=n$, Equation (2.35) becomes $(\lambda_n-\lambda_n^*)\mathbf{x}^{(n)\dagger}\mathbf{x}^{(n)}=0$. Now [Eq. (2.26)]

$$\mathbf{x}^{(n)\dagger}\mathbf{x}^{(n)}=|x_1^{(n)}|^2+|x_2^{(n)}|^2+\cdots>0$$

since the eigenvector $\mathbf{x}^{(n)}$ is by definition not a null vector. Since $\mathbf{x}^{(n)\dagger}\mathbf{x}^{(n)}$ is nonzero, we must have $(\lambda_n-\lambda_n^*)=0$, or $\lambda_n=\lambda_n^*$. We have proved that *eigenvalues of a Hermitian matrix are real.*

For $m\neq n$, Equation (2.35) becomes $(\lambda_m-\lambda_n)\mathbf{x}^{(n)\dagger}\mathbf{x}^{(m)}=0$. If $\lambda_m\neq\lambda_n$, then $\mathbf{x}^{(n)\dagger}\mathbf{x}^{(m)}=0$. We have proved that *eigenvectors of a Hermitian matrix that correspond to different eigenvalues are orthogonal.* For the case $\lambda_m=\lambda_n$, we can take linear combinations of $\mathbf{x}^{(m)}$ and $\mathbf{x}^{(n)}$ to form orthogonal eigenvectors. Thus the vectors

$$\mathbf{x}^{(m)} \qquad \text{and} \qquad \mathbf{x}^{(n)}-\frac{(\mathbf{x}^{(m)\dagger}\mathbf{x}^{(n)})\mathbf{x}^{(m)}}{(\mathbf{x}^{(m)\dagger}\mathbf{x}^{(m)})}$$

are orthogonal, as the conscientious reader will verify. [Note the analogy to (1.26); note also that $\mathbf{x}^{(m)\dagger}\mathbf{x}^{(n)}$ and $\mathbf{x}^{(m)\dagger}\mathbf{x}^{(m)}$ are scalars.] Of course, the eigenvectors can each be normalized. Hence *the eigenvectors of a Hermitian matrix can be chosen to form an orthonormal set.*

How do we actually find the eigenvalues and eigenvectors of a matrix? The matrix equation (2.32) can be rewritten as

$$(\mathbf{A}-\lambda_n\mathbf{I})\mathbf{x}^{(n)}=\mathbf{0} \tag{2.36}$$

Using (2.12) and (2.16) to form the ith element of the column vectors $\mathbf{Ax}^{(n)}$ and $\mathbf{Ix}^{(n)}$, we find that (2.36) is equivalent to the scalar equations

$$\sum_k (a_{ik}-\lambda_n\delta_{ik})x_k^{(n)}=0, \qquad i=1,2,\ldots \tag{2.37}$$

This is a set of simultaneous linear homogeneous equations for the unknowns $x_1^{(n)}$, $x_2^{(n)}$, ..., and for a nontrivial solution, the coefficient determinant must vanish (Section 1.2):

$$\det(a_{ik}-\lambda_n\delta_{ik})=\det(\mathbf{A}-\lambda_n\mathbf{I})=0 \tag{2.38}$$

$$\begin{vmatrix} a_{11}-\lambda_n & a_{12} & \cdots \\ a_{21} & a_{22}-\lambda_n & \cdots \\ \vdots & \vdots & \vdots\vdots\vdots \end{vmatrix}=0 \tag{2.39}$$

This *secular equation* is a polynomial in λ_n of degree equal to the order of the square matrix **A**. Note from Theorem 6 on determinants in Section 1.2 and (2.39) that for a diagonal matrix, the eigenvalues are equal to the diagonal elements.

Having found the eigenvalues from (2.39), we substitute each λ_n in turn into the simultaneous equations (2.37) and solve for $x_1^{(n)}, x_2^{(n)}, \ldots,$ which make up the eigenvector $\mathbf{x}^{(n)}$. (This sort of procedure should be pretty familiar by now.) A numerical example is given in the next section.

Let us take the column eigenvectors $\mathbf{x}^{(n)}$ and place them side by side to form a square matrix **X** of eigenvectors:

$$\mathbf{X}=\left(\mathbf{x}^{(1)} \mid \mathbf{x}^{(2)} \mid \cdots\right)=\begin{pmatrix} x_1^{(1)} & x_1^{(2)} & \cdots \\ x_2^{(1)} & x_2^{(2)} & \cdots \\ \vdots & \vdots & \vdots\vdots\vdots \end{pmatrix} \tag{2.40}$$

Let $\boldsymbol{\Lambda}$ be the diagonal matrix whose diagonal elements are the eigenvalues of **A**:

$$\boldsymbol{\Lambda}=\begin{pmatrix} \lambda_1 & 0 & \cdots \\ 0 & \lambda_2 & \cdots \\ \vdots & \vdots & \vdots\vdots\vdots \end{pmatrix} \tag{2.41}$$

We assert that the set of matrix equations (2.32) with $n=1,2,\ldots$ can be written as the single matrix equation

$$\mathbf{AX}=\mathbf{X}\boldsymbol{\Lambda} \tag{2.42}$$

To verify (2.42) we show that corresponding matrix elements of **AX** and of $\mathbf{X}\boldsymbol{\Lambda}$ are equal. Using (2.37) we have

$$(\mathbf{AX})_{in}=\sum_k a_{ik}x_k^{(n)}=\sum_k \lambda_n \delta_{ik} x_k^{(n)}=\lambda_n x_i^{(n)}$$

$$(\mathbf{X}\boldsymbol{\Lambda})_{in}=\sum_k x_i^{(k)}(\boldsymbol{\Lambda})_{kn}=\sum_k x_i^{(k)}\delta_{kn}\lambda_n=x_i^{(n)}\lambda_n$$

If **A** is a Hermitian matrix, then as proved above, the eigenvectors $\mathbf{x}^{(n)}$ can be chosen to be orthonormal. Hence [Equation (2.24)] the **X** matrix can be chosen as unitary; therefore, **X** has an inverse (namely, its Hermitian conjugate $\mathbf{X}^\dagger$). Application of $\mathbf{X}^{-1}$ on the left of each side of

(2.42) gives $\mathbf{X}^{-1}\mathbf{AX}=\mathbf{X}^{-1}\mathbf{X\Lambda}=\mathbf{I\Lambda}=\mathbf{\Lambda}$; hence

$$\mathbf{X}^{-1}\mathbf{AX}=\mathbf{\Lambda} \qquad \text{or} \qquad \mathbf{X}^{\dagger}\mathbf{AX}=\mathbf{\Lambda} \tag{2.43}$$

so that a Hermitian matrix **A** is transformed to its diagonal eigenvalue matrix by a similarity transformation with the unitary matrix of its eigenvectors. This process is called *diagonalization* of **A**. Instead of finding the eigenvalues of **A** by solving the algebraic secular equation (2.39), we can try and find a matrix **X** that will diagonalize **A**; the eigenvalues are then found as the diagonal elements of $\mathbf{X}^{\dagger}\mathbf{AX}=\mathbf{\Lambda}$. Since the trace is invariant under a similarity transformation, the sum of the eigenvalues of **A** equals the trace of **A**. For the special case of a real Hermitian matrix, all the eigenvectors must be real (since their elements are found by the solution of real linear simultaneous equations), and **X** is a real unitary matrix, that is, an orthogonal matrix. Hence *a real symmetric matrix can be diagonalized by an orthogonal similarity transformation; a Hermitian matrix can be diagonalized by a unitary similarity transformation.*

A *real quadratic form f* is a scalar algebraic expression that has the form

$$f=\sum_{i=1}^{n}\sum_{k=1}^{n} a_{ik}x_i x_k \tag{2.44}$$

$$a_{ik}=a_{ki} \quad \text{for all } i \text{ and } k$$

where the x's are real variables and the a_{ik}'s are real constants. An example with $n=2$ is

$$f=a_{11}x_1^2+a_{12}x_1x_2+a_{21}x_2x_1+a_{22}x_2^2=a_{11}x_1^2+2a_{12}x_1x_2+a_{22}x_2^2$$

Let **A** be the real symmetric n by n matrix whose elements are the a_{ik}'s, and let **x** be the n by 1 column vector with elements $x_1,\ldots,x_n$. Then **Ax** is an n by 1 column vector whose ith element is given by (2.12) with b_k replaced by x_k. The matrix $\tilde{\mathbf{x}}$ is a 1 by n row vector. The product $\tilde{\mathbf{x}}(\mathbf{Ax})$ is the product of a row vector with a column vector and [as in (2.6)] gives a scalar; this scalar is readily seen to be equal to the right side of (2.44). Hence (2.44) may be written as

$$f=\tilde{\mathbf{x}}\mathbf{Ax} \tag{2.45}$$

Suppose we want to carry out a change of variables that will eliminate all cross products (those with $i\neq k$) from f, so as to have f expressed as a sum of squares. Let **C** be the square matrix of eigenvectors of **A**. We previously used **X** to denote this matrix, but to avoid confusion with the variables x_i in (2.44) we now switch to **C**. Equations (2.42) and (2.43) give

$$\mathbf{AC}=\mathbf{C\Lambda},\qquad \mathbf{C}^{-1}\mathbf{AC}=\tilde{\mathbf{C}}\mathbf{AC}=\mathbf{\Lambda} \tag{2.46}$$

where $\mathbf{\Lambda}$ is given by (2.41) and the λ_i's are the eigenvalues of **A**. Since **A** is real and symmetric, **C** can be chosen as orthogonal. We now define a new set of variables $y_1,\ldots,y_n$ as the following linear combinations of the original variables (the x_i's):

$$y_i=\sum_k c_{ki}x_k=\sum_k \tilde{c}_{ik}x_k,\qquad i=1,\ldots,n$$

$$\mathbf{y}=\tilde{\mathbf{C}}\mathbf{x} \tag{2.47}$$

where **y** is the column vector with elements $y_1,\ldots,y_n$. Since **C** is orthogonal, (2.47) reads $\mathbf{y}=\mathbf{C}^{-1}\mathbf{x}$; left multiplication by **C** gives $\mathbf{Cy}=\mathbf{CC}^{-1}\mathbf{x}$ or

$$\mathbf{x}=\mathbf{Cy} \tag{2.48}$$

Substitution into (2.45) and use of (2.21) and (2.46) give

$$f=\tilde{\mathbf{x}}\mathbf{Ax}=\widetilde{(\mathbf{Cy})}\mathbf{ACy}=\tilde{\mathbf{y}}\tilde{\mathbf{C}}\mathbf{ACy}=\tilde{\mathbf{y}}\mathbf{\Lambda y}$$

The right side of this last equation has the same form as the right side of (2.45), and by analogy to (2.44) we have

$$f=\sum_{i=1}^{n}\sum_{k=1}^{n} y_i(\Lambda)_{ik}y_k=\sum_i\sum_k y_i\delta_{ik}\lambda_i y_k=\sum_{i=1}^{n}\lambda_i y_i^2$$

Hence use of **C** (the square matrix of eigenvectors of the coefficient matrix **A**) to carry out the change of variables (2.47) reduces a real quadratic form to a sum of squares. Since **C** is orthogonal, (2.47) is called an *orthogonal transformation* of variables. A simple application is in finding the coordinate rotation that brings the coordinate axes into coincidence with the principal axes of an ellipse. The equation of an ellipse can be expressed in terms of a real quadratic form. The cross products can be eliminated by an orthogonal transformation of variables, which, as noted in Section 2.1, corresponds to a rotation of coordinates. (See Problem 2.17.) More exciting applications of quadratic forms will be found in Chapters 5 and 6.

2.3 MATRICES AND QUANTUM MECHANICS

Historically, the first formulation of quantum mechanics was given by Heisenberg a few months before Schrödinger discovered his equation. The

Heisenberg formulation was in terms of certain arrays of numbers, for which Heisenberg deduced the rules of combination. Heisenberg (like most scientists in 1925) was unfamiliar with matrices and did not realize that the rules he found were those of matrix algebra. (In a letter, Heisenberg wrote: "Now the learned Göttingen mathematicians talk so much about Hermitian matrices, but I do not even know what a matrix is.") Max Born recognized Heisenberg's arrays as matrices and together with Heisenberg and Jordan he further developed Heisenberg's matrix mechanics. (Jordan happened to be traveling on the same train as Born and overheard Born telling a colleague of his difficulties with matrix calculations. Jordan introduced himself and volunteered his services.) In this section, we shall show how quantum mechanics can be formulated in terms of matrices.

We have referred to integrals of the form

$$\int \varphi_i^* \hat{F} \varphi_j \, d\tau = F_{ij} \tag{2.49}$$

as *matrix elements*; we now justify this terminology. Let

$$\varphi_1, \varphi_2, \ldots, \varphi_k, \ldots \tag{2.50}$$

be some complete orthonormal set of well-behaved functions (not necessarily eigenfunctions of $\hat{F}$). We form the integrals (2.49) for all i and j and arrange them in a square matrix, which we call $\mathbf{F}$:

$$\mathbf{F} = \begin{pmatrix} F_{11} & F_{12} & F_{13} & \cdots \\ F_{21} & F_{22} & F_{23} & \cdots \\ F_{31} & F_{32} & F_{33} & \cdots \\ \cdot & \cdot & \cdot & \cdots \end{pmatrix} \tag{2.51}$$

The complete set (2.50) will usually have an infinite number of functions, and the matrix (2.51) will be of infinite order.

In any quantum-mechanical problem, we have a set of operators $\hat{F}, \hat{G}, \ldots$ which obey certain relations. (An example is the set of angular-momentum commutation relations.) What we want to show is that the matrices formed from these operators obey the same relations as the operators, so that we can, if we like, work with the matrices instead of with the operators. There are three ways of combining operators: we can add two operators; we can multiply an operator by a constant; we can multiply two operators. We shall examine each process in turn.

First, suppose that the sum of $\hat{F}$ and $\hat{G}$ is the operator $\hat{K}$: $\hat{K} = \hat{F} + \hat{G}$. The

(i,j)th element of the matrix **K** is given by the definition (2.49) as

$$K_{ij} = \int \varphi_i^* \hat{K}\varphi_j \, d\tau = \int \varphi_i^* (\hat{F} + \hat{G})\varphi_j \, d\tau = \int \varphi_i^* \left(\hat{F}\varphi_j + \hat{G}\varphi_j\right) d\tau$$

$$= \int \varphi_i^* \hat{F}\varphi_j \, d\tau + \int \varphi_i^* \hat{G}\varphi_j \, d\tau = F_{ij} + G_{ij}$$

Since $K_{ij} = F_{ij} + G_{ij}$ for all elements, then $\mathbf{K} = \mathbf{F} + \mathbf{G}$, which is the same relation satisfied by the operators $\hat{K}$, $\hat{F}$, and $\hat{G}$.

Next suppose that $k\hat{F} = \hat{B}$, where k is a constant. Then

$$B_{ij} = \int \varphi_i^* \hat{B}\varphi_j \, d\tau = \int \varphi_i^* k\hat{F}\varphi_j \, d\tau = k\int \varphi_i^* \hat{F}\varphi_j \, d\tau = kF_{ij}$$

Since kF_{ij} is the (i,j)th element of the matrix $k\mathbf{F}$, we have $\mathbf{B} = k\mathbf{F}$, the same relation as the operators obey.

Finally, suppose that $\hat{F}\hat{G} = \hat{A}$. We have

$$A_{ij} = \int \varphi_i^* \hat{A}\varphi_j \, d\tau = \int \varphi_i^* \hat{F}\hat{G}\varphi_j \, d\tau \tag{2.52}$$

The expansion of any function f in terms of the complete set (2.50) is given by (1.33). Replacement of f in (1.33) by the function $\hat{G}\varphi_j$ gives

$$\hat{G}\varphi_j = \sum_k \left[\int \varphi_k^* \hat{G}\varphi_j \, d\tau \right] \varphi_k = \sum_k G_{kj}\varphi_k \tag{2.53}$$

Substitution into (2.52) gives (G_{kj} is a constant)

$$A_{ij} = \sum_k G_{kj} \int \varphi_i^* \hat{F}\varphi_k \, d\tau$$

$$A_{ij} = \sum_k F_{ik} G_{kj} \tag{2.54}$$

We recognize (2.54) as the matrix-multiplication rule (2.11). Hence the matrix **A** is the product of the matrices **F** and **G**: $\mathbf{A} = \mathbf{FG}$, the same relation obeyed by the corresponding operators. Actually, we proved this earlier—see (1.52).

Thus the matrices formed according to (2.49) and (2.51) obey the same equations as the corresponding operators. For example, we know that [Equation (1.60)] $\hat{x}\hat{p}_x - \hat{p}_x\hat{x} = i\hbar\hat{1}$. Hence, if we form matrices $\mathbf{x}$ and $\mathbf{p}_x$ with matrix elements

$$x_{ij} = \langle \varphi_i | \hat{x} | \varphi_j \rangle, \qquad (p_x)_{ij} = \langle \varphi_i | \hat{p}_x | \varphi_j \rangle \tag{2.55}$$

where φ_j is any complete orthonormal set, we will find that $\mathbf{xp}_x - \mathbf{p}_x\mathbf{x} = i\hbar\mathbf{I}$, where $\mathbf{I}$ is the unit matrix. (Because of orthonormality, the matrix elements of the unit operator $\hat{1}$ are δ_{ij}.)

The functions (2.50) are called *basis functions*. The matrices **F**, **G**, ... are called *matrix representatives* of the operators $\hat{F}$, $\hat{G}$, ... in the φ basis. Of course, the specific form of the matrix representation of a set of operators depends on the basis chosen. Equation (2.53) shows that the effect of the operator $\hat{G}$ on the basis functions is determined by the matrix elements G_{kj}. Since an arbitrary well-behaved function can be expanded using the complete set (2.50), knowledge of the matrix **G** allows one to determine the effect of the operator $\hat{G}$ on an arbitrary function. Thus, knowledge of the square matrix **G** is fully equivalent to knowledge of the corresponding operator $\hat{G}$. Since $\hat{G}$ is a Hermitian operator, its matrix elements satisfy $G_{ij} = (G_{ji})^*$. Hence the matrix **G** representing $\hat{G}$ is a Hermitian matrix (Section 2.1).

As an example, suppose we want matrix representatives for the electron-spin operators $\hat{S}_x, \hat{S}_y, \hat{S}_z$. An obvious choice of basis functions for the representation is the set of eigenfunctions of $\hat{S}_z$, the functions α and β. (This is a complete set since it consists of the eigenfunctions of the Hermitian operator $\hat{S}_z$.) Equations (1.244), (1.245), and the orthonormality of α and β [(1.237) and (1.238)] give for the elements of the $\mathbf{S}_x$ matrix

$$(\mathbf{S}_x)_{11} = \langle \alpha | \hat{S}_x | \alpha \rangle = \tfrac{1}{2}\hbar \langle \alpha | \beta \rangle = 0$$

$$(\mathbf{S}_x)_{12} = \langle \alpha | \hat{S}_x | \beta \rangle = \tfrac{1}{2}\hbar \langle \alpha | \alpha \rangle = \tfrac{1}{2}\hbar = [(\mathbf{S}_x)_{21}]^*$$

$$(\mathbf{S}_x)_{22} = \langle \beta | \hat{S}_x | \beta \rangle = \tfrac{1}{2}\hbar \langle \beta | \alpha \rangle = 0$$

where we used the fact that $\hat{S}_x$ is Hermitian. The $\mathbf{S}_x$ matrix is then

$$\mathbf{S}_x = \tfrac{1}{2}\hbar \begin{pmatrix} 0 & 1 \\ 1 & 0 \end{pmatrix} \tag{2.56}$$

Similarly, we find (Problem 2.16)

$$\mathbf{S}_y = \tfrac{1}{2}\hbar \begin{pmatrix} 0 & -i \\ i & 0 \end{pmatrix}, \qquad \mathbf{S}_z = \tfrac{1}{2}\hbar \begin{pmatrix} 1 & 0 \\ 0 & -1 \end{pmatrix} \tag{2.57}$$

These (without the $\tfrac{1}{2}\hbar$ factor) are the *Pauli spin matrices*.

Having established the correspondence between square matrices and operators, we now consider how to represent a function with a column matrix. If f is an arbitrary well-behaved function, we can expand it

[Equation (1.30)] using the complete set (2.50):

$$f = \sum_k c_k \varphi_k \tag{2.58}$$

We shall represent the function f by the column vector $\mathbf{f}$ whose elements are the coefficients c_k in the expansion (2.58). Since each basis function can be "expanded" as $\varphi_k = \Sigma_j \delta_{jk} \varphi_j$, each φ_k is represented by a column vector with one element unity and all others zero. Thus

$$\mathbf{f} = \begin{pmatrix} c_1 \\ c_2 \\ c_3 \\ \vdots \end{pmatrix} \quad \boldsymbol{\varphi}_1 = \begin{pmatrix} 1 \\ 0 \\ 0 \\ \vdots \end{pmatrix} \quad \boldsymbol{\varphi}_2 = \begin{pmatrix} 0 \\ 1 \\ 0 \\ \vdots \end{pmatrix} \quad \cdots \tag{2.59}$$

where $\boldsymbol{\varphi}_k$ is the column vector representing φ_k. Equation (2.58) corresponds to the matrix equation

$$\mathbf{f} = \sum_k c_k \boldsymbol{\varphi}_k \tag{2.60}$$

as is readily verified. The vectors $\boldsymbol{\varphi}_k$ are *basis vectors*. Equation (2.60) is analogous to writing an ordinary three-dimensional vector as a linear combination of the unit vectors **i**, **j**, and **k**. The set of all vectors formed as linear combinations of the basis vectors $\boldsymbol{\varphi}_k$ constitutes what mathematicians call a *vector space*.

As an example, the most general one-electron spin function $c_1\alpha + c_2\beta$ has as its representative in the α, β basis

$$\begin{pmatrix} c_1 \\ c_2 \end{pmatrix} = c_1 \begin{pmatrix} 1 \\ 0 \end{pmatrix} + c_2 \begin{pmatrix} 0 \\ 1 \end{pmatrix} \tag{2.61}$$

The basis vectors are the two vectors on the right of (2.61).

If f in (2.58) is the state function Ψ, then the column vector of expansion coefficients representing Ψ is called the *state vector*.

To show the validity of representing a function by the column vector of expansion coefficients, suppose that the operator $\hat{A}$ acting on the function f converts f to the function g: $g = \hat{A}f$. We can prove that the corresponding matrix equation $\mathbf{g} = \mathbf{A}\mathbf{f}$ holds, where $\mathbf{g}$ and $\mathbf{f}$ are the column vectors representing g and f, and $\mathbf{A}$ is the square matrix representing $\hat{A}$. The proof is outlined in Problem 2.21. As a special case, suppose that $\hat{A}f = \lambda f$ (i.e., $g = \lambda f$). Then $\mathbf{A}\mathbf{f} = \lambda\mathbf{f}$, showing that $\mathbf{f}$ is an eigenvector of $\mathbf{A}$ with eigenvalue λ if f is an eigenfunction of $\hat{A}$ with eigenvalue λ.

Let $\mathbf{f}$ and $\mathbf{g}$ be the column vectors representing the functions f and g in some orthonormal basis and let $\mathbf{A}$ represent the operator $\hat{A}$ in that basis. The reader can verify (Problem 2.22) that the integrals $\langle f | g \rangle$ and $\langle f | \hat{A} | g \rangle$

are given by

$$\langle f|g\rangle = \mathbf{f}^\dagger\mathbf{g} \tag{2.62}$$

$$\langle f|\hat{A}|g\rangle = \mathbf{f}^\dagger\mathbf{A}\mathbf{g} \tag{2.63}$$

The quantity $\mathbf{f}^\dagger\mathbf{g}$ is called the *inner product* (or *scalar product*) of the column vectors **f** and **g**. The inner product is a generalization of the dot product (1.55) to vectors with an arbitrary number of complex components. Since a Hermitian operator satisfies (1.13), then (2.63) shows that for a Hermitian matrix **A**

$$\mathbf{f}^\dagger\mathbf{A}\mathbf{g} = (\mathbf{g}^\dagger\mathbf{A}\mathbf{f})^* = \tilde{\mathbf{g}}\mathbf{A}^*\mathbf{f}^*$$

Note from (2.62) that orthonormal functions have orthonormal column vector representatives.

Because of the correspondence between matrices and operators and between column vectors and functions, all the theorems we proved about operators and eigenfunctions have their matrix counterparts. We have already seen examples of this in Section 2.2.

We now show how the fundamental problem of quantum mechanics, the finding of the eigenvalues and eigenfunctions of $\hat{F}$, $\hat{G}, \ldots$ is formulated using matrix algebra.

Let Ω_n and λ_n be the eigenfunctions and eigenvalues of $\hat{F}$:

$$\hat{F}\Omega_n = \lambda_n\Omega_n \tag{2.64}$$

We expand the unknown eigenfunctions using the complete set (2.50):

$$\Omega_n = \sum_j c_j^{(n)}\varphi_j \tag{2.65}$$

where the superscripts on the expansion coefficients $c_j^{(n)}$ indicate which eigenfunction is being expanded. Substitution of (2.65) into (2.64), multiplication by φ_i^*, integration over all space, and use of the orthonormality of the functions (2.50) gives

$$\sum_j c_j^{(n)}\hat{F}\varphi_j = \lambda_n \sum_j c_j^{(n)}\varphi_j$$

$$\sum_j c_j^{(n)}\langle\varphi_i|\hat{F}|\varphi_j\rangle = \lambda_n \sum_j c_j^{(n)}\langle\varphi_i|\varphi_j\rangle = \lambda_n \sum_j c_j^{(n)}\delta_{ij}$$

$$\sum_j F_{ij}c_j^{(n)} = \lambda_n c_i^{(n)}, \qquad i = 1, 2, 3, \ldots \tag{2.66}$$

$$\sum_j (F_{ij} - \lambda_n\delta_{ij})c_j^{(n)} = 0, \qquad i = 1, 2, 3, \ldots \tag{2.67}$$

The equations of (2.67) form a set of simultaneous linear homogeneous equations in the unknowns $c_1^{(n)}$, $c_2^{(n)}, \ldots$; for a nontrivial solution to exist, the determinant of the coefficients must vanish:

$$\det(F_{ij} - \lambda_n \delta_{ij}) = 0 \tag{2.68}$$

The roots λ_n of the *secular equation* (2.68) give the eigenvalues of $\hat{F}$. Substitution of each root in turn into the equations of (2.67) allows the expansion coefficients $c_j^{(n)}$ to be found as the solutions of the resulting simultaneous equations; from the $c_j^{(n)}$'s, the eigenfunctions of $\hat{F}$ are found according to (2.65).

If **F** is the matrix representative of $\hat{F}$ in the φ basis, then the equations (2.66) can be written as the matrix equation:

$$\mathbf{F}\mathbf{c}^{(n)} = \lambda_n \mathbf{c}^{(n)} \tag{2.69}$$

where $\mathbf{c}^{(n)}$ is the column vector with elements $c_1^{(n)}$, $c_2^{(n)}, \ldots$. But (2.69) is the equation satisfied by the eigenvalues λ_n and eigenvectors $\mathbf{c}^{(n)}$ of the matrix **F**. [Compare (2.32).] Hence finding the eigenfunctions and eigenvalues of the operator $\hat{F}$ is equivalent to finding the eigenvectors and eigenvalues of the matrix **F**. The operator and matrix eigenvalues are the same, and (2.65) shows the column eigenvector $\mathbf{c}^{(n)}$ to be the representative of the eigenfunction Ω_n in the φ basis used to form **F**. [Actually, we pointed all this out earlier—see the discussion following (2.61).]

Since $\hat{F}$ is a Hermitian operator, **F** is a Hermitian matrix and its eigenvectors can be chosen as orthonormal. Let **C** be the unitary square matrix of column eigenvectors of **F**:

$$\mathbf{C} = \left(\mathbf{c}^{(1)} \;\vdots\; \mathbf{c}^{(2)} \;\vdots\; \mathbf{c}^{(3)} \;\vdots\; \cdots \right) \tag{2.70}$$

Equation (2.42) shows that the set of equations (2.69) with $n = 1, 2, \ldots$ can be written as

$$\mathbf{FC} = \mathbf{C}\boldsymbol{\Lambda} \tag{2.71}$$

$$\boldsymbol{\Lambda} = \mathbf{C}^{-1}\mathbf{FC} = \mathbf{C}^{\dagger}\mathbf{FC} \tag{2.72}$$

where $\boldsymbol{\Lambda}$ is the diagonal eigenvalue matrix (2.41).

As an example, suppose we want the eigenfunctions and eigenvalues of the electron spin operator $\hat{S}_x$. We shall use the basis functions α and β for the calculation. Equation (2.56) gives the $\mathbf{S}_x$ matrix in this basis. The secular equation (2.68) is then

$$\begin{vmatrix} 0-\lambda_n & \frac{1}{2}\hbar \\ \frac{1}{2}\hbar & 0-\lambda_n \end{vmatrix} = 0 \tag{2.73}$$

$$\lambda_n^2 - \tfrac{1}{4}\hbar^2 = 0$$

$$\lambda_1 = +\tfrac{1}{2}\hbar, \qquad \lambda_2 = -\tfrac{1}{2}\hbar$$

The eigenvalues are, of course, the same as for $\hat{S}_z$ since all directions of space are equivalent. As a check, the sum of the eigenvalues equals zero, which equals the trace of $\mathbf{S}_x$. To find the eigenvector corresponding to λ_1, we substitute λ_1 into (2.67) to get

$$-\tfrac{1}{2}\hbar c_1^{(1)} + \tfrac{1}{2}\hbar c_2^{(1)} = 0$$

$$\tfrac{1}{2}\hbar c_1^{(1)} - \tfrac{1}{2}\hbar c_2^{(1)} = 0$$

Thus $c_1^{(1)} = c_2^{(1)}$. Normalization of the λ_1 eigenvector gives $|c_1^{(1)}|^2 + |c_2^{(1)}|^2 = 1$. Hence $c_1^{(1)} = c_2^{(1)} = 2^{-1/2}$. The normalized $\hat{S}_x$ eigenfunction corresponding to the eigenvalue $+\frac{1}{2}\hbar$ is then $2^{-1/2}(\alpha+\beta)$. Similarly the λ_2 eigenfunction is found to be $2^{-1/2}(\alpha-\beta)$. The column vector representatives of the $\hat{S}_x$ eigenfunctions in the α, β basis are

$$\begin{pmatrix} 2^{-1/2} \\ 2^{-1/2} \end{pmatrix} \quad \text{and} \quad \begin{pmatrix} 2^{-1/2} \\ -2^{-1/2} \end{pmatrix}$$

As a check, we verify that a similarity transformation with the unitary square matrix of eigenvectors diagonalizes $\mathbf{S}_x$:

$$\begin{pmatrix} 2^{-1/2} & 2^{-1/2} \\ 2^{-1/2} & -2^{-1/2} \end{pmatrix} \begin{pmatrix} 0 & \frac{1}{2}\hbar \\ \frac{1}{2}\hbar & 0 \end{pmatrix} \begin{pmatrix} 2^{-1/2} & 2^{-1/2} \\ 2^{-1/2} & -2^{-1/2} \end{pmatrix} = \begin{pmatrix} \frac{1}{2}\hbar & 0 \\ 0 & -\frac{1}{2}\hbar \end{pmatrix}$$

At this point the reader may be wondering why we bothered to introduce the notion of matrices, since the eigenvalues and eigenfunction coefficients can be found by solving the algebraic equations (2.68) and (2.67), and we need not utter the word "matrix" at all. Actually, there are both practical and theoretical reasons for introducing matrices. Most eigenvalue problems are solved on computers. It is generally much faster to have the computer calculate the matrix **C** that diagonalizes **F** under a similarity transformation according to (2.72) than to have it solve the algebraic equations (2.68) and (2.67). For a real Hermitian (i.e., real symmetric) matrix, one commonly used procedure (the *Jacobi method*) for finding **C** is as follows: The computer finds the largest off-diagonal element in **F** and then carries out a similarity transformation to reduce this element to zero. The process is repeated for the transformed matrix; the process is repeated over and over until all the off-diagonal elements are reduced to sufficiently small values. The eigenvalues are then obtained (to the desired degree of accuracy) as the diagonal elements of the resulting matrix; the matrix of eigenvectors is the product of the matrices used to carry out the successive similarity transformations. (Further details and a computer program are given in *Offenhartz*, pp. 338–346.)

Matrices are also a great convenience in the theoretical development of quantum chemistry. They make possible an economy of notation, and use of matrix-algebra theorems simplifies derivations considerably. Much of the quantum chemistry literature is formulated in matrix language. The vector-space formulation of quantum mechanics (which we have just touched on) is very fruitful for advanced applications; see *Merzbacher*, Chapter 14 and later chapters.

For a second-order real Hermitian matrix **F** there is a simple formula for its eigenvector matrix **C**. In fact

$$\mathbf{C}=\begin{pmatrix} \cos\theta & -\sin\theta \\ \sin\theta & \cos\theta \end{pmatrix} \tag{2.74}$$

where $\tan 2\theta \equiv 2f_{12}/(f_{11}-f_{22})$. The matrix **C** is readily verified to be orthogonal and we find (Problem 2.23) $\tilde{\mathbf{C}}\mathbf{F}\mathbf{C}$ to be a diagonal matrix with diagonal elements

$$\begin{gathered} f_{11}\cos^2\theta + f_{22}\sin^2\theta + 2f_{12}\sin\theta\cos\theta \\ f_{11}\sin^2\theta + f_{22}\cos^2\theta - 2f_{12}\sin\theta\cos\theta \end{gathered} \tag{2.75}$$

These are the eigenvalues.

In the Jacobi method, a series of similarity transformations is carried out. It is easily proven that similar matrices have the same eigenvalues. Let $\mathbf{A}=\mathbf{P}^{-1}\mathbf{B}\mathbf{P}$. The eigenvalues of **A** satisfy the secular equation (2.38):

$$0=\det(\mathbf{A}-\lambda_n\mathbf{I})=\det(\mathbf{P}^{-1}\mathbf{B}\mathbf{P}-\lambda_n\mathbf{I})=\det[\mathbf{P}^{-1}(\mathbf{B}\mathbf{P}-\lambda_n\mathbf{P})]$$

It follows from (2.17) that $\det(\mathbf{AB})=\det(\mathbf{BA})$; hence

$$0=\det[(\mathbf{BP}-\lambda_n\mathbf{P})\mathbf{P}^{-1}]=\det(\mathbf{B}-\lambda_n\mathbf{I}) \tag{2.76}$$

so that the eigenvalues of **A** satisfy the secular equation of **B**. Q.E.D.

When $\hat{F}$ in (2.64) is a time-independent Hamiltonian, then (2.64) is the time-independent Schrödinger equation $\hat{H}\psi_n=E_n\psi_n$ and the energies satisfy (2.68):

$$\det\left[\langle\varphi_i|\hat{H}|\varphi_j\rangle-\delta_{ij}E_n\right]=0 \tag{2.77}$$

This equation is fully equivalent to the time-independent Schrödinger equation and its roots give the exact energies. Since the set (2.50) generally has an infinite number of functions, the secular determinant generally is of infinite order. In certain cases, however, one need only deal with a finite-order secular determinant. For example, in electron-spin (and nuclear-spin) problems, the number of basis spin functions is finite. Applications will be found in later chapters.

Sometimes one uses a set of basis functions φ_i that are not orthogonal. In this case, (2.68) and (2.71) become (Problem 2.24)

$$\det(F_{ij}-\lambda_n S_{ij})=0, \qquad S_{ij}\equiv\langle\varphi_i|\varphi_j\rangle \tag{2.78}$$

$$\mathbf{FC}=\mathbf{SC\Lambda} \tag{2.79}$$

Nonorthogonality complicates the matrix diagonalization process; see *Offenhartz*, pp. 338–341 for the procedure used. (Note that we can, if we like, use the Schmidt or some other orthogonalization procedure to form orthogonal linear combinations of the nonorthogonal basis functions and then use these orthogonalized basis functions to form the secular equation.)

If the basis set used is finite and incomplete, solution of the secular equation yields approximate, rather than exact, eigenvalues. An example is the linear variation method; note that (2.78) and (1.190) have the same form, except that (1.190) uses an incomplete basis set. An important application of the linear variation method is the Hartree–Fock–Roothaan secular equation (1.298); here, basis AOs centered on different nuclei are nonorthogonal. *Ab initio* and semiempirical SCF methods use matrix-diagonalization procedures to solve the Roothaan equations.

Perturbation theory can be conveniently formulated using matrices; see *Bates*, Vol. I, pp. 199–201.

As an example of the application of matrix methods to MO theory, consider the transformation between the set of delocalized canonical Hartree–Fock MOs $\varphi_{i,\text{can}}$ and the set of energy-localized MOs $\varphi_{i,\text{loc}}$ (Section 1.20). The localized MOs are certain linear combinations of the canonical MOs:

$$\varphi_{j,\text{loc}} = \sum_k \varphi_{k,\text{can}} b_{kj}, \qquad j = 1, 2, \ldots \tag{2.80}$$

The set of equations (2.80) is equivalent to the matrix equation

$$\mathbf{\Phi}_{\text{loc}} = \mathbf{\Phi}_{\text{can}} \mathbf{B} \tag{2.81}$$

where $\mathbf{\Phi}_{\text{loc}}$ and $\mathbf{\Phi}_{\text{can}}$ are the following row matrices:

$$\mathbf{\Phi}_{\text{loc}} = (\varphi_{1,\text{loc}} \quad \varphi_{2,\text{loc}} \quad \cdots), \qquad \mathbf{\Phi}_{\text{can}} = (\varphi_{1,\text{can}} \quad \varphi_{2,\text{can}} \quad \cdots) \tag{2.82}$$

and $\mathbf{B}$ is the square matrix with elements b_{kj}. To verify the equivalence of (2.80) and (2.81), just drop the row index i from the matrix-multiplication rule (2.11).

We require the localized MOs to be orthonormal:

$$\delta_{ij} = \langle \varphi_{i,\text{loc}} | \varphi_{j,\text{loc}} \rangle = \sum_m \sum_k \langle \varphi_{m,\text{can}} | \varphi_{k,\text{can}} \rangle b_{mi}^* b_{kj} = \sum_m \sum_k \delta_{mk} b_{mi}^* b_{kj}$$

$$\delta_{ij} = \sum_k b_{ki}^* b_{kj} \tag{2.83}$$

Since the column vectors of $\mathbf{B}$ are orthonormal, $\mathbf{B}$ is a unitary matrix [Equation (2.24)]. The localized and canonical MOs are related by a unitary transformation. Since $\mathbf{B}$ is unitary, we can write (2.81) as $\mathbf{\Phi}_{\text{can}} = \mathbf{\Phi}_{\text{loc}} \mathbf{B}^\dagger$.

We now prove that any unitary transformation of the orbitals leaves a closed-shell SCF Slater-determinant wave function unchanged, thereby showing the validity of transforming to localized MOs. Let ψ_{loc} be the SCF wave function written using localized orbitals. We wish to prove that ψ_{loc} equals ψ_{can}, where ψ_{can} uses the canonical MOs. In the notation of (1.260), we have

$$\psi_{\text{loc}} = |\varphi_{1,\text{loc}} \overline{\varphi_{1,\text{loc}}} \varphi_{2,\text{loc}} \overline{\varphi_{2,\text{loc}}} \cdots | = |\varphi_{1,\text{loc}} \varphi_{2,\text{loc}} \cdots \overline{\varphi_{1,\text{loc}}} \; \overline{\varphi_{2,\text{loc}}} \cdots | \tag{2.84}$$

where we used the fact that column interchanges can, at most, change the sign of ψ, which is physically irrelevant. Let $\mathbf{L}$ be the matrix whose determinant is ψ_{loc} ($\det\mathbf{L}=\psi_{\text{loc}}$) and let $\mathbf{C}$ be the matrix whose determinant is ψ_{can} ($\det\mathbf{C}=\psi_{\text{can}}$). We partition $\mathbf{L}$ into n rows (where n is the number of electrons) and two columns. We have from (2.84) and (2.82)

$$\mathbf{L}=\left(\begin{array}{c|c} \boldsymbol{\Phi}_{\text{loc}}(1)\alpha(1) & \boldsymbol{\Phi}_{\text{loc}}(1)\beta(1) \\ \hline \boldsymbol{\Phi}_{\text{loc}}(2)\alpha(2) & \boldsymbol{\Phi}_{\text{loc}}(2)\beta(2) \\ \hline \vdots & \vdots \end{array}\right)=\left(\begin{array}{c|c} \boldsymbol{\Phi}_{\text{can}}(1)\mathbf{B}\alpha(1) & \boldsymbol{\Phi}_{\text{can}}(1)\mathbf{B}\beta(1) \\ \hline \boldsymbol{\Phi}_{\text{can}}(2)\mathbf{B}\alpha(2) & \boldsymbol{\Phi}_{\text{can}}(2)\mathbf{B}\beta(2) \\ \hline \vdots & \vdots \end{array}\right)$$

where the numbers in parentheses indicate the coordinates of electrons $1,2,\ldots$, and (2.81) was used. The matrix on the right of this last equation can be expressed as the product of two partitioned matrices as follows:

$$\mathbf{L}=\begin{pmatrix} \boldsymbol{\Phi}_{\text{can}}(1)\alpha(1) & \boldsymbol{\Phi}_{\text{can}}(1)\beta(1) \\ \boldsymbol{\Phi}_{\text{can}}(2)\alpha(2) & \boldsymbol{\Phi}_{\text{can}}(2)\beta(2) \\ \vdots & \vdots \end{pmatrix}\begin{pmatrix} \mathbf{B} & \mathbf{0} \\ \mathbf{0} & \mathbf{B} \end{pmatrix}=\mathbf{C}\begin{pmatrix} \mathbf{B} & \mathbf{0} \\ \mathbf{0} & \mathbf{B} \end{pmatrix} \qquad (2.85)$$

The determinant of a block-diagonal matrix equals the product of the block determinants (Section 1.2). Therefore (2.17) gives

$$\det\mathbf{L}=\det\mathbf{C}(\det\mathbf{B})^2$$

Since $\mathbf{B}$ is unitary, it follows (Problem 2.12) that $|\det\mathbf{B}|=1$. Hence $\det\mathbf{L}=\det\mathbf{C}$, that is, $\psi_{\text{loc}}=\psi_{\text{can}}$ (except possibly for an irrelevant factor of absolute value 1). ψ_{loc} and ψ_{can} are the same wave function.

In open-shell electronic states, the orbitals are not all doubly occupied, and the preceding procedure is not applicable. However, *if* the wave function can be written as a single Slater determinant, one can use a modified procedure to obtain energy-localized MOs here also. The procedure is to deal with the α spin-orbitals and the β spin-orbitals separately, using two different unitary transformation matrices $\mathbf{B}_\alpha$ and $\mathbf{B}_\beta$ in (2.85). Energy-localized MOs for the O_2 triplet ground state have been calculated in this manner. [See D. M. Hirst and M. E. Linington, *Theor. Chim. Acta*, **16**, 55 (1970).] This procedure can also be applied to radicals with a single unpaired electron (e.g., vinyl).

PROBLEMS

2.1 Compute $\mathbf{A}+\mathbf{B}$, $2\mathbf{A}$, $\mathbf{AB}$, $\mathbf{BA}$, $\mathbf{AD}$, $\mathbf{CA}$, $\mathbf{CD}$, and $\mathbf{DC}$, where

$$\mathbf{A}=\begin{pmatrix} 1 & 2 & 3 \\ -1 & -3 & 0 \\ 0 & 1 & 2 \end{pmatrix} \qquad \mathbf{B}=\begin{pmatrix} 0 & 5 & 1 \\ 3 & 5 & -2 \\ -1 & 0 & 5 \end{pmatrix} \qquad \mathbf{D}=\begin{pmatrix} 0 \\ 4 \\ i \end{pmatrix}$$

$$\mathbf{C}=(1 \quad i \quad -1)$$

2.2 Let $\mathbf{A}$, $\mathbf{B}$, and $\mathbf{C}$ be $j\times k$, $k\times l$, and $l\times m$ matrices, respectively. Prove that $(\mathbf{AB})\mathbf{C}=\mathbf{A}(\mathbf{BC})$.

2.3 Let $\mathbf{R}$, $\mathbf{S}$, and $\mathbf{A}$ be square matrices of order n, with $\mathbf{R}$ and $\mathbf{S}$ diagonal and $\mathbf{A}$ nondiagonal. (a) Must $\mathbf{RS}$ equal $\mathbf{SR}$? (b) Must $\mathbf{RA}$ equal $\mathbf{AR}$?

2.4 Prove that (2.18) gives the elements of $\mathbf{A}^{-1}$. *Hint*: use Cramer's rule and (2.28).

2.5 Verify (2.24).

2.6 Verify (2.31).

2.7 (a) Use (2.18) to find $\mathbf{A}^{-1}$, where $\mathbf{A}$ is given in Problem 2.1. Verify your answer by calculating $\mathbf{AA}^{-1}$. (b) Write down a second-order square matrix that has all nonzero elements but has no inverse.

2.8 Which of the following matrices are (a) real; (b) symmetric; (c) Hermitian?

$$\mathbf{A}=\begin{pmatrix} i & i \\ i & i \end{pmatrix} \qquad \mathbf{B}=\begin{pmatrix} i & 1+i \\ 1-i & i \end{pmatrix} \qquad \mathbf{C}=\begin{pmatrix} 1 & 2 \\ 2 & -4 \end{pmatrix}$$

2.9 Which of these matrices are (a) orthogonal; (b) unitary? (It is not necessary to use (2.18) here.)

$$\mathbf{A}=\begin{pmatrix} \cos\alpha & -\sin\alpha & 0 \\ \sin\alpha & \cos\alpha & 0 \\ 0 & 0 & 1 \end{pmatrix} \qquad \mathbf{B}=\begin{pmatrix} i & i \\ i & i \end{pmatrix} \qquad \mathbf{C}=\begin{pmatrix} 1 & 0 \\ 0 & i \end{pmatrix}$$

2.10 Show that $(\mathbf{AB})^{-1}=\mathbf{B}^{-1}\mathbf{A}^{-1}$, where $\mathbf{A}$ and $\mathbf{B}$ are nonsingular square matrices.

2.11 (a) Prove the second and third equations in (2.21). (b) Show $\widetilde{(\mathbf{ABC})}=\tilde{\mathbf{C}}\tilde{\mathbf{B}}\tilde{\mathbf{A}}$.

2.12 (a) Use (2.17) to show that the determinant of an orthogonal matrix must equal +1 or −1. (b) What is true about the determinant of a unitary matrix?

2.13 Let $\mathbf{AB}=\mathbf{AC}$, where **A**, **B**, and **C** are nonnull square matrices of the same order. Explain why **B** and **C** are not necessarily equal.

2.14 State whether each of the following concepts is applicable to all matrices or to only square matrices: (a) real matrix; (b) symmetric matrix; (c) diagonal matrix; (d) null matrix; (e) unit matrix; (f) Hermitian matrix; (g) orthogonal matrix; (h) transpose; (i) inverse; (j) Hermitian conjugate; (k) eigenvalues.

2.15 (a) Find the eigenvalues and the normalized eigenvectors of the following matrices:

$$\mathbf{A}=\begin{pmatrix} 1 & 3 \\ 4 & 2 \end{pmatrix} \qquad \mathbf{B}=\begin{pmatrix} 5 & -2 \\ -2 & 2 \end{pmatrix} \qquad \mathbf{C}=\begin{pmatrix} 8 & 4i \\ -4i & 2 \end{pmatrix}$$

(b) The matrix of eigenvectors of **A** is called the *modal matrix* of **A**. Verify that the modal matrix of the symmetric matrix **B** is orthogonal and that the modal matrix of the Hermitian matrix **C** is unitary. What theorem about Hermitian operators corresponds to these results for matrices? (c) Verify directly that a similarity transformation using the appropriate modal matrix transforms **A**, **B**, and **C** to diagonal form, with the diagonal elements being the eigenvalues.

2.16 Verify (2.57).

2.17 A certain ellipse has the equation $5x^2+2y^2-4xy=1$. The left side of this equation is a real quadratic form. Find the eigenvalues and eigenvectors of the coefficient matrix. Use the results to find the lengths of the principal axes of the ellipse and the angles they make with the x,y coordinate axes.

2.18 Find the formulas for the matrix elements of the matrix representatives of $\hat{x}$ and $\hat{p}_x$ in the particle-in-a-box energy representation. (This is the representation with particle-in-a-box wave functions as basis functions.) Use the general formulas to write down the first several elements in the northwest corner of each matrix.

2.19 (a) Suppose we want the eigenfunctions of $\hat{L}_x$ for quantum number $l=1$. Choose a convenient basis set, and find the $\mathbf{L}_x$ matrix (just the portion with $l=1$); the results of Problem 1.10 will be useful for this purpose. (b) Now find the eigenvalues and eigenvectors of $\mathbf{L}_x$, and the eigenfunctions of $\hat{L}_x$.

2.20 Which of the following are equal to Tr(**ABC**)? (a) Tr(**BCA**); (b) Tr(**CBA**); (c) Tr(**BAC**).

2.21 Prove that if $g = \hat{A}f$, then $\mathbf{g} = \mathbf{Af}$, where **f** and **g** are the column vector representatives of the functions f and g, and **A** is the matrix representative of $\hat{A}$, all in the φ basis. *Hint*: use (2.53) with $\hat{G}$ replaced by $\hat{A}$ to aid in showing that corresponding elements of **g** and **Af** are equal.

2.22 Verify (2.62) and (2.63).

2.23 (a) Verify that **C** in (2.74) diagonalizes **F** and that the eigenvalues are (2.75). (b) Check that (2.74) and (2.75) give the correct eigenvalues and eigenvectors for $\mathbf{S}_x$.

2.24 Verify that (2.78) and (2.79) result from use of a nonorthogonal basis set.

2.25 Prove that similar matrices have equal determinants.

2.26 Show that the sum of the n first-order energy corrections $E_j^{(1)}$ for an n-fold degenerate energy level equals the sum of the n matrix elements H'_{ii} in (1.207).

2.27 Show that the product of two orthogonal matrices is an orthogonal matrix.

2.28 Which of the following matrices are unitary? (a) The matrix of HMO coefficients of a planar molecule. (b) The matrix of coefficients relating the real and complex $2p$ hydrogen orbitals. (c) The coefficient matrix relating MOs to AOs in an SCF MO calculation. (d) The inverse of a unitary matrix. (e) The transpose of a unitary matrix.

2.29 Explain why the following equation holds for the HMO coefficients of a planar conjugated organic compound with n_{C} carbon atoms.

$$\sum_{i=1}^{n_{\text{C}}} |c_{ir}|^2 = 1$$

where the sum is over the orbitals, and not over the carbon atoms.

2.30 Use the theorem of Problem 2.25 to show that the product of the eigenvalues of a matrix equals its determinant.

2.31 What is the sum of the HMO energies of a planar conjugated hydrocarbon with n_{C} carbon atoms?

2.32 We have seen the correspondence between Hermitian operators and Hermitian matrices. In this problem, we extend the correspondence between matrices and operators. The operator $\hat{G}$ is said to be the *adjoint* of $\hat{F}$, if

$$\langle\varphi|\hat{G}|\varphi\rangle = \langle\varphi|\hat{F}|\varphi\rangle^* \tag{2.86}$$

for all well-behaved functions φ; we write $\hat{G}=\hat{F}^\dagger$. By a procedure like that of Problem 1.9, it follows from (2.86) that

$$\langle\varphi_1|\hat{F}^\dagger|\varphi_2\rangle = \langle\varphi_2|\hat{F}|\varphi_1\rangle^* \tag{2.87}$$

for all well-behaved functions φ_1 and φ_2; this is sometimes called "the turnover rule." (a) If $\hat{G}=\hat{F}^\dagger$, find $\hat{G}^\dagger$. (b) An operator $\hat{A}$ is self-adjoint if $\hat{A}=\hat{A}^\dagger$. "Self-adjoint" is a synonym for what term? (c) Find the adjoints of the following operators: (1) multiplication by $f(x)+ig(x)$, where f and g are real functions; (2) id/dx; (3) d/dx; (4) $\hat{O}_{C_4}$. (d) The *inverse* $\hat{F}^{-1}$ of the operator $\hat{F}$ is the operator that satisfies $\hat{F}\hat{F}^{-1}=\hat{1}$. Find the inverses of the following operators: (1) multiplication by $f(x)$; (2) $\hat{O}_{C_4}$. (e) A *normal operator* $\hat{N}$ is one that commutes with its adjoint: $[\hat{N},\hat{N}^\dagger]=\hat{0}$. A *unitary* operator $\hat{U}$ is one for which $\hat{U}\hat{U}^\dagger=\hat{1}=\hat{U}^\dagger\hat{U}$. Prove that Hermitian operators are normal operators. Prove that unitary operators are normal operators. (f) Prove that the eigenvalues of a unitary operator each have absolute value 1.

2.33 Let f in (2.58) be the state function Ψ and let $\mathbf{c}$ be the column vector of expansion coefficients (i.e., the state vector). Let $\boldsymbol{\rho}\equiv\mathbf{c}\mathbf{c}^\dagger$. The matrix $\boldsymbol{\rho}$ is a square matrix called the *density matrix* for the state Ψ. Let A be some quantum-mechanical property. Show that $\langle A\rangle=\mathrm{Tr}(\boldsymbol{\rho}\mathbf{A})$, where $\mathbf{A}$ is the matrix representative of $\hat{A}$ in the φ basis. *Hint*: use (2.63).

2.34 It can be proven that a necessary and sufficient condition for there to exist a matrix that will simultaneously diagonalize the Hermitian matrices **A** and **B** is for **AB** to equal **BA**. To which theorem in Chapter 1 does this matrix-algebra theorem correspond?

2.35 (a) From (2.56) find $\mathbf{S}_x^2$. What are its eigenvalues? (b) From (2.56) and (2.57) find $\mathbf{S}_+$ and $\mathbf{S}_-$.

2.36 For a one-electron spin problem, let the operator $\hat{G}$ have the matrix representative

$$\mathbf{G}=\begin{pmatrix} 1 & i \\ 0 & 2 \end{pmatrix}$$

in the α,β basis. Express $\hat{G}$ in terms of $\hat{S}_x$, $\hat{S}_y$, and $\hat{S}_z$.

2.37 Write a computer program to multiply two rectangular matrices of maximum size 20 by 20. Check it out on some simple examples.

2.38 Prove that $\mathbf{f}^\dagger\mathbf{A}^\dagger\mathbf{g}=\tilde{\mathbf{g}}\mathbf{A}^*\mathbf{f}^*$, where $\mathbf{f}$ and $\mathbf{g}$ are n by 1 column vectors and $\mathbf{A}$ is an n by n square matrix. [When $\mathbf{A}$ is Hermitian, this identity reduces to the equation following (2.63).] *Hint*: take the Hermitian conjugate of $\mathbf{g}^\dagger\mathbf{A}\mathbf{f}$, noting that this product is a complex scalar.

CHAPTER 3

TIME-DEPENDENT STATES AND SPECTROSCOPY

3.1 TIME DEPENDENT PERTURBATION THEORY

Spectroscopy involves transitions between states. To discuss transitions, we must use the time-dependent Schrödinger equation, which is harder to deal with than the time-independent equation. Rather than attempt an exact solution, we will use an approximate treatment called time-dependent perturbation theory.

In spectroscopy, we have a system (atom or molecule) that starts in some stationary state of definite energy, is exposed to electromagnetic radiation for a limited time, and is then found to be in some other stationary state. Let $\hat{H}^0$ be the time-independent Hamiltonian of the system in the absence of radiation. We have

$$\hat{H}^0\psi_k^{(0)}(x) = E_k^{(0)}\psi_k^{(0)}(x) \tag{3.1}$$

where $\psi_k^{(0)}$ and $E_k^{(0)}$ are the unperturbed stationary-state wave functions and energies, and x stands for the $3n$ spatial and n spin coordinates of the n-particle system. Let $\hat{H}'(t)$ be the additional term(s) in the Hamiltonian due to the interaction between the system and the radiation or other perturbing influence. (Our treatment will be applicable to any time-dependent perturbation, not just radiation.) Since $\hat{H}$ depends on time, we must use the time-dependent Schrödinger equation $i\hbar\partial\Psi/\partial t = \hat{H}\Psi$ to investigate how the state changes with time. During the time interval that the perturbation acts, we have

$$i\hbar\frac{\partial\Psi}{\partial t} = \left[\hat{H}^0 + \hat{H}'(t)\right]\Psi \tag{3.2}$$

At any particular time, say $t = t'$, the state function is a function of the $4n$ coordinates of the system:

$$\Psi = \Psi(x, t') \tag{3.3}$$

where t' is some particular numerical value. Since the unperturbed time-independent wave functions form a complete set, we can expand $\Psi(x, t')$ in terms of them:

$$\Psi(x, t') = \sum_k a'_k \psi_k^{(0)}(x) \tag{3.4}$$

where the a'_k are constants. For some other time, say t'', we can make a similar expansion:

$$\Psi(x, t'') = \sum_k a''_k \psi_k^{(0)}(x) \tag{3.5}$$

In general, we have

$$\Psi(x, t) = \sum_k a_k(t) \psi_k^{(0)}(x) \tag{3.6}$$

where the expansion coefficients change with time.

Let us now define new expansion coefficients c_k:

$$c_k(t) \equiv a_k(t) \exp\left(iE_k^{(0)} t/\hbar\right) \tag{3.7}$$

This step will simplify the work. Equation (3.6) becomes

$$\Psi(x, t) = \sum_k c_k(t) \exp\left(-iE_k^{(0)} t/\hbar\right) \psi_k^{(0)}(x) \tag{3.8}$$

The function $\exp(-iE_k^{(0)} t/\hbar)\psi_k^{(0)}(x)$ is what the state function would be if the perturbation $\hat{H}'$ were absent and the system were in the kth stationary state. [See Equation (1.19).] Using (3.8) in (3.2), we have

$$i\hbar \sum_k \frac{dc_k}{dt} \exp\left(-iE_k^{(0)} t/\hbar\right) \psi_k^{(0)} + \sum_k E_k^{(0)} c_k \exp\left(-iE_k^{(0)} t/\hbar\right) \psi_k^{(0)}$$

$$= \sum_k c_k \exp\left(-iE_k^{(0)} t/\hbar\right) \hat{H}^0 \psi_k^{(0)} + \sum_k c_k \exp\left(-iE_k^{(0)} t/\hbar\right) \hat{H}' \psi_k^{(0)} \tag{3.9}$$

From (3.1), the first sum on the right of (3.9) equals the second sum on the left; we have

$$i\hbar \sum_k \frac{dc_k}{dt} \exp\left(-iE_k^{(0)} t/\hbar\right) \psi_k^{(0)} = \sum_k c_k \exp\left(-iE_k^{(0)} t/\hbar\right) \hat{H}' \psi_k^{(0)} \tag{3.10}$$

Now we multiply by $\psi_m^{(0)*}$ and integrate over all space:

$$i\hbar \sum_k \frac{dc_k}{dt} \exp(-iE_k^{(0)}t/\hbar) \int \psi_m^{(0)*}\psi_k^{(0)}\, d\tau$$

$$= \sum_k c_k \exp(-iE_k^{(0)}t/\hbar) \int \psi_m^{(0)*}\hat{H}'\psi_k^{(0)}\, d\tau \qquad (3.11)$$

Using orthonormality of the unperturbed wave functions, and then multiplying by $\exp(iE_m^{(0)}t/\hbar)$, we simplify (3.11) to

$$i\hbar \frac{dc_m}{dt} = \sum_k c_k \exp[i(E_m^{(0)} - E_k^{(0)})t/\hbar] \int \psi_m^{(0)*}\hat{H}'\psi_k^{(0)}\, d\tau \qquad (3.12)$$

$$\frac{dc_m}{dt} = -\frac{i}{\hbar}\sum_k c_k e^{i\omega_{mk}t}\langle \psi_m^{(0)}|\hat{H}'|\psi_k^{(0)}\rangle, \qquad m = 1, 2, 3, \ldots \qquad (3.13)$$

where

$$\omega_{mk} \equiv (E_m^{(0)} - E_k^{(0)})/\hbar \qquad (3.14)$$

Let us emphasize that we have made no approximations yet. Equation (3.13) is a set of simultaneous differential equations for the coefficients c_m that determine the state function; (3.13) is fully equivalent to the time-dependent Schrödinger equation. [The column vector $\mathbf{c}(t)$ whose elements are the coefficients c_k in (3.8) is the state vector in the representation that uses the $\Psi_k^{(0)}$'s as basis functions. Thus (3.13) is a matrix formulation of the time-dependent Schrödinger equation and can be written as the matrix equation $i\hbar d\mathbf{c}/dt = \mathbf{Gc}$, where $d\mathbf{c}/dt$ has elements dc_m/dt and $\mathbf{G}$ is the square matrix with elements $\exp(i\omega_{mk}t)H'_{mk}$.]

Now suppose that the perturbation is applied at time $t = 0$ to a system that is in stationary state n. At $t = 0$, we have $\Psi(x, 0) = \psi_n^{(0)}$. Comparing this equation with (3.8), we have the initial conditions:

$$c_n(0) = 1$$
$$c_k(0) = 0, \qquad k \neq n \qquad (3.15)$$

If the perturbation $\hat{H}'$ is zero, (3.13) says that the c_m remain constant in time, and thus the system will stay in stationary state n indefinitely. If $\hat{H}'$ is small and does not act for too long, then dc_m/dt will be small, and as a first approximation we can use the initial values of the coefficients c_m in computing dc_m/dt. With this approximation,[1] (3.13) becomes

$$\frac{dc_m}{dt} = -\frac{i}{\hbar} e^{i\omega_{mn}t} \langle \psi_m^{(0)} | \hat{H}' | \psi_n^{(0)} \rangle, \qquad m = 1, 2, 3, \ldots \tag{3.16}$$

Let the perturbation $\hat{H}'(t)$ act from $t=0$ to $t=t_1$. Taking the definite integral of (3.16), we have

$$c_m(t_1) = c_m(0) - \frac{i}{\hbar} \int_0^{t_1} e^{i\omega_{mn}t} \langle \psi_m^{(0)} | \hat{H}'(t) | \psi_n^{(0)} \rangle \, dt \tag{3.17}$$

$$c_m(0) = \delta_{mn} \tag{3.18}$$

For times after t_1, we have $\hat{H}'(t) = 0$, and (3.13) tells us that

$$\frac{dc_m}{dt} = 0, \qquad t > t_1 \tag{3.19}$$

Hence the state function after t_1 is, according to (3.8),

$$\Psi(x,t) = \sum_m c_m(t_1) \exp(-iE_m^{(0)}t/\hbar) \psi_m^{(0)}(x), \qquad t \geqslant t_1 \tag{3.20}$$

where the $c_m(t_1)$ are constants.

The state function (3.20) is a superposition of the energy eigenfunctions $\psi_m^{(0)}$ of $\hat{H}^0$, the expansion coefficients being

$$c_m(t_1) \exp(-iE_m^{(0)}t/\hbar) \tag{3.21}$$

From (1.43), if we make an energy measurement on a system with Hamiltonian $\hat{H}^0$ and state function (3.20), the quantity

$$|c_m(t_1) \exp(-iE_m^{(0)}t/\hbar)|^2 = |c_m(t_1)|^2 \tag{3.22}$$

gives the probability of finding the result $E_m^{(0)}$. If we had carried out an energy measurement at any time before $t=0$, when

$$\Psi = \exp(-iE_n^{(0)}t/\hbar) \psi_n^{(0)}, \qquad t \leqslant 0 \tag{3.23}$$

we would certainly have gotten $E_n^{(0)}$. If we carry out an energy measurement at any time after t_1, we can, in general, find any one of the possible

[1]Just as with time-independent perturbation theory, we can go to higher-order approximations if necessary. See *Fong*, pp. 234–244.

stationary-state energies $E_m^{(0)}$; the probability of each result is given by (3.22), where $c_m(t_1)$ is approximated by (3.17). We say that the perturbation has induced a transition from state n to one of the states m; the probability for this transition is $|c_m(t_1)|^2$. If $E_m^{(0)}$ is the value we find when we carry out the energy measurement, then the state function is changed from (3.20) to $\psi_m^{(0)}$ by the measurement (reduction of the state function—Section 1.1).

3.2 ABSORPTION AND EMISSION OF RADIATION

We shall apply the time-dependent perturbation theory of the last section to a system exposed to electromagnetic radiation. Before doing so, we review the classical wave theory of light.[2]

The theory of light developed by James Clerk Maxwell (1831–1879) views light as a transverse electromagnetic wave. By *transverse*, we mean that the vibrating electric-field and magnetic-field vectors are at right angles to the direction of propagation of the wave. The magnetic vector **B** is always perpendicular to the electric vector **E** at any point in the wave. Light waves may be plane-polarized or unpolarized. By *plane-polarized*, we mean that for all points in the wave, the electric vector points in the same direction. An instantaneous "snapshot" of a plane-polarized light wave traveling in the z direction is shown in Fig. 3.1; the plane of polarization is the xz plane, the plane of the **E** vectors. Unpolarized light is a superposition of plane-polarized waves with different planes of polarization.

For a light wave traveling in the z direction and polarized in the xz plane, the space and time variations of **E** and **B** are:

$$\mathbf{E}=\mathbf{i}\mathcal{E}_x=\mathbf{i}\mathcal{E}_x^0\cos(2\pi\nu t-2\pi z/\lambda) \tag{3.24}$$

$$\mathbf{B}=\mathbf{j}B_y=\mathbf{j}B_y^0\cos(2\pi\nu t-2\pi z/\lambda) \tag{3.25}$$

$\mathcal{E}_x^0$ and B_y^0 are the maximum magnitudes of **E** and **B**. The wavelength λ is the distance between successive crests of **E**. The frequency ν is the number of crests of **E** that pass a fixed point in space each second. We have

$$\lambda\nu=c=2.9979\times10^{10}\text{ cm/sec} \tag{3.26}$$

where c is the speed of light. The ratio of the magnitudes of **E** and **B** is of interest. Application of Maxwell's equations of electromagnetism shows

[2] *Halliday and Resnick*, Sections 39–5, 39–6, and 46–1.

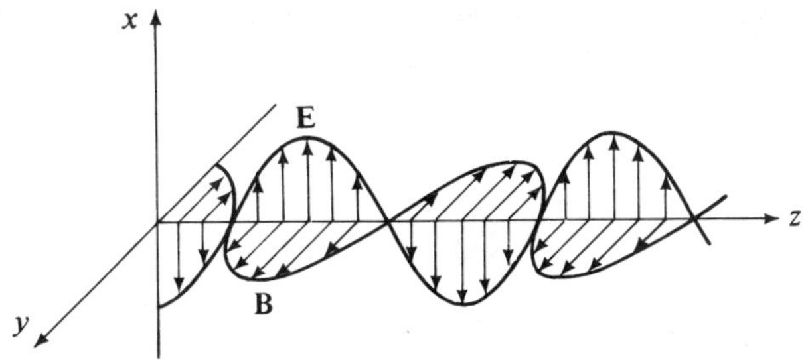

Fig. 3.1 A plane-polarized light wave traveling in the z direction. The **E** and **B** vectors at various points on the z axis are indicated.

that in Gaussian units

$$\mathcal{E}_x^0 = B_y^0$$

For convenience, the electromagnetic spectrum is divided into several regions, according to the frequency involved; see Fig. 3.2. Of course, there is no sharp division between adjacent regions.

An electromagnetic wave transports energy. The energy flow per second per unit area (i.e., the *intensity*) is given by the *Poynting vector* **S**; for an electromagnetic wave moving in vacuum,

$$\mathbf{S} = \frac{c}{4\pi}\mathbf{E} \times \mathbf{B} \tag{3.27}$$

in Gaussian units. Note that **S** points in the direction of propagation, which is the direction of energy flow. Since **E** and **B** are mutually

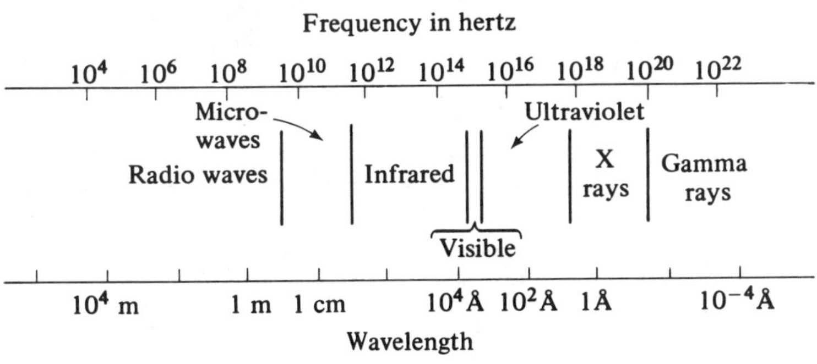

Fig. 3.2 The electromagnetic spectrum. The scale is logarithmic. 1 hertz (Hz) $\equiv$ 1 cycle/sec.

perpendicular and are equal in magnitude, we have for the magnitude of **S**

$$S=\frac{c}{4\pi}\mathcal{E}_x^2 \tag{3.28}$$

where we have assumed polarization in the xz plane. From (3.24), the average value of $\mathcal{E}_x^2$ over one cycle is $\frac{1}{2}(\mathcal{E}_x^0)^2$. Thus

$$\langle S\rangle=\frac{c}{8\pi}(\mathcal{E}_x^0)^2 \tag{3.29}$$

We now consider the effect of exposing a system to electromagnetic radiation. Our treatment will involve approximations beyond that of replacing (3.13) with (3.16). A proper treatment of the interaction of radiation with matter must treat both the atom and the radiation field quantum-mechanically; this gives what is called *quantum field theory* (or *quantum electrodynamics*). However, the quantum theory of radiation is beyond the scope of this book. We will treat the atom quantum-mechanically, but will treat the radiation field as a classical wave, ignoring its photon aspect. Thus our treatment is *semiclassical.*

The electric and magnetic fields of the radiation will interact with the atomic or molecular electrons giving a time-dependent perturbation. We can estimate the relative magnitudes of these two interactions as follows: the force on a particle with charge q and velocity **v** moving through electric and magnetic fields **E** and **B** is[3] (in Gaussian units)

$$\mathbf{F}=q\mathbf{E}+\frac{q}{c}\mathbf{v}\times\mathbf{B} \tag{3.30}$$

Since the magnitudes of the electric and magnetic fields of electromagnetic radiation are equal, the ratio of the magnetic to the electric force is v/c. For the ground-state hydrogen-atom electron, we have (Problem 3.22) $\langle v^2\rangle^{1/2}/c=1/137$. This is also the order of magnitude for valence electrons of other atoms. Thus to a good approximation, we can ignore the interaction between **B** and the electrons, and consider only the effect of **E**.

If (3.16) is to be a good approximation, the effect of the radiation's electric field must be a small addition to the atomic Hamiltonian. For a radiation intensity of 1 watt/cm^2, we have for the radiation's electric field, using (3.29),

$$\left[\frac{8\pi}{c}10^7\frac{\text{ergs}}{\text{sec}\cdot\text{cm}^2}\right]^{1/2}\approx 10^{-1}\,\frac{\text{statvolts}}{\text{cm}}$$

[3] *Halliday and Resnick*, Section 33–2.

To estimate the internal electric field acting on an electron in an atom, we will calculate the field of a proton at a distance of 1 bohr:

$$\frac{4.8\times10^{-10}}{(0.53)^2\times10^{-16}}\ \frac{\text{statcoul}}{\text{cm}^2}\approx10^7\,\frac{\text{statvolts}}{\text{cm}}$$

Thus we are justified in considering the radiation as a small perturbation (provided we do not use a high-power laser).

The potential energy of interaction between a system of charged particles and the electric field (3.24) is [Equation (1.285)]

$$-\sum_i\int q_i\mathscr{E}_x\,dx_i=-\sum_i q_i x_i\mathscr{E}_x \tag{3.31}$$

where x_i and q_i are the x coordinates and charges of the particles. We thus have as the perturbation part of the Hamiltonian

$$\hat{H}'(t)=-\mathscr{E}_x^0\sum_i q_i x_i\cos\left(2\pi\nu t-\frac{2\pi z_i}{\lambda}\right) \tag{3.32}$$

For transitions between different atomic or molecular electronic states, the wavelengths usually lie in the ultraviolet; typically, $\lambda\approx10^3$ Å. For vibrational and rotational transitions, λ is even larger. The size of atoms and molecules is about 1 Å. Hence λ is usually much greater than the size of the molecule; as far as electrons confined to move within the molecule are concerned, the spatial variation of the radiation's electric field is negligible: $z_i/\lambda\approx0$. With this further approximation, we have

$$\hat{H}'(t)=-\mathscr{E}_x^0\cos\omega t\sum_i q_i x_i \tag{3.33}$$

$$\omega\equiv2\pi\nu \tag{3.34}$$

The sum in (3.33) is the operator for the x component of the system's dipole moment [Equation (1.286)]

$$\hat{d}_x=\sum_i q_i x_i \tag{3.35}$$

Hence

$$\hat{H}'(t)=-\hat{d}_x\mathscr{E}_x^0\tfrac{1}{2}(e^{i\omega t}+e^{-i\omega t}) \tag{3.36}$$

where the identity $\cos\theta = \frac{1}{2}(e^{i\theta} + e^{-i\theta})$ was used. Equation (3.17) becomes

$$c_m(t_1) = \delta_{mn} + \frac{i\mathcal{E}_x^0}{2\hbar}\langle\psi_m^{(0)}|\hat{d}_x|\psi_n^{(0)}\rangle\int_0^{t_1}[e^{i(\omega_{nm}+\omega)t} + e^{i(\omega_{mn}-\omega)t}]\,dt \qquad (3.37)$$

$$c_m(t_1) = \delta_{mn} + \frac{i\mathcal{E}_x^0}{2\hbar}\langle\psi_m^{(0)}|\hat{d}_x|\psi_n^{(0)}\rangle\left[\frac{e^{i(\omega_{mn}+\omega)t_1}-1}{\omega_{mn}+\omega} + \frac{e^{i(\omega_{mn}-\omega)t_1}-1}{\omega_{mn}-\omega}\right] \qquad (3.38)$$

Under what conditions is the probability of a transition from state n to state m of appreciable magnitude? One possibility is

$$\omega_{mn} = \omega \qquad (3.39)$$

which makes the denominator of the second fraction in (3.38) zero. This makes $c_m(t_1)$ large, but not infinite, since

$$\lim_{a\to 0}\frac{e^{it_1 a}-1}{a} = it_1 \qquad (3.40)$$

Using (3.34) and (3.14), we have for (3.39)

$$E_m^{(0)} - E_n^{(0)} = h\nu \qquad (3.41)$$

Thus we have an atom exposed to light of frequency ν making a transition from state n to m, such that (3.41) holds. If we assume that the energy needed for the transition came from the radiation field, we have the familiar phenomenon of *light absorption*.

If $\omega_{mn} = -\omega$, that is, if

$$E_m^{(0)} - E_n^{(0)} = -h\nu \qquad (3.42)$$

the denominator of the first fraction in (3.38) vanishes, and the transition probability is large. In this case, the final state m has an energy that is $h\nu$ less than the initial-state energy. What happened to this energy? It is reasonable to suppose that it was emitted from the atom as a photon of frequency ν. (Our semiclassical treatment cannot prove this; to reach rigorous conclusions about the radiation field, we must use quantum field theory.) Thus an atom in an excited state exposed to electromagnetic radiation of frequency ν may emit radiation of frequency ν, thereby falling to a lower state. This phenomenon is called *stimulated emission*.

If ω is not close to either ω_{mn} or $-\omega_{mn}$, the transition probability is negligible (Problem 3.7).

The radiation's electric-field amplitude $\mathcal{E}_x^0$ in (3.38) is related to the radiation intensity. Consider a cross section of area A perpendicular to the radiation beam. The radiation energy passing through this cross section in time t is given by (3.29) as $c(\mathcal{E}_x^0)^2At/8\pi$. During time t the beam has traveled a distance $l = ct$, and the energy that passed through area A in time t is spread through a volume $Al = Act$. The radiation density ρ_x, defined as the electromagnetic radiation energy per unit volume, is therefore (the subscript reminds us that we are considering radiation plane-polarized in the x direction)

$$\rho_x = (\mathcal{E}_x^0)^2/8\pi \tag{3.43}$$

Let us assume $E_m^{(0)} > E_n^{(0)}$ and let ω be close to ω_{mn}, corresponding to absorption. We can neglect the first fraction in (3.38) in comparison to the second. From (3.22) the probability of a transition to state m from state n is $c_m c_m^*$. Using (3.43) and

$$e^{i\theta} - 1 = e^{i\theta/2}(e^{i\theta/2} - e^{-i\theta/2}) = 2ie^{i\theta/2}\sin\tfrac{1}{2}\theta$$

we find for the probability of an absorptive transition from state n to state m

$$|c_m(t_1)|^2 = 8\pi\rho_x|\langle\psi_m^{(0)}|\hat{d}_x|\psi_n^{(0)}\rangle|^2\sin^2[\tfrac{1}{2}(\omega_{mn} - \omega)t_1]\hbar^{-2}(\omega_{mn} - \omega)^{-2} \tag{3.44}$$

So far we have assumed a beam of strictly monochromatic radiation [Equation 3.24)]. Actually, spectroscopic work at optical frequencies is usually done with a beam containing a continuous range of frequencies. (This isn't true when lasers are used.) The amount of radiation at any one precise frequency is infinitesimally small. Let $u_x(\nu)\,d\nu$ be the electromagnetic radiation energy per unit volume with frequency between ν and $\nu + d\nu$. (u is an energy per unit volume per unit frequency, and has different dimensions than ρ, which is an energy per unit volume.) The probability of a transition to state m being induced by radiation with frequency in the range ν to $\nu + d\nu$ is gotten by replacing ρ_x in (3.44) by $u_x(\nu)\,d\nu$. The total probability of a transition to state m is found by summing the infinitesimal probabilities over the various frequencies; this amounts to an integration. Therefore, for radiation of continuous frequency,

$$|c_m|^2 = 8\pi|\langle m|\hat{d}_x|n\rangle|^2\int_0^\infty \frac{\sin^2[(E_m^{(0)} - E_n^{(0)} - h\nu)\pi t_1/h]}{[E_m^{(0)} - E_n^{(0)} - h\nu]^2}u_x(\nu)\,d\nu$$

As noted above, the integrand is only of significant size for ν close to ν_{mn}, where

$$\nu_{mn} \equiv \omega_{mn}/2\pi = (E_m^{(0)} - E_n^{(0)})/h \tag{3.45}$$

Hence there is no harm done if we change the lower limit to $-\infty$. (Between 0 and $-\infty$ the integrand is negligible.) Similarly, although $u_x(\nu)$ is a function of frequency, we can replace it with $u_x(\nu_{mn})$, its value at ν_{mn}. Making these changes, and using $\int_{-\infty}^{\infty}[(\sin^2 x)/x^2]dx = \pi$, we get

$$|c_m(t_1)|^2 = 2\pi\hbar^{-2}t_1|\langle m|\hat{d}_x|n\rangle|^2 u_x(\nu_{mn}) \tag{3.46}$$

Equation (3.46) gives the probability that any one molecule will make a transition to state m after having been illuminated for a time t_1. The number of transitions to m per second (i.e., the *transition rate*) is given by (3.46) multiplied by N_n and divided by t_1, where N_n is the number of molecules in state n.

To find the probability for stimulated emission from state m to state n, we consider the molecule to have originally been in state m. Thus we interchange m and n in (3.38) and again assume ω near ω_{mn}. We end up with an expression for $|c_n|^2$ that is the same as (3.46), except that $|\langle m|\hat{d}_x|n\rangle|$ is replaced by $|\langle n|\hat{d}_x|m\rangle|$. Hence the probability of stimulated emission from m to n equals the probability of absorption from n to m.

Equation (3.46) applies for radiation plane-polarized in the x direction. If the radiation is isotropic, we also get contributions from terms involving the matrix elements of $\hat{d}_y$ and $\hat{d}_z$. For isotropic radiation, we have $u_x = u_y = u_z = u/3$, where u is the total radiation density. Hence for isotropic radiation, (3.46) becomes

$$|c_m|^2 = (2\pi t_1/3\hbar^2)\left[|\langle m|\hat{d}_x|n\rangle|^2 + |\langle m|\hat{d}_y|n\rangle|^2 + |\langle m|\hat{d}_z|n\rangle|^2\right]u(\nu_{mn})$$

$$|c_m|^2 = (2\pi t_1/3\hbar^2)|\langle m|\hat{\mathbf{d}}|n\rangle|^2 u(\nu_{mn}) \tag{3.47}$$

where the electric dipole-moment operator $\hat{\mathbf{d}}$ is $\mathbf{i}\hat{d}_x + \mathbf{j}\hat{d}_y + \mathbf{k}\hat{d}_z$ [Equation (1.286)] and $\langle m|\hat{\mathbf{d}}|n\rangle$ is a vector whose components are the matrix elements of $\hat{d}_x$, $\hat{d}_y$, and $\hat{d}_z$. Thus

$$|\langle m|\hat{\mathbf{d}}|n\rangle|^2 = |\langle m|\hat{d}_x|n\rangle|^2 + |\langle m|\hat{d}_y|n\rangle|^2 + |\langle m|\hat{d}_z|n\rangle|^2 \tag{3.48}$$

We call $\langle m|\hat{\mathbf{d}}|n\rangle$ the *transition* (*dipole*) *moment*.

The semiclassical treatment just given has the defect of not predicting *spontaneous emission*. According to (3.13), if there is no outside perturbation, that is, if $\hat{H}'(t)=0$, then $dc_m/dt=0$ for all m; if the atom is in the nth stationary state at $t=0$, it will persist in that state forever. However, experimentally we find that unperturbed atoms in excited states spontaneously radiate energy and drop to lower states. Quantum field theory does predict spontaneous emission. Since quantum field theory is beyond us, we shall use an argument given by Einstein in 1917 to find the spontaneous-emission probability.

Consider a system in which matter and radiation are in equilibrium in a closed cavity at temperature T. (This equilibrium situation does not generally hold in spectroscopy, but the transition probabilities are fundamental properties of the interaction between radiation and matter and cannot be affected by the presence or absence of equilibrium.) As before, let $E_m^{(0)}$ be greater than $E_n^{(0)}$. The probability of absorption from state n to state m is proportional to the number of photons with frequency near ν_{mn}; the number of photons is proportional to the radiation density $u(\nu_{mn})$. Hence the rate of absorption is given by $B_{n\to m}N_n u(\nu_{mn})$, where N_n is the number of molecules in state n and $B_{n\to m}$ is a proportionality constant called the *Einstein coefficient* for absorption. From the discussion following (3.46) and from (3.47), it follows that

$$B_{n\to m}=(2\pi/3\hbar^2)|\langle m|\hat{\mathbf{d}}|n\rangle|^2 \tag{3.49}$$

Similarly, the rate of stimulated emission from state m to n equals $B_{m\to n}N_m u(\nu_{mn})$, where $B_{m\to n}$ is another constant. The *spontaneous* emission probability is independent of the presence or absence of radiation. Hence the rate of spontaneous emission from m to n is $A_{m\to n}N_m$, where $A_{m\to n}$ is still another constant.

At equilibrium, the rates of upward and downward transitions balance. Hence

$$B_{n\to m}N_n u(\nu_{mn})=B_{m\to n}N_m u(\nu_{mn})+A_{m\to n}N_m$$

$$N_m/N_n=B_{n\to m}u(\nu_{mn})/[B_{m\to n}u(\nu_{mn})+A_{m\to n}] \tag{3.50}$$

The equilibrium values of N_m and N_n satisfy the Boltzmann distribution law:

$$N_m/N_n=\exp[-(E_m^{(0)}-E_n^{(0)})/kT]=e^{-h\nu_{mn}/kT} \tag{3.51}$$

where k is Boltzmann's constant. The density of blackbody radiation in equilibrium with matter is given by the radiation law derived by Planck in 1900; this law is [for Planck's derivation of (3.52), see *Gatz*, pp. 17–19]

$$u(\nu_{mn}) = \frac{8\pi h\nu_{mn}^3}{c^3} \frac{1}{e^{h\nu_{mn}/kT} - 1} \tag{3.52}$$

If we combine (3.50) and (3.51) and solve for u, we get

$$u(\nu_{mn}) = \frac{A_{m\to n}}{B_{n\to m} e^{h\nu_{mn}/kT} - B_{m\to n}} \tag{3.53}$$

For (3.53) to be consistent with Planck's law (3.52), we must have

$$B_{n\to m} = B_{m\to n} \tag{3.54}$$

$$A_{m\to n} = (8\pi h\nu_{mn}^3/c^3) B_{n\to m} = (64\pi^4\nu_{mn}^3/3hc^3)|\langle m|\hat{\mathbf{d}}|n\rangle|^2 \tag{3.55}$$

We derived (3.54) earlier from our semiclassical treatment, but (3.55) is new information. Quantum electrodynamics (*Merzbacher*, Chapter 22) confirms the correctness of (3.55).

The coefficient $A_{m\to n}$ is related to the lifetime of state m. Suppose we initially have N_{m0} molecules in the excited state m and let no radiation be present. Assume that state n is the only available state to which the excited molecules can decay. Then the rate of decay is $dN_m/dt = -A_{m\to n}N_m$. This equation integrates to

$$N_m = N_{m0} e^{-A_{m\to n}t} \tag{3.56}$$

After a time $\tau = 1/A_{m\to n}$, the number of molecules in state m will be reduced to $1/e$ of its initial value; τ is the *average lifetime* of the state m (Problem 3.11).

Finally, consider the observed intensity of the absorption transition $n\to m$. Spontaneous emission from molecules in state m is sent out in random directions and can be ignored. However, emission from m stimulated by the incident light beam can be shown to travel in the same direction as the beam and thus decreases the observed intensity of the absorption line. Stimulated emission and absorption are each proportional to the population of the initial state of the transition and the $n\to m$ and $m\to n$ transition moments are equal. Hence the observed absorption intensity is proportional to the population *difference* $N_n - N_m$ [and to $|\mathbf{d}_{mn}|^2$ and $u(\nu_{mn})$].

3.3 SELECTION RULES

According to the results of the last section, if the integrals in (3.48) all vanish, then the probability for a transition between states m and n is zero. Actually, (3.47) is the result of several approximations, and even if the electric dipole-moment matrix elements vanish, there still might be some probability for the transition to occur.

Thus suppose we had included the interaction of the radiation's magnetic field **B** with the atomic or molecular electrons and nuclei. The Hamiltonian for this interaction is [Equation (1.268)] $-\mathbf{B}\cdot\hat{\boldsymbol{\mu}}$, where $\hat{\boldsymbol{\mu}}$ is the magnetic dipole-moment operator for the system. This gives additional terms in c_m that are proportional to

$$\langle \psi_m^{(0)} | \hat{\mu}_x | \psi_n^{(0)} \rangle, \qquad \langle \psi_m^{(0)} | \hat{\mu}_y | \psi_n^{(0)} \rangle, \qquad \text{and} \qquad \langle \psi_m^{(0)} | \hat{\mu}_z | \psi_n^{(0)} \rangle \tag{3.57}$$

where the magnetic dipole-moment operators can be written in terms of spin and orbital angular-momentum operators, as in (1.269). We showed that the interaction between **B** and the charges was about $\frac{1}{137}$th the interaction between **E** and the charges. Since the transition probability is given by $|c_m|^2$, the contribution of the magnetic dipole-moment matrix element $\boldsymbol{\mu}_{mn}$ is only $1/10^4$ or $1/10^5$ that of the electric dipole-moment matrix element $\mathbf{d}_{mn}$. If $\mathbf{d}_{mn}$ is nonzero, we can neglect $\boldsymbol{\mu}_{mn}$; a transition due to a nonvanishing $\mathbf{d}_{mn}$ is called an *electric-dipole transition*. If $\mathbf{d}_{mn}$ vanishes and $\boldsymbol{\mu}_{mn}$ does not, we have the possibility of a *magnetic-dipole transition* between states n and m.

We can also remove the approximation of neglecting the spatial variation of the radiation's electric field. This will give contributions to c_m involving matrix elements of the operators (see *Kauzmann*, p. 654)

$$\sum_i q_i x_i^2, \qquad \sum_i q_i x_i y_i, \qquad \sum_i q_i x_i z_i, \qquad \text{etc.} \tag{3.58}$$

Transitions due to such matrix elements are called *electric-quadrupole transitions*. Since the wavelength of ultraviolet light is about 10^3 times the size of atoms, electric-quadrupole transitions are about $1/10^6$ the intensity of electric-dipole transitions.

In the remainder of this section, we will consider only electric-dipole transitions. These are the strongest transitions, and account for most of the observed atomic and molecular spectroscopic transitions. (Magnetic-dipole transitions occur in magnetic-resonance spectroscopy.) When the integral $\mathbf{d}_{mn}$ vanishes, we say that a transition between states n and m is *forbidden*.

The conditions for which $\mathbf{d}_{mn} \neq 0$ give rise to the *selection rules*, which specify the allowed transitions.

As an example, consider the electric-dipole selection rules for a particle of charge q in a one-dimensional box. We have

$$\langle \psi_m^{(0)} | \hat{d}_x | \psi_n^{(0)} \rangle = q \int_0^l \left(\frac{2}{l} \right) x \sin\left(\frac{m\pi x}{l} \right) \sin\left(\frac{n\pi x}{l} \right) dx \tag{3.59}$$

Evaluation of this integral (Problem 3.3) shows that it vanishes for $m-n$ even, and is nonzero for $m-n$ odd. The selection rule for the particle in a box is the following: electric-dipole transitions are allowed if and only if the change in quantum number is odd.

Now consider a one-dimensional harmonic oscillator of charge q. We must evaluate $q\langle m|x|n \rangle$. The harmonic-oscillator wave functions are given by (1.133). Using (1.133) and the Hermite-polynomial identity (1.138) with $z = \alpha^{1/2}x$, we have

$$\alpha^{1/2} x \psi_n = n(2^n n!)^{-1/2} \alpha^{1/4} \pi^{-1/4} e^{-\alpha x^2/2} H_{n-1}(\alpha^{1/2}x)$$
$$+ \tfrac{1}{2}(2^n n!)^{-1/2} \alpha^{1/4} \pi^{-1/4} e^{-\alpha x^2/2} H_{n+1}(\alpha^{1/2}x)$$

$$x\psi_n = (n/2\alpha)^{1/2} \psi_{n-1} + [(n+1)/2\alpha]^{1/2} \psi_{n+1} \tag{3.60}$$

where (1.133) with $v = n-1$ and $v = n+1$ was used. Multiplication by ψ_m^* and integration gives

$$\langle \psi_m | x | \psi_n \rangle = (n/2\alpha)^{1/2} \langle \psi_m | \psi_{n-1} \rangle + [(n+1)/2\alpha]^{1/2} \langle \psi_m | \psi_{n+1} \rangle$$

$$\langle m|x|n \rangle = (n/2\alpha)^{1/2} \delta_{m,n-1} + [(n+1)/2\alpha]^{1/2} \delta_{m,n+1} \tag{3.61}$$

where orthonormality was used. Thus the allowed electric-dipole transitions for the harmonic oscillator involve a change of ± 1 in the quantum number. A harmonic oscillator of vibration frequency ν_{vib} will emit or absorb light of frequency ν_{light} given by [Equations (3.41), (3.42), and (1.140)]

$$\nu_{\text{light}} = (E_2 - E_1)/h = [(v_2 + \tfrac{1}{2})h\nu_{\text{vib}} - (v_1 + \tfrac{1}{2})h\nu_{\text{vib}}]/h = \nu_{\text{vib}}$$

Consider the electric dipole selection rules for the hydrogen atom. We must evaluate

$$\langle \psi_j^{(0)} | \hat{\mathbf{d}} | \psi_k^{(0)} \rangle = -e \langle \psi_j^{(0)} | \mathbf{r} | \psi_k^{(0)} \rangle \tag{3.62}$$

where $-e$ is the electronic charge, and $\mathbf{r}$ is the vector from the origin (nucleus) to the electron:

$$\mathbf{r}=\mathbf{i}x+\mathbf{j}y+\mathbf{k}z=\mathbf{i}r\sin\theta\cos\varphi+\mathbf{j}r\sin\theta\sin\varphi+\mathbf{k}r\cos\theta \tag{3.63}$$

Using primes to indicate quantum numbers of state j, we want to evaluate

$$\int z\psi^*_{n'l'm'm_s'}\psi_{nlmm_s}\,d\tau \tag{3.64}$$

where $\int d\tau$ means integration over the space variables and summation over the spin variable. The factor z in (3.64) is a constant as far as the spin variable is concerned, and since the spin functions α and β are orthogonal [(1.238)], the integral (3.64) vanishes unless $m_s'=m_s$. We have the selection rule

$$\Delta m_s=0 \tag{3.65}$$

We now have for (3.64)

$$\delta_{m_s,m_s'}\int_0^\infty rR_{n'l'}R_{nl}r^2\,dr\int_0^{2\pi}\int_0^\pi[Y_{l'}^{m'}]^*Y_l^m\cos\theta\sin\theta\,d\theta\,d\varphi \tag{3.66}$$

The phi integral in (3.66) is [Equation (1.157)]

$$\frac{1}{2\pi}\int_0^{2\pi}e^{-im'\varphi}e^{im\varphi}\,d\varphi=\frac{1}{2\pi}\int_0^{2\pi}e^{i(m-m')\varphi}\,d\varphi \tag{3.67}$$

which vanishes unless $m=m'$. We have the selection rule

$$\Delta m=0 \qquad \text{for } z\text{-polarized light} \tag{3.68}$$

The r integration is involved, and we simply state that the r integral is nonzero for all combinations of quantum numbers; we have

$$\Delta n=0,\ \pm1,\ \pm2,\ \pm3,\ \ldots \tag{3.69}$$

The theta integration involves the associated Legendre functions. Using[4]

$$\cos\theta\,P_l^{|m|}(\cos\theta)=\frac{(l+|m|)}{(2l+1)}P_{l-1}^{|m|}(\cos\theta)+\frac{(l-|m|+1)}{(2l+1)}P_{l+1}^{|m|}(\cos\theta) \tag{3.70}$$

and the orthogonality of the associated Legendre functions

$$\int_0^\pi[P_{l'}^{|m|}(\cos\theta)]^*P_l^{|m|}(\cos\theta)\sin\theta\,d\theta=0 \qquad \text{if } l'\neq l \tag{3.71}$$

[4] *Kauzmann*, p. 98.

[(3.71) follows from (1.163) with $m = m'$] we deduce the selection rule

$$\Delta l = \pm 1 \qquad \text{for } z\text{-polarized light} \tag{3.72}$$

For x- and y-polarized light, a similar treatment[5] gives

$$\left.\begin{aligned} \Delta l &= \pm 1 \quad (3.73) \\ \Delta m &= \pm 1 \quad (3.74) \end{aligned}\right\} \text{for } x\text{- or } y\text{-polarized light}$$

The above selection rules for m can be verified by applying a weak external magnetic field to define the z direction and remove the m degeneracy.

Combining these results, we have the hydrogen-atom electric-dipole selection rules:

$$\left.\begin{aligned} \Delta n &= 0, \pm 1, \pm 2, \ldots \quad (3.75) \\ \Delta l &= \pm 1 \quad (3.76) \\ \Delta m &= 0, \pm 1 \quad (3.77) \\ \Delta m_s &= 0 \quad (3.78) \end{aligned}\right\} \text{for isotropic radiation}$$

The selection rules (3.76)–(3.78) were derived without reference to the form of the radial functions; hence they are valid for any one-particle central-field problem. For example, they hold well for the sodium valence electron, which moves outside a closed-shell structure.

From (1.226), (1.224), (3.83), and (3.75), the reciprocal wavelengths of the H-atom spectral lines are

$$\frac{1}{\lambda} = \frac{\nu}{c} = \frac{E_2 - E_1}{hc} = \frac{e^2}{2ahc}\left(\frac{1}{n_1^2} - \frac{1}{n_2^2}\right) \equiv R_{\mathrm{H}}\left(\frac{1}{n_1^2} - \frac{1}{n_2^2}\right)$$

where $R_{\mathrm{H}} = 109{,}677.6\ \mathrm{cm}^{-1}$ is the *Rydberg constant* for H.

We can use parity to aid in determining selection rules. Recall (Section 1.8) that the integral $\langle m|\hat{\mathbf{d}}|n\rangle$ vanishes if the integrand is an odd function of the Cartesian coordinates. The operator $\hat{\mathbf{d}}$ [Equation (1.286)] is an odd function. If the wave functions are of definite parity, as is usually true, then if states m and n have the same parity, the integrand in $\mathbf{d}_{mn}$ will be odd. Hence electric-dipole transitions are forbidden between states of the same parity; we have the selection rule: *parity changes*. (This is the *Laporte rule.*)

For example, consider the one-dimensional harmonic-oscillator states. These states are either even or odd, depending on whether the quantum

[5] *Kauzmann*, pp. 658–659.

number v is even or odd. Therefore, electric-dipole transitions are forbidden between harmonic-oscillator states whose quantum numbers are both even, or both odd. This agrees with our previous result that the quantum number changes by ± 1. Note, however, that parity considerations do not tell us all the forbidden transitions in this case, but only some of them.

Another example is the particle in a box. With the origin at the center of the box, the potential energy is an even function, and the wave functions are of definite parity, determined by whether the quantum number is odd or even. Hence for electric-dipole transitions, the quantum number must go from even to odd, or vice versa, as concluded previously.

For the hydrogen atom, the parity is determined by l. (See Section 1.17.) The selection rule that $\Delta l = \pm 1$ is in agreement with the rule that parity changes in electric-dipole transitions.

What about parity in electric-quadrupole and magnetic-dipole transitions? The quantities (3.58) are even functions. Hence for electric-quadrupole transitions, parity remains the same. Magnetic-dipole transitions involve angular momentum operators. For example, consider $\hat{L}_z = -i\hbar(x\partial/\partial y - y\partial/\partial x)$. Inversion of coordinates leaves this operator unchanged. Hence for magnetic-dipole transitions, parity remains the same.

We now consider many-electron atoms. We will assume Russell–Saunders coupling, so that an atomic state can be characterized by total electronic orbital and spin angular-momentum quantum numbers L and S, and total electronic angular-momentum quantum numbers J and M_J. (See Section 1.17.) The electric-dipole selection rules for L, J, and M_J can be shown to be (*Bethe and Jackiw*, p. 224)

$$\Delta L = 0, \pm 1 \tag{3.79}$$

$$\Delta J = 0, \pm 1 \tag{3.80}$$

$$\Delta M_J = 0, \pm 1 \tag{3.81}$$

except that transitions between states both of which have $J = 0$ are forbidden. [The selection rule (3.81) is significant only when an external magnetic field has removed the M_J degeneracy.] Now consider S. The atomic wave functions are eigenfunctions of $\hat{S}^2$ with eigenvalues $S(S+1)\hbar^2$. The operator $\hat{\mathbf{d}}$ involves only spatial coordinates and commutes with $\hat{S}^2$. Now recall the following theorem [Equation (1.51)]: if $\hat{A}$ is a Hermitian operator that commutes with $\hat{B}$, and if φ_1 and φ_2 are eigenfunctions of $\hat{A}$ with different eigenvalues, then $\langle \varphi_1 | \hat{B} | \varphi_2 \rangle = 0$. From this theorem, it follows that the integral $\langle m | \hat{\mathbf{d}} | n \rangle$ vanishes unless states m and n have the same values of S. Thus

$$\Delta S = 0 \tag{3.82}$$

The parity of the state of a many-electron atom is even or odd according to whether the sum of the l values of the individual electrons is even or odd. (See Section 1.17.) For example, the $1s2p$ configuration of helium gives rise to 12 states (Section 1.16), all of odd parity since $0+1=1$ is odd. (Often one adds a degree sign as a right superscript to the level symbol if the states of the level are of odd parity. Thus the lowest level of boron, which arises from the configuration $1s^2 2s^2 2p$, is designated ${}^2P^{\circ}{}_{1/2}$.) Since all atomic states belonging to the same configuration have the same parity, electric-dipole transitions are forbidden between such states. The selection rule (3.79) allows $\Delta L = 0$. It might be thought that this violates the Laporte rule, but this is not so. The parity is determined by the *algebraic* sum of the l_i's, whereas L is determined by the *vector* sum of the l_i's.

Most atomic transitions are due to one electron changing its orbital. Using the central-field approximation, we have the angular part of the orbital function being a spherical harmonic, for which the selection rule is $\Delta l = \pm 1$ [(3.76)]. Hence for a one-electron atomic transition, the l value of the electron making the jump changes by ± 1.

The selection rules for atoms give rise to the existence of *metastable states*. Consider, for example, the $1s2s\ {}^3S_1$ level in helium. This is the lowest triplet level, and the only lower level is the singlet $1s^2$ ground state. The selection rule $\Delta S = 0$ forbids an atom in this triplet level from decaying to the ground state by emission of radiation; hence the triplet state has a relatively long lifetime, and is said to be metastable. Actually, the selection rule $\Delta S = 0$ is not strictly obeyed. If spin–orbit interaction (Section 1.17) is included in the Hamiltonian, then the wave functions are no longer exact eigenfunctions of $\hat{S}^2$, and radiative transitions between states of different multiplicity can occur. The relatively slow (10^{-3} sec or longer) decay from a metastable triplet state of an atom or molecule to a lower singlet state gives *phosphorescence*. The rapid (10^{-8} sec, typically) decay from a higher to a lower state of the same multiplicity gives *fluorescence*. The $1s2s\ {}^3S_1$ helium level is especially long-lived because decay to the ground state also violates the $\Delta l = \pm 1$ electric-dipole parity selection rule; this level has the exceedingly long lifetime of 8000 sec (*Physics Today*, March 1971, p. 19). The $1s2s\ {}^1S_0$ He level has a lifetime of 0.02 sec; its metastability is due to the violation of the $\Delta l = \pm 1$ selection rule. $\Delta l = 0$ magnetic-dipole and electric-quadrupole transitions are allowed, but occur with very low probability.

There are other ways for a system to lose excitation energy besides the emission of light. Excitation energy of an atom or molecule can be transferred to another atom or molecule during a collision; such collisions are called collisions of the *second kind*. The energy transferred can go into translational (kinetic) energy, or to excite the second atom or molecule, or to break a chemical bond. Similarly, excitation energy can be gained by processes besides absorption of light. A molecule can gain excitation energy in a collision of the second kind with a molecule that is excited. Also, a molecule can be excited by what is called a collision of the *first kind*; in such a collision, translational kinetic energy is converted to excitation energy. The selection rules derived above apply only for radiative transitions. A long-lived excited state like He $1s2s\ ^3S_1$ will far more likely decay to the ground state during a collision with another atom or with the container wall than by emitting radiation; the radiation is said to be *quenched*. When the gas density is very low, substantial emission of electric-dipole-forbidden lines may occur. The aurora borealis and aurora australis occur in the earth's upper atmosphere (whose density is low) and involve a forbidden transition of oxygen atoms.

3.4 SPECTROSCOPY

The relation $\Delta E = h\nu$ allows the separations between the energy levels of a system to be determined by measurement of the frequencies of light absorbed or emitted by the system, giving rise to *absorption* and *emission spectroscopy*. The techniques used in spectroscopy vary with the region of the electromagnetic spectrum. In the following paragraphs, we survey the various kinds of spectroscopy.

The smallest energy differences in atoms and molecules are associated with the possible orientations of electron and nuclear spin magnetic moments in magnetic fields, and lead to transitions in the radio-frequency portion of the electromagnetic spectrum. In Section 1.17, we mentioned atomic hyperfine structure, which is due to the interaction of the nuclear spin magnetic moment with the magnetic field associated with the electronic spin and orbital angular momenta; radio astronomers tune their radio telescopes to the 1420 Mc/sec line arising from the transition between the hyperfine levels of the ground-state hydrogen atom. In most molecular ground states, the total electronic spin is zero, and therefore does not give a hyperfine structure. Application of an external magnetic field to a molecule splits the ground state into a number of levels corresponding to the possible orientations of the nuclear spin magnetic moments in the external field; in nuclear magnetic resonance, we observe transitions between these levels. For molecules with one or more unpaired electrons, application of an external magnetic field allows study of transitions between levels corresponding to different orientations of the electronic spin (electron spin resonance). Magnetic resonance will be discussed in Chapter 8.

Differences between rotational energy levels of molecules correspond to frequencies lying in the microwave region of the electromagnetic spectrum; see Chapter 5.

Differences in molecular vibration energies correspond to infrared frequencies; see Chapter 6.

Differences in atomic and molecular electronic energies involving excitation of a valence electron correspond to visible and ultraviolet frequencies. See Chapter 7 for molecular electronic spectroscopy.

Investigation of atomic spectra yields atomic energy levels. An important chemical application of atomic spectroscopy is in elemental analysis. Atomic absorption spectroscopy and emission spectroscopy are used for rapid, accurate quantitative analysis of most metals and some nonmetals, and have replaced the older, wet methods of analysis in many applications. One compares the intensity of a spectral line of the element being analyzed with a standard line of known intensity. In atomic absorption spectroscopy, a flame is used to vaporize the sample; in emission spectroscopy, one passes a powerful electric discharge through the sample or uses a flame to produce the spectrum. Atomic spectroscopy is used clinically in the determination of Ca, Mg, K, Na, and Pb in blood samples. For details, see *Robinson*.

Differences in atomic and molecular electronic energies involving inner-shell electrons correspond to x-ray frequencies. (Orbital energies are proportional to the square of the effective atomic number [(1.226)]; inner-shell electrons see a nucleus that is not well screened.) X rays are produced by high-energy electrons striking a target. An incident electron can knock an inner-shell electron out of a target atom. When an outer electron drops down to fill the inner-shell vacancy, the atom emits an x-ray photon. X rays provide a means of determining crystal structure. X-ray wavelengths are about 1 Å, which is also the magnitude of interatomic distances in crystals. Hence we can use a crystal as a diffraction grating for x rays; the resulting diffraction pattern provides information on the crystal structure.

Differences between nuclear energy levels correspond to γ-ray frequencies. The emission of an α or β particle from a nucleus often produces a nucleus in an excited state, which then decays to a lower state by emission of a γ-ray photon. The 1958 discovery of the Mössbauer effect, in which certain nuclei emit γ rays of sharply defined frequency, allows nuclear energy-level differences to be accurately determined. The energy of a transition between nuclear levels depends very slightly on the chemical environment of the nucleus. The finite size of the nucleus causes a slight change in the potential energy of electrostatic interaction between the electrons and the nucleus, as compared to the assumption of a point nucleus; when an electron penetrates the nucleus, its potential energy is decreased. Different compounds of, say, Fe have different electron probability densities at the Fe nucleus; moreover, the nuclear radius varies from one nuclear state to another, so that the electron–nucleus interaction energy is different for the two nuclear levels involved in the transition. Thus the frequency of the Mössbauer-effect line differs for different compounds of the same element, and can provide information on the chemical bonding. Iron occurs in many biological compounds and Fe^{57} Mössbauer spectroscopy has been useful for investigating them. For example, the binding of O_2 to Fe in oxygenated hemoglobin has been studied. (For more on Mössbauer spectroscopy, see *Chang*, Chapter 5; N. N. Greenwood and T. C. Gibb, *Mössbauer Spectroscopy*, Harper and Row, 1971.)

We now consider the energy units used in spectroscopy. For electronic energies, the electron volt (eV), the kinetic energy acquired by an electron accelerated through a potential difference of 1 volt, is often used; 1 eV $= 1.60 \times 10^{-12}$ erg. Using

$$\Delta E = h\nu = hc/\lambda \tag{3.83}$$

we can relate energy differences to frequencies and to reciprocal wavelengths. For a mole of substance, a convenient unit is the kcal/mole; a molecular energy difference of 1 eV corresponds to a molar energy difference of about 23 kcal. Table A.3 in the Appendix gives energy conversion factors; note that the unit cm^{-1} is not actually a unit of energy —a photon with $1/\lambda = 1\ cm^{-1}$ has an energy 1.99×10^{-16} erg, as listed in the table.

In ultraviolet, visible, and infrared spectroscopy, one usually measures the spectral line wavelengths (rather than frequencies), using a prism or diffraction grating. Until 1972, the speed of light was not known to very high accuracy, so it became traditional to convert measured wavelengths to reciprocal wavelengths $1/\lambda$ (rather than to frequencies); the quantity $1/\lambda$ is called the *wave number*.

For wavelengths below 2000 Å, the O_2 molecules of the air absorb strongly, and ultraviolet work below this wavelength is done with the apparatus evacuated. For ultraviolet work above 2000 Å, and for visible and infrared work, wavelengths are generally measured in air. To have a constant proportionality factor *hc* between energy and wave number, we must use the *vacuum* wave number. Thus before calculating the wave number of a spectral line, the wavelength in air, λ_{air}, must be converted to the vacuum wavelength, λ_{vac}. We have $\lambda_{air}\nu = c_{air}$ and $\lambda_{vac}\nu = c$, where c_{air} and c are the speeds of light in air and in vacuum, respectively. The index of refraction n of air is $n \equiv c/c_{air}$, so that

$$\lambda_{vac} = n\lambda_{air} = \lambda_{air} + (n-1)\lambda_{air}$$

The index of refraction of air varies with temperature and pressure; "standard air" is defined as dry air at 15°C and 1 atmosphere pressure, with 0.03 percent CO_2. To correct wavelengths measured in air to vacuum, we add $(n-1)\lambda_{air}$. Some values of $n-1$ for standard air are[6]

λ_{air}	2000 Å	4000 Å	6000 Å	1 μm	2 μm	1000 μm
$10^4(n-1)$	3.24	2.83	2.77	2.74	2.730	2.726

The symbol for the wave number of a spectral line is σ:

$$\sigma = 1/\lambda_{vac} \tag{3.84}$$

(Infrared spectroscopists often use ν as the symbol for wave number and ν' for frequency.)

In microwave work, the frequency is measured, rather than the wavelength, and no correction is needed.

Units for wavelengths include the *angstrom* (Å), the *micrometer* (μm), and the *nanometer* (nm):

$$1\ \text{Å} = 10^{-8}\ \text{cm}, \qquad 1\ \mu\text{m} = 10^{-6}\ \text{m} = 10^4\ \text{Å}, \qquad 1\ \text{nm} = 10^{-9}\ \text{m} = 10\ \text{Å}$$

[The micrometer was formerly called the micron (μ).] Frequency units are the hertz (Hz) and its multiples, such as the kilohertz (kHz), megahertz (MHz), and gigahertz (GHz):

$$1\ \text{Hz} \equiv 1\ \text{cycle/sec}, \quad 1\ \text{kHz} = 10^3\ \text{Hz}, \quad 1\ \text{MHz} = 10^6\ \text{Hz}, \quad 1\ \text{GHz} = 10^9\ \text{Hz}$$

Wavelengths below 2000 Å are generally reported as vacuum wavelengths, whereas those above 2000 Å are generally reported as wavelengths in standard air. Wave numbers are always reported as vacuum wave numbers. The cgs unit of wave number is the reciprocal centimeter, cm^{-1}.

[6]For extensive tables, see C. D. Coleman, W. R. Bozman, and W. F. Meggers, *Table of Wavenumbers*, National Bureau of Standards Monograph 3 (1960).

It is a common, but illogical, practice to refer to this unit as a wave number; correctly used, the term *wave number* refers to the physical quantity $1/\lambda_{vac}$, and not to the unit of this quantity.

3.5 LINE SHAPES AND WIDTHS

A spectral absorption or emission line does not occur at one precise frequency; rather, the line has a measurable width. A convenient measure of this width is $\Delta\nu$, the *half-width at half maximum*; if the line's peak occurs at frequency ν_0, then at $\nu_0+\Delta\nu$ and at $\nu_0-\Delta\nu$ the line's intensity has fallen to half its value at ν_0. There are several causes of the finite width of lines.

The energy–time uncertainty relation (1.50) shows that a state with a finite lifetime does not have a precisely defined energy; rather, its energy has a spread or uncertainty, which increases with decreasing lifetime. Since all states but the ground state show spontaneous emission, an excited state does not have a sharply defined energy. The finite lifetime of excited states and consequent energy uncertainty then gives a breadth to spectral lines. [Note that (3.38) might seem to imply a violation of conservation of energy in that it gives some probability for a transition to occur with absorption or emission of a photon whose energy $h\nu$ differs somewhat from the stationary-state energy difference $E_m^{(0)}-E_n^{(0)}$; the lack of sharpness of excited-state energies resolves this apparent violation of energy conservation.] Weisskopf and Wigner in 1930 used quantum field theory to derive the following theoretical line shape for a transition between states n and m having finite lifetimes:

$$\frac{a}{(\nu-\nu_{mn})^2+(\tau_n^{-1}+\tau_m^{-1})^2/16\pi^2} \tag{3.85}$$

where τ_n and τ_m are the average lifetimes of states n and m, the frequency ν_{mn} is given by (3.45), and a is a constant characteristic of the transition.

The line shape (3.85) is an example of a *Lorentzian* shape, whose general form is

$$\frac{a}{(\nu-\nu_0)^2+b^2} \tag{3.86}$$

For the Lorentzian line shape, the maximum occurs at $\nu=\nu_0$ and equals a/b^2. At half maximum, $\nu=\nu_0\pm\Delta\nu$ and (3.86) gives

$$\tfrac{1}{2}(a/b^2)=a/\left[(\Delta\nu)^2+b^2\right];$$

solving, we get $\Delta\nu = b$. The half width at half maximum is b for the Lorentzian shape (3.86).

For the Weisskopf–Wigner shape (3.85), b is $(\tau_n^{-1} + \tau_m^{-1})/4\pi$. If state n is the ground state, then $\tau_n = \infty$ and the width of the spectral line measures the uncertainty ΔE_m in the energy of the excited state. We can take $h\Delta\nu$ as giving ΔE_m, where $\Delta\nu$ is the half-width at half maximum of the line. Hence $\Delta E_m = h\Delta\nu = h/4\pi\tau_m$, or

$$\Delta E_m \tau_m = \tfrac{1}{2}\hbar \tag{3.87}$$

in agreement with the uncertainty principle (1.50). From the discussion following (3.56), τ_m is the reciprocal of the Einstein coefficient $A_{m\to n}$. Hence from (3.55), we have for the half-width at half maximum

$$\Delta\nu = 1/4\pi\tau_m = A_{m\to n}/4\pi = (16\pi^3\nu_{mn}^3/3hc^3)|\mathbf{d}_{mn}|^2 \tag{3.88}$$

Besides the *uncertainty broadening* just discussed, there are other causes of line broadening which make line widths generally considerably greater than the *natural width* (3.88). The Doppler effect causes an apparent change in radiation frequency for molecules with a component of velocity v_{obs} in the direction of observation of the radiation. Different molecules have different values of v_{obs} and we get a *Doppler broadened* line.

A third contributor to line widths is *pressure broadening* (or *collision broadening*). We can look at this in two ways. The stationary-state energies are derived for an isolated molecule. However, in bulk matter, molecules interact and the varying intermolecular forces continually shift the stationary-state energy levels, giving line broadening. Alternatively, we can look at the intermolecular collisions as shortening the lifetimes of excited states, thereby broadening spectral lines. Much theoretical work has been done on calculating pressure-broadened line shapes.

A fourth cause of broadening is *saturation broadening*. In our semiclassical treatment of radiation, we assumed the radiation intensity was sufficiently weak that a first-order perturbation theory could be used. If the radiation is very intense, this causes substantial depletion of the population of the lower level. A more accurate perturbation treatment is needed in this case, and the result is that the absorption line is appreciably broadened. (See *Townes and Schawlow*, Section 13-15 and references cited therein.)

To reduce pressure broadening, one uses a gaseous sample at low pressure. Doppler broadening can be reduced by using a molecular beam and making observations at right angles to the beam. Saturation broadening is avoided by not using excessive power in the light beam. When these causes of broadening are eliminated, we are left with the natural line shape (3.85).

The two most commonly observed line shapes are the Lorentzian (3.86) and the *Gaussian*:

$$ce^{-d(\nu-\nu_0)^2} \tag{3.89}$$

Figure 1.1*a* shows the Gaussian function. The Lorentzian shape is similar to the Gaussian, but falls off more slowly. The Doppler shift of radiation from an emitting molecule is proportional to its velocity component in the direction of observation. The one-dimensional distribution of speeds in a gas is a Gaussian function. (See any physical-chemistry text.) Hence when Doppler broadening is dominant, we get a Gaussian-shaped line.

3.6 LASERS

The word *laser* is an acronym for light amplification by stimulated emission of radiation. The first laser was produced in 1960, and since then a wealth of applications have followed. We now consider the theory of the laser.

Let state m have higher energy than state n. As noted at the end of Section 3.2, the absorption intensity is proportional to $N_n - N_m$, since stimulated emission from state m to n diminishes the observed absorption. Ordinarily, the system levels have equilibrium populations, which are given by the Boltzmann distribution law (3.51). Thus N_m is usually less than N_n, resulting in net absorption of light when radiation of frequency ν_{mn} passes through the system. Suppose, however, that by some means we achieve a nonequilibrium population distribution with $N_m > N_n$. If we have achieved such a *population inversion*, then when light of frequency ν_{mn} is applied to the system, stimulated emission from m to n will exceed absorption from n to m, and we will get amplification of the light at frequency ν_{mn}. This is the principle behind lasers.

A laser has three components: (1) a system with a set of energy levels; (2) a means for achieving population inversion between two of these levels; (3) a resonant optical cavity in which the radiation is amplified.

The first laser produced was the *ruby laser*. Ruby is a crystal of Al_2O_3 in which a fraction of 1 percent of the Al^{3+} ions is replaced by Cr^{3+} ions. A diagram of the lowest Cr^{3+} levels in the ruby crystal is given in Fig. 3.3. The symbols A_2, E, T_2 are the symmetry-species symbols (Section 1.19). (These are derived using ligand-field theory.) The ruby laser consists of a

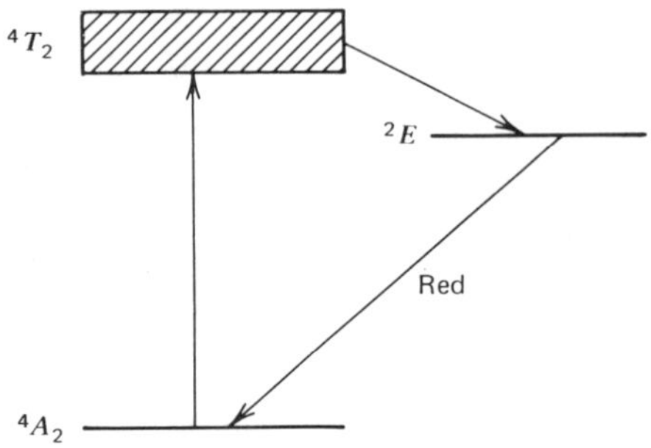

Fig. 3.3 Simplified diagram of Cr^{3+} levels in ruby.

ruby rod surrounded by a spiral high-energy flash lamp; the ends of the rod are silvered to form a reflecting surface. The majority of the Cr^{3+} ions are pumped from the $^4A_2$ ground level to the $^4T_2$ level by discharging the surrounding flash lamp. (This is an example of *optical pumping*.) The $^4T_2$ level is shown as broad; this is because the crystal lattice vibrations of surrounding ions appreciably affect the energy of this level, and its energy changes during the course of lattice vibrations. (However, ligand-field theory shows that the ^2E level is not significantly affected by lattice vibrations.) The broad $^4A_2 \rightarrow {}^4T_2$ absorption of flash-lamp radiation occurs in the blue and green regions of the spectrum. (That's why rubies are red.) The most probable means of de-excitation of the $^4T_2$ Cr^{3+} ions is for them to give up some electronic energy to vibrations of the crystal lattice and thereby make radiationless transitions to the ^2E level. Because of the $\Delta S = 0$ selection rule (3.82), this level is metastable with a relatively long lifetime of 3 msec with respect to decay to the ground $^4A_2$ level. (This transition does have a finite probability because of spin–orbit interaction.) Because of the metastability, the population of the ^2E level builds up, and we get a population inversion between ^2E and $^4A_2$. Some of the ^2E Cr^{3+} ions spontaneously emit red photons in random directions and jump to the ground state. Photons emitted at angles to the axis of the ruby rod simply pass out of the crystal and are not heard from again. Photons emitted parallel to the axis are reflected back and forth between the silvered ends of the rod (which must be accurately parallel). As they go back and forth, they stimulate ^2E chromium ions to radiate the red line. [Since $N(^2E) \gg N(^4A_2)$, stimulated emission from ^2E is more likely than absorption from $^4A_2$.] The photons emitted by stimulated emission travel in the same direction (along the axis) and in phase with the photons that stimulated them; traveling back and forth, they stimulate further emission. Thus we get a very intense beam. One end of the rod is only partially silvered, so that photons hitting it have some probability to be transmitted rather than reflected. This allows part of the laser beam to emerge from the rod. (Of course, if this end is made too transmitting, losses of photons will be too great to allow laser action.)

The first stimulated-emission amplification device was produced at microwave frequencies by Townes and co-workers in 1954. (This was the *maser*—microwave amplification....) In 1958 Schawlow and Townes described the conditions needed to produce light amplification at optical frequencies. In 1960 Maiman built the first laser, the ruby laser described above. (Amusingly, his paper reporting the laser was rejected by *Physical Review Letters*, the first journal he sent it to, on the grounds that the paper did not merit rapid publication.)

Maiman's ruby laser emits a pulse of laser light, and then must be pumped again before another pulse is emitted. The first continuously operating laser, called a CW (continuous wave) laser, was the *helium–neon gas laser*, invented in 1961 by Javan. This laser consists of a glass tube containing He at 1 torr and Ne at 0.1 torr. Electrodes are sealed in the tube and a continuous electric discharge is passed through the gas mixture. Parallel mirrors, one of which is partially transmitting, are placed at the ends of the tube. Electrons of the discharge collide with helium (and neon) atoms, exciting them to a variety of excited states. Many of the excited He atoms will make transitions to the 2^3S_1 level. (In this notation, the "2" is the principal quantum number of the excited electron.) As noted in Section 3.3, this metastable level has an 8000-sec lifetime and will most likely be de-excited by collisions with other atoms (and with the wall) rather than by emitting a photon. It happens that the four levels of the Ne $1s^22s^22p^54s$ configuration lie at energies above the Ne ground state which are nearly equal to the 2^3S_1 He excitation energy. (Russell–Saunders coupling happens not to be a good approximation for Ne, so we avoid the usual term symbols for the Ne levels.) Hence the 2^3S_1 helium atoms can get back to the 1^1S_0 ground state by exciting ground-state neon atoms with which they collide to these four levels. Thus a significant population of the neon $2p^54s$ levels is achieved. These levels decay mainly by photon emission to the 10 levels of the $2p^53p$ configuration. However, these $2p^53p$ levels have short lifetimes and are rapidly depopulated by emission to $2p^53s$ levels. We thus get a population inversion between the neon $2p^54s$ and $2p^53p$ levels. Spontaneously emitted photons from $4s$ to $3p$ transitions are reflected back and forth, and we get laser action.

Thousands of different lasers have been built. Chemical lasers use the energy of chemical reactions to achieve population inversion. For example, flash photolysis of a mixture of H_2 and Cl_2 initiates a chain reaction that produces HCl molecules in vibrationally excited states, thereby giving a population inversion with respect to the ground vibrational state and laser action. Both pulse and CW chemical lasers have been produced.

Early lasers emitted light at one (or a few) fixed frequencies. If one wanted to do spectroscopy using a laser light source, one was limited to transitions whose frequencies happened to match those of existing lasers. However, in the late 1960s, the development of *tunable dye lasers* began. These use an organic dye (e.g., fluorescein, eosin, polymethines) dissolved in a solvent such as ethanol. The dye is optically pumped to excited states with light from another laser (e.g., a CW gas laser). The vibration–rotation levels of the dye are collision-broadened in solution to give continuous bands of energy, and we can get laser emission over a broad frequency range. To select the frequency, one of the end mirrors is replaced with a diffraction grating placed at an angle to the cell axis. The first-order diffraction line of the grating is reflected back along the axis. Changing the grating angle changes the wavelength that is reflected back and thus changes the wavelength that will resonate in the cavity. As an example, a commercially available CW rhodamine 6G dye laser is continuously tunable over the range 530 to 640 nm with an output line width of 0.02 nm. Various tunable dye lasers produce light of frequencies covering the region from 2500 to 10,000 Å. Tunable CW semiconductor lasers produce light of infrared frequencies; tuning is accomplished by varying pressure, temperature, or magnetic field.

What's so special about laser light? Laser light is (1) highly directional, (2) nearly monochromatic, (3) coherent, and (4) intense. The directionality of the beam results from the resonance in the optical cavity. (Because of diffraction at the transmitting mirror, the directionality is not perfect, and the beam does show a very slight spread.) The monochromaticity also results from the resonance in the optical cavity. A standing wave pattern is set up in the cavity; for resonance, an integral number of half wavelengths of the light must fit along the length of the cavity. Light whose wavelength does not satisfy this condition is eliminated by destructive interference. Hence the output is virtually monochromatic. Actually, the laser resonant cavity is a three-dimensional structure, and its resonant modes are quite similar to the wave functions of a particle in a three-dimensional box. We can characterize these modes by three numbers that give the number of nodes in the standing-wave pattern in each direction. The number of nodes in the axial direction is some very large integer n, such that $n\lambda/2$ equals the cavity length. The simplest mode has zero nodes in each of the two directions perpendicular to the axis; here, the radiation electric-field amplitude pattern in a plane perpendicular to the axis resembles the wave-function contours of the lowest state ($n_x = n_y = 1$) of a particle in a two-dimensional box. This is called the TEM_{00} mode, where the letters stand for "transverse electromagnetic" and the subscripts give the mode node numbers. Several modes in a laser cavity can be excited at once, and a theoretical treatment reveals that the resonant wavelength differs slightly for different modes. Special techniques allow one to select just the TEM_{00} mode, thereby giving the fullest monochromaticity. To give some figures, for a line with $\lambda = 10{,}000$ Å and $\nu = 3 \times 10^{14}$ Hz, a typical Doppler half-width is 500×10^6 Hz, and a typical natural half-width is 30×10^6 Hz. These can be compared with an observed half-width as low as 3 Hz for the He–Ne gas laser. This laser line width is only $1/10^7$ of the natural width. (Solid and liquid lasers do not give as good a monochromaticity as gas lasers; here, widths are measured in kHz or MHz.) The coherence and intensity result from the coherence of stimulated emission plus the resonance conditions in the cavity. (The term *coherent* indicates that all along the beam, the phase of the light wave varies in a smooth, continuous, nonrandom fashion.)

For extreme intensities, lasers can only be operated in pulsed rather than CW fashion, due to the heating of the cavity and the high pumping power required for the population inversion. Laser pulses with peak powers of 2.5×10^{13} watts have been achieved; of course, such a pulse lasts for only a rather short time (10^{-12} sec).

Before the laser, the light sources used in infrared, visible, and ultraviolet spectroscopy were heated solids or gas discharge tubes. These sources are based on *spontaneous* emission. Such light is emitted in random directions with random phases. Even "lines" of gas discharge tubes lack true monochromaticity due to pressure-, Doppler-, and natural-broadening. Thus light from these traditional sources does not possess the above-mentioned four qualities of laser light, which is based on *stimulated* emission.

Actually, coherent, monochromatic electromagnetic radiation has been around for many years. Radio waves emitted from oscillating electric circuits are coherent and monochromatic, as are the microwaves produced by klystron tubes (Section 5.8). However, coherent electromagnetic radiation of the high frequencies of infrared, visible, and ultraviolet light was not available until 1960. The use of lasers as light sources in spectroscopy research is growing, and within the next few years they may largely replace conventional sources, thus bringing much greater precision and resolution to optical spectroscopy.

The high intensity of laser light (lots of photons per cm^3) gives significant probability to transitions in which *two* photons, rather than one, are absorbed. Such two-photon processes are a field of active study.

In 1972 a 100-fold increase in the accuracy of the known speed of light was achieved by measuring the frequency of a helium–neon laser line in terms of the Cs time standard (Appendix Table A.2) and measuring its wavelength in terms of the Kr length standard. [K. M. Evenson *et al.*, *Phys. Rev. Letters*, **29**, 1346 (1972).] The improved value is

$$c = (299{,}792{,}458 \pm 1)\ \text{m/sec}$$

Lasers have applications in studying chemical reactions. One can use a monochromatic laser pulse to populate a specific vibrationally excited state of a molecular species, and then follow the kinetics of the reactions the excited species undergoes.

Other uses of lasers include eye surgery on detached retinas, spot welding, holography, isotope separation, accurate determination of the moon's orbit by reflection of laser light off a reflector placed on the moon's surface, and laser-guided bombs and missiles. Possible future uses include terrestrial and extraterrestrial communication, applications to computers, and production of the high temperatures needed for controlled nuclear-fusion reactions.

PROBLEMS

3.1 A sample is irradiated with sodium D light ($\lambda = 5900$ Å) of intensity 0.10 watt/cm^2. (a) What is the amplitude of the radiation's electric field strength? (b) What is the amplitude of the radiation's magnetic induction B? (c) How many photons pass through a 1-cm^2 cross section per second?

3.2 Does the state function (3.20) represent a stationary state?

3.3 Evaluate the integral in (3.59).

3.4 What are the electric-dipole selection rules for a particle in a three-dimensional box exposed to isotropic radiation?

3.5 What are the electric-dipole selection rules for a three-dimensional harmonic oscillator exposed to isotropic radiation?

3.6 Consider the integral $\langle m|x^2|n\rangle$, where the wave functions are one-dimensional harmonic-oscillator wave functions. For what values of $m-n$ is this integral nonzero?

3.7 Let ω_{mn} in (3.38) be 2π times the frequency of sodium D light ($\lambda = 5900$ Å). Consider two values of ω, one equal to ω_{mn} and one equal to $1.001\ \omega_{mn}$. Calculate the ratio of the two values of $|c_m|^2$ for these two values of ω. Assume an irradiation time of 1 sec. An order-of-magnitude calculation is sufficient.

3.8 Verify (3.46).

3.9 In deriving (3.56), we assumed that state m decayed only to state n. Suppose there are three states n, p, and s to which state m decays. What form does (3.56) have in this case?

3.10 Consider a one-dimensional harmonic oscillator with vibrational frequency 5×10^{13} sec^{-1} and mass 1×10^{-23} g. (These are typical values for a diatomic molecule.) Find the average lifetime of the $v=1$ vibrational state.

3.11 Prove that $1/A_{m\to n}$ in (3.56) is the average lifetime of state m. *Hint*: Let $-dN_m$ molecules decay between t and $t+dt$. Then the lifetime of these $-dN_m$ molecules in state m was t. The average lifetime is then found by summing (i.e., integrating) $-t\,dN_m/N_{m0}$; see (1.42).

3.12 Verify the selection rule (3.74).

3.13 What are the electric-dipole selection rules for a two-particle rigid rotor exposed to isotropic radiation?

3.14 Between which pairs of levels of the helium $1s^2$, $1s2s$, and $1s2p$ configurations are electric-dipole transitions allowed?

3.15 Early investigators of the spectrum of helium believed that there were two kinds of helium, which they called orthohelium and parahelium; each kind had its own set of energy levels, and transitions between the two kinds of levels did not occur. Explain these observations.

3.16 Derive the conversion factor between eV/molecule and kcal/mole.

3.17 Calculate the vacuum wave numbers of the sodium D lines; the wavelengths in standard air are 5890.0 Å and 5895.9 Å. What is the separation in eV between the levels of the $1s^22s^22p^63p$ sodium configuration?

3.18 Show that if there were no spin–orbit interaction in atoms, the probability of a magnetic-dipole transition between any two atomic states would be zero.

3.19 Find the natural half-width at half maximum for the $v=0\to1$ absorption transition of the harmonic oscillator of Problem 3.10.

3.20 What is the half-width at half maximum for the Gaussian shape (3.89)?

3.21 Plot the Gaussian and Lorentzian functions (3.89) and (3.86) against $\nu-\nu_0$. Use the same axes for both plots and take the peak heights and half-widths to be equal.

3.22 For a $1s$ H atom, find $\langle T\rangle$, and use this to calculate $\langle v^2\rangle^{1/2}/c$, where c is the speed of light.

CHAPTER 4

VIBRATION AND ROTATION OF DIATOMIC MOLECULES

4.1 NUCLEAR MOTION IN DIATOMIC MOLECULES

In Section 1.19, we showed how the Born–Oppenheimer approximation separates the electronic motions and nuclear motions in a molecule. We considered the electronic Schrödinger equation in Sections 1.20 and 1.21; we now study the nuclear Schrödinger equation. In this chapter, we consider diatomic molecules. Unless otherwise stated, we will restrict ourselves to diatomic molecules in $^1\Sigma$ electronic states; the presence of electronic spin or orbital angular momentum complicates things, and will be discussed in Section 4.11. Most diatomic molecules have $^1\Sigma$ electronic ground states, notable exceptions being O_2 and NO. As usual, we will be considering gas-phase molecules, neglecting intermolecular interactions.

The Schrödinger equation for the nuclear motion of a diatomic molecule is, according to (1.280) and (1.281),

$$\left[-\frac{\hbar^2}{2m_a}\nabla_a^2-\frac{\hbar^2}{2m_b}\nabla_b^2+U(R)\right]\psi_{N,\mathrm{tot}}=E_{\mathrm{tot}}\psi_{N,\mathrm{tot}} \tag{4.1}$$

The first two terms in brackets are the kinetic-energy operators for nuclei a and b; the variable R is the internuclear distance; the function $U(R)$ is the potential-energy function (including nuclear repulsion) for nuclear motion, and is found by solving the electronic Schrödinger equation; $\psi_{N,\mathrm{tot}}$ is the wave function for nuclear motion; E_{tot} is the total energy of the molecule, including translational energy.

Equation (4.1) is a two-particle problem. In Section 1.12 we showed that we can separate a two-particle problem into two one-particle problems, provided that the potential energy is a function of only the relative coordinates of the two particles. Since R involves only the relative coordinates of the nuclei, this condition is met. We conclude that

$$\psi_{N,\mathrm{tot}} = \psi_{N,\mathrm{trans}} \psi_{N,\mathrm{int}} \tag{4.2}$$

$$E_{\mathrm{tot}} = E_{\mathrm{trans}} + E_{\mathrm{int}} \tag{4.3}$$

where $\psi_{N,\mathrm{trans}}$ is the wave function for translational motion of the nuclei as a whole, and $\psi_{N,\mathrm{int}}$ is the wave function for the internal motion of the nuclei relative to each other. The translational motion of the molecule simply adds the constant quantity E_{trans} to the molecular energy. Writing ψ_N for $\psi_{N,\mathrm{int}}$ and E for E_{int}, we have (Section 1.12)

$$\left[-\frac{\hbar^2}{2\mu} \nabla^2 + U(R) \right] \psi_N = E \psi_N \tag{4.4}$$

where the reduced mass μ of the nuclei is

$$\mu = \frac{m_a m_b}{m_a + m_b} \tag{4.5}$$

E is the total energy of the molecule (nuclei *and* electrons), excluding translational energy. The independent variables in (4.4) are the relative spherical polar coordinates of the nuclei: (R, θ_N, ϕ_N); the angles θ_N and ϕ_N give the orientation of the internuclear axis relative to the coordinate system shown in Fig. 4.1. This system has its origin at the molecular center of mass, and therefore translates through space with the molecule; however, it does not rotate with the molecule. Since the molecular translational motion is of no interest, we call the nonrotating XYZ axes the *space-fixed* axes.

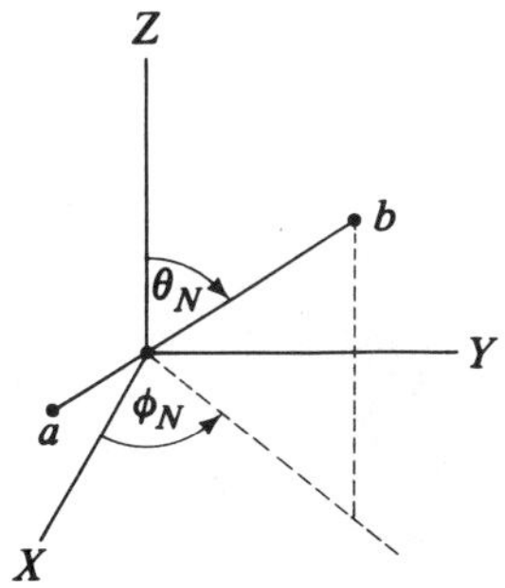

Fig. 4.1 Space-fixed coordinate system for internal nuclear motion.

Since the potential energy in (4.4) depends only on R, we have a central-force problem, and the results of Section 1.11 show that

$$\psi_N = F(R) Y_J^M(\theta_N, \phi_N) \tag{4.6}$$

where Y_J^M is a spherical harmonic, and F is some function of R. We use J and M instead of l and m for the angular-momentum quantum numbers to conform with the standard notation for molecules. We have (Section 1.6)

$$J = 0, 1, 2, 3, \ldots \tag{4.7}$$

$$M = -J, -J+1, \ldots, J-1, J \tag{4.8}$$

The rotational angular momentum of the nuclei is $\sqrt{J(J+1)}\,\hbar$; the component of this angular momentum along the space-fixed Z axis is $M\hbar$. [Recall the two-particle rigid rotor, treated in Section 1.13. The difference between (4.4) and the rigid rotor is that we are allowing the internuclear distance to vary, and therefore have vibrational energy in addition to rotational energy.]

$F(R)$ is found by solving the differential equation (1.213):

$$-\frac{\hbar^2}{2\mu}\left[F''(R) + \frac{2}{R}F'(R)\right] + \frac{J(J+1)\hbar^2}{2\mu R^2}F(R) + U(R)F(R) = EF(R) \tag{4.9}$$

Equation (4.9) is simplified by the following change of dependent variable:

$$G(R) \equiv RF(R) \tag{4.10}$$

$$F' = G'R^{-1} - GR^{-2}$$

$$F'' = G''R^{-1} - 2G'R^{-2} + 2GR^{-3}$$

Substitution into (4.9) gives

$$-\frac{\hbar^2}{2\mu}G''(R) + \left[\frac{J(J+1)\hbar^2}{2\mu R^2} + U(R) - E\right]G(R) = 0 \tag{4.11}$$

To proceed further, we need an expression for $U(R)$. The potential energy U (which is found by solving the molecular electronic Schrödinger equation) is a different function not only for each molecule, but for each electronic state of each molecule. Rather than deal with a different nuclear

Schrödinger equation for each molecular electronic state, we seek an approximation that will represent U reasonably well for most diatomics. For a bound electronic state, we know that U has the general appearance of the solid curve in Fig. 4.2. We expect the nuclei to vibrate about the position of minimum potential energy; therefore, we expand U in a Taylor series (Section 1.2) about R_e, the equilibrium internuclear separation:

$$U(R) = U(R_e) + U'(R_e)(R - R_e) + U''(R_e)(R - R_e)^2/2!$$
$$+ U'''(R_e)(R - R_e)^3/3! + \cdots \quad (4.12)$$

At the position of minimum potential energy, we have

$$U'(R_e) = 0 \quad (4.13)$$

For the region near R_e, we note that $(R - R_e)^3$ and higher powers of $(R - R_e)$ are small, and we shall neglect them; we justify this by saying that for small vibrations the internuclear separation will be close to R_e most of the time. Thus (4.12) becomes

$$U(R) \approx U(R_e) + \tfrac{1}{2} U''(R_e)(R - R_e)^2 \quad (4.14)$$

This amounts to replacing the true curve by a parabola (Fig. 4.2); the fit is good only near R_e.

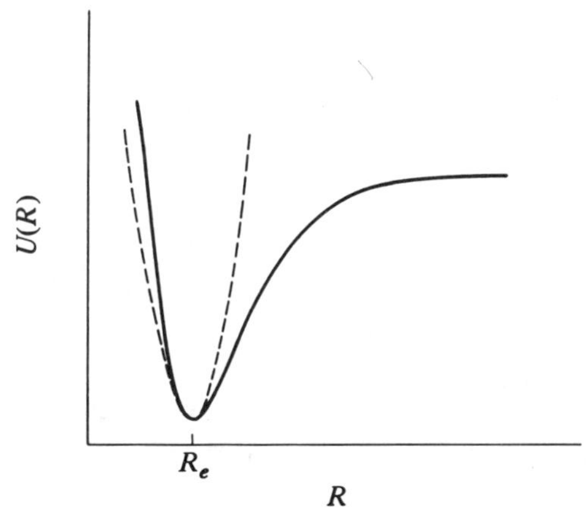

Fig. 4.2 Approximation of $U(R)$ by a parabola. Because $U'''(R_e)$ is generally negative, the true curve (solid line) lies below the parabola (dotted line) for $R > R_e$ and above the parabola for $R < R_e$.

We now change from R to a new independent variable q, which is the deviation of R from its equilibrium value:

$$q \equiv R - R_e \tag{4.15}$$

Defining

$$k_e \equiv U''(R_e) \tag{4.16}$$

we write (4.14) as

$$U \approx U(R_e) + \tfrac{1}{2}k_e q^2 \tag{4.17}$$

$U(R_e)$ is the value of the electronic energy (including nuclear repulsion) at the equilibrium internuclear separation. k_e is called the *equilibrium molecular force constant*. Making the change of variable (4.15) in (4.11), we have

$$S(q) \equiv G(R) \tag{4.18}$$

$$\frac{dS}{dq} = \frac{dG}{dR}\frac{dR}{dq} = \frac{dG}{dR}, \qquad \frac{d^2S}{dq^2} = \frac{d^2G}{dR^2} \tag{4.19}$$

and (4.11) becomes

$$-\frac{\hbar^2}{2\mu}S''(q) + \left[\frac{J(J+1)\hbar^2}{2\mu(q+R_e)^2} + U(R_e) + \tfrac{1}{2}k_e q^2 - E\right]S(q) = 0 \tag{4.20}$$

where we have used the approximation (4.17).

The first term in brackets is troublesome, and we therefore expand it in a Taylor series (see Problem 4.1):

$$\frac{J(J+1)\hbar^2}{2\mu(R_e+q)^2} = \frac{J(J+1)\hbar^2}{2\mu R_e^2}\frac{1}{(1+q/R_e)^2}$$

$$= \frac{J(J+1)\hbar^2}{2\mu R_e^2}\left[1 - 2\frac{q}{R_e} + 3\frac{q^2}{R_e^2} - \cdots\right] \tag{4.21}$$

For R near R_e, the terms in brackets in (4.21) become successively smaller, and we shall neglect all terms beyond the first. Finally, defining the constant W as

$$W \equiv E - U(R_e) - \frac{J(J+1)\hbar^2}{2\mu R_e^2} \tag{4.22}$$

we have for (4.20)

$$-\frac{\hbar^2}{2\mu}S''(q)+[\tfrac{1}{2}k_e q^2 - W]S(q)=0 \tag{4.23}$$

These manipulations have brought us to a familiar equation; we recognize (4.23) as the Schrödinger equation (1.132) for a one-dimensional harmonic oscillator with force constant k_e. Before we can conclude that (4.23) and (1.132) have the same solutions, we must verify that the boundary conditions are the same. For quadratic integrability, we require that $S(q)$ vanish for $q=\infty$. Also, since the radial factor $F(R)$ in the nuclear wave function is

$$F(R)=G(R)/R=S(q)/R \tag{4.24}$$

we require that $S(q)$ vanish for $q=-R_e$, which corresponds to $R=0$. (A radial function that behaves as $1/R$ for small R is unacceptable; see *Levine*, Section 6.4.) The harmonic-oscillator wave functions satisfy the boundary conditions of vanishing at $+\infty$ and $-\infty$, instead of at $+\infty$ and $-R_e$, as is desired for the solutions of (4.23). However, the harmonic-oscillator wave functions decrease very rapidly outside the classically allowed region, and, provided the vibrational quantum number is not too high, they are nearly zero at $-R_e$ (see Problem 4.2). Hence there is no serious error introduced by ignoring the difference in boundary conditions between (4.23) and (1.132). We therefore take $S(q)$ to be a harmonic-oscillator wave function. For the state with vibrational quantum number v, we have from (1.133):

$$S_v(q)=\left(\frac{\alpha}{\pi}\right)^{1/4}\frac{1}{(2^v v!)^{1/2}}H_v(\alpha^{1/2}q)e^{-\alpha q^2/2} \tag{4.25}$$

where H_v is a Hermite polynomial, and [Eq. (1.134)]

$$\alpha \equiv 4\pi^2\nu_e\mu/h \tag{4.26}$$

The equilibrium vibration frequency ν_e is [Eq. (1.131)]

$$\nu_e \equiv \frac{1}{2\pi}\left(\frac{k_e}{\mu}\right)^{1/2}=\frac{1}{2\pi}\left[\frac{U''(R_e)}{\mu}\right]^{1/2} \tag{4.27}$$

Combining (4.6) and (4.24), we have as the approximate wave function for the internal nuclear motion of a diatomic molecule

$$\psi_N=\frac{S_v(q)}{R}Y_J^M(\theta_N,\phi_N) \tag{4.28}$$

W in (4.23) corresponds to E in (1.132), and we have [(1.140) and (4.22)]

$$W=(v+\tfrac{1}{2})h\nu_e, \qquad v=0,1,2,\ldots \tag{4.29}$$

$$E=U(R_e)+(v+\tfrac{1}{2})h\nu_e+J(J+1)hB_e \tag{4.30}$$

where the *equilibrium rotational constant* B_e is defined as

$$B_e \equiv \frac{h}{8\pi^2 I_e} \tag{4.31}$$

and where the *equilibrium moment of inertia* I_e is defined by

$$I_e \equiv \mu R_e^2 \tag{4.32}$$

The first term on the right side of (4.30) is the value of the electronic energy including nuclear repulsion at R_e, and depends on which electronic state we are considering; we call this term the (equilibrium) electronic energy, U_e. The "electronic energy" U_e is to be distinguished from the purely electronic energy $E_{\text{el}}(R)$ and from the electronic energy including nuclear repulsion $U(R)$ of Section 1.19. The quantity U_e is a constant for a given electronic state and is given by

$$U_e \equiv U(R_e)=E_{\text{el}}(R_e)+Z_aZ_be^2/R_e \tag{4.33}$$

The second term in (4.30) is the energy of a harmonic oscillator with force constant $U''(R_e)$; we call this term the vibrational energy, E_{vib}. The last term is the energy [Eq. (1.219)] of a two-particle rigid rotor with moment of inertia μR_e^2; we call this the rotational energy, E_{rot}. Thus we write (4.30) as

$$E=U_e+E_{\text{vib}}+E_{\text{rot}} \tag{4.34}$$

This equation is not exact; its derivation involved several approximations. If we include the translational motion of the molecule, then (4.3) gives for the total energy of the molecule

$$E_{\text{tot}}=U_e+E_{\text{vib}}+E_{\text{rot}}+E_{\text{trans}}$$

The relative magnitudes of these energies are important. For the ground state of H_2, the electronic energy U_e equals -31.9 eV. (The hydrogen molecule has $D_e=4.7$ eV, and it dissociates to two hydrogen atoms with total energy -27.2 eV.) The lowest stable excited state of H_2 has $U_e=-20.6$ eV. For the H_2 ground state, $h\nu_e=0.5$ eV and $hB_e=8\times10^{-3}$ eV. The separation between electronic energy levels is much greater than the separation between vibrational levels, which, in turn, is much greater than the separation between rotational levels.

The Born–Oppenheimer approximation separates the molecular wave function into a product of electronic and nuclear wave functions:

$$\psi = \psi_{\text{el}}\psi_{N,\text{tot}}$$

Our treatment of the nuclear Schrödinger equation for diatomic molecules has shown that the wave function for nuclear motion can be separated into rotational, vibrational, and translational wave functions:

$$\psi_{N,\text{tot}} = \psi_{\text{vib}}\psi_{\text{rot}}\psi_{\text{trans}} \tag{4.35}$$

$$\psi_{\text{vib}} = \frac{S_v(R-R_e)}{R}, \qquad \psi_{\text{rot}} = Y_J^M(\theta_N, \phi_N)$$

Molecular vibration is much faster than molecular rotation, so that we can use an average internuclear separation R_e in calculating the rotational moment of inertia.

In calculating the moment of inertia, $I_e = \mu R_e^2$, one customarily evaluates μ from the masses of the *neutral* atoms, considering this mass to be concentrated at the nucleus.[1] Of course, this is an approximation, but it is a very good one, since electrons are so light. The atomic masses m_a and m_b in (4.5) cannot usually be gotten from tables of atomic weights, since most elements are mixtures of different isotopes. Each isotopic species has its own vibration–rotation levels. Masses for some common isotopes are given in the Appendix.

With each electronic state, we have a series of vibration–rotation levels. See Fig. 4.3. The total molecular energy (excluding translation) for a given state of electronic and nuclear motion is the sum of the equilibrium electronic energy U_e, the vibrational energy E_{vib}, and the rotational energy E_{rot}.

4.2 ANHARMONICITY, VIBRATION–ROTATION INTERACTION, AND CENTRIFUGAL DISTORTION

In the last section, we obtained an approximate solution to the nuclear Schrödinger equation for diatomic molecules. We now improve the accuracy of that solution. In going from (4.11) to (4.23), we neglected the smaller terms in the expansions (4.12) and (4.21). We shall deal with the

[1] See, however, P. R. Bunker, *J. Mol. Spectrosc.*, **35**, 306 (1970).

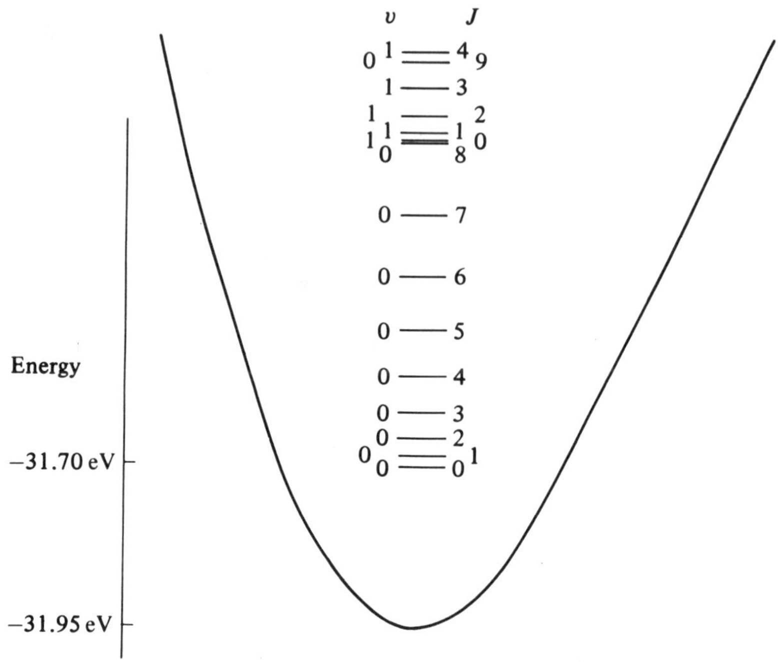

Fig. 4.3 Lowest energy levels of the ground electronic state of H_2, according to (4.30). For this case we have $\nu_e \approx 72B_e$. For diatomic molecules of intermediate mass, a typical value is $\nu_e = 10^3 B_e$, and there are perhaps 30 levels between the $v=0$, $J=0$ level and the $v=1$, $J=0$ level. The lower part of the $U(R)$ curve is sketched in.

neglected terms by considering them as a perturbation. The calculation of the energy corrections is rather involved and can be skimmed if desired—begin reading again at (4.67).

Considering only the first two previously neglected terms in (4.12) and (4.21), we have as the perturbation

$$\hat{H}' = aq + bq^2 + cq^3 + dq^4 \tag{4.36}$$

where $q = R - R_e$, and [(4.31)]

$$a \equiv -\frac{2J(J+1)hB_e}{R_e}, \qquad b \equiv \frac{3J(J+1)hB_e}{R_e^2}$$
$$c \equiv (1/6)U'''(R_e), \qquad d \equiv (1/24)U^{iv}(R_e) \tag{4.37}$$

The zeroth-order (unperturbed) nuclear wave functions are [Eq. (4.28)]

$$\psi_{vJM}^{(0)} = \frac{S_v(q)}{R} Y_J^M(\theta_N, \phi_N) \tag{4.38}$$

The zeroth-order energies are [Equation (4.30)]

$$E^{(0)} = U_e + (v + \tfrac{1}{2})h\nu_e + J(J+1)hB_e \tag{4.39}$$

Since the value of M does not affect $E^{(0)}$, the unperturbed levels are $(2J+1)$-fold degenerate. Hence, before forging ahead, we must be sure that we have the correct zeroth-order wave functions for the perturbation (4.36). (See Section 1.10.) It was noted in Section 1.10 that when the secular determinant is diagonal, we have the right zeroth-order functions. We now show that the functions (4.38) give a diagonal secular determinant. An off-diagonal element has the form

$$\langle \psi^{(0)}_{vJM'}(q,\theta_N,\phi_N)|\hat{H}'(q)|\psi^{(0)}_{vJM}(q,\theta_N,\phi_N)\rangle, \qquad M' \neq M \tag{4.40}$$

Since $\hat{H}'$ involves only q, the triple integral (4.40) can be written as the product of an integral over q, and the integral

$$\int_0^{2\pi}\int_0^{\pi}[Y_J^{M'}]^* Y_J^M \sin\theta_N \, d\theta_N \, d\phi_N, \qquad M' \neq M$$

This integral vanishes because of the orthogonality of the spherical harmonics. Hence (4.38) are the correct zeroth-order functions.

The first-order correction to the energy of the state with quantum numbers vJM is [Equation (1.201)]

$$E^{(1)} = \langle \psi^{(0)}_{vJM}|\hat{H}'|\psi^{(0)}_{vJM}\rangle$$

$$E^{(1)} = \int_0^{\infty} \frac{S_v^*}{R}\frac{S_v}{R}(aq + bq^2 + cq^3 + dq^4)R^2 \, dR \int_0^{2\pi}\int_0^{\pi} |Y_J^M|^2 \sin\theta_N \, d\theta_N \, d\phi_N$$

$$E^{(1)} = \int_{-R_e}^{\infty} |S_v(q)|^2 (aq + bq^2 + cq^3 + dq^4)\, dq \tag{4.41}$$

(The spherical harmonics are normalized.) The harmonic-oscillator function $S(q)$ is very small for q less than $-R_e$, so that we can write the lower limit in (4.41) as $-\infty$ without serious error. Also S is real. Equation (4.41) becomes

$$E^{(1)} = \int_{-\infty}^{\infty}(aq + cq^3)S_v^2 \, dq + \int_{-\infty}^{\infty}(bq^2 + dq^4)S_v^2 \, dq \tag{4.42}$$

The harmonic-oscillator functions are either even or odd, so that S_v^2 is even. Thus the integrand of the first integral in (4.42) is an odd function, and the first integral is zero. We have

$$E^{(1)} = b\int_{-\infty}^{\infty} S_v q^2 S_v \, dq + d\int_{-\infty}^{\infty} q^4 S_v^2 \, dq \tag{4.43}$$

$$E^{(1)} = b\langle v|q^2|v\rangle + d\langle v|q^4|v\rangle \tag{4.44}$$

To evaluate these integrals, we start with the harmonic-oscillator matrix element (3.61), which reads in the notation of the present chapter

$$\langle v'|q|v\rangle = (v/2\alpha)^{1/2}\delta_{v',v-1} + [(v+1)/2\alpha]^{1/2}\delta_{v',v+1} \tag{4.45}$$

The operator $\hat{q}^2$ is the square of the operator $\hat{q}$. Since matrix representatives of operators obey the same relations as the operators (Section 2.3), the $\mathbf{q}^2$ matrix [whose elements are $\langle v'|q^2|v\rangle$] is the square of the $\mathbf{q}$ matrix [whose elements are (4.45)]. Hence, using the matrix-multiplication rule (2.11), we have

$$\langle v'|q^2|v\rangle = \sum_k \langle v'|q|k\rangle\langle k|q|v\rangle \tag{4.46}$$

$$= \sum_k \left[\left(\frac{k}{2\alpha}\right)^{1/2}\delta_{v',k-1} + \left(\frac{k+1}{2\alpha}\right)^{1/2}\delta_{v',k+1}\right]$$

$$\times\left[\left(\frac{v}{2\alpha}\right)^{1/2}\delta_{k,v-1} + \left(\frac{v+1}{2\alpha}\right)^{1/2}\delta_{k,v+1}\right]$$

Multiplying out the terms, we get four sums. The first sum is

$$(1/2\alpha)\sum_k (kv)^{1/2}\delta_{v',k-1}\delta_{k,v-1} \tag{4.47}$$

This sum will vanish unless the values of v' and v are such that a single value of k can satisfy both $k = v'+1$ and $k = v-1$. Thus v' must equal $v-2$, or (4.47) will vanish. Hence (4.47) equals

$$(1/2\alpha)[(v-1)v]^{1/2}\delta_{v',v-2} \tag{4.48}$$

Evaluating the remaining sums similarly, we find (Problem 4.3)

$$\langle v'|q^2|v\rangle = \frac{[v(v-1)]^{1/2}}{2\alpha}\delta_{v',v-2} + \frac{2v+1}{2\alpha}\delta_{v',v} + \frac{[(v+1)(v+2)]^{1/2}}{2\alpha}\delta_{v',v+2} \tag{4.49}$$

Setting $v' = v$, we have for the first integral in (4.44)

$$\langle v|q^2|v\rangle = (v + \tfrac{1}{2})/\alpha \tag{4.50}$$

To find $\langle v|q^4|v\rangle$, we square the $\mathbf{q}^2$ matrix and take the diagonal elements of $\mathbf{q}^4$. The result is (Problem 4.4)

$$\langle v|q^4|v\rangle = (6v^2 + 6v + 3)/4\alpha^2 \tag{4.51}$$

Combining (4.44), (4.50), (4.51), (4.37), (4.31), and (4.26), we have for the first-order energy correction

$$E^{(1)} = \left(\frac{6hB_e^2}{\nu_e}\right) J(J+1)(v+\tfrac{1}{2}) + \frac{B_e^2 R_e^4 U^{iv}(R_e)}{8\nu_e^2}(2v^2 + 2v + 1) \tag{4.52}$$

The energy correction (4.52) takes account of the effect of the terms $(bq^2 + dq^4)$ in the perturbation (4.36). However, since the first-order correction due to the terms $(aq + cq^3)$ vanished, we must go to the second-order energy correction to take these terms into account. The contribution of $(aq + cq^3)$ to $E^{(2)}$ will turn out to be of the same order of magnitude as the contribution of $(bq^2 + dq^4)$ to $E^{(1)}$. We have the correct zeroth-order wave functions (4.38), and we can use (1.202) for the second-order energy correction for state n:

$$E^{(2)} = \sum_{k > n} \frac{|\langle \psi_k^{(0)}|\hat{H}'|\psi_n^{(0)}\rangle|^2}{E_n^{(0)} - E_k^{(0)}} \tag{4.53}$$

The sum in (4.53) is over all unperturbed states except those having energy $E_n^{(0)}$. In our problem, the state is defined by the quantum numbers v, J, and M, and the energy depends only on J and v; hence (4.53) becomes

$$E^{(2)} = \sum_{v', J', M'}{}' \frac{|\langle \psi_{v'J'M'}^{(0)}|\hat{H}'(q)|\psi_{vJM}^{(0)}\rangle|^2}{E_{vJM}^{(0)} - E_{v'J'M'}^{(0)}} \tag{4.54}$$

The prime on the summation sign indicates that the triple summation does not include the terms with both $v' = v$ and $J' = J$. From (4.38), the integrals occurring in (4.54) are

$$\int_{-\infty}^{\infty} S_{v'}\hat{H}'(q) S_v \, dq \int_0^{2\pi} \int_0^{\pi} [Y_{J'}^{M'}]^* Y_J^M \sin\theta_N \, d\theta_N \, d\phi_N \tag{4.55}$$

The angular integral in (4.55) equals $\delta_{J'J}\delta_{M'M}$ [Equation (1.163)]. Hence the only nonvanishing terms in (4.54) are those for which J' and M' are equal to J and M, respectively; therefore

$$E^{(2)}=\sum_{v'\neq v}\frac{\left|\int_{-\infty}^{\infty}S_{v'}\hat{H}'S_v\,dq\right|^2}{E^{(0)}_{vJM}-E^{(0)}_{v'JM}} \tag{4.56}$$

Using (4.39) for the zeroth-order energies, and including only the $(aq+cq^3)$ part of $\hat{H}'$, we have

$$E^{(2)}=\sum_{v'\neq v}\frac{[aq_{v'v}+c(q^3)_{v'v}]^2}{h\nu_e(v-v')} \tag{4.57}$$

$$q_{v'v}\equiv\int_{-\infty}^{\infty}S_{v'}qS_v\,dq, \qquad (q^3)_{v'v}\equiv\int_{-\infty}^{\infty}S_{v'}q^3S_v\,dq \tag{4.58}$$

(We dropped the absolute-value signs because all quantities involved are real.) Breaking (4.57) up into three sums, we have

$$E^{(2)}=\frac{a^2}{h\nu_e}\sum_{v'\neq v}\frac{[q_{v'v}]^2}{v-v'}+\frac{c^2}{h\nu_e}\sum_{v'\neq v}\frac{[(q^3)_{v'v}]^2}{v-v'}+\frac{2ac}{h\nu_e}\sum_{v'\neq v}\frac{q_{v'v}(q^3)_{v'v}}{v-v'} \tag{4.59}$$

Equation (4.45) gives $q_{v'v}$. To evaluate $(q^3)_{v'v}$, we multiply the $\mathbf{q}$ and $\mathbf{q}^2$ matrices [(4.45) and (4.49)]. The result is (Problem 4.5)

$$(q^3)_{v'v}=\left[\frac{(v+1)(v+2)(v+3)}{8\alpha^3}\right]^{1/2}\delta_{v',v+3}+3\left[\frac{v+1}{2\alpha}\right]^{3/2}\delta_{v',v+1}$$

$$+3\left[\frac{v}{2\alpha}\right]^{3/2}\delta_{v',v-1}+\left[\frac{v(v-1)(v-2)}{8\alpha^3}\right]^{1/2}\delta_{v',v-3} \tag{4.60}$$

Using (4.45), (4.60), and Kronecker-delta identities, we have

$$[q_{v'v}]^2=\frac{v}{2\alpha}\delta_{v',v-1}+\frac{v+1}{2\alpha}\delta_{v',v+1} \tag{4.61}$$

$$[(q^3)_{v'v}]^2=\frac{(v+1)(v+2)(v+3)}{8\alpha^3}\delta_{v',v+3}+\frac{9(v+1)^3}{8\alpha^3}\delta_{v',v+1}$$

$$+\frac{9v^3}{8\alpha^3}\delta_{v',v-1}+\frac{v(v-1)(v-2)}{8\alpha^3}\delta_{v',v-3} \tag{4.62}$$

$$q_{v'v}(q^3)_{v'v} = \frac{3v^2}{4\alpha^2}\delta_{v',v-1} + \frac{3(v+1)^2}{4\alpha^2}\delta_{v',v+1} \tag{4.63}$$

Equation (4.61) gives for the first sum in (4.59)

$$\frac{a^2}{h\nu_e}\left[\frac{v}{(+1)(2\alpha)} + \frac{v+1}{(-1)(2\alpha)}\right] = -\frac{a^2}{2\alpha h\nu_e} \tag{4.64}$$

Evaluating the remaining two sums, we find

$$E^{(2)} = -\frac{a^2}{2\alpha h\nu_e} - \frac{c^2}{8\alpha^3 h\nu_e}(30v^2+30v+11) - \frac{3ac}{2\alpha^2 h\nu_e}(2v+1) \tag{4.65}$$

Substituting in a, c, and α from (4.37) and (4.26), we have

$$E^{(2)} = -\left(\frac{4hB_e^3}{\nu_e^2}\right)J^2(J+1)^2 - \frac{5B_e^3R_e^6[U'''(R_e)]^2}{6h\nu_e^4}(v^2+v+\tfrac{11}{30})$$

$$+\frac{4B_e^3R_e^3U'''(R_e)}{\nu_e^3}J(J+1)(v+\tfrac{1}{2}) \tag{4.66}$$

We now add (4.39), (4.52), and (4.66) to obtain the following improved expression for the total energy (excluding translational energy) of a diatomic molecule:

$$E \approx U(R_e) + h\nu_e(v+\tfrac{1}{2}) + hB_eJ(J+1) - h\nu_e x_e(v+\tfrac{1}{2})^2$$

$$-h\alpha_e(v+\tfrac{1}{2})J(J+1) - h\bar{D}_eJ^2(J+1)^2 + hY_{00} \tag{4.67}$$

where

$$B_e \equiv \frac{h}{8\pi^2 I_e}, \qquad I_e \equiv \mu R_e^2, \qquad \nu_e \equiv \frac{1}{2\pi}\left[\frac{U''(R_e)}{\mu}\right]^{1/2} \tag{4.68}$$

$$\nu_e x_e \equiv \frac{B_e^2 R_e^4}{4h\nu_e^2}\left[\frac{10B_eR_e^2[U'''(R_e)]^2}{3h\nu_e^2} - U^{iv}(R_e)\right] \tag{4.69}$$

$$\alpha_e \equiv -\frac{2B_e^2}{\nu_e}\left[\frac{2B_eR_e^3U'''(R_e)}{h\nu_e^2} + 3\right] \tag{4.70}$$

$$\overline{D}_e \equiv \frac{4B_e^3}{\nu_e^2} \tag{4.71}$$

$$Y_{00} \equiv \frac{B_e^2 R_e^4}{16h\nu_e^2}\left[U^{iv}(R_e) - \frac{14R_e^2 B_e [U'''(R_e)]^2}{9h\nu_e^2} \right] \tag{4.72}$$

Consider the physical significance of the additional terms in (4.67) as compared to (4.39). The fourth term on the right side of (4.67) represents a shift in the vibrational levels. The constant $\nu_e x_e$ involves the third and fourth derivatives of U evaluated at R_e, and is therefore a consequence of the deviation of the potential energy function from the (quadratic) harmonic-oscillator potential:

$$U(R) - U(R_e) \approx \tfrac{1}{2}k_e(R-R_e)^2 \tag{4.73}$$

Hence we call $\nu_e x_e$ an *anharmonicity constant.* For almost all molecules, $\nu_e x_e$ is positive (the main exceptions are the alkali hydrides, all of whose first excited electronic states have a negative $\nu_e x_e$, and also a negative α_e) so that anharmonicity lowers all the vibrational levels. From (4.67), this lowering increases rapidly with increasing v, thereby decreasing the spacing between vibrational levels with increasing v. For example, Fig. 4.4 shows the 15 vibrational levels ($J=0$) of the H_2 ground electronic state. The higher vibrational levels deviate greatly from harmonic-oscillator levels. Above $v=14$, the vibrational energy exceeds the dissociation energy D_e, and (as in the hydrogen atom) we have a continuum of levels. [For a molecule that dissociates into ions, the potential energy U at large distances is given by the Coulomb's law attraction between the ions:

$$U(R) \sim U(\infty) - c^2e^2/R, \qquad R \text{ large} \tag{4.74}$$

where c is the charge on the ions. This $-1/R$ form of the potential energy is like the hydrogen-atom potential energy, where we get an infinite number of discrete energy levels below the continuum. Similarly, for a molecule that dissociates into ions, the behavior (4.74) gives an infinite number of discrete vibrational levels below the dissociation limit. For a molecule that dissociates into neutral atoms, the potential energy at large distances consists of the van der Waals attraction, which is usually proportional to the inverse sixth power of R (*Kauzmann*, Chapter 13). For this potential, it turns out[2] that there are a finite number of discrete levels

[2]*Landau and Lifshitz*, Section 18.

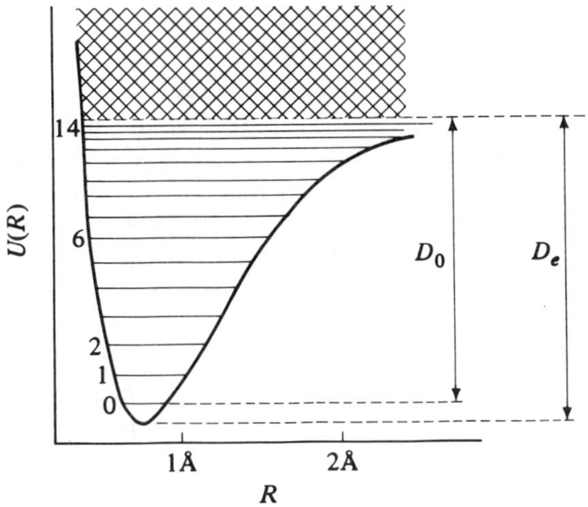

Fig. 4.4 The vibrational levels of the H_2 ground electronic state. The levels are drawn between the classical limits of vibration, but there is a small probability for vibration to extend into the classically forbidden region.

below the continuum. Since almost all neutral molecules dissociate into neutral atoms, we usually have a finite number of vibrational levels below the dissociation limit.]

The fifth term in (4.67) represents an interaction between vibration and rotation, and α_e is called a *vibration–rotation coupling constant*. [Do not confuse α_e with α in (4.26).] As the vibrational quantum number increases, the average internuclear distance increases, because of the anharmonicity of the potential-energy curve (Fig. 4.4). This increases the effective moment of inertia, and therefore decreases the rotational energy. We can define a mean rotational constant B_v for states with vibrational quantum number v by

$$B_v = B_e - \alpha_e(v + \tfrac{1}{2}) \tag{4.75}$$

Interestingly, even if the potential had no anharmonicity, α_e would not vanish; setting $U'''(R_e)$ equal to zero in (4.70), we find that this gives α_e the value $-6B_e^2/\nu_e$. This is because it is the average value of $1/R^2$ that enters into the mean rotational constant B_v. Averaging $1/(R_e+\epsilon)^2$ with $1/(R_e-\epsilon)^2$, where ϵ is a small quantity, gives a number greater than $1/R_e^2$, so that B_v is increased by this effect. Generally $U'''(R_e)$ is sufficiently negative to outweigh the $-6B_e^2/\nu_e$ term, and α_e is positive for nearly all molecules. Because of the vibration–rotation interaction, writing the energy of nuclear motion as the sum of vibrational and rotational energies is only an approximation.

The sixth term in (4.67) represents the effects of centrifugal distortion. As the rotational quantum number increases, the mean value of R increases, since the molecule is not completely rigid; this increases the effective moment of inertia, and decreases the rotational energy, as compared to that of a rigid rotor. The *centrifugal-distortion* constant $\bar{D}_e$ can be a source of confusion. $\bar{D}_e$ has nothing to do with the dissociation energy D_e. Unfortunately, most papers in the literature use the same symbol D_e for both the centrifugal-distortion constant and the dissociation energy. We have arbitrarily added the bar to avoid confusion. Many early workers defined the centrifugal-distortion constant as the negative of (4.71). From (4.71), $\bar{D}_e$ is always positive. The ratio of $\bar{D}_e$ to B_e is $4B_e^2/\nu_e^2$. Since rotational energies are typically 10^{-3} times as small as vibrational energies, the centrifugal distortion usually amounts to an extremely small correction to the rotational energy. For light molecules (hydrides), centrifugal distortion is greater.

The final term in (4.67) is a constant, and affects all vibrational levels equally. We see from (4.72) that Y_{00} is a result of anharmonicity in U. The molecular zero-point vibrational energy is

$$\tfrac{1}{2}h\nu_e - \tfrac{1}{4}h\nu_e x_e + hY_{00} \tag{4.76}$$

which is slightly different from the zero-point energy of a harmonic oscillator. An equivalent expression for Y_{00} is

$$Y_{00} = \frac{B_e}{4} + \frac{\alpha_e \nu_e}{12B_e} + \frac{\alpha_e^2 \nu_e^2}{144B_e^3} - \frac{\nu_e x_e}{4} \tag{4.77}$$

Often Y_{00} is omitted, either deliberately or inadvertently. Although Y_{00} is small, it is not completely negligible for light molecules and affects their zero-point energies; this must be taken into account when isotopic-exchange equilibrium constants are calculated; see M. Wolfsberg, *Advances in Chemistry Series*, No. 89, p. 185 (1969).

We have defined ν_e, B_e, $\nu_e x_e$, α_e, $\bar{D}_e$, and Y_{00} in a manner that gives them units of cycles per second. The vibrational and rotational constants are often listed in units of cm^{-1}; we shall use a tilde to indicate the corresponding constants in units of cm^{-1}:

$$\tilde{\nu}_e = \nu_e/c, \qquad \tilde{B}_e = B_e/c, \qquad \text{etc.} \tag{4.78}$$

where c is the speed of light. In the literature the tilde is not used, and B_e can mean the rotational constant in units of cm^{-1} or Hz. Instead of $\tilde{\nu}_e$, it is common practice to write ω_e. This notation is unfortunate, in that ω is usually used to represent 2π times a vibrational frequency, whereas ω_e means the vibrational frequency divided by c.

We can calculate a better approximation to the energy by taking further terms in (4.12) and (4.21), and by going to higher-order perturbation theory. This introduces further terms into the energy expression, the most important such term being

$$+\nu_e y_e(v+\tfrac{1}{2})^3 \tag{4.79}$$

which is a further anharmonicity correction to the vibrational energy levels. The anharmonicity constant $\nu_e y_e$ can be positive or negative. The term (4.79) adds $\nu_e y_e/8$ to the expression (4.76) for the zero-point energy.

The higher-order perturbations also add small corrections to the coefficients in (4.67). For example, the term $h\nu_e(v+1/2)$ becomes $h\nu_e(v+1/2)[1+(B_e^2/4\nu_e^2)A]$, where A is a complicated constant involving k_e, R_e, $U'''(R_e)$, $U^{iv}(R_e)$, $U^{v}(R_e)$, and $U^{vi}(R_e)$, and where ν_e and B_e are defined by (4.68). The coefficients of the third, fourth, fifth, and sixth terms on the right of (4.67) are also affected. Because these corrections involve B_e^2/ν_e^2, they are negligible except for light molecules (hydrides), and are usually omitted. However, they must be included in accurate work on hydrides. The accurate expression for the total energy of a diatomic molecule then has the form

$$E = U(R_e) + \sum_l \sum_m Y_{lm} h(v+\tfrac{1}{2})^l J^m (J+1)^m \tag{4.80}$$

where the coefficients Y_{lm} (no connection with the spherical harmonics) are the *Dunham coefficients*. (For their explicit forms, see *Townes and Schawlow*, p. 10.) Comparison with (4.67) and (4.79) shows that $Y_{10} \approx \nu_e$, $Y_{20} \approx -\nu_e x_e$, $Y_{01} \approx B_e$, $Y_{11} \approx -\alpha_e$, $Y_{02} \approx -\bar{D}_e$, $Y_{30} \approx \nu_e y_e$.

The perturbation-theory derivation of (4.67) and (4.80) depends on the expansions (4.12) and (4.21), which are not convergent for large R (Problem 4.1). Although (4.67) is valid for the lower levels of an electronic state, it is invalid for the higher levels, which involve large values of R.

Table 4.1 lists rotational and vibrational constants for some diatomic-molecule states. The most extensive compilation of such data is *Bourcier*.

For H_2, accurate theoretical calculations[3] of the vibration–rotation energy levels have been done by solving the radial differential equation (4.11) using numerical integration. The potential-energy function used is that found from a 100-term variational electronic wave function.

[3] W. Kolos and L. Wolniewicz, *J. Chem. Phys.*, **49**, 404 (1968).

Table 4.1 Molecular Constants for Some Diatomic Molecules[a]

Molecule	Electronic Term	R_e (Å)	k_e (mdyn/Å)	$\tilde{\nu}_e$ (cm^{-1})	D_e/hc (cm^{-1})	$\tilde{B}_e$ (cm^{-1})	$\tilde{\alpha}_e$ (cm^{-1})	$\tilde{\nu}_e x_e$ (cm^{-1})	$\tilde{D}_e$ (cm^{-1})
H_2	$X\,^1\Sigma_g^+$	0.7412	5.756	4403.2	38,297	60.85	3.06	121.3	4.7×10^{-2}
HF	$X\,^1\Sigma^+$	0.9168	9.659	4138.7	49,314	20.956	0.796	90.0	2.2×10^{-3}
N_2	$X\,^1\Sigma_g^+$	1.0977	22.94	2358.0	79,868	1.998	0.0177	14.1	5.7×10^{-6}
CO	$X\,^1\Sigma^+$	1.1283	19.018	2169.8	90,542	1.93127	0.01751	13.294	6.20×10^{-6}
CO	$d\,^3\Delta$	1.3700	5.366	1152.6	29,522	1.3099	0.0168	7.281	5.8×10^{-6}
O_2	$X\,^3\Sigma_g^-$	1.2074	11.766	1580.2	42,046	1.4456	0.0158	12.0	4.8×10^{-6}
I_2	$X\,^1\Sigma_g^+$	2.667	1.720	214.52	12,560	0.03739	0.00012	0.61	4.5×10^{-9}

[a] Data mainly from *Bourcier*. The isotopic species are ^{1}H, ^{19}F, ^{14}N, ^{12}C, ^{16}O, ^{127}I.

4.3 POTENTIAL-ENERGY FUNCTIONS FOR DIATOMIC MOLECULES

We have expanded the potential energy $U(R)$ in a series about R_e, taking the first few terms; this approximation is good in the region near R_e. However, for large values of R, the Taylor series (4.12) is divergent. To represent the potential energy over the whole range of R, many empirical potential-energy functions have been proposed. These functions contain parameters (usually three) that are evaluated from experimental quantities for the molecular electronic state of interest. Since the detailed shape of the potential-energy curve is peculiar to each molecule, we cannot hope to construct a "universal" potential-energy function which will give a good representation of $U(R)$ for all bound electronic states of all molecules.

A bound-state potential-energy function should go to infinity as R goes to zero; it should pass through a minimum at some value of R; it should approach a finite limiting value as R goes to infinity. To evaluate the parameters in a three-parameter function, we use the relations

$$U(\infty)-U(R_e)=D_e \tag{4.81}$$

$$U'(R_e)=0 \tag{4.82}$$

$$U''(R_e)=k_e \tag{4.83}$$

where R_e, D_e, and k_e are the equilibrium internuclear distance, dissociation energy, and force constant, respectively [Equation (4.16)].

The most famous three-parameter empirical potential-energy function is the Morse function:

$$U = D_e[1 - e^{-a(R - R_e)}]^2 \tag{4.84}$$

The Morse function approaches a very large, finite value as R goes to zero, rather than becoming infinite. Since the behavior of U near $R=0$ is unimportant, this "defect" is of no significance. From (4.84), we have

$$U(\infty) = D_e, \qquad U(R_e) = 0 \tag{4.85}$$

so that (4.81) is satisfied. (For convenience, the zero of energy has been taken at the bottom of the potential-energy curve.) Differentiating (4.84), we verify that $U'(R_e)=0$. To evaluate the parameter a, we use (4.83); we find

$$U''(R_e) = 2a^2 D_e \tag{4.86}$$

$$a = \left(\frac{k_e}{2D_e}\right)^{1/2} = \frac{\Delta^{1/2}}{R_e} \tag{4.87}$$

where the Sutherland parameter Δ is defined as

$$\Delta \equiv \frac{k_e R_e^2}{2D_e} = \frac{h\nu_e^2}{4B_e D_e} \tag{4.88}$$

To test the accuracy of a three-parameter potential-energy function, we want to compare the predicted values of U with the experimental values for a variety of molecular electronic states. How can we construct an experimental $U(R)$ curve? This can be done using the experimentally observed vibrational energy levels, by a numerical method developed by Rydberg, Klein, and Rees (the RKR method). (The RKR method is the reverse of what one usually does in quantum mechanics; we usually start with some potential-energy function U and use it to calculate the allowed energy levels from the Schrödinger equation.) The RKR method is limited to molecular states for which sufficient vibrational levels have been spectroscopically determined, and the number of RKR calculations performed is not large. Steele *et al.* have compared the points on the RKR potential curves of 19 diatomic electronic states with the predictions of several potential-energy functions.[4] Of the eight three-parameter functions tested, the Morse function was the third poorest; the average value of

[4] D. Steele, E. R. Lippincott, and J. T. Vanderslice, *Rev. Mod. Phys.*, **34**, 239 (1962).

$|U_{\text{Morse}} - U_{\text{RKR}}|/D_e$ being 3.68 percent. Thus the Morse function generally gives a rather poor fit to $U(R)$, but because of its simplicity and early invention, it has been widely used.

A considerably more accurate empirical potential-energy function is the Varshni function:

$$U = D_e\left\{1 - (R_e/R)\exp\left[-\beta(R^2 - R_e^2)\right]\right\}^2 \tag{4.89}$$

for which $|U - U_{\text{RKR}}|/D_e$ averages 2.31 percent. A modification of (4.89) that has an average error of 1.99 percent for the above-mentioned 19 states is the following three-parameter function[5]:

$$U = D_e\left\{1 - (R_e/R)\exp\left[-b(R^p - R_e^p)\right]\right\}^2$$

$$p \equiv 2 + \frac{(\Delta^{1/2} - 4)(\Delta^{1/2} - 2)}{4(\Delta^{1/2} - 1)}$$

where Δ is defined by (4.88). Since p is a function of k_e, R_e, and D_e, this function has three parameters.

4.4 SELECTION RULES FOR ROTATIONAL AND VIBRATIONAL TRANSITIONS

To investigate the spectra of diatomic molecules, we need the selection rules for radiative transitions. We now investigate the electric-dipole selection rules for transitions between vibration–rotation levels belonging to the same $^1\Sigma$ electronic state. (Transitions in which the electronic state changes will be considered in Chapter 7.)

To deduce whether a transition is allowed between two stationary states, we investigate the matrix element of the electric dipole-moment operator between those states (Section 3.2). We will use the Born–Oppenheimer approximation of writing the stationary-state molecular wave functions as products of electronic and nuclear wave functions:

$$\psi = \psi_{\text{el}}\psi_N \tag{4.90}$$

For a transition between states ψ and ψ' in which the electronic state doesn't change, we are interested in the integral

$$\langle\psi|\hat{\mathbf{d}}|\psi'\rangle = \int\int \psi_{\text{el}}^*\psi_N^*\hat{\mathbf{d}}\psi_{\text{el}}\psi_N'\,d\tau_{\text{el}}\,d\tau_N \tag{4.91}$$

The dipole-moment operator is given by (1.287).

[5] I. N. Levine, *J. Chem. Phys.*, **45**, 827 (1966).

We can write (4.91) as

$$\int \psi_N^* \psi_N' \left[\int \psi_{\text{el}}^* \hat{\mathbf{d}} \psi_{\text{el}} \, d\tau_{\text{el}} \right] d\tau_N \tag{4.92}$$

The quantity in brackets is the permanent electric dipole moment $\mathbf{d}$ corresponding to the state ψ_{el} [see (1.289)]; of course, since both ψ_{el} and the operator $\hat{\mathbf{d}}$ depend on the nuclear configuration, the dipole-moment vector $\mathbf{d}$ depends on the nuclear configuration. Symmetry requires that $\mathbf{d}$ be oriented along the internuclear axis in Fig. 4.1; therefore,

$$\mathbf{d} = \mathbf{i} d_X + \mathbf{j} d_Y + \mathbf{k} d_Z \tag{4.93}$$

$$\mathbf{d} = \mathbf{i} d \sin\theta_N \cos\phi_N + \mathbf{j} d \sin\theta_N \sin\phi_N + \mathbf{k} d \cos\theta_N \tag{4.94}$$

where the magnitude of $\mathbf{d}$ depends on the internuclear separation R:

$$d = d(R) \tag{4.95}$$

The integral (4.92) becomes

$$\int_0^{2\pi} \int_0^{\pi} \int_0^{\infty} \psi_N^* \psi_N' \, \mathbf{d} R^2 \sin\theta_N \, dR \, d\theta_N \, d\phi_N \tag{4.96}$$

For the nuclear wave functions, we use the approximate expression (4.28). We have for (4.96)

$$\int_0^{\infty} S_v(q) S_{v'}(q) d(R) \, dR \int_0^{2\pi} \int_0^{\pi} Y_{J'}^{M'} [Y_J^M]^* (\mathbf{i} \sin\theta_N \cos\phi_N$$

$$+ \mathbf{j} \sin\theta_N \sin\phi_N + \mathbf{k} \cos\theta_N) \sin\theta_N \, d\theta_N \, d\phi_N \tag{4.97}$$

We have three angular integrals, the third of which is

$$\int_0^{2\pi} \int_0^{\pi} [Y_J^M]^* Y_{J'}^{M'} \cos\theta_N \sin\theta_N \, d\theta_N \, d\phi_N \tag{4.98}$$

The integral (4.98) has the identical form as the angular integral in the integral (3.66). Hence from (3.72) and (3.68), we conclude that (4.98) vanishes unless $M' = M$ and $J' = J \pm 1$. Similarly, we see that the other two angular integrals in (4.97) have the same forms as the angular integrals that arise for a hydrogen atom exposed to x- and y-polarized light [compare (4.94) and (3.63)]. Thus from (3.76) and (3.77), the selection rules for changes in the rotational quantum numbers of a diatomic molecule for an electric-dipole transition with no change in electronic state are

$$\Delta J = \pm 1, \qquad \Delta M = 0, \pm 1 \tag{4.99}$$

(These selection rules are applicable only to $^1\Sigma$ electronic states.)

Now consider the radial integral in (4.97). For a homonuclear diatomic molecule, the electric dipole moment is zero for all internuclear separations, $d(R)=0$, so that the radial integral vanishes. Hence for homonuclear diatomic molecules, electric-dipole transitions with no change in electronic state cannot occur. For heteronuclear diatomic molecules, we must consider the function $d(R)$. For $R=0$, the nuclei have come together to form the corresponding united atom; since atoms do not have electric dipole moments, we have $d(0)=0$. For $R=\infty$, we have the separated atoms; since almost all neutral diatomic molecules dissociate into neutral atoms, rather than into ions, we have $d(\infty)=0$. Not too much is known about the detailed behavior of $d(R)$ for R between zero and infinity, but a plausible curve might have $d(R)$ rise from 0 at $R=0$ to a maximum at an internuclear distance greater than R_e and then fall back to 0 as R goes to infinity. As is usual when we don't know very much about a function, we expand it in a Taylor series about R_e:

$$d(R) = d(R_e) + d'(R_e)(R-R_e) + \tfrac{1}{2}d''(R_e)(R-R_e)^2 + \cdots \tag{4.100}$$

Substituting this expansion into the radial integral and changing to the variable $q = R - R_e$, we have for the radial integral

$$d(R_e)\int_{-\infty}^{\infty} S_{v'}(q)S_v(q)\,dq + d'(R_e)\int_{-\infty}^{\infty} S_{v'}(q)qS_v(q)\,dq + \frac{d''(R_e)}{2}\int_{-\infty}^{\infty} S_{v'}(q)q^2S_v(q)\,dq + \cdots \tag{4.101}$$

where we have made the usual approximation of extending the lower integration limit to minus infinity. The functions $S_v(q)$ are harmonic-oscillator wave functions. Because of orthogonality, the first integral in (4.101) is nonzero only when $v'=v$. The second integral is given by (4.45), and is nonzero only when $v'=v\pm1$. From (4.49), the third integral is nonzero only for $v'=v\pm2$ or $v'=v$. The next integral is given by (4.60), and is nonzero for $v'=v\pm1$ and $v'=v\pm3$. And so on. The little evidence available indicates that the contributions to (4.100) from $d''(R_e)$ and higher derivatives at R_e are usually small compared to that from $d'(R_e)$. For example, theoretical analysis of observed line intensities in the HBr vibration–rotation spectrum gave $R_e d'(R_e) = 0.64$ D as compared to

$\frac{1}{2}R_e^2 d''(R_e) = 0.17$ D. [R. H. Tipping and R. M. Herman, *J. Mol. Spectrosc.*, **36**, 404 (1970). For HI, however, $R_e d'(R_e) = 0.08$ D and $\frac{1}{2}R_e^2 d''(R_e) = 0.50$ D; here it seems that the maximum value of d occurs for R considerably smaller than R_e; see R. H. Tipping and A. Forbes, *ibid.*, **39**, 65 (1971).] If we neglect the second and higher derivatives, then the vibrational quantum number can change by 0 or ± 1; including these derivatives, we have the possibility of v changing by ± 2, ± 3, etc. We write

$$\Delta v = 0, \pm 1 \ (\pm 2, \pm 3, \pm \cdots) \tag{4.102}$$

where the changes in parentheses are less probable.

A transition with no change in electronic state, and with $\Delta v = 0$ gives a line in the pure rotation spectrum of the molecule. Transitions with $\Delta v = \pm 1, \pm 2, \ldots$ give the vibration–rotation spectrum. These two types of transitions will be discussed separately in the next two sections.

It should be noted that (4.28) is only an approximation for the nuclear wave function. The perturbation terms (4.36) will mix into the nuclear wave function small contributions from harmonic-oscillator functions with quantum numbers other than v. These anharmonicity corrections to the vibrational wave function will add further to the probability of transitions with $|\Delta v| > 1$.

The vibration–rotation energy expression (4.67) contains the rotational constant B_e, which depends on the equilibrium internuclear separation R_e; hence investigation of vibrational and rotational spectra of diatomic molecules allows their structure to be determined.

4.5 ROTATIONAL SPECTRA OF DIATOMIC MOLECULES

We now consider the pure-rotation spectra of gaseous diatomic molecules. As noted in the previous section, a diatomic molecule with an electric dipole moment can undergo a pure-rotational radiative transition with $\Delta J = +1$ or -1, corresponding to absorption or emission. Rotational spectra are studied as absorption spectra. For a transition with the initial and final rotational quantum numbers J and $J+1$, respectively, and with no change in electronic or vibrational levels, we find from (4.67)

$$\nu(J \to J+1) = (E_{J+1} - E_J)/h$$

$$= 2(J+1)[B_e - \alpha_e(v + \tfrac{1}{2})] - 4\bar{D}_e(J+1)^3 \tag{4.103}$$

$$\nu(J \to J+1) = 2(J+1)B_v - 4\bar{D}_e(J+1)^3 \tag{4.104}$$

We see from (4.104) that, although the vibrational quantum number is not changing, the frequency of a pure-rotational transition depends on the vibrational quantum number of the molecule undergoing the transition. (Recall that vibration changes the effective moment of inertia, and thus affects the rotational energies.) For a collection of diatomic molecules at temperature T, the relative populations of the energy levels are given by the Boltzmann distribution law; the ratio of the number of molecules with vibrational quantum number v to the number with vibrational quantum number zero is

$$\frac{n(v)}{n(0)}=\frac{e^{-(v+1/2)h\nu_e/kT}}{e^{-(h\nu_e/2)/kT}}=e^{-vh\nu_e/kT} \tag{4.105}$$

where we have neglected anharmonicity corrections. At room temperature, kT/hc is about 200 cm^{-1}. For most diatomic molecules, $\tilde{\nu}_e$ is of the order of 2000 cm^{-1} (Table 4.1), and

$$\frac{n(v=1)}{n(v=0)}=e^{-10}=5\times10^{-5}$$

which is quite small. However, for heavy diatomic molecules, the vibrational frequency may be sufficiently small to have a considerable population in excited vibrational levels. (Recall that ν_e is proportional to $\mu^{-1/2}$.) Thus for ICl^{35}, the vibrational constant $\tilde{\nu}_e$ is about 400 cm^{-1}, and at room temperature: $n(v=1)/n(v=0)=e^{-2}=0.14$.

Considering first the case where the population of excited vibrational levels is negligible, we have for the rotational frequencies

$$\nu(J\to J+1, v=0)=2(J+1)B_0-4\overline{D}_e(J+1)^3, \qquad J=0,1,2,\ldots \tag{4.106}$$

$$B_0=B_e-\tfrac{1}{2}\alpha_e \tag{4.107}$$

If centrifugal distortion were neglected, we would have a series of equally spaced lines at frequencies $2B_0, 4B_0, 6B_0, \ldots$. The rotational constant B is proportional to $1/\mu$. For light diatomic molecules, we see from Table 4.1 that $\tilde{B}$ is roughly 20 cm^{-1}. The "boundary" between the infrared and microwave regions of the spectrum is at, say, 10 cm^{-1}; hence the pure-rotation spectra of light diatomic molecules lie in the far infrared region. For heavier diatomic molecules, $\tilde{B}$ runs about 1 cm^{-1}, and the rotational spectrum lies in the microwave region.

The centrifugal distortion constant $\overline{D}_e$ is small compared to B_0, and causes small shifts from the hypothetical frequencies $2B_0, 4B_0, \ldots$. Figure 4.5 gives the pure-rotation spectrum of diatomic molecules having $v=0$.

Note the increasing effect of centrifugal distortion as J increases.

Now consider the case with appreciable population of excited vibrational levels. The set of molecules with vibrational quantum number v will have its own value of B_v and will give rise to its own pure-rotation spectrum. Thus each line in Fig. 4.5 will have a series of vibrational satellites. B_v is given by (4.75), where α_e is small compared to B_e, so that the vibrational satellites lie near the main line. These satellites are shown for the $J=0\rightarrow1$ transition in Fig. 4.6. Note the rapid decrease in intensity as v increases, due to the decreasing population of the vibrational levels.

Different isotopic species of a molecule have different reduced masses and different rotational constants. Hence each such species has its own pure-rotation spectrum, the intensity of the spectrum being determined by the relative isotopic abundance.

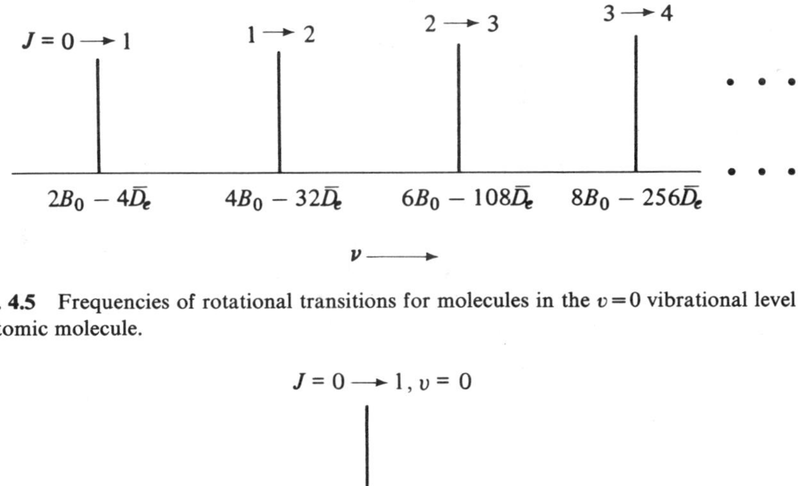

Fig. 4.5 Frequencies of rotational transitions for molecules in the $v=0$ vibrational level of a diatomic molecule.

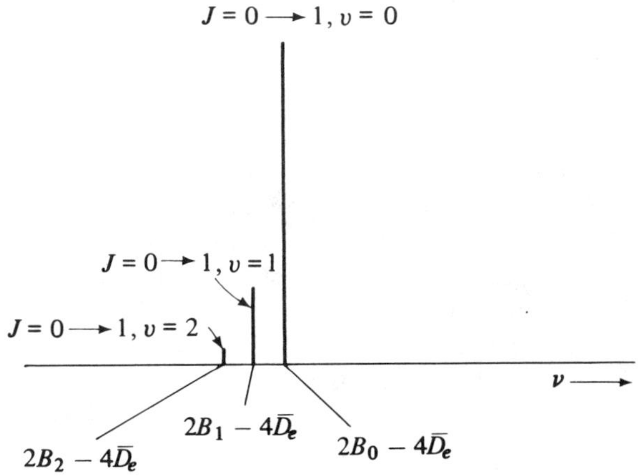

Fig. 4.6 The $J=0\rightarrow1$ transition of a diatomic molecule with appreciable population of excited vibrational levels.

Table 4.2 lists some of the pure-rotation transitions of CO.

Table 4.2 Rotational Transitions[a] of CO

Molecule	$J \rightarrow J'$	Frequency (MHz)
$C^{12}O^{16}$	$0 \rightarrow 1$	115,271.20
	$1 \rightarrow 2$	230,537.97
	$2 \rightarrow 3$	345,795.9
	$3 \rightarrow 4$	461,040.7
	$4 \rightarrow 5$	576,267.8
$C^{13}O^{16}$	$0 \rightarrow 1$	110,201.37
$C^{12}O^{18}$	$0 \rightarrow 1$	109,782.18

[a] P. H. Krupenie, "The Band Spectrum of Carbon Monoxide," National Bureau of Standards publication NSRDS-NBS 5 (1966). All transitions are for $v=0$.

4.6 VIBRATION–ROTATION SPECTRA OF DIATOMIC MOLECULES

We now consider radiative transitions for which both v and J change, but the electronic state does not; these transitions give the vibration–rotation spectra of diatomic molecules. The selection rules for these transitions were found in Section 4.4 to be ($^1\Sigma$ states only)

$$\Delta v = \pm 1 \ (\pm 2, \pm 3, \ldots), \qquad \Delta J = \pm 1 \tag{4.108}$$

Vibrational energies are much greater than rotational energies. Hence the vibration–rotation spectrum consists of a series of bands; the lines of a given band all correspond to transitions with the same initial and the same final vibrational quantum numbers, but different initial and/or final rotational quantum numbers. To get an overall view of the spectrum, we will first neglect the rotational structure of the bands, and will calculate the frequencies of hypothetical lines for which the J value of both the initial and final vibration–rotation levels is zero. This gives the position of the *band origin*. Since $\Delta J = 0$ is forbidden, the actual vibration–rotation band will not have a line at the band origin, but will have lines on either side, corresponding to $\Delta J = \pm 1$. (See Fig. 4.9) Vibration–rotation spectra are usually observed as absorption rather than emission spectra, so that we will consider only positive changes in v.

We will calculate the wave number σ, rather than the frequency ν, of each transition. We have

$$\sigma \equiv \frac{1}{\lambda} = \frac{\nu}{c} = \frac{\Delta E}{hc} \tag{4.109}$$

Setting $J=0$ in (4.67), we calculate the wave number of the band origin σ_0 for a transition from vibrational level v to level v', as

$$\sigma_0(v \to v') = \tilde{\nu}_e(v'-v) - \tilde{\nu}_e x_e[v'(v'+1) - v(v+1)] \tag{4.110}$$

As previously mentioned, for most diatomic molecules at room temperature, the population of excited vibrational levels is negligible. We therefore first consider transitions for which the initial vibrational level is $v=0$. The selection rules (4.108) allow the transitions $v=0 \to 1, 0 \to 2, 0 \to 3, \ldots$, where the higher transitions are of decreasing intensity. The $0 \to 1$ transition is the *fundamental band*; the $0 \to 2$ transition is the *first overtone*, and so on. The wave numbers of the band origins for these bands are

$$\sigma_0(0 \to v') = \tilde{\nu}_e v' - \tilde{\nu}_e x_e v'(v'+1) \tag{4.111}$$

The anharmonicity constant $\tilde{\nu}_e x_e$ is small compared to $\tilde{\nu}_e$, but its effect increases as v' increases, and the overtones deviate more and more from simple multiples of the fundamental frequency with increasing v'. See Fig. 4.7. The infrared region extends from 10 to 14,000 cm^{-1} (7000 Å). Molecular vibrational frequencies run from 100 to 4000 cm^{-1}, so that the fundamental and lower overtones lie in the infrared region.

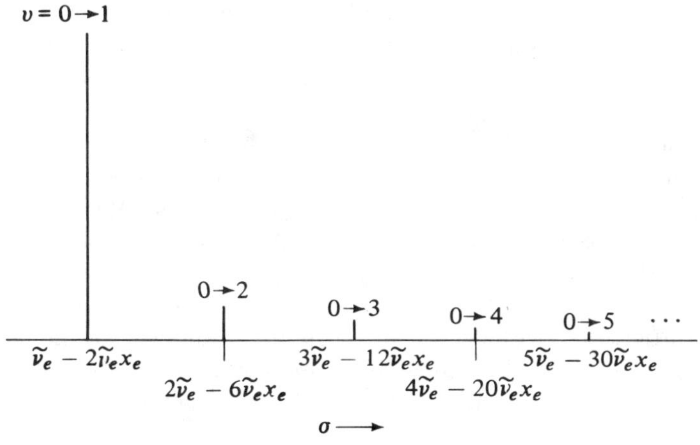

Fig. 4.7 Positions of the band origins of the infrared absorption spectrum of a diatomic molecule.

We can test (4.111) as follows. We define the *first differences* $\Delta\sigma_0(v')$ by

$$\Delta\sigma_0(v') \equiv \sigma_0(0\rightarrow v'+1) - \sigma_0(0\rightarrow v') \tag{4.112}$$

From (4.111) we find

$$\Delta\sigma_0(v') = \tilde{\nu}_e - 2\tilde{\nu}_e x_e(v'+1) \tag{4.113}$$

We define the *second differences* $\Delta^2\sigma_0(v')$ by

$$\Delta^2\sigma_0(v') \equiv \Delta\sigma_0(v'+1) - \Delta\sigma_0(v') \tag{4.114}$$

From (4.113) we find

$$\Delta^2\sigma_0(v') = -2\tilde{\nu}_e x_e \tag{4.115}$$

which says that the second differences of the band-origin wave numbers should all be equal to the constant $-2\tilde{\nu}_e x_e$.

The entity Δ is the *difference operator*,[6] and is defined by

$$\Delta f(x) = f(x+1) - f(x) \tag{4.116}$$

The square of the difference operator has the following effect:

$$\Delta^2 f(x) = \Delta[\Delta f(x)] = \Delta[f(x+1) - f(x)] = \Delta f(x+1) - \Delta f(x) \tag{4.117}$$

where we have used the linearity of Δ. It is easily shown (Problem 4.18) that differencing a polynomial once reduces its degree by one. (Differencing is similar to differentiation in this respect.) Thus if $f(x)$ is a polynomial of degree n, then $\Delta^n f(x)$ is a constant. For example, $\sigma_0(v')$ is a polynomial of degree 2 [Eq. (4.111)] and $\Delta^2\sigma_0(v')$ is a constant. The rotational energy of a diatomic molecule is a quartic function of J (in the approximation we are using), and the difference between adjacent rotational levels is a cubic function of J [Eq. (4.104)].

The values of $\sigma_0(0\rightarrow v')$ for HCl^{35} are listed in Table 4.3, which also gives the first and second differences. We see that $\Delta^2\sigma_0$ is nearly constant, and that $\tilde{\nu}_e x_e$ is about $51\frac{1}{2}$ cm^{-1}. The small variation in the $\Delta^2\sigma_0$ values can be attributed to the cubic term (4.79) in the vibrational energy, and to experimental error, perhaps.

If the diatomic molecule has a low vibrational frequency, or if the gas is heated sufficiently, we will have substantial numbers of molecules in excited vibrational levels, and we will observe infrared absorption bands for which the initial vibrational level is not $v=0$. Such bands are called *hot bands*. See Fig. 4.8.

[6] *Sokolnikoff and Redheffer*, pp. 663–667.

Table 4.3 Band Origins of HCl^{35} Infrared Bands[a]

Transition	σ_0 (cm^{-1})	$\Delta\sigma_0$ (cm^{-1})	$\Delta^2\sigma_0$ (cm^{-1})
$v=0\rightarrow0$	0		
		2885.98	
$v=0\rightarrow1$	2885.98		-103.98
		2782.00	
$v=0\rightarrow2$	5667.98		-103.20
		2678.80	
$v=0\rightarrow3$	8346.78		-102.77
		2576.03	
$v=0\rightarrow4$	10,922.81		-102.65
		2473.38	
$v=0\rightarrow5$	13,396.19		

[a] D. H. Rank, D. P. Eastman, B. S. Rao, and T. A. Wiggins, *J. Opt. Soc. Am.*, **52**, 1 (1962).

Now consider the rotational structure of vibration–rotation absorption bands. Since the rotational energy is small compared to the vibrational energy, we can have $\Delta J = -1$, as well as $\Delta J = +1$, in an infrared absorption transition. Vibration–rotation absorption transitions with $\Delta J = +1$ form the *R branch* of a vibration–rotation band, while $\Delta J = -1$ vibration–rotation transitions form the *P branch* of the band.

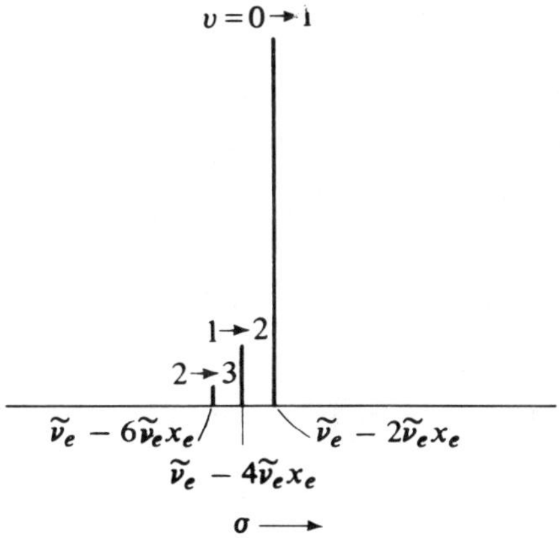

Fig. 4.8 Band origins of infrared hot bands of a diatomic molecule.

For the wave numbers σ_R of R-branch lines, we find from (4.67), after some manipulation,

$$\sigma_R(v\to v',J\to J+1)=\sigma_0(v\to v')+\left[2\tilde{B}_e-\tilde{\alpha}_e(v'+v+1)\right](J+1)$$
$$-\tilde{\alpha}_e(v'-v)(J+1)^2-4\tilde{\bar{D}}_e(J+1)^3$$
$$J=0,1,2,\ldots \qquad (4.118)$$

where σ_0 is given by (4.110). If we neglect $\tilde{\alpha}_e$ and $\tilde{\bar{D}}_e$ in comparison to $\tilde{B}_e$, we get a series of equally spaced lines on the high-frequency side of σ_0 (just as we got equally spaced lines in the pure-rotation spectrum if we neglected $\bar{D}_e$). Because of the terms quadratic and cubic in $(J+1)$ in (4.118), the spacing between successive R-branch lines slowly decreases as we go away from the band origin. In fact, if we go to a high enough value of J, the quadratic and cubic terms will overpower the linear term, and we will observe a *reversal* of the R-branch lines, with the frequencies of the lines decreasing as J increases; this reversal is said to form a *band head* at the R-branch line of maximum frequency. In actual practice, because of the low intensities of R-branch lines for high J, band-head formation is usually not observed in infrared absorption spectra.

For the P-branch lines, we find similarly

$$\sigma_P(v\to v',J\to J-1)=\sigma_0(v\to v')-\left[2\tilde{B}_e-\tilde{\alpha}_e(v'+v+1)\right]J$$
$$-\tilde{\alpha}_e(v'-v)J^2+4\tilde{\bar{D}}_eJ^3 \qquad J=1,2,3,\ldots \qquad (4.119)$$

Neglecting $\tilde{\alpha}_e$ and $\tilde{\bar{D}}_e$, we would have a series of equally spaced lines on the low-frequency side of σ_0. The effect of the term quadratic in J is to increase the spacing between P-branch lines as J increases, while the term cubic in J tends to decrease the spacing as J increases. Since $\tilde{\alpha}_e$ is much larger than $\bar{D}_e$ (Table 4.1), the spacing between P-branch lines increases as we go away from the band origin.

We can combine (4.118) and (4.119) by introducing the number m, which takes on all nonzero integral values. We have for the wave numbers of the lines of an infrared band of a diatomic molecule

$$\sigma(v\to v',J\to J\pm 1)=\sigma_0(v\to v')+\left[2\tilde{B}_e-\tilde{\alpha}_e(v'+v+1)\right]m$$
$$-\tilde{\alpha}_e(v'-v)m^2-4\tilde{\bar{D}}_em^3 \quad m=\pm 1,\pm 2,\pm 3,\ldots \qquad (4.120)$$

Positive values of m give the R-branch lines; negative values give the

P-branch lines. (The *running number* m is not to be confused with the magnetic quantum number.)

The appearance of a vibration–rotation band is shown in Fig. 4.9. Because $\Delta J = 0$ is forbidden, we have a gap at the wave number of the band origin σ_0. (We are considering only diatomic molecules in $^1\Sigma$ electronic states; for electronic states in which the electronic orbital angular-momentum quantum number Λ is nonzero, transitions with $\Delta J = 0$ are allowed, giving a Q *branch* in each band. An example is NO, which has a $^2\Pi$ electronic ground term. Q-branch transitions also occur for vibration–rotation bands of polyatomic molecules; see Chapter 6.) Under low resolution, an infrared band of a diatomic molecule looks like Fig. 4.10.

The pattern of intensities in Fig. 4.9 deserves mention. The intensities of absorption lines are proportional to the population of the lower level, and to the square of the dipole-moment matrix element (4.97). It turns out that for vibration–rotation transitions in the same band, the integral (4.97)

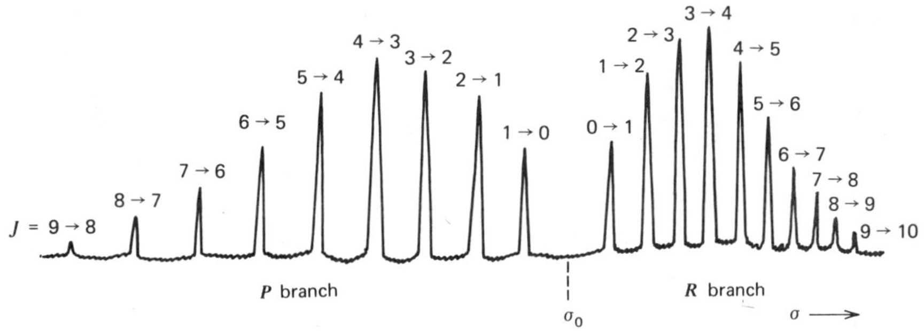

Fig. 4.9 Rotational fine structure of a vibration–rotation band of a diatomic molecule. Note the decreasing spacing with increasing J in the R branch, and the increasing spacing with increasing J in the P branch.

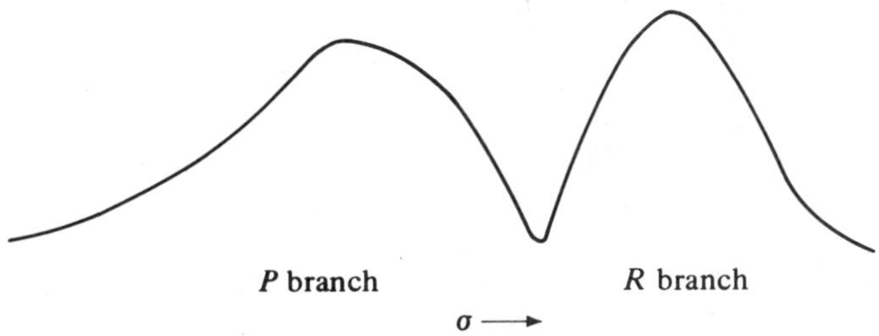

Fig. 4.10 Appearance of a vibration–rotation band of a diatomic molecule under low resolution.

varies rather slowly with J, so that the major factor in determining the relative intensities of the rotational fine-structure lines is the relative populations of the rotational levels. In using the Boltzmann distribution law to determine the populations, we must remember that there are $2J+1$ rotational states (each with a different value of M) for each rotational energy level. Hence the ratio of the population of the level with vibrational and rotational quantum numbers v and J to the population of the level with quantum numbers v and 0 is

$$\frac{N_{v,J}}{N_{v,0}} = (2J+1)e^{-B_e hJ(J+1)/kT} \tag{4.121}$$

where centrifugal distortion and vibration–rotation interaction have been neglected. The exponential factor decreases as J increases, but for low J, the $2J+1$ degeneracy factor overcomes the exponential factor, causing the population of low J levels to increase with increasing J. For high J, the exponential factor overpowers the degeneracy factor, and the population decreases with increasing J. Thus we get a maximum population for some intermediate value of J, which explains the pattern of intensities in a vibration–rotation band.

Table 4.4 lists the wave numbers of the lines of the fundamental band of CO.

Table 4.4 Wave Numbers of Lines in the $C^{12}O^{16}$ $v=0 \rightarrow 1$ Infrared Band[a]

J	$R(J)$	$P(J)$
0	2147.084 cm^{-1}	—
1	2150.858	2139.427 cm^{-1}
2	2154.599	2135.548
3	2158.301	2131.633
4	2161.971	2127.684
5	2165.602	2123.700
6	2169.200	2119.681
7	2172.759	2115.632
8	2176.287	2111.545

[a] Taken from D. H. Rank, D. P. Eastman, B. S. Rao, and T. A. Wiggins, *J. Opt. Soc. Am.*, **51**, 929 (1961); E. K. Plyler, L. R. Blaine, and W. S. Conner, *ibid.*, **45**, 102 (1955).

4.7 PARITY OF DIATOMIC-MOLECULE WAVE FUNCTIONS

The complete, nonrelativistic Hamiltonian for a diatomic molecule is given by (1.272). If one inverts the Cartesian coordinates of all particles (nuclei and electrons), then $\hat{H}$ in (1.272) is unchanged, since all interparticle distances are unchanged. Thus the parity operator $\hat{\Pi}$ commutes with this Hamiltonian, and we can characterize the overall wave function of a diatomic molecule by its parity. (This statement applies to both homonuclear and heteronuclear diatomics.)

In the Born–Oppenheimer approximation, the molecular wave function ψ is the product of electronic and nuclear wave functions; see (4.90). We now examine the behavior of ψ with respect to inversion. We must, however, exercise some care. In finding the nuclear wave functions ψ_N we have used a set of axes fixed in space (except for translation with the molecule). However, in dealing with ψ_{el} (Sections 1.19 and 1.20) we defined the electronic coordinates with respect to a set of axes fixed in the molecule, with the z axis being the internuclear axis. To find the effect on ψ of inversion of all nuclear and electronic coordinates, we must use the set of space-fixed axes for both ψ_N and ψ_{el}. We shall call the space-fixed axes X, Y, and Z, and the molecule-fixed axes x, y, and z. The nuclear wave function of a diatomic molecule has the (approximate) form (4.28) for $^1\Sigma$ electronic states, where $q = R - R_e$, and where the angles are defined with respect to space-fixed axes. When we replace each nuclear coordinate in ψ_N by its negative, the internuclear distance R is unaffected, so that the vibrational wave function has even parity. The parity of the spherical harmonic Y_J^M is even or odd according to whether J is even or odd (Section 1.17). Thus the parity eigenvalue of ψ_N is $(-1)^J$.

Now consider ψ_{el}. We set up the space-fixed and molecule-fixed coordinate systems with a common origin on the internuclear axis, midway between the nuclei, as in Fig. 4.11. (Previously in this chapter, we put the origin at the center of mass, but the difference is of no consequence.) The electronic wave function depends on the electronic spatial and spin coordinates and parametrically on R. The parity operator does not affect spin coordinates, and we shall only be considering transformations of spatial coordinates in this section.

It might be thought that inverting the molecule-fixed coordinates of all particles is equivalent to inverting the space-fixed coordinates of all particles, but this is not so. The direction of the z axis is defined as going from nucleus a to nucleus b, since the xyz axes are rigidly connected to the nuclear framework; when we invert the coordinates of electrons and nuclei, we interchange nuclei a and b, and thereby reverse the direction of

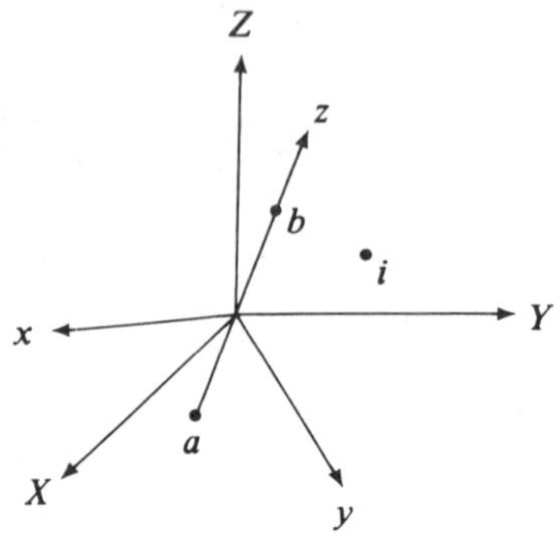

Fig. 4.11 Space-fixed and molecule-fixed axes in a diatomic molecule. The nuclei are at *a* and *b*, and *i* represents the location of a typical electron.

the (molecule-fixed) z axis. For coordinate axes that are defined independently of the locations of the particles, a transformation of particle coordinates leaves the coordinate axes unmoved; thus the XYZ set of axes remains fixed for all transformations of particle coordinates. Therefore, the inversion of the space-fixed coordinates of all particles is *not* equivalent to the inversion of the molecule-fixed coordinates of all particles.

To determine the effect on ψ_{el} of inversion of all space-fixed coordinates, we carry out the inversion in two steps: (a) We rotate all the particles (nuclei and electrons) by 180° about the y axis. (b) We then reflect all the electrons in the xz plane. The net effect of these steps is the inversion of all space-fixed nuclear and electronic coordinates. Step (a) rotates the x and z axes (which are rigidly attached to the nuclei) by 180° and has no effect on the xyz coordinates of the electrons or the nuclei; see Fig. 4.12*a*. Step (b) does nothing to the nuclei and therefore nothing to the xyz axes; it does convert the y coordinate of each electron to $-y$ (Fig. 4.12*b*).

Thus the net effect of inversion of the space-fixed coordinates of all particles is to convert each (molecule-fixed) y electronic coordinate to $-y$. This amounts to reflecting the electronic coordinates in the xz plane, which contains the nuclei. But we noted in Section 1.19 that we could classify diatomic-molecule electronic wave functions as having either eigenvalue $+1$ or -1 with respect to this reflection. We used a right superscript + or − on the electronic term symbol to indicate this eigenvalue. We noted that Σ^+ and Σ^- terms have different energies, and that for $\Pi, \Delta, \Phi, \ldots$ terms the + and − components have the same energy (omitting Λ-type doubling). Thus the parity of ψ_{el} with respect to inversion of all particles is even for $\Sigma^+, \Pi^+, \ldots$ states and odd for $\Sigma^-, \Pi^-, \ldots$ states.

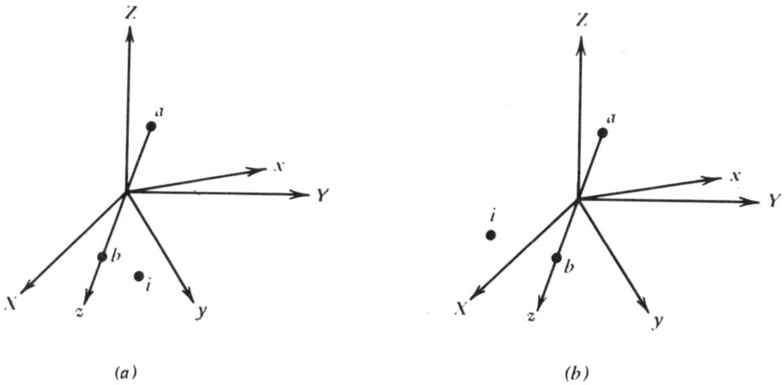

Fig. 4.12 Two-step inversion of space-fixed coordinates in Fig. 4.11.

For the effect of $\hat{\Pi}$ on the overall wave function, we have

$$\hat{\Pi}\psi=\hat{\Pi}(\psi_{\text{el}}\psi_N)=(\hat{\Pi}\psi_{\text{el}})(\hat{\Pi}\psi_N)=(-1)^J\psi_N(\hat{\Pi}\psi_{\text{el}}) \tag{4.122}$$

ψ is thus an eigenfunction of the parity operator with eigenvalue $(-1)^J$ for the levels of ${}^1\Sigma^+$ electronic states and with eigenvalue $(-1)^{J+1}$ for the levels of ${}^1\Sigma^-$ electronic states. It is conventional to denote vibration–rotation levels as + if ψ is an even function and − if ψ is an odd function. (This notation is unfortunate in that it invites confusion with the + and − superscripts giving the behavior of ψ_{el} with respect to reflection of electronic coordinates in a plane containing the nuclei.) Figure 4.13 shows the parity of the rotational levels for ${}^1\Sigma^+$ and ${}^1\Sigma^-$ electronic states (the parity is independent of the vibrational quantum number of the levels). Each level is $(2J+1)$-fold degenerate, but since M does not affect the parity, the $2J+1$ states all have the same parity.

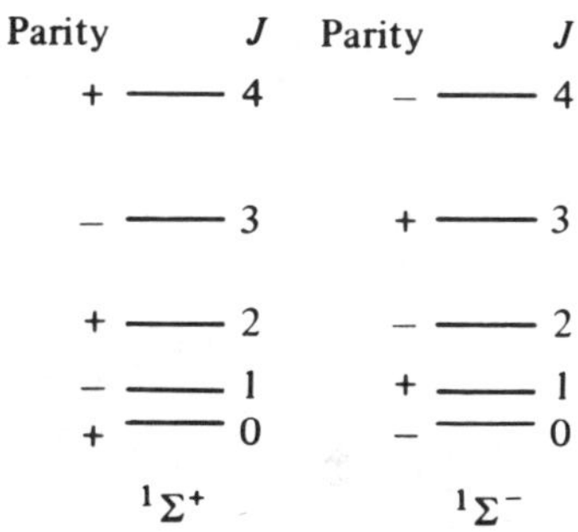

Fig. 4.13 Parity of the lowest rotational levels of ${}^1\Sigma^+$ and ${}^1\Sigma^-$ electronic states.

Recall the selection rule (Section 3.3) that parity changes in an electric-dipole transition. Hence we have the electric-dipole selection rule that forbids a transition between two + or two − rotational levels. This rule is symbolized by

$$+ \leftarrow\!|\!\rightarrow + \qquad - \leftarrow\!|\!\rightarrow - \tag{4.123}$$

where $\leftarrow\!|\!\rightarrow$ indicates a forbidden transition. The selection rule (4.123) is consistent with the previously found selection rule $\Delta J = \pm 1$ for transitions with no change in ψ_{el}.

In Section 1.19 we classified the electronic wave functions of homonuclear diatomic molecules as g or u, according to whether they were even or odd with respect to inversion; g and u refer to inversion of the *electronic* coordinates with respect to the *molecule-fixed* axes. This is to be distinguished from the inversion of electronic *and* nuclear coordinates with respect to *space-fixed* axes, which was discussed in this section. The electronic Hamiltonian for a diatomic molecule is

$$-\frac{\hbar^2}{2m}\sum_i \nabla_i^2 - \sum_i \frac{Z_a e^2}{r_{ia}} - \sum_i \frac{Z_b e^2}{r_{ib}} + \sum_j \sum_{i>j} \frac{e^2}{r_{ij}} \tag{4.124}$$

The electronic Hamiltonian (4.124) is invariant under inversion of electronic coordinates with respect to the molecule-fixed axes if and only if the molecule is homonuclear. However, (4.124) is invariant under inversion of electronic *and* nuclear coordinates with respect to space-fixed axes, for both homonuclear and heteronuclear diatomic molecules.

4.8 NUCLEAR SPIN AND THE PAULI PRINCIPLE

So far, we have not worried about nuclear spin—the wave function (4.28) for nuclear motion contains only the nuclear spatial coordinates. However, nuclei possess an intrinsic or "spin" angular momentum of magnitude

$$\sqrt{I(I+1)}\,\hbar \tag{4.125}$$

where the nuclear spin quantum number I can be $0, \frac{1}{2}, 1, \frac{3}{2}, \ldots$. Some values of I are given in the Appendix.

The spin of a proton is, like an electron, $\frac{1}{2}$; a neutron also has spin $\frac{1}{2}$. In addition to spin angular momentum, the nucleons in a nucleus have orbital angular momentum; the orbital angular-momentum quantum number of a nucleon can be $0, 1, 2, \ldots$. The spin angular momentum of each nucleon adds vectorially [i.e., according to (1.265)] to its orbital angular momentum to give a total angular momentum for that nucleon. The total angular momenta of the individual nucleons then combine vectorially to give a total nuclear angular momentum of magnitude (4.125). (This scheme resembles $j-j$ coupling in atoms.) The total nuclear angular momentum is thus made up of both spin and orbital angular momentum of the nucleons, but it is customary to call this total angular momentum "the nuclear spin." We will treat the nuclei in a diatomic molecule as if they were elementary particles with spin I, and not worry about the details of nuclear structure. I is generally small; this results from the pairing of nucleonic angular momenta. Nuclei have a shell structure resembling the electronic shell structure of atoms. The angular momentum of a filled nuclear shell is zero. The spins in the Appendix are for the ground states of the nuclei, which is usually what interests the chemist.

For a *heteronuclear* diatomic molecule, we can take nuclear spin into account simply by multiplying the spatial nuclear wave function (4.28) by two spin functions—one for each nucleus. The overall molecular wave function (4.90) then has the (approximate) form

$$\psi = \psi_{el}\psi_{vib}\psi_{rot}\psi_{ns} \tag{4.126}$$

where ψ_{ns} is the nuclear-spin factor. Since the nuclei are distinguishable, the Pauli principle is irrelevant.

Now consider *homonuclear* diatomic molecules. To satisfy the Pauli principle, the overall molecular wave function must have the following behavior with respect to interchange of the spin and spatial coordinates of the identical nuclei: it must be symmetric if the nuclei are bosons ($I = 0, 1, 2, \ldots$); it must be antisymmetric if the nuclei are fermions ($I = \frac{1}{2}, \frac{3}{2}, \frac{5}{2}, \ldots$). We therefore investigate the behavior of the factors in ψ with respect to interchange of the nuclei. (In the last section, we considered the parity of ψ, which is determined by its behavior on inversion of the spatial coordinates of all electrons and nuclei; since inversion does not affect the spin coordinates, the omission of ψ_{ns} does not affect the conclusions we reached.)

First consider ψ_{ns}. For a nucleus with spin I, there are $2I+1$ possible orientations of the spin: $M_I = -I, \ldots, +I$. Thus for a diatomic molecule there are $(2I_a+1)(2I_b+1)$ possible nuclear-spin functions, where I_a and I_b are the spins of the two nuclei. For a homonuclear diatomic, $I_a = I_b = I$, and we have $(2I+1)^2$ nuclear-spin functions. We will use $|M_I(a)\rangle$ to denote the spin function for which nucleus a has $M_I\hbar$ as the z component of its nuclear spin.

Of the $(2I+1)^2$ possible spin functions for the homonuclear diatomic molecule, $2I+1$ functions have equal M_I values for nuclei a and b; these functions have the form

$$|M_I(a)\rangle|M_I(b)\rangle \tag{4.127}$$

Interchanging a and b, we see that these functions are symmetric with respect to exchange of nuclear coordinates. Now consider the remaining $(2I+1)^2-(2I+1)$ spin functions. We can write functions like

$$|M_I'(a)\rangle|M_I''(b)\rangle \qquad \text{and} \qquad |M_I'(b)\rangle|M_I''(a)\rangle$$

where $M_I'' \neq M_I'$. However, these are not acceptable functions since they distinguish between the indistinguishable nuclei. Instead, we take the symmetric and antisymmetric linear combinations [recall (1.252) and (1.254)]:

$$2^{-1/2}[|M_I'(a)\rangle|M_I''(b)\rangle+|M_I'(b)\rangle|M_I''(a)\rangle] \tag{4.128}$$

$$2^{-1/2}[|M_I'(a)\rangle|M_I''(b)\rangle-|M_I'(b)\rangle|M_I''(a)\rangle] \tag{4.129}$$

We have $\frac{1}{2}[(2I+1)^2-(2I+1)]$ functions like (4.128) and and equal number of functions like (4.129). Thus the total number of symmetric nuclear-spin functions is

$$(2I+1)+\tfrac{1}{2}\left[(2I+1)^2-(2I+1)\right]=(I+1)(2I+1) \tag{4.130}$$

while there are $I(2I+1)$ antisymmetric nuclear-spin functions. As a check, for $I=\frac{1}{2}$ we get three symmetric functions and one antisymmetric function, in agreement with (1.251)–(1.254).

Now consider ψ_{rot} which for $^1\Sigma$ states is

$$\psi_{\text{rot}}=Y_J^M(\theta_N,\phi_N) \tag{4.131}$$

In the last section, we showed ψ_{rot} to have the eigenvalue $(-1)^J$ for inversion of the nuclei. Since inversion amounts to interchanging the nuclear spatial coordinates, and since the nuclear spin coordinates do not occur in ψ_{rot}, this is also the eigenvalue of ψ_{rot} for the operator $\hat{P}_{ab}$, which interchanges the space and spin coordinates of the two nuclei.

For the vibrational wave function, we have

$$\psi_{\text{vib}}=S(R-R_e)/R \tag{4.132}$$

Since R does not change on interchange of the nuclei, ψ_{vib} is symmetric with respect to such interchange.

Finally, consider ψ_{el}:

$$\psi_{el} = \psi_{el}(x_i, m_{si}; R) \tag{4.133}$$

Interchanging the nuclear coordinates does not affect R, but it *does* affect the electronic spatial coordinates x_i, since they are defined with respect to the molecule-fixed *xyz* axes, which are rigidly attached to the nuclei. To find the effect on ψ_{el} of interchanging the nuclear coordinates, we will first invert the space-fixed coordinates of the nuclei *and* the electrons, and then carry out a second inversion of the space-fixed electronic coordinates only; the net effect will be the interchange of the space-fixed coordinates of the two nuclei. We found in the last section that inversion of the space-fixed coordinates of all particles left ψ_{el} unchanged for $\Sigma^+, \Pi^+, \ldots$ electronic states, but multiplied it by -1 for $\Sigma^-, \Pi^-, \ldots$ states. Consider now the effect of the second step, reinversion of the electronic space-fixed coordinates. Since the nuclei are unaffected by this step, the molecule-fixed axes remain fixed for this inversion, so that inversion of the space-fixed coordinates of the electrons also inverts their molecule-fixed coordinates. But we noted in Section 1.19 that the electronic wave functions of homonuclear diatomics could be classified as g or u, according to whether inversion of molecule-fixed electronic coordinates multiplies ψ_{el} by $+1$ or -1. We conclude that for $\Sigma_g^+, \Sigma_u^-, \Pi_g^+, \Pi_u^-, \ldots$ electronic states, ψ_{el} is symmetric with respect to interchange of nuclear coordinates, whereas for $\Sigma_g^-, \Sigma_u^+, \Pi_g^-, \Pi_u^+, \ldots$ states, ψ_{el} is antisymmetric with respect to such interchange.

From the preceding discussion, we conclude that the function

$$\psi/\psi_{ns} = \psi_{el}\psi_{vib}\psi_{rot} \tag{4.134}$$

has the following behavior with respect to interchange of the nuclear coordinates of a homonuclear diatomic molecule: symmetric for $J = 0, 2, 4, \ldots$ rotational levels of ${}^1\Sigma_g^+$ and ${}^1\Sigma_u^-$ electronic states; antisymmetric for $J = 1, 3, 5, \ldots$ rotational levels of ${}^1\Sigma_g^+$ and ${}^1\Sigma_u^-$ electronic states; symmetric for $J = 1, 3, 5, \ldots$ rotational levels of ${}^1\Sigma_g^-$ and ${}^1\Sigma_u^+$ electronic states; antisymmetric for $J = 0, 2, 4, \ldots$ rotational levels of ${}^1\Sigma_g^-$ and ${}^1\Sigma_u^+$ electronic states. (We have not considered ψ_{rot} for non-${}^1\Sigma$ electronic states.) We use the letters s or a to indicate that ψ/ψ_{ns} is symmetric or antisymmetric with respect to interchange of identical nuclei. See Fig. 4.14.

For homonuclear diatomic molecules, the overall ψ (with ψ_{ns} included) must be antisymmetric with respect to interchange of nuclear space and spin coordinates if the nuclei are fermions. For such molecules, the s rotational levels must have an antisymmetric ψ_{ns}, whereas the a rotational levels must have a symmetric ψ_{ns}. If the nuclei are bosons, the situation is

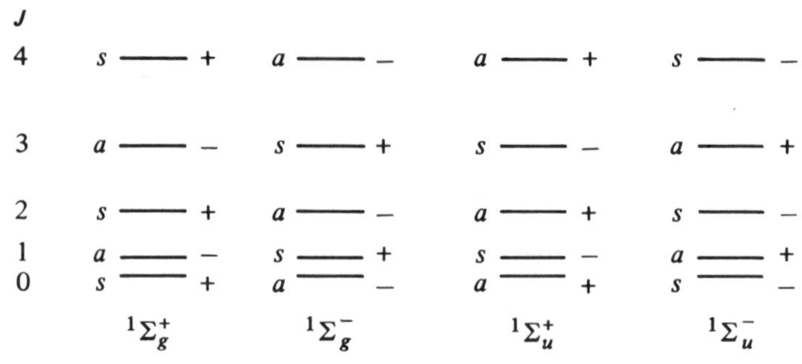

Fig. 4.14 Symmetry of rotational levels of a homonuclear diatomic molecule. The letters *s* and *a* refer to the nuclear-interchange symmetry of the wave function with the nuclear-spin factor omitted. The signs + and − refer to the parity of the wave function with respect to inversion of all particles.

reversed. We derived the number of symmetric and antisymmetric nuclear spin functions as

$$\text{number of symmetric } \psi_{\text{ns}} = (I+1)(2I+1) \tag{4.135}$$

$$\text{number of antisymmetric } \psi_{\text{ns}} = I(2I+1) \tag{4.136}$$

To a very good approximation, the molecular Hamiltonian does not involve nuclear spin, so that these different nuclear spin functions do not affect the energy. Thus, in addition to the previously mentioned $(2J+1)$-fold degeneracy of the rotational levels, there is a nuclear-spin degeneracy (which is often called a nuclear statistical weight). Since *s* and *a* rotational levels alternate, the nuclear statistical weights alternate. This gives an alternation in the population of adjacent rotational levels, the ratio being $I/(I+1)$. (This alternation is in addition to the $2J+1$ factor.) Although homonuclear diatomics don't have rotation or vibration–rotation spectra, the alternation can be investigated by observing electronic transitions; this is one way of experimentally determining nuclear spins. The nuclear statistical weights affect the molecular partition function.[7]

Suppose the spin of each nucleus of the homonuclear diatomic molecule is zero. The nuclei are then bosons and ψ must be symmetric with respect to nuclear exchange. For example, the C^{12} nucleus has $I=0$, and we will consider the rotational levels of the ground electronic state of C_2, which is a $^1\Sigma_g^+$ state (*Levine*, Section 13.6). The rotational levels with $J=0,2,4,\ldots$ have ψ/ψ_{ns} symmetric (Fig. 4.14) and require a symmetric ψ_{ns}. With $I=0$, there is one symmetric nuclear-spin function [Eq. (4.135)]. The rotational

[7] See *Davidson*.

levels with $J=1,3,5,\ldots$ require an antisymmetric ψ_{ns}, but with $I=0$, we do not have any antisymmetric nuclear-spin functions. Thus, the Pauli principle leads to the conclusion that half the rotational levels of a homonuclear diatomic molecule composed of nuclei with zero spin *do not exist*. This is verifiable by observing the rotational structure of electronic transitions.

4.9 ORTHO AND PARA MODIFICATIONS OF HOMONUCLEAR DIATOMIC MOLECULES

The selection rules for radiative transitions generally do not hold for nonradiative transitions. For example, suppose we start with a gas of homonuclear diatomic molecules at an extremely low temperature, so that only the lowest $(J=0, v=0)$ vibration–rotation level is populated. We then warm the gas by placing it in contact with a hot body. This transfers energy to the gas through collisions between the hot-body molecules and the gas molecules. Suppose that the temperature to which we warm the gas is not high enough to cause significant population of excited electronic states. We saw in Section 4.4 that radiative transitions between different vibration–rotation levels of the same electronic state of a homonuclear diatomic molecule are forbidden. Hence, if these radiative selection rules also held for nonradiative transitions, collisions between molecules would not lead to excitation of the higher vibration–rotation levels of a homonuclear diatomic. Actually, the radiative selection rules are not applicable to transitions induced by collisions, since the form of the time-dependent perturbation operator differs from (3.33); molecular collisions cause nonradiative transitions to the higher vibration–rotation levels at the expense of translational kinetic energy, thereby populating the levels in accord with the Boltzmann distribution law.

The preceding discussion has not taken into account a selection rule that results from the existence of nuclear spin. We now consider the effect of nuclear spin on selection rules.

For an electric-dipole transition between two molecular states, the transition-moment integral is

$$\langle\psi'|\hat{\mathbf{d}}|\psi\rangle = \int\int \psi_{\text{el}}'^{*}\psi_{\text{vib}}'^{*}\psi_{\text{rot}}'^{*}\hat{\mathbf{d}}\psi_{\text{el}}\psi_{\text{vib}}\psi_{\text{rot}}\,d\tau_{\text{el}}\,dV_N \sum_{M_{I_a}}\sum_{M_{I_b}} \psi_{\text{ns}}'^{*}\psi_{\text{ns}} \qquad (4.137)$$

where $\int d\tau_{\text{el}}$ indicates integration over the electronic spatial coordinates and summation over the electronic spin coordinates, and $\int dV_N$ indicates integration over the nuclear spatial coordinates. Having included the nuclear-spin factor, we must also sum over the nuclear spin coordinates. In

(4.137), we have used the fact that $\hat{\mathbf{d}}$ does not involve nuclear spin. For *homonuclear* diatomic molecules the nuclear spin functions are eigenfunctions of the Hermitian operator $\hat{P}_{ab}$, the operator that interchanges the space and spin coordinates of nuclei a and b; therefore, if ψ_{ns} and ψ'_{ns} correspond to different eigenvalues of $\hat{P}_{ab}$, they are orthogonal [Eq. (1.25)], and the summation over nuclear spin coordinates in (4.137) vanishes. We thus have the selection rule that forbids transitions between s and a rotational levels of a homonuclear diatomic molecule:

$$s \leftarrow\!\mid\!\rightarrow a \tag{4.138}$$

We previously found the selection rule $\Delta J = \pm 1$ for a $^1\Sigma$ diatomic-molecule vibration–rotation or pure-rotation transition. The rule (4.138) forbids $\Delta J = \pm 1$ for homonuclear diatomics; this gives us no new information as far as vibration–rotation spectra are concerned, since the absence of a dipole moment insures the absence of a vibration–rotation or pure-rotation spectrum, anyway.

The selection rule (4.138) differs from previously discussed selection rules in that it holds well for nonradiative transitions, as well as for radiative transitions. In deriving (4.138), we made no reference to the operator $\hat{\mathbf{d}}$, beyond the statement that it did not involve the nuclear spin coordinates. For *any* time-dependent perturbation that does not involve nuclear spin, the selection rule (4.138) will hold. Thus molecular collisions will not cause nonradiative transitions between symmetric and antisymmetric rotational levels of a homonuclear diatomic molecule. If we somehow start with all the molecules in symmetric levels, the collisions will not populate the antisymmetric levels.

As an example, consider H_2. The nuclear spin of H is $\frac{1}{2}$, and we have three symmetric nuclear spin functions and one antisymmetric function. The symmetric spin functions are of the form (1.251)–(1.253), and correspond to the two nuclear spins being parallel. Designating the quantum number of the vector sum of the two nuclear spins as T, we have $T = 1$ for the symmetric spin functions. The antisymmetric spin function has the form (1.254), and corresponds to $T = 0$. The ground electronic state of H_2 is a $^1\Sigma_g^+$ state, and the nuclei are fermions; hence the symmetric ($T = 1$) nuclear spin functions go with the $J = 1, 3, 5, \ldots$ rotational levels, whereas the $T = 0$ spin function goes with the $J = 0, 2, 4, \ldots$ levels.

Now suppose we cool H_2 to a very low temperature. Because of the selection rule (4.138), molecules originally in levels with odd values of J will all fall to the $J = 1, v = 0$ level, whereas molecules originally in levels with even values of J will all fall to the $J = 0, v = 0$ level. Actually, (4.138) is not absolutely inviolable. There are very weak interactions between the nuclear spins and the electrons; as a result, the separation of the molecular wave function into a factor depending only on nuclear spin coordinates times a factor independent of these coordinates is an approximation, albeit an extremely good approximation. Because of these weak interactions, collisions can lead to transitions between symmetric and antisymmetric rotational levels; however, equilibration between the two kinds of levels takes many months. If we wait a year or two after having cooled the H_2, the $J=1$ molecules will all eventually fall to the $J=0$ rotational level. If we lack the patience to wait, a small amount of some paramagnetic substance can be added as a catalyst to equilibrate the H_2. If we now warm the $J=0$ H_2 (in the absence of any catalyst), we will get a gas for which only the $J=0,2,4,\ldots$ rotational levels are populated, and this condition will persist for weeks. This gas will have different properties from ordinary H_2; for example, its specific heat will differ and the rotational structure of its electronic absorption spectrum will have half the lines missing. This modified form of H_2 is called *para hydrogen*; it has been prepared by the process just discussed. The modification of H_2 with only $J=1,3,5,\ldots$ rotational levels populated is called *ortho hydrogen*. Ortho H_2 cannot be prepared by the method used to isolate para H_2. However, it is possible to cleanly separate ortho and para H_2 by gas chromatography, passing ordinary H_2 through an alumina-packed column.[8] There are slight differences in the melting points of the two forms; ortho—13.94°K, para—13.81°K. The specific heats differ widely. Ordinary H_2 is essentially a 3:1 mixture of ortho to para hydrogen.

Now consider D_2. The nuclear spin of a deuteron is 1, and we can have $T=0$, 1, or 2. The $T=2$ and $T=0$ nuclear spin functions are symmetric and the $T=1$ functions are antisymmetric (Problem 4.24). The ground electronic state must be ${}^1\Sigma_g^+$ (as in H_2), and the nuclei are bosons. Hence the $T=0$ and $T=2$ spin functions go with the $J=0,2,4,\ldots$ rotational levels. Cooling D_2 to low temperature in the presence of a paramagnetic catalyst gives molecules with $J=0$ and $T=0$ and 2. Warming this gas in the absence of catalyst gives $J=0,2,4,\ldots$ molecules; this modification of D_2 is called *ortho deuterium*. (The convention is to use *ortho* for the modification with symmetric nuclear spin functions.)

For any homonuclear diatomic molecule whose nuclei have nonzero spin, it should, in principle, be possible to isolate a modified form with only even-numbered rotational levels populated. (For $I=0$, half the rotational levels do not exist.) However, only for H_2 and D_2 has this been achieved. The small moments of inertia of these light molecules give a relatively large spacing between $J=0$ and $J=1$ rotational levels, so that we can get nearly all the molecules into the $J=0$ level at a temperature above the freezing point of the substance.

[8] W. R. Moore and H. R. Ward, *J. Am. Chem. Soc.*, **80**, 2909 (1958).

4.10 THE RAMAN EFFECT

When a photon collides with a molecule, several things can happen. If the photon's frequency corresponds to a difference between energy levels of the system, the photon may be absorbed. There is also the possibility of stimulated emission. If the photon's frequency does not correspond to such an energy difference, it may be *scattered*; that is, it may undergo a change in direction of motion. Of the scattered photons, most will have undergone no change in frequency (Rayleigh scattering); however, during the time the photon that is being scattered is interacting with the molecule, some of the photon's energy can be transferred to the molecule, or vice versa. This then leads to a change in frequency of the scattered photon, known as the *Raman effect*, after the Indian physicist who observed it in 1928. The quantum mechanics of Raman scattering is involved, and we shall simply summarize the results.

The change in frequency of the scattered photon must correspond to a difference between two energy levels of the molecule doing the scattering. The Raman spectrum of a gas of diatomic molecules is shown in Fig. 4.15. The frequency of the exciting radiation is ν_{exc}. The lines on either side of ν_{exc} and close to it correspond to pure-rotational transitions in the molecule. To the low-frequency side of the pure-rotation lines are lines corresponding to $v=0\rightarrow 1$ vibration–rotation transitions of the gas molecules; a gain of energy by the molecules means a decrease in frequency of the Raman-scattered photons. To the high-frequency side of the pure-rotation lines is the $v=1\rightarrow 0$ vibration–rotation band, which results from $v=1$ molecules giving up energy to the scattered photons and thereby falling to the $v=0$ vibrational level. Since the population of the rotational levels of the $v=1$ vibrational level is negligible for most diatomic molecules at room temperature, this $v=1\rightarrow 0$ band is usually not observed. Lines of lower frequency than ν_{exc} are called *Stokes* lines, whereas lines of higher frequency than ν_{exc} are called *anti-Stokes* lines. The displacements of the Raman spectral lines from ν_{exc} are, of course, independent of the value of ν_{exc}, and give the differences between energy levels of the molecules doing the scattering.

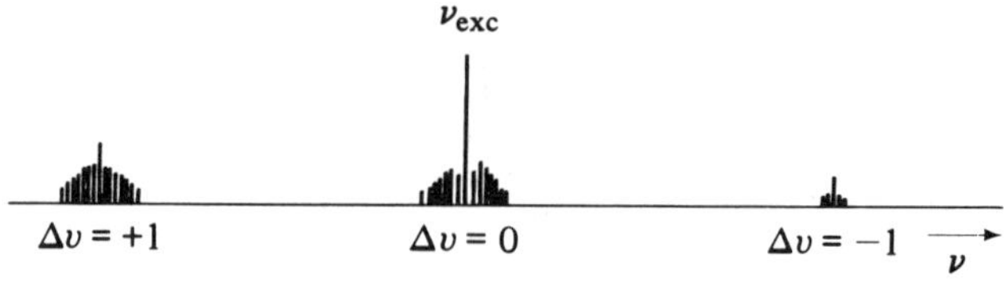

Fig. 4.15 Raman spectrum of a diatomic gas. The central line corresponds to photons scattered with no frequency change (Rayleigh scattering).

The selection rules for the Raman spectrum turn out to depend not on the matrix elements of the electric dipole moment, but on the matrix elements of the molecular polarizability, which we now define. The application of an electric field **E** to a molecule gives rise to an induced electric dipole moment $\mathbf{d}_{\text{ind}}$ (which is in addition to the permanent dipole moment **d**). If $\mathbf{E} = \mathscr{E}_x\mathbf{i} + \mathscr{E}_y\mathbf{j} + \mathscr{E}_z\mathbf{k}$, then the induced dipole moment has the components

$$\begin{pmatrix} d_{x,\text{ind}} \\ d_{y,\text{ind}} \\ d_{z,\text{ind}} \end{pmatrix} = \begin{pmatrix} \alpha_{xx} & \alpha_{xy} & \alpha_{xz} \\ \alpha_{yx} & \alpha_{yy} & \alpha_{yz} \\ \alpha_{zx} & \alpha_{zy} & \alpha_{zz} \end{pmatrix} \begin{pmatrix} \mathscr{E}_x \\ \mathscr{E}_y \\ \mathscr{E}_z \end{pmatrix} \tag{4.139}$$

where α_{xx} etc. are constants that depend parametrically on the nuclear configuration and on the choice of coordinate axes x, y, z. The molecular *polarizability* $\boldsymbol{\alpha}$ is thus a matrix, and the induced dipole-moment vector is not necessarily in the same direction as the applied electric field.

(Theoretical calculation of $\boldsymbol{\alpha}$ is not as straightforward as for, say, the dipole moment; to find $\boldsymbol{\alpha}$ one must include the applied electric field in the molecular Hamiltonian and calculate the resulting change in the molecular dipole moment. Theoretical values of the components of $\boldsymbol{\alpha}$ have been calculated for several diatomic and small polyatomic molecules using a Hartree–Fock approach; the calculated values are reasonably accurate. See *Schaefer*, pp. 243–250, for details.)

We have considered scalar, vector, and matrix molecular properties. A scalar is a zero-dimensional array; a vector is a one-dimensional array; a matrix is a two-dimensional array. In general, an s-dimensional array is called a *tensor* of *rank* (or *order*) s; a tensor of order s has n^s components, where n is the number of dimensions of the coordinate system (usually 3). Thus the dipole moment is a first-order tensor with $3^1 = 3$ components; the polarizability is a second-order tensor with $3^2 = 9$ components. The molecular first hyperpolarizability (which we will not define) is a third-order tensor.

Consideration of the matrix elements $\langle m|\boldsymbol{\alpha}|n\rangle$ of the polarizability shows that the selection rule for a pure-rotational Raman transition of a $^1\Sigma$ diatomic molecule is (see *Wilson, Decius, and Cross*)

$$\Delta J = 0, \pm 2 \tag{4.140}$$

which differs from (4.99). The selection rules for vibration–rotation Raman transitions of a $^1\Sigma$ diatomic molecule are

$$\Delta v = \pm 1, \qquad \Delta J = 0, \pm 2 \tag{4.141}$$

(As in the infrared spectrum, overtone bands with $\Delta v > 1$ are possible, but have much weaker intensity and are usually not observed.) The $\Delta J = -2$, 0, and $+2$ branches of a vibration–rotation Raman band are called O, Q, and S branches, respectively, in an extension of the P, Q, R notation used in infrared spectra.

We saw that homonuclear diatomic molecules exhibit no pure-rotation or vibration–rotation spectra, because they have zero electric dipole moment for all internuclear separations. The Raman effect depends on the polarizability and not the electric dipole moment; homonuclear diatomic molecules do have a nonzero polarizability which varies with varying internuclear separation. Hence they exhibit pure-rotation and vibration–rotation Raman spectra. Raman spectroscopy provides information on the vibrational and rotational constants of homonuclear diatomic molecules.

The Raman effect should be distinguished from the phenomenon of light absorption. In absorption, a photon whose frequency corresponds to an energy difference in the molecule is absorbed, thereby raising the molecule to an excited state. After some time has elapsed, the excited molecule may reradiate the photon, or lose its excitation energy in other ways. In the Raman effect, the frequency of the incident photon bears no relation to energy differences in the molecule; the photon is not absorbed, but is scattered, with its frequency being changed in the process.

4.11 ROTATIONAL ENERGIES OF NON-$^1\Sigma$ DIATOMIC MOLECULES

Earlier in this chapter, we considered nuclear motion for diatomic molecules in $^1\Sigma$ electronic states. When the molecule is not in a $^1\Sigma$ state there is an interaction between the rotational angular momentum of the molecule and the electronic spin and/or orbital angular momenta; this interaction changes the expression for the rotational energy. The details of coupling the angular momenta are involved, and we shall simply present a summary of the results for the important cases.[9] The coupling of angular momenta in non-$^1\Sigma$ molecules was first treated by Hund.

Most molecules with $\Lambda > 0$ obey Hund's *case (a)*. Here, the axial components of electronic orbital angular momentum ($\Lambda\hbar$) and of electronic spin angular momentum ($\Sigma\hbar$) combine to give a resultant axial component of total electronic angular momentum $(\Lambda + \Sigma)\hbar$. The quantum number Ω is defined as (Section 1.19)

$$\Omega = |\Lambda + \Sigma| \tag{4.142}$$

[9] For further details, see *King*, Chapter 6; *Townes and Schawlow*, Chapter 7.

The angular momentum $\mathbf{\Omega}$ then combines with the angular momentum $\mathbf{O}$ due to the rotation of the molecule to give a total angular momentum (apart from nuclear spin) $\mathbf{J}$. (In the older literature, the rotational angular momentum of the nuclei is called $\mathbf{N}$ instead of $\mathbf{O}$.) The magnitude of $\mathbf{J}$ is

$$\sqrt{J(J+1)}\,\hbar \tag{4.143}$$

where J has the possible values

$$J=\Omega, \Omega+1, \Omega+2, \ldots \tag{4.144}$$

(For $^1\Sigma$ molecules, we used J as the rotational angular-momentum quantum number; since $^1\Sigma$ molecules have no electronic spin or orbital angular momentum, J is also the total angular-momentum quantum number, exclusive of nuclear spin, for such molecules. Recall that for atoms J is also used as the total angular-momentum quantum number apart from nuclear spin.) The rotational energy in case (a) is given approximately by

$$hB_v[J(J+1)-\Omega^2] \tag{4.145}$$

where B_v is the rotational constant.

As an example, consider NO. The ground electronic term is $^2\Pi$, meaning $\Lambda=1$ and $S=\frac{1}{2}$. The possible values of $\Lambda+\Sigma$ are $\frac{3}{2}$ and $\frac{1}{2}$, and we have $^2\Pi_{3/2}$ and $^2\Pi_{1/2}$ electronic levels. $^2\Pi_{1/2}$ is the ground electronic level. For this level, we have $\Omega=|\Lambda+\Sigma|=\frac{1}{2}$. The possible values of J are $J=\frac{1}{2}, \frac{3}{2}, \frac{5}{2}, \ldots$. Note that J is half-integral; J does not represent just rotational angular momentum, but rotational and electronic angular momentum. As discussed in Section 1.19, Π (and $\Delta, \Phi, \ldots$) levels are doubly degenerate in the zeroth approximation; however, the interaction between electronic and rotational motions leads to the splitting of the degenerate Π electronic level into the nondegenerate Π^+ and Π^- electronic states (Λ-type doubling). See Fig. 4.16.

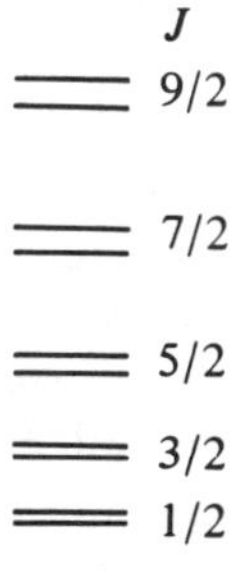

Fig. 4.16 Λ-type doubling in a $^2\Pi_{1/2}$ electronic level. Half the rotational levels belong to the $^2\Pi^+_{1/2}$ electronic state, half to the $^2\Pi^-_{1/2}$ state. The doubling is exaggerated; in practice, it is small and often not resolved.

For nearly all molecules with $\Lambda=0$, we have *case (b)* coupling; there are also a few molecules with $\Lambda\neq 0$ that fall in this case. In case (b), the angular momentum $\mathbf{\Lambda}$ combines with the molecular rotational angular momentum $\mathbf{O}$ to give a total angular momentum apart from electronic and nuclear spin, which is called $\mathbf{N}$. The magnitude of $\mathbf{N}$ is

$$\sqrt{N(N+1)}\,\hbar \tag{4.146}$$

and the quantum number N takes on the values

$$N=\Lambda,\Lambda+1,\Lambda+2,\ldots \tag{4.147}$$

(The older literature uses K instead of N.) For the most common case-(b) case, we have $\Lambda=0$; here N has the possible values $0,1,2,\ldots$ and represents just rotational angular momentum. The angular momentum $\mathbf{N}$ then adds to the electronic spin angular momentum $\mathbf{S}$ to give a total angular momentum apart from nuclear spin, which, as usual, is called $\mathbf{J}$. The quantum number J has the possible values [Equation (1.265)]

$$J=N+S,N+S-1,\ldots,|N-S| \tag{4.148}$$

The rotational energy for case (b) is given approximately by

$$hB_v[N(N+1)-\Lambda^2]+f(J,N,S) \tag{4.149}$$

where $f(J,N,S)$ is a small term whose value depends on J, N, and S. (See also Section 5.5.)

As an example of case (b) coupling, consider $O^{16}O^{18}$, whose ground electronic level is (like O_2^{16}) $^3\Sigma_g^-$. Although $O^{16}O^{18}$ is a heteronuclear diatomic, the nuclear masses are absent from the electronic Schrödinger equation, and we can still classify the electronic states as g or u. In the Born–Oppenheimer approximation, isotopic substitution does not affect the electronic wave functions. (Actually, there are very small deviations from this approximation; for example, the molecule HD has a small permanent electric dipole moment of about[10] 10^{-3} debye. The amplitudes of the H and D zero-point vibrations differ, and the electrons do not adjust their wave function quite fast enough to follow perfectly the nuclear motion; together with the anharmonicity of the vibrations, this gives a dipole moment.) We have $\Lambda=0$, so that N takes the values $0,1,2,\ldots$; also $S=1$, and there are three possible values of J for each N, except $N=0$, where $J=1$. For $^1\Sigma$ heteronuclear and homonuclear diatomic molecules, we showed that the parity eigenvalue with respect to inversion of all particles is $(-1)^J$ for $^1\Sigma^+$ states, and $(-1)^{J+1}$ for $^1\Sigma^-$ states. For Σ states that are not singlets, N gives the rotational angular momentum, and thus

[10] L. Wolniewicz and T. Kowalski, *Chem. Phys. Letters*, **18**, 55 (1973).

the parity is $(-1)^N$ for Σ^+ states and $(-1)^{N+1}$ for Σ^- states. (Of course, for $^1\Sigma$ states, $N=J$.) Thus, Fig. 4.13 can be applied to all Σ states if J is replaced by N. Figure 4.17 gives the parities of $O^{16}O^{18}$ rotational levels.

Now consider O_2^{16}. For $^1\Sigma$ homonuclear diatomic molecules, we showed how to determine whether ψ/ψ_{ns} is symmetric or antisymmetric with respect to interchange of the identical nuclei. Figure 4.14 summarizes the results. For Σ states that are not singlets, this figure can be used, except that J must be replaced by N, since it is N that represents the rotational angular momentum of Σ molecules in case (b) coupling. Thus Fig. 4.14 with N replacing J tells us that for the $^3\Sigma_g^-$ electronic ground level of O_2^{16}, the $N=0,2,4,\ldots$ rotational levels are a. Since O^{16} has zero nuclear spin, we conclude (Section 4.8) that the $N=0,2,4,\ldots$ rotational levels *do not exist* for O_2^{16}. Figure 4.18 gives the symmetry and parity of the rotational levels of the ground electronic state of O_2^{16}. Just as Λ-type doubling resolves the electronic orbital angular-momentum degeneracy in $\Pi,\Delta,\ldots$ states, we can view the term $f(J,N,S)$ in (4.149) as resolving the electronic spin degeneracy in $^2\Sigma, {}^3\Sigma,\ldots$ electronic levels; for example, we can consider each of the three rotational levels for a given N in Fig. 4.18 as belonging to a different electronic state, characterized by a different orientation of **S**.

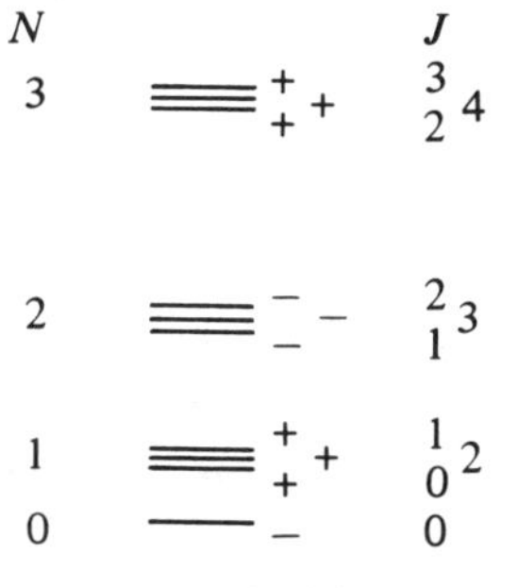

Fig. 4.17 Parities of rotational levels of the $O^{16}O^{18}$ $^3\Sigma_g^-$ ground electronic level.

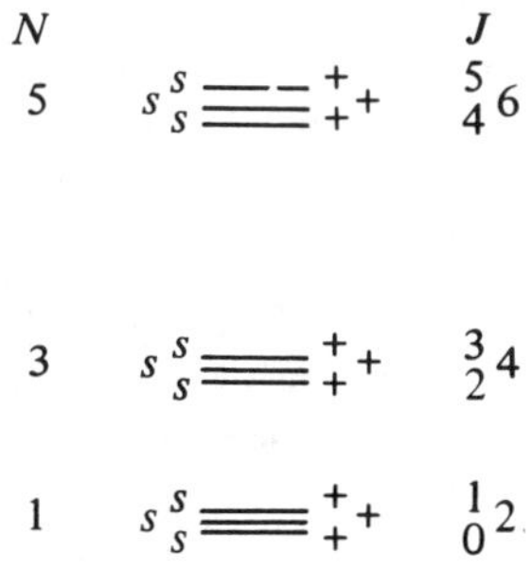

Fig. 4.18 Symmetry and parity of the rotational levels of the O_2^{16} Σ_g^- ground electronic level.

PROBLEMS

4.1 (a) Verify (4.21) by finding the Taylor-series expansion about $x=0$ for the function $(1+x)^{-2}$. (See Section 1.2) (b) What is the region of convergence for the expansion in part (a)? (c) For what range of values of R does the expansion (4.21) converge?

4.2 Compute $S_0(-R_e)/S_0(0)$ for the hydrogen-molecule ground electronic state, where S_0 is the ground-state wave function of a one-dimensional harmonic oscillator.

4.3 Verify (4.49).

4.4 Verify (4.51).

4.5 Verify (4.60).

4.6 Verify (4.67)–(4.72).

4.7 For the ground electronic state of H_2^1, compute $\tilde{Y}_{00}$ and the zero-point "energy" in cm^{-1}; use $\tilde{\nu}_e y_e = +0.7\ cm^{-1}$.

4.8 Verify (4.65) and (4.66).

4.9 (a) From the average of the $\Delta^2\sigma_0$ values in Table 4.3, calculate $\tilde{\nu}_e x_e$ for HCl^{35}. (b) Add $2\tilde{\nu}_e x_e(v'+1)$ to the $\Delta\sigma_0$ values of Table 4.3 and calculate $\tilde{\nu}_e$ for HCl^{35} as the average of the values obtained.

4.10 Verify (4.87).

4.11 Show that the Morse function predicts

$$\alpha_e = 6(\Delta^{1/2}-1)B_e^2/\nu_e \qquad \text{and} \qquad \nu_e x_e = B_e\Delta$$

4.12 Using the results of Problem 4.11, compute the numerical values of α_e and $\nu_e x_e$ predicted by the Morse function for the H_2^1 ground electronic state; what is the percent error for each quantity, as compared to the experimental values?

4.13 Express the parameter β in the Varshni function (4.89) in terms of Δ and R_e, where Δ is given by (4.88).

4.14 The vibrational levels of the H_2 ground electronic state are listed in S. Weissman, J. T. Vanderslice, and R. Battino, *J. Chem. Phys.*, **39**, 2226 (1963). Graph the separation between adjacent vibrational levels against the vibrational quantum number. What do you observe?

4.15 (a) Why do we expect the potential-energy curves for H_2 and HD to be virtually identical? (b) Use the value of $\tilde{\nu}_e$ in Table 4.1 for H_2 to predict the value of $\tilde{\nu}_e$ for HD. (c) Would D_e for H_2 and HD be virtually identical? (d) Would D_0 for H_2 and HD be virtually identical?

4.16 Verify that the result (4.45) is consistent with the fact that multiplication by q is a Hermitian operator.

4.17 (a) Use the data of Table 4.2 to calculate B_0 and $\overline{D}_e$ for $C^{12}O^{16}$. (b) Suppose the $J=0\rightarrow 1$ transition of $C^{12}O^{16}$ with $v=1$ occurs at 114,221.2 MHz; calculate α_e for $C^{12}O^{16}$. (Actually, this transition is too weak to be observed and α_e must be found from the infrared spectrum.) (c) Calculate B_e for $C^{12}O^{16}$. (d) Show from (4.70) and (4.71) that the following proportionalities hold:

$$\alpha_e \propto \mu^{-3/2}, \qquad \overline{D}_e \propto \mu^{-2}$$

(e) Calculate α_e and $\overline{D}_e$ for $C^{13}O^{16}$. (f) From Table 4.2, find B_0 and B_e for $C^{13}O^{16}$. (g) Calculate R_e and R_0 for $C^{12}O^{16}$ and for $C^{13}O^{16}$; comment on the results. (By definition, $R_0^2 = h/8\pi^2\mu B_0$.)

4.18 Given that $f(x)$ is a polynomial of degree n, show that $\Delta f(x)$ is a polynomial of degree $n-1$.

4.19 (a) From the average of the second differences of the R-branch and P-branch lines of Table 4.4, calculate $\tilde{\alpha}_e$ for $C^{12}O^{16}$, assuming that $\overline{D}_e$ is negligible. (b) Use the value of $\tilde{\alpha}_e$ found in part (a) to eliminate the J-dependent part of the first differences of the wave numbers of Table 4.4 and then calculate $\tilde{B}_e$ for $C^{12}O^{16}$ (assume $\overline{D}_e$ is negligible). Compare the infrared value with the microwave value (Problem 4.17). (The method of this problem is only approximate; the best procedure would be a least-squares fit of the observed lines.)

4.20 (a) From the data of Table 4.1, calculate the J value that corresponds to the band head of the R branch of the $v=0\rightarrow 1$ transition of $C^{12}O^{16}$. (b) Calculate the ratio at 300°K of the population of the band-head J level to the $J=0$ level. (Do not forget the degeneracy factor.)

4.21 Use the virial theorem (*Levine*, Chapter 14) to show that for a diatomic molecule

$$\left.\frac{\partial\langle V\rangle}{\partial R}\right|_{R_e} = R_e k_e = -\left.\frac{\partial\langle T_{\text{el}}\rangle}{\partial R}\right|_{R_e}$$

where $\langle V\rangle = \langle\psi_{\text{el}}|V|\psi_{\text{el}}\rangle$, with V and $\hat{T}_{\text{el}}$ being the operators for the electronic energy including nuclear repulsion and the electronic kinetic energy, respectively. Thus the average electronic potential energy must be decreasing with decreasing R at R_e, whereas the average electronic kinetic energy must be increasing with decreasing R at R_e.

4.22 Up to about 1930, it was believed that the nucleus of an atom with atomic number Z and mass number A consisted of A protons and $A-Z$ electrons. Which of the first 13 nuclear spins in the Appendix Table A.6 are incompatible with this hypothesis of nuclear structure? Explain your answer.

4.23 Much effort is being devoted to an attempt to produce elements of atomic numbers 114 and 126, even though many elements of lower atomic number have not been made. Why?

4.24 Write down the nine nuclear spin functions of D_2. Show that the three antisymmetric spin functions are eigenfunctions of the operator for the square of the magnitude of the total nuclear spin with the eigenvalue $2\hbar^2$. Find the corresponding eigenvalues for the symmetric spin functions.

4.25 Calculate the temperature for which the population of the $J=1$ rotational level is reduced to 1 percent of the population of the $J=0$ rotational level for each of the molecules H_2, D_2, and N_2. Compare these temperatures with the freezing points of each compound. What is the significance of the results?

4.26 What is the total degeneracy of the $J=0$, 1, and 2 levels of the ground electronic state of (a) H_2; (b) D_2; (c) HD; (d) C_2?

4.27 When liquid H_2 is to be stored for any period of time, a paramagnetic substance is added to it. Why?

4.28 Calculate ΔH at absolute zero for the reaction

$$\text{para } H_2 \rightarrow \text{ortho } H_2$$

4.29 What are the cgs units of the molecular polarizability components?

CHAPTER 5

ROTATION OF POLYATOMIC MOLECULES

5.1 NUCLEAR MOTION OF POLYATOMIC MOLECULES

We now consider the nuclear motions of polyatomic molecules. We are using the Born–Oppenheimer approximation, writing the Hamiltonian $\hat{H}_N$ for nuclear motion as the sum of the nuclear kinetic-energy $\hat{T}_N$ and a potential-energy term U derived from solving the electronic Schrödinger equation. We then solve the nuclear Schrödinger equation

$$\hat{H}_N \psi_N = E\psi_N \tag{5.1}$$

We found it easy to deal with the nuclear kinetic-energy terms for a diatomic molecule; but for a polyatomic molecule, the necessary manipulations to bring the classical-mechanical kinetic-energy expression into a form suitable for the quantum-mechanical treatment are complex. We omit these manipulations[1] and simply state their result. The classical expression for the kinetic energy T_N of the nuclei is the sum of several terms:

$$T_N = T_{\text{trans}} + T_{\text{rot}} + T_{\text{vib}} + T_{\text{vib-rot}} \tag{5.2}$$

The first term on the right is the translational kinetic energy of the molecule as a whole; this simply adds a constant to the total energy, and we shall omit this term. The second and third terms are the rotational and vibrational kinetic energies of the molecule. The final term is the energy of interaction between rotation and vibration. To get the classical-mechanical Hamiltonian function, we add the potential energy U to (5.2), where U is a function of the relative positions of the nuclei.

[1] See *Wilson, Decius, and Cross*, Chapter 11; *Margenau and Murphy*, Chapter 9.

The coordinate systems used are as follows: The translation of the molecule is described by giving the coordinates of the center of mass using a system of space-fixed Cartesian coordinates. The rotation is described by specifying the orientation of the molecule relative to a nonrotating set of axes XYZ with origin at the center of mass; these axes translate with the molecule. The vibration of the molecule is described relative to a set of axes with origin at the center of mass and rotating with the molecule; these are called *molecule-fixed* axes. Molecule-fixed axes are also used in solving the electronic Schrödinger equation for U.

The vibration–rotation interaction term makes the Hamiltonian for nuclear motion of a polyatomic molecule difficult to deal with. Frequently, this term is small compared to the other terms. We shall make the initial approximation of omitting $T_{\text{vib-rot}}$. The rotational kinetic energy T_{rot} involves the moments of inertia of the molecule, which in turn depend on the instantaneous nuclear configuration. However, the vibrational motions are much faster than the rotational motions, so that we can make the approximation of calculating the moments of inertia averaged over the vibrational motions.

With these approximations, the classical-mechanical Hamiltonian function for nuclear motion (omitting translation) is

$$H_N = T_{\text{rot}} + [T_{\text{vib}} + U] \tag{5.3}$$

We have bracketed the potential energy U with T_{vib} because both quantities depend on the coordinates of the nuclei relative to one another, whereas T_{rot} averaged over the vibrations does not. The quantum-mechanical Hamiltonian operator is obtained by replacing the classical quantities with operators:

$$\hat{H}_N = \hat{T}_{\text{rot}} + \left[\hat{T}_{\text{vib}} + U\right] \tag{5.4}$$

$$\hat{H}_N = \hat{H}_{\text{rot}} + \hat{H}_{\text{vib}} \tag{5.5}$$

where the rotational Hamiltonian involves the kinetic energy of rotation, and the vibrational Hamiltonian involves the bracketed terms in (5.4). Because of the (approximate) separation of the nuclear Hamiltonian into two parts, each involving different coordinates, we conclude that (Section 1.2) the nuclear wave function is approximately a product of rotational and vibrational wave functions, and the molecular energy is (approximately) the sum of rotational and vibrational energies. To find the rotational energies and wave functions, we use $\hat{H}_{\text{rot}}$ to form a rotational Schrödinger equation; similarly, we use $\hat{H}_{\text{vib}}$ to find the vibrational energies and wave functions. In this chapter, we will consider the rotational motion of polyatomic molecules, and in the next chapter we will deal with their vibrational motions.

Whereas for diatomic molecules the vibration–rotation interaction added only a small correction to the energy, for a number of polyatomic molecules the vibration–rotation interaction leads to relatively large corrections. Similarly, although the Born–Oppenheimer separation of electronic and nuclear motions holds extremely well for diatomic molecules, it occasionally breaks down for polyatomic molecules, leading to substantial interactions between electronic and nuclear motions.

For electronic states of polyatomic molecules for which the net electronic spin **S** is nonzero, we get an interaction between rotational and spin angular momenta. We shall not consider this case, but will restrict ourselves to rotational energies of singlet electronic states. Polyatomic molecules with nonsinglet ground electronic states are rare.

5.2 CLASSICAL MECHANICS OF A RIGID ROTOR

To consider the quantum mechanics of rotation of a polyatomic molecule, we first need the classical-mechanical expression for the rotational energy. We are considering the molecule to be a rigid rotor, with dimensions obtained by averaging over the vibrational motions. The classical mechanics of rotation of a rigid body in three dimensions is involved, and we shall simply summarize the results.[2]

We have a collection of n point masses (the nuclei) held fixed relative to one another, but free to rotate as a whole. The rotational energy expression involves the moments of inertia of the body. The moment of inertia I_α of a rigid body about axis α is defined as

$$I_\alpha \equiv \sum_{i=1}^{n} m_i r_i^2 \tag{5.6}$$

where r_i is the perpendicular distance from the particle of mass m_i to the axis α. Clearly, I_α depends on the choice of axis α. We will restrict ourselves to axes that pass through the molecular center of mass. We can make a sort of three-dimensional plot of the corresponding moments of inertia, as follows. On every possible axis α through the center of mass, we lay off on each side of the center of mass a distance numerically equal to

[2] See R. A. Becker, *Introduction to Theoretical Mechanics*, McGraw-Hill, New York, 1954, Chapter 12.

$1/\sqrt{I_\alpha}$. The result will be a set of points forming a surface in three-dimensional space. It can be shown that this surface is an ellipsoid with center at the center of mass. The ellipsoid is called the *ellipsoid of inertia*, or the *momental ellipsoid*. The ellipsoid of inertia has three mutually perpendicular principal axes, which we designate a, b, and c. These axes are the *principal axes of inertia* of the molecule; the corresponding moments of inertia about these three axes, I_a, I_b, and I_c, are the *principal moments of inertia*. The axes are labeled so that

$$I_a \leqslant I_b \leqslant I_c \tag{5.7}$$

Figure 5.1 shows the ellipsoid of inertia for $CH_2{=}O$.

The symmetry elements (Section 1.18) of a molecule aid in locating its principal axes. Clearly, a molecular symmetry operation must carry the momental ellipsoid to an orientation indistinguishable from its original position. Consider a $\hat{C}_n$ rotation ($n \neq 1$); such a rotation about any axis other than one of the three principal axes of the momental ellipsoid will send the ellipsoid to a nonequivalent spatial orientation. Hence a C_n symmetry axis of the molecule must coincide with one of its principal axes. (The converse is, of course, not true; every molecule has three principal

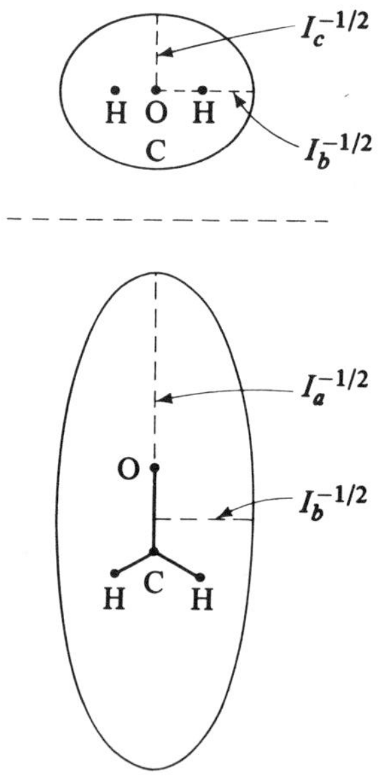

Fig. 5.1 The ellipsoid of inertia of CH_2O. Each of the two cross sections shown contains two principal axes.

axes, but need not have any symmetry axes.) The cross sections perpendicular to a principal axis are ellipses; for $n=2$, a $\hat{C}_2$ (180°) rotation brings these ellipses to an equivalent orientation. However, for $n \geqslant 3$, a $\hat{C}_n$ rotation will not bring these ellipses to an equivalent orientation unless they are circles. Thus for a molecule with a C_3 or higher-order rotational axis, the momental ellipsoid becomes an ellipsoid of revolution. Suppose the threefold or higher symmetry axis is the c principal axis. Since the cross section perpendicular to the C_n axis and containing the other two principal axes is a circle, we have $1/\sqrt{I_a} = 1/\sqrt{I_b}$, and $I_a = I_b$; hence a molecule with a C_n axis, $n \geqslant 3$, has two principal moments of inertia equal. The same argument shows that a molecule with two or more noncoincident C_n axes, $n \geqslant 3$, has all three principal moments equal; in this case, the momental ellipsoid is a sphere, and any axis through the center of mass is a principal axis.

Rotors are classified as follows: If all three principal moments of inertia are equal, we have a *spherical top*. If two principal moments are equal, we have a *symmetric top*. A *prolate* symmetric top has $I_b = I_c$ and the a axis (the axis of smallest principal moment) is the symmetry axis; an *oblate* symmetric top has $I_a = I_b$ and the c axis is the symmetry axis. A football (American, not British) is a prolate symmetric top; a discus is an oblate symmetric top. If all three principal moments are unequal, we have an *asymmetric top*. Thus

$$\text{spherical top:} \qquad I_a = I_b = I_c \tag{5.8}$$

$$\text{prolate symmetric top:} \qquad I_a \neq I_b = I_c \tag{5.9}$$

$$\text{oblate symmetric top:} \qquad I_a = I_b \neq I_c \tag{5.10}$$

$$\text{asymmetric top:} \qquad I_a \neq I_b \neq I_c \tag{5.11}$$

We have shown that (1) any C_n axis with $n \geqslant 2$ is a principal axis; (2) a molecule with one C_n axis with $n \geqslant 3$ is a symmetric top; (3) a molecule with two or more noncoincident C_n axes with $n \geqslant 3$ is a spherical top.

Thus CH_4 and SF_6 are spherical tops; C_6H_6 and CH_3Br are oblate and prolate symmetric tops, respectively; $CH_2{=}CHF$ and H_2O are asymmetric tops.

What information does an S_n axis give? We noted in Section 1.18 that an S_n axis for n odd implies the presence of a C_n axis of the same order; hence $S_3, S_5, \ldots$ axes give no new information. Since $\hat{S}_6^2 = \hat{C}_3$, $\hat{S}_8^2 = \hat{C}_4, \ldots$, such axes give no information that cannot be deduced from the $C_{n/2}$ axis. Now consider an S_4 axis. Since $\hat{S}_4^2 = \hat{C}_2$, the S_4 axis must be a principal axis. We have $\hat{S}_4 = \hat{C}_4\hat{\sigma}_h$. Reflection in a plane perpendicular to a principal axis and passing through the center of mass has no effect on the momental ellipsoid; the $\hat{S}_4$ operation thus has the same effect on the momental ellipsoid as a $\hat{C}_4$ rotation. Hence a molecule with an S_4 axis must be a symmetric top. Since the S_4 axis need not be a C_4 axis, this is new information. For example, allene, $CH_2{=}C{=}CH_2$, is a symmetric top, although it has no C_n axis with $n \geqslant 3$.

Next consider S_1 and S_2 axes. Since the only kind of reflection that sends the momental ellipsoid into an equivalent configuration is one in a plane containing two principal axes, a molecular symmetry plane contains two of the principal axes and is perpendicular to the third. A molecular center of symmetry coincides with the center of mass, which is the point of intersection of the three principal axes.

As an example, consider $CH_2{=}O$, formaldehyde. The three principal axes intersect at the center of mass, which lies on the C_2 axis passing through the C and O nuclei. This C_2 axis is one of the principal axes. A second principal axis must lie in the molecular plane, intersecting the C_2 axis at right angles. The third principal axis is perpendicular to the molecular plane, intersecting it at the center of mass. Having located the three principal axes, we can readily calculate the three principal moments from the molecular geometry. Since the moment of inertia about the C_2 axis has contributions from the light hydrogen atoms only, it is the smallest, making the C_2 axis the a principal axis. The molecule is an asymmetric rotor, since it has no threefold or higher symmetry axis.

For a molecule with little or no symmetry, finding the principal axes and principal moments is more involved. For any molecule, the equation of the ellipsoid of inertia can be shown to be[3]

$$I_{xx}x^2 + I_{yy}y^2 + I_{zz}z^2 + 2I_{xy}xy + 2I_{xz}xz + 2I_{yz}yz = 1 \qquad (5.12)$$

where the xyz coordinate system is a set of molecule-fixed axes with origin at the center of mass and an arbitrary orientation, and where the moments of inertia I_{xx}, I_{yy}, I_{zz} and products of inertia I_{xy}, I_{xz}, I_{yz} for the xyz system are defined as

$$I_{xx} \equiv \sum_i m_i(y_i^2 + z_i^2) \quad \text{etc.,} \qquad I_{xy} \equiv -\sum_i m_i x_i y_i \quad \text{etc.} \qquad (5.13)$$

(Note the minus sign.) The terms in xy, xz, and yz occur because the coordinate axes do not coincide with the principal axes. The left side of (5.12) is a real quadratic form (Section 2.2) and (5.12) can be written as the

[3] Becker, *op. cit.*, pp. 282–283.

matrix equation [see (2.44) and (2.45)]:

$$\tilde{\mathbf{x}}\mathbf{I}_{xyz}\mathbf{x}=1 \tag{5.14}$$

where $\mathbf{x}$ is the column vector with elements x, y, z, and $\mathbf{I}_{xyz}$ is the symmetric matrix

$$\mathbf{I}_{xyz}=\begin{pmatrix} I_{xx} & I_{xy} & I_{xz} \\ I_{xy} & I_{yy} & I_{yz} \\ I_{xz} & I_{yz} & I_{zz} \end{pmatrix} \tag{5.15}$$

$\mathbf{I}_{xyz}$ is called the *inertia tensor*. Let $\mathbf{L}$ be the square matrix of eigenvectors of $\mathbf{I}_{xyz}$; since $\mathbf{I}_{xyz}$ is symmetric, $\mathbf{L}$ can be chosen as orthogonal. Let I_a, I_b, and I_c denote the eigenvalues of $\mathbf{I}_{xyz}$. Then [Equation (2.46)]

$$\tilde{\mathbf{L}}\mathbf{I}_{xyz}\mathbf{L}=\mathbf{I}_{abc} \tag{5.16}$$

where $\mathbf{I}_{abc}$ is the diagonal eigenvalue matrix with diagonal elements I_a, I_b, I_c. We can reduce the quadratic form in (5.12) to a sum of squares by defining new variables a, b, c as [*cf.* (2.47) and (2.48)]

$$\mathbf{a}\equiv\tilde{\mathbf{L}}\mathbf{x}=\mathbf{L}^{-1}\mathbf{x}, \qquad \mathbf{x}=\mathbf{L}\mathbf{a} \tag{5.17}$$

where $\mathbf{a}$ is the column vector with elements a, b, c. Thus $a=l_{11}x+l_{21}y+l_{31}z$, with similar equations for b and c. Substitution into (5.14) gives

$$1=\tilde{\mathbf{x}}\mathbf{I}_{xyz}\mathbf{x}=(\widetilde{\mathbf{La}})\mathbf{I}_{xyz}\mathbf{La}=\tilde{\mathbf{a}}\tilde{\mathbf{L}}\mathbf{I}_{xyz}\mathbf{La}=\tilde{\mathbf{a}}\mathbf{I}_{abc}\mathbf{a} \tag{5.18}$$

Equation (5.18) reads [*cf.* the two equations following (2.48)]

$$I_a a^2+I_b b^2+I_c c^2=1 \tag{5.19}$$

In two dimensions, when the coordinate axes coincide with the principal axes of an ellipse, the ellipse has the equation $x^2/s^2+y^2/t^2=1$, where s and t are half the lengths of the principal axes. Extension to three dimensions shows that (5.19) is the equation of an ellipsoid with principal axes coincident with the coordinate axes. The orthogonal transformation of coordinates (5.17) is a rotation from the arbitrary axes xyz to the principal axes abc, and I_a, I_b, I_c are, therefore, the principal moments of inertia. The components of the orthogonal matrix $\tilde{\mathbf{L}}$ are the direction cosines of the abc axes in the xyz system [see (1.120) and (2.29)].

The steps in finding the principal axes and principal moments of a molecule with no or low symmetry are, therefore, the following: (1) Find the center of mass. (2) Set up an arbitrary coordinate system *xyz* with origin at the center of mass; calculate the moments and products of inertia in this system to give the matrix $\mathbf{I}_{xyz}$. (3) Find the eigenvalues and normalized eigenvectors of $\mathbf{I}_{xyz}$. The eigenvalues are the principal moments of inertia I_a, I_b, I_c. The transpose of the orthogonal matrix of eigenvectors gives the direction cosines of the *abc* principal-axis system with respect to the *xyz* axes.

[Like the inertia tensor $\mathbf{I}_{xyz}$, the polarizability $\boldsymbol{\alpha}$ in (4.139) is a matrix property. Moreover, $\boldsymbol{\alpha}$ can be shown to be a symmetric matrix (*Atkins*, pp. 398–399). Thus much of what we said about the inertia tensor applies to $\boldsymbol{\alpha}$. It follows from (4.139) that the component of $\mathbf{d}_{\text{ind}}$ in the direction of the applied field $\mathbf{E}$ is proportional to the magnitude $\mathcal{E}$ of the field—if $\mathbf{E}$ in $\mathbf{d}_{\text{ind}} = \boldsymbol{\alpha}\mathbf{E}$ is multiplied by a constant c, then $\mathbf{d}_{\text{ind}}$ and every component of $\mathbf{d}_{\text{ind}}$ is multiplied by c. Hence $d_{\text{ind},E} = \alpha_E \mathcal{E}$, where the proportionality constant α_E depends on the components of $\boldsymbol{\alpha}$ and the direction of $\mathbf{E}$. If we now plot distances $1/\alpha_E^{1/2}$ in each direction from the origin, we obtain the *polarizability ellipsoid* with equation $\tilde{\mathbf{x}}\boldsymbol{\alpha}_{xyz}\mathbf{x} = 1$, where the subscripts on $\boldsymbol{\alpha}$ emphasize the coordinate system. A rotation of coordinate axes to the principal axes *abc* of the polarizability ellipsoid gives a diagonal polarizability matrix $\boldsymbol{\alpha}_{abc}$ with diagonal elements α_a, α_b, α_c, the *principal values* of the polarizability. The same symmetry arguments used with the inertia ellipsoid apply to the polarizability ellipsoid. Hence a molecular symmetry axis must be a principal axis of the polarizability ellipsoid; a symmetric top must have two of the three principal values of $\boldsymbol{\alpha}$ equal; a spherical top has a spherical polarizability ellipsoid with $\alpha_a = \alpha_b = \alpha_c$. Note, however, that $\boldsymbol{\alpha}$ depends on the distribution of *charges*, whereas the inertia tensor depends on the distribution of *masses*; thus CH_3D is a symmetric top, but its polarizability ellipsoid is a sphere, as in CH_4.]

We now consider the classical-mechanical expression for the rotational energy of a rigid body. At any instant, the body can be considered to be rotating with angular velocity $\boldsymbol{\omega}$ about some axis in space, where $\boldsymbol{\omega}$ is a vector whose magnitude is the time rate of change of the angle about the axis of rotation, and whose direction is along the axis of rotation. The relation between $\boldsymbol{\omega}$ and the velocity $\mathbf{v}_i$ of the *i*th particle of the body is $\mathbf{v}_i = \boldsymbol{\omega} \times \mathbf{r}_i$, where $\mathbf{r}_i$ is a vector from any point on the rotation axis to particle *i*. It turns out that if we refer the rotation to the principal axes, the expression for the rotational kinetic energy takes the simple form

$$T_{\text{rot}} = \tfrac{1}{2}I_a\omega_a^2 + \tfrac{1}{2}I_b\omega_b^2 + \tfrac{1}{2}I_c\omega_c^2 \tag{5.20}$$

where I_a, I_b, and I_c are the principal moments of inertia, and ω_a, ω_b, and ω_c are the components of the angular velocity $\boldsymbol{\omega}$ along the three principal axes.

To form the Hamiltonian operator, we need the energy in terms of angular momenta, not angular velocities. Let **P** be the rotational angular momentum of the rigid body. We have [Equation (1.145)]

$$\mathbf{P} = \sum_{i=1}^{n} m_i \mathbf{r}_i \times \mathbf{v}_i \tag{5.21}$$

If we use principal axes, then the components of **P** along these axes can be shown to be

$$P_a = I_a \omega_a, \qquad P_b = I_b \omega_b, \qquad P_c = I_c \omega_c \tag{5.22}$$

and (5.20) becomes

$$T_{\text{rot}} = \frac{P_a^2}{2I_a} + \frac{P_b^2}{2I_b} + \frac{P_c^2}{2I_c} \tag{5.23}$$

This is the classical-mechanical Hamiltonian for the rotation of a rigid body. [Note the similarity of (5.20), (5.22), and (5.23) to the corresponding equations for linear motion; we get the equations for rotational motion by replacing the velocity v with the angular velocity ω, the linear momentum p with the angular momentum P, and the mass with the principal moments of inertia.]

5.3 THE ROTATIONAL HAMILTONIAN OPERATOR

To form the quantum-mechanical Hamiltonian operator for a rigid rotating body, we replace P_a, P_b, P_c in (5.23) by operators. These will be differential operators involving the coordinates describing the rotation. The appropriate coordinates are those describing the orientation of the molecule relative to a nonrotating coordinate system XYZ at the center of mass. We shall call this nonrotating system the *space-fixed* coordinate system, although, strictly speaking, it is translating with the molecule. To specify the orientation of the molecule-fixed principal axes a, b, and c with respect to the space-fixed axes X, Y, and Z, we must specify three angles. Placing the origin at the center of mass, point O, we first specify the orientation of one of the principal axes, say the c axis, by giving the angles θ and φ, where θ is the angle between OZ and Oc, and φ is the angle between OX and the projection of Oc in the XY plane (Fig. 5.2). (If the

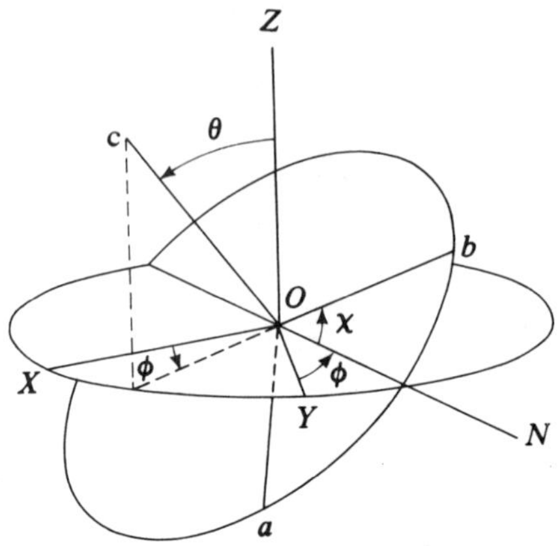

Fig. 5.2 The Eulerian angles. ON is the line of intersection of the XY and ab planes and is therefore perpendicular to both OZ and Oc. θ is the angle of rotation about ON; φ is the angle of rotation about OZ; χ is the angle of rotation about Oc.

molecule possesses a symmetry axis, we specify the orientation of the symmetry axis, which might or might not be the c axis. For simplicity, we will refer to θ as the angle between OZ and Oc, although it is more accurately described as the angle between OZ and the symmetry axis.) Having specified the orientation of the c axis, we specify the orientations of the a and b axes by giving the angle of rotation about the c axis; we shall call this angle χ (chi). (In classical mechanics, this third angle is usually called ψ, which is inappropriate for quantum mechanics.) The angles θ, φ, and χ are called the *Eulerian angles*, after the mathematician Euler (pronounced "oiler"). Their ranges are

$$0 \leqslant \theta \leqslant \pi, \qquad 0 \leqslant \varphi \leqslant 2\pi, \qquad 0 \leqslant \chi \leqslant 2\pi$$

The line of intersection of the XY and ab planes is called the *line of nodes*; in Fig. 5.2, this is line ON. (Unfortunately, there is no uniformity in the definition of the Eulerian angles.)

We now need the operators for P_a, P_b, and P_c in (5.23). Since $\hat{L}_z = -i\hbar\partial/\partial\varphi$ [Equation (1.154)], and there is nothing special about the z axis, the operator corresponding to the component of angular momentum along some axis is $-i\hbar\partial/\partial\beta$, where β is the angle of rotation about that axis. The angles of rotation about the OZ, the Oc, and the ON axes are φ, χ, and θ, respectively (Fig. 5.2). Hence the operators for the components of $\mathbf{P}$

along these axes are

$$\hat{P}_z = -i\hbar\frac{\partial}{\partial\varphi}, \qquad \hat{P}_c = -i\hbar\frac{\partial}{\partial\chi}, \qquad \hat{P}_N = -i\hbar\frac{\partial}{\partial\theta} \tag{5.24}$$

To derive the expressions for $\hat{P}_a$, $\hat{P}_b$, $\hat{P}_c$, we must relate these operators to the operators in (5.24). We therefore express the physical quantities corresponding to the operators in (5.24) in terms of P_a, P_b, P_c.

The component P_N is, from the definition of the dot product, the dot product of **P** with the unit vector $\mathbf{e}_N$ directed along ON:

$$P_N = \mathbf{e}_N \cdot \mathbf{P} \tag{5.25}$$

Writing **P** as

$$\mathbf{P} = \mathbf{e}_a P_a + \mathbf{e}_b P_b + \mathbf{e}_c P_c \tag{5.26}$$

where $\mathbf{e}_a$, $\mathbf{e}_b$, and $\mathbf{e}_c$ are unit vectors along the mutually perpendicular a, b, and c axes, we have

$$P_N = (\mathbf{e}_N \cdot \mathbf{e}_a)P_a + (\mathbf{e}_N \cdot \mathbf{e}_b)P_b + (\mathbf{e}_N \cdot \mathbf{e}_c)P_c \tag{5.27}$$

The quantity $(\mathbf{e}_N \cdot \mathbf{e}_a)$ is the cosine of the angle between the N and a axes. From Fig. 5.2, angle $NOa = (\frac{1}{2}\pi - \chi)$, so that

$$\mathbf{e}_N \cdot \mathbf{e}_a = \sin\chi$$

Similarly

$$\mathbf{e}_N \cdot \mathbf{e}_b = \cos\chi, \qquad \mathbf{e}_N \cdot \mathbf{e}_c = 0$$

Hence

$$P_N = \sin\chi\ P_a + \cos\chi\ P_b \tag{5.28}$$

For P_Z we have $P_Z = \mathbf{e}_Z \cdot \mathbf{P}$, and a similar treatment (the geometrical relations are harder to see in this case) gives

$$P_Z = -\sin\theta\cos\chi\ P_a + \sin\theta\sin\chi\ P_b + \cos\theta\ P_c \tag{5.29}$$

We can regard (5.28) and (5.29) as two equations in the two unknowns P_a and P_b; solving for them, we get

$$P_a = -\cos\chi\csc\theta\ P_Z + \cos\chi\cot\theta\ P_c + \sin\chi\ P_N \tag{5.30}$$

$$P_b = \sin\chi\csc\theta\ P_Z - \sin\chi\cot\theta\ P_c + \cos\chi\ P_N \tag{5.31}$$

Replacing physical quantities by operators, and using (5.24), we have the desired operators:

$$\hat{P}_a = i\hbar\left[\cos\chi\csc\theta\frac{\partial}{\partial\varphi} - \cos\chi\cot\theta\frac{\partial}{\partial\chi} - \sin\chi\frac{\partial}{\partial\theta}\right] \tag{5.32}$$

$$\hat{P}_b = i\hbar\left[-\sin\chi\csc\theta\frac{\partial}{\partial\varphi} + \sin\chi\cot\theta\frac{\partial}{\partial\chi} - \cos\chi\frac{\partial}{\partial\theta}\right] \tag{5.33}$$

$$\hat{P}_c = -i\hbar\frac{\partial}{\partial\chi} \tag{5.34}$$

From (5.23), the Hamiltonian operator for rotation is

$$\hat{H}_{\text{rot}} = \frac{\hat{P}_a^2}{2I_a} + \frac{\hat{P}_b^2}{2I_b} + \frac{\hat{P}_c^2}{2I_c} \tag{5.35}$$

We could now square the operators (5.32)–(5.34), substitute into (5.35), and form the Schrödinger equation for rotation. Instead of solving the Schrödinger equation directly, we will find the rotational-energy eigenvalues by using properties of the various operators. For this purpose, we will also need the operators $\hat{P}_X$, $\hat{P}_Y$, $\hat{P}_Z$ for the angular-momentum components along the space-fixed X, Y, Z axes. These are derived from (5.24) the same way the operators (5.32)–(5.34) for the molecule-fixed components were derived; they are found to be

$$\hat{P}_X = i\hbar\left[\cos\varphi\cot\theta\frac{\partial}{\partial\varphi} - \cos\varphi\csc\theta\frac{\partial}{\partial\chi} + \sin\varphi\frac{\partial}{\partial\theta}\right] \tag{5.36}$$

$$\hat{P}_Y = i\hbar\left[\sin\varphi\cot\theta\frac{\partial}{\partial\varphi} - \sin\varphi\csc\theta\frac{\partial}{\partial\chi} - \cos\varphi\frac{\partial}{\partial\theta}\right] \tag{5.37}$$

$$\hat{P}_Z = -i\hbar\frac{\partial}{\partial\varphi} \tag{5.38}$$

The operator for the total angular momentum of the n-particle rigid rotor is

$$\hat{P}^2 = \hat{P}_a^2 + \hat{P}_b^2 + \hat{P}_c^2 = \hat{P}_X^2 + \hat{P}_Y^2 + \hat{P}_Z^2 \tag{5.39}$$

and using either (5.32)–(5.34) or (5.36)–(5.38), we get

$$\hat{P}^2 = -\hbar^2\left[\csc^2\theta\frac{\partial^2}{\partial\varphi^2} + \csc^2\theta\frac{\partial^2}{\partial\chi^2} - 2\cot\theta\csc\theta\frac{\partial^2}{\partial\varphi\,\partial\chi} + \frac{\partial^2}{\partial\theta^2} + \cot\theta\frac{\partial}{\partial\theta}\right] \tag{5.40}$$

We now evaluate some commutators. We find

$$\left[\hat{P}_X, \hat{P}_Y\right] = i\hbar \hat{P}_Z \qquad \text{etc.} \tag{5.41}$$

$$\left[\hat{P}_a, \hat{P}_b\right] = -i\hbar \hat{P}_c \qquad \text{etc.} \tag{5.42}$$

where *etc.* indicates two other commutation relations formed by cyclic permutation of the subscripts. From (5.39), (5.41) and (5.42), together with commutator identities such as (1.68), we find

$$\left[\hat{P}^2, \hat{P}_c\right] = 0 \qquad \text{etc.} \tag{5.43}$$

$$\left[\hat{P}^2, \hat{P}_Z\right] = 0 \qquad \text{etc.} \tag{5.44}$$

We also find that each space-fixed component of angular momentum commutes with each molecule-fixed component:

$$\left[\hat{P}_Z, \hat{P}_c\right] = 0 \qquad \text{etc., etc.} \tag{5.45}$$

Equation (5.45) comprises nine commutation relations.

Now consider the commutators of the Hamiltonian with the angular-momentum operators. We have

$$\left[\hat{H}_{\text{rot}}, \hat{P}^2\right] = \frac{1}{2I_a}\left[\hat{P}_a^2, \hat{P}^2\right] + \frac{1}{2I_b}\left[\hat{P}_b^2, \hat{P}^2\right] + \frac{1}{2I_c}\left[\hat{P}_c^2, \hat{P}^2\right] \tag{5.46}$$

For the first commutator on the right, we have

$$\left[\hat{P}_a^2, \hat{P}^2\right] = \hat{P}_a\left[\hat{P}_a, \hat{P}^2\right] + \left[\hat{P}_a, \hat{P}^2\right]\hat{P}_a = 0$$

where (5.43) has been used. By cyclic permutation, the other two commutators on the right in (5.46) are zero, so that

$$\left[\hat{H}_{\text{rot}}, \hat{P}^2\right] = 0 \tag{5.47}$$

We also have

$$\left[\hat{H}_{\text{rot}}, \hat{P}_Z\right] = \frac{1}{2I_a}\left[\hat{P}_a^2, \hat{P}_Z\right] + \frac{1}{2I_b}\left[\hat{P}_b^2, \hat{P}_Z\right] + \frac{1}{2I_c}\left[\hat{P}_c^2, \hat{P}_Z\right] \tag{5.48}$$

$$\left[\hat{H}_{\text{rot}}, \hat{P}_Z\right] = 0 \qquad \text{etc.} \tag{5.49}$$

where (5.45) was used. Also

$$[\hat{H}_{\mathrm{rot}},\hat{P}_c]=\frac{1}{2I_a}[\hat{P}_a^2,\hat{P}_c]+\frac{1}{2I_b}[\hat{P}_b^2,\hat{P}_c]+\frac{1}{2I_c}[\hat{P}_c^2,\hat{P}_c]$$

$$=\frac{1}{2I_a}\hat{P}_a[\hat{P}_a,\hat{P}_c]+\frac{1}{2I_a}[\hat{P}_a,\hat{P}_c]\hat{P}_a+\frac{1}{2I_b}\hat{P}_b[\hat{P}_b,\hat{P}_c]$$

$$+\frac{1}{2I_b}[\hat{P}_b,\hat{P}_c]\hat{P}_b$$

$$[\hat{H}_{\mathrm{rot}},\hat{P}_c]=i\hbar\left(\frac{1}{2I_a}-\frac{1}{2I_b}\right)(\hat{P}_a\hat{P}_b+\hat{P}_b\hat{P}_a) \tag{5.50}$$

Now consider the eigenfunctions ψ of the rotational Hamiltonian: $\hat{H}\psi = E\psi$. (This equation is more fully written as $\hat{H}_{\mathrm{rot}}\psi_{\mathrm{rot}}=E_{\mathrm{rot}}\psi_{\mathrm{rot}}$, but since we are considering only rotation in this chapter, we omit the subscripts.) Because $\hat{H}$ commutes with $\hat{P}^2$ and $\hat{P}_Z$, the rotational eigenfunctions ψ can be chosen as eigenfunctions of these two operators.

It was shown in Section 1.7 that when the operators $\hat{P}_X$, $\hat{P}_Y$, $\hat{P}_Z$ obey the general angular-momentum commutation relations, as in (5.41), then the eigenvalues of $\hat{P}^2$ and $\hat{P}_Z$ are $J(J+1)\hbar^2$ and $M\hbar$, respectively, where M ranges from $-J$ to J, and J is integral or half-integral. However, we exclude the half-integral values of the rotational quantum number, since these occur only when spin is involved.

We conclude that for any rotor, the following relations hold:

$$\hat{H}\psi=E\psi \tag{5.51}$$

$$\hat{P}^2\psi=J(J+1)\hbar^2\psi, \qquad J=0,1,2,\ldots \tag{5.52}$$

$$\hat{P}_Z\psi=M\hbar\psi, \qquad M=0,\pm1,\ldots,\pm J \tag{5.53}$$

where $\sqrt{J(J+1)}\,\hbar$ is the magnitude of the total rotational angular momentum and $M\hbar$ is its component along a space-fixed axis. Since the eigenfunctions of $\hat{P}_Z=-i\hbar\partial/\partial\varphi$ are $e^{iM\varphi}/(2\pi)^{1/2}$, the rotor eigenfunctions have the form

$$\psi=F(\theta,\chi)(2\pi)^{-1/2}e^{iM\varphi} \tag{5.54}$$

5.4 THE SPHERICAL TOP

From (5.8), (5.39), and (5.35), the spherical-top Hamiltonian is

$$\hat{H} = \frac{\hat{P}_a^2}{2I} + \frac{\hat{P}_b^2}{2I} + \frac{\hat{P}_c^2}{2I} = \frac{\hat{P}^2}{2I} \tag{5.55}$$

where I is the moment of inertia about any axis through the center of mass. The Schrödinger equation (5.51) is

$$(1/2I)\hat{P}^2\psi = E\psi \tag{5.56}$$

Using (5.52), we can find the energy eigenvalues:

$$(1/2I)J(J+1)\hbar^2\psi = E\psi$$

$$E = \frac{J(J+1)\hbar^2}{2I}, \qquad J = 0, 1, 2, \ldots \tag{5.57}$$

The rotational Hamiltonian does not generally commute with $\hat{P}_c$ [(5.50)]; for the spherical top, however, the Hamiltonian involves only $\hat{P}^2$, which commutes with $\hat{P}_c$. Hence $[\hat{H}, \hat{P}_c] = 0$ for the spherical top. [This also follows from (5.50) on setting $I_a = I_b$.] Therefore, the spherical-top energy eigenfunctions can be chosen as eigenfunctions of $\hat{P}_c$. The angular-momentum components along molecule-fixed axes obey the general angular-momentum commutation relations, except that i is replaced by $-i$ [Eq. (5.42)]. If we replace i by $-i$ in all the equations of Section 1.7, the resulting eigenvalues are unchanged; we conclude that the eigenvalues of $\hat{P}_c$ are

$$K\hbar, \qquad K = 0, \pm 1, \ldots, \pm J \tag{5.58}$$

The quantum number K gives the rotational angular-momentum component along a molecule-fixed axis of the spherical top. From (5.34) and (5.58), we see that the eigenfunctions of $\hat{P}_c$ are

$$\left(1/\sqrt{2\pi}\right)e^{iK\chi} \tag{5.59}$$

Hence the spherical-top eigenfunctions have the form[4]

$$H_{JKM}(\theta)(1/2\pi)e^{iM\varphi}e^{iK\chi} \tag{5.60}$$

[4] Some workers restrict K to nonnegative values, in which case K is replaced by $\pm K$ in (5.60).

We have three quantum numbers J, K, and M, but the energy depends only on J. Since K and M each can have $2J+1$ different values for each value of J, the spherical-top rotational levels are $(2J+1)^2$-fold degenerate.

5.5 THE SYMMETRIC TOP

For the symmetric top, we have either $I_a = I_b$ or $I_b = I_c$. We shall initially assume that $I_a = I_b$, so that the c principal axis is the symmetry axis. The Hamiltonian is

$$\hat{H} = \frac{\hat{P}_a^2 + \hat{P}_b^2}{2I_b} + \frac{\hat{P}_c^2}{2I_c} = \frac{\hat{P}^2 - \hat{P}_c^2}{2I_b} + \frac{\hat{P}_c^2}{2I_c} \tag{5.61}$$

Since $\hat{P}_c$ commutes with $\hat{P}^2$ and with $\hat{P}_c^2$, the symmetric-top Hamiltonian commutes with $\hat{P}_c$ [this also follows from (5.50) with $I_a = I_b$], and the energy eigenfunctions can be chosen as eigenfunctions of $\hat{P}_c$. We thus have [Equation (5.58)]

$$\hat{P}_c \psi = K\hbar\psi$$

just as for the spherical top. Also

$$\hat{P}_c^2 \psi = \hat{P}_c\left(\hat{P}_c \psi\right) = K\hbar \hat{P}_c \psi = K^2\hbar^2\psi \tag{5.62}$$

The energy eigenvalues are easily found using (5.62) and (5.52):

$$\hat{H}\psi = E\psi$$

$$\frac{\hat{P}^2 - \hat{P}_c^2}{2I_b}\psi + \frac{\hat{P}_c^2}{2I_c}\psi = E\psi$$

$$\frac{J(J+1)\hbar^2 - K^2\hbar^2}{2I_b}\psi + \frac{K^2\hbar^2}{2I_c}\psi = E\psi$$

$$E = \frac{J(J+1)\hbar^2}{2I_b} + K^2\hbar^2\left(\frac{1}{2I_c} - \frac{1}{2I_b}\right) \tag{5.63}$$

Analogous to (4.31) for the diatomic molecule, the *rotational constants* of a polyatomic molecule are defined as

$$A \equiv \frac{h}{8\pi^2 I_a} \quad \geqslant \quad B \equiv \frac{h}{8\pi^2 I_b} \quad \geqslant \quad C \equiv \frac{h}{8\pi^2 I_c} \tag{5.64}$$

where (5.7) was used. The symmetric-top energies (5.63) are

$$E/h = BJ(J+1) + (C-B)K^2 \qquad \text{(oblate)} \tag{5.65}$$

$$J = 0, 1, 2, \ldots \qquad K = 0, \pm 1, \pm 2, \ldots, \pm J \tag{5.66}$$

Note that $(C-B)$ is negative. If the symmetry axis is the a axis, we must interchange a and c in the above derivation, and the energies are

$$E/h = BJ(J+1) + (A-B)K^2 \qquad \text{(prolate)} \tag{5.67}$$

Figure 5.3 shows the rotational levels for the two kinds of symmetric tops. If we set $A = B = C$ in (5.65) and (5.67), these formulas reduce to the spherical-top energies (5.57), as they should.

Since the symmetric-top energy eigenfunctions are eigenfunctions of $\hat{P}_c$ and of $\hat{P}_Z$, they have the form

$$G_{JKM}(\theta)(1/2\pi)e^{iM\varphi}e^{iK\chi} \tag{5.68}$$

The theta factor (which turns out to be a rather complicated function) can be found[5] by direct solution of the Schrödinger equation or by the use of ladder operators. The work is involved and we omit it. The spherical-top wave functions (5.60) are identical with the symmetric-top wave functions (5.68)—see Problem 5.14.

The symmetric-top energy depends on J and on K^2. There is a $(2J+1)$-fold degeneracy associated with M. In addition for $K \neq 0$, there is a twofold degeneracy associated with the values $+|K|$ and $-|K|$, which give the same value of K^2. The symmetric-top degeneracies are then $4J+2$ for $K \neq 0$ and $2J+1$ for $K = 0$.

What about linear molecules? Here the internuclear axis is a C_∞ axis, and a linear molecule is a symmetric top. Since the nuclei all lie on the symmetry axis, we have $I_a = 0$; also $I_b = I_c$. The ellipsoid of inertia is an infinite right circular cylinder. For a linear molecule, the nuclei cannot have any angular momentum about the a axis, so that K must be zero. Thus for a linear molecule, (5.67) reduces to

$$E = hBJ(J+1) \qquad \text{(linear)} \tag{5.69}$$

as found in Chapter 4 for $^1\Sigma$ diatomic molecules.

However, for a linear molecule, we *can* have *electronic* angular momentum about the internuclear axis; thus in a sense, we can view a linear molecule that is in a non-$^1\Sigma$ electronic state as a symmetric top with $K\hbar$

[5] *Pauling and Wilson*, p. 277; W. H. Shaffer, *J. Mol. Spectry*., **1**, 69 (1957); D. G. Burkhard, *ibid*., **2**, 187 (1958).

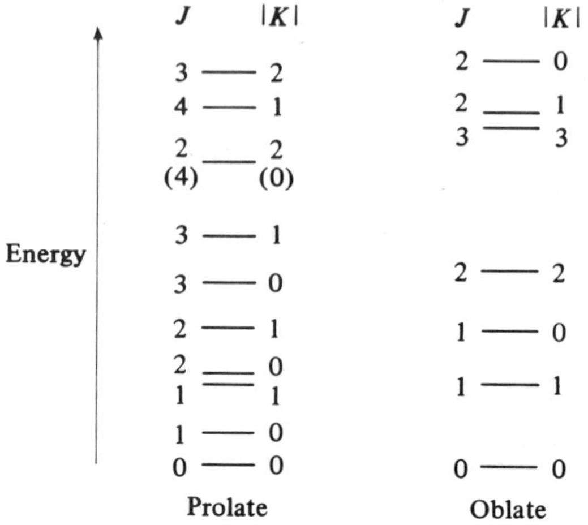

Fig. 5.3 Rotational levels for a prolate symmetric top with $A=4.5B$ and for an oblate symmetric top with $B=4.5C$. Note the accidental degeneracy of the $J=2$, $|K|=2$, and $J=4$, $|K|=0$ prolate-top levels.

being replaced by the component of electronic angular momentum along the internuclear axis. For case (a) coupling, the absolute value of the axial electronic angular momentum is $\Omega\hbar$; replacing $|K|$ by Ω in (5.67), we get

$$E/h=B[J(J+1)-\Omega^2]+a\Omega^2 \tag{5.70}$$

where B is calculated from the nuclear positions and masses, and a is some constant that can be viewed roughly as $h/8\pi^2 I_{el}$, where I_{el} is some sort of average moment of inertia of the electrons about the internuclear axis. Since electrons are much lighter than nuclei, it follows that $a \gg B$. The quantity $a\Omega^2$ is really electronic energy, not rotational energy of the molecule as a whole, and is best included in E_{el} rather than E_{rot}. This then gives the expression (4.145) for the rotational energy in case (a) coupling. The condition (5.66) that $J \geqslant |K|$ becomes $J \geqslant \Omega$ [Eq. (4.144)].

For case (b) coupling, if we first neglect electronic spin angular momentum, then N is analogous to J and Λ is analogous to $|K|$, so that (5.67) becomes

$$E/h=B[N(N+1)-\Lambda^2]+a\Lambda^2 \tag{5.71}$$

Considering $a\Lambda^2$ as part of the electronic energy and adding a term to take account of the electronic spin angular momentum, we get the expression (4.149). The condition $J \geqslant |K|$ becomes $N \geqslant \Lambda$ [Equation (4.147)].

Up to this point, the molecule has been considered to be a *rigid* rotor, but the work in Chapter 4 on diatomics shows that we must add corrections for rotation–vibration interaction and centrifugal distortion. For a polyatomic molecule, there are several normal modes of vibration, each with its own vibrational quantum number v_i (see Chapter 6). By analogy to (4.75), we write for polyatomic molecules

$$B_{[v]} = B_e - \sum_i \alpha_i^B (v_i + \tfrac{1}{2}) \tag{5.72}$$

where B_e is the equilibrium B rotational constant, $B_{[v]}$ is[6] the effective rotational constant for a given vibrational state, and where the sum is over all the vibrational modes of the molecule. (A separate term is included for each mode of a degenerate vibration.) We have similar expressions for the A and C rotational constants. For a diatomic molecule, the centrifugal-distortion correction is $-\bar{D}_e J^2(J+1)^2$; for polyatomic symmetric tops, we must include centrifugal-distortion terms for both J and K, and (5.65) becomes

$$E/h = B_{[v]} J(J+1) + [C_{[v]} - B_{[v]}]K^2 \\ - D_J J^2(J+1)^2 - D_{JK} J(J+1)K^2 - D_K K^4 \tag{5.73}$$

with a similar expression with A replacing C for a prolate symmetric top. The centrifugal-distortion constants D_J, D_{JK}, and D_K are generally very small compared to B and C.

5.6 THE ASYMMETRIC TOP

The asymmetric-top Hamiltonian is

$$\hat{H} = \frac{\hat{P}_a^2}{2I_a} + \frac{\hat{P}_b^2}{2I_b} + \frac{\hat{P}_c^2}{2I_c} = \frac{hA}{\hbar^2}\hat{P}_a^2 + \frac{hB}{\hbar^2}\hat{P}_b^2 + \frac{hC}{\hbar^2}\hat{P}_c^2 \tag{5.74}$$

The wave functions have the form (5.54), but since $\hat{P}_c$ does not commute with $\hat{H}$, we cannot separate out a chi factor; the Schrödinger equation is not separable, and we will try another method of dealing with the problem. We saw in Section 2.3 that the eigenvalues of an operator $\hat{H}$ can be found by expanding the unknown eigenfunctions ψ_i in terms of some known complete orthonormal set φ_i and solving the resulting secular equation (2.77):

$$\det\left[\langle \varphi_m | \hat{H} | \varphi_n \rangle - E_i \delta_{mn}\right] = 0 \tag{5.75}$$

[6] The standard notation for the spectra of polyatomic molecules is given in R. S. Mulliken, *J. Chem. Phys.*, **23**, 1997 (1955).

Let ψ_i be an *asymmetric*-top wave function. A convenient complete orthonormal set to use here is the *symmetric*-top wave functions, which are functions of the same coordinates (the Eulerian angles) and satisfy the same boundary conditions as the asymmetric-top functions:

$$\psi_i = \sum_{J'} \sum_{K'} \sum_{M'} c_{i,J'M'K'} \varphi_{J'M'K'} \tag{5.76}$$

where the $\varphi_{J'M'K'}$ are given by (5.68).

Since (Section 5.3) the asymmetric-top function ψ_i is an eigenfunction of $\hat{P}^2$ with eigenvalue $J(J+1)\hbar^2$, we only include in the sum those symmetric-top functions that have the same J value as ψ_i. [See the paragraph after (1.37).] Similarly, since ψ_i is an eigenfunction of $\hat{P}_z$ with eigenvalue $M\hbar$, we only include φ's that have the same M value as ψ_i. The infinite sum (5.76) over the quantum numbers J', K', and M' therefore reduces to a finite sum over the $2J+1$ possible values of K:

$$\psi_i = \sum_{K'=-J}^{J} c_{i,JMK'} \varphi_{JMK'} \tag{5.77}$$

where J and M are the asymmetric-top quantum numbers for the total rotational angular momentum and its component along a space-fixed axis. Correspondingly, the secular determinant in (5.75) reduces to a finite determinant of order $2J+1$:

$$\det(H_{K'K''} - E_i \delta_{K'K''}) = 0 \tag{5.78}$$

$$H_{K'K''} \equiv \int \varphi^*_{JMK'} \hat{H} \varphi_{JMK''}\, d\tau \tag{5.79}$$

where K' and K'' each range from $-J$ to $+J$. From the known symmetric-top wave functions (5.68) and the asymmetric-top Hamiltonian (5.74), the integral $H_{K'K''}$ can be evaluated. We omit the derivation, and simply give the result:

$$\begin{aligned} H_{K'K''} = {} & \delta_{K'',K'}(\tfrac{1}{2}h)\left[(2C-A-B)(K')^2 + (A+B)J(J+1)\right] \\ & + \delta_{K'',K'+2}(\tfrac{1}{4}h)(B-A)[J(J+1)-K'(K'+1)]^{1/2} \\ & \times [J(J+1)-(K'+1)(K'+2)]^{1/2} \\ & + \delta_{K'',K'-2}(\tfrac{1}{4}h)(B-A)[J(J+1)-K'(K'-1)]^{1/2} \\ & \times [J(J+1)-(K'-1)(K'-2)]^{1/2} \end{aligned} \tag{5.80}$$

In general, the roots of (5.78) are all different, so that for each value of J and M we have $2J+1$ different energies. For a given J and M, we arrange these $2J+1$ energies in order of increasing value, and add the label τ to each asymmetric-top wave function to indicate which energy it goes with:

$$\psi_i = \psi_{JM\tau}$$

The convention is that τ ranges from $-J$ to J as E increases. The index τ is a bookkeeping number, rather than a true quantum number. The degeneracy of the asymmetric-top energy levels is $2J+1$, corresponding to the $2J+1$ values of M, which do not affect the energy. Each asymmetric-top wave function ψ_i is a linear combination of the $2J+1$ symmetric-top wave functions with the same value of J and of M as ψ_i.

Let us consider some examples. For $J=0$, the only value of K is 0, and the secular equation (5.78) is

$$|H_{00}-E|=0 \quad \text{or} \quad E=H_{00}$$

Equation (5.80) then gives $E=0$ for $J=0$. The $J=0$ asymmetric-top wave function is, of course, identical with the symmetric-top $J=0$ wave function.

For $J=1$, we have $K=-1,0$, and $+1$. We can write the φ functions in the expansion (5.77) in any order we please; we choose to write the functions with odd K' first, so that the secular equation (5.78) is

$$\begin{vmatrix} H_{-1,-1}-E & H_{-1,1} & H_{-1,0} \\ H_{1,-1} & H_{1,1}-E & H_{1,0} \\ H_{0,-1} & H_{0,1} & H_{0,0}-E \end{vmatrix} = 0 \qquad (5.81)$$

(The motive for taking the order as $-1,1,0$ will be found in Problem 5.8.) Equation (5.80) is used to evaluate the matrix elements, and the secular equation (5.81) becomes

$$\begin{vmatrix} \frac{1}{2}h(2C+A+B)-E & \frac{1}{2}h(B-A) & 0 \\ \frac{1}{2}h(B-A) & \frac{1}{2}h(2C+A+B)-E & 0 \\ 0 & 0 & h(A+B)-E \end{vmatrix} = 0 \qquad (5.82)$$

which factors into a 2×2 and 1×1 determinant, giving

$$[\tfrac{1}{2}h(2C+A+B)-E]^2-[\tfrac{1}{2}h(B-A)]^2=0$$

$$h(A+B)-E=0$$

In order of increasing energy, the roots are

$$E = h(B+C), \quad h(A+C), \quad h(A+B) \tag{5.83}$$

Note the permutation symmetry in (5.83).

For large J, the secular equation is an algebraic equation of high degree in E, and we have to resort to approximation procedures for hand calculations; see *Townes and Schawlow*, Chapter 4. Computer programs that diagonalize the asymmetric-top Hamiltonian matrix to yield accurate energy eigenvalues (for the assumed moments of inertia) are available.

The parameter B in the asymmetric-top Hamiltonian (5.74) can take all values between A and C. For $B=A$, we have an oblate symmetric top; for $B=C$, we have a prolate symmetric top; for $A>B>C$, we have an asymmetric top. A correlation diagram can be drawn showing the energy variation as a function of B, keeping A and C fixed. Figure 5.4 shows the correlation between the lowest asymmetric-rotor energy levels and the corresponding oblate and prolate symmetric-rotor levels. Note the splitting of the twofold K degeneracy. Instead of using the index τ to label the asymmetric-top energies and wave functions, we can give the values of $|K|$ for the prolate- and oblate-top levels that correlate with the asymmetric-top level; these values are written as subscripts to the J value: $J_{K_{pr}K_{ob}}$. Both notations are used in Fig. 5.4. (Of course, neither K_{pr} nor K_{ob} is a true quantum number for the asymmetric top.)

As an example of determining the correlations, consider the $J=1$ levels (5.83). The lowest $J=1$ level is 1_{-1} with energy $h(B+C)$. For a prolate symmetric top, $B=C$, and the 1_{-1} energy becomes $2Bh$ in the prolate symmetric-top limit. Equation (5.67) then tells us that the 1_{-1} asymmetric-top level must correlate with the $J=1$, $K=0$ prolate-top level (as shown in Fig. 5.4). For an oblate top, $A=B$, and the 1_{-1} energy is $h(B+C) = h(2B+C-B)$ in the oblate limit. Hence from (5.65), the 1_{-1} level correlates with the $J=1$, $K=1$ oblate-top level. Similarly, we find the 1_0 level to be the 1_{11} level, and the 1_1 level to be the 1_{10} level.

Vibration–rotation interaction causes the rotational constants to vary with the vibrational quantum numbers [Eq. (5.72)]. Correction terms for centrifugal distortion can also be added to the energy expressions for asymmetric tops.

5.7 PURE-ROTATION SPECTRA OF POLYATOMIC MOLECULES

The electric-dipole selection rules for pure-rotation transitions will be considered in Section 6.5. Here we will simply give the results.

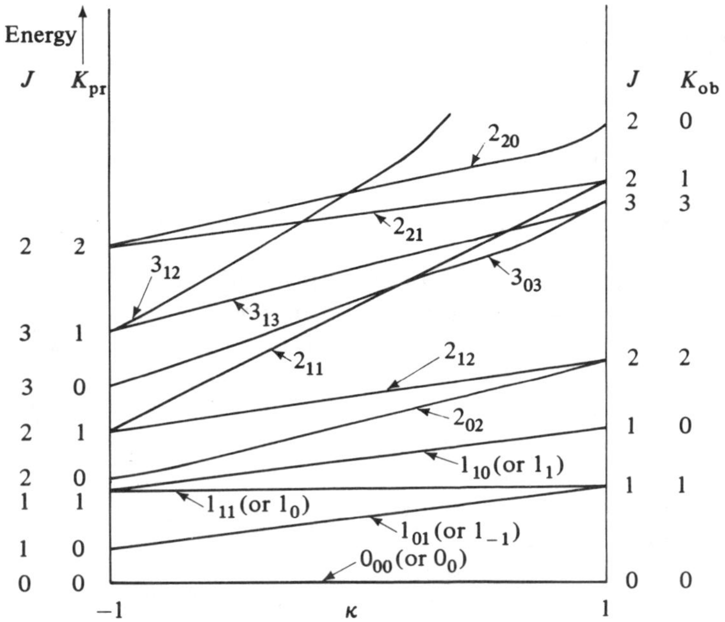

Fig. 5.4 Correlation diagram for asymmetric-top levels. *Ray's asymmetry parameter* κ is defined as $\kappa = (2B - A - C)/(A - C)$. The symmetric-top levels have been taken from Fig. 5.3; thus it is assumed that $A = 4.5C$. Note that the 1_{11} energy is independent of B and hence gives a horizontal line.

First of all, a molecule must possess a permanent electric dipole moment to exhibit electric-dipole pure-rotation transitions.

Spherical-top molecules have two or more noncoincident C_n axes and therefore (Section 1.18) have no permanent electric dipole moment. Thus spherical tops show a particularly simple microwave spectrum, namely, no lines at all.

Symmetric tops with no dipole moment have no microwave spectrum. For example, planar symmetric-top molecules have a C_n axis and a σ_h symmetry plane; such molecules cannot have a dipole moment. Thus benzene has no microwave spectrum. For a symmetric top with a permanent electric dipole moment, the selection rules for pure-rotation transitions are

$$\Delta J = 0, \pm 1, \qquad \Delta K = 0, \qquad \Delta M = 0, \pm 1 \tag{5.84}$$

(Since K is forbidden to change, a $\Delta J = 0$ transition with M changing leads to no absorption or emission of radiation.) Microwave transitions are studied in absorption. From (5.73), the frequencies of the microwave absorption lines of a symmetric top are

$$\nu(J \to J+1) = 2(B_{[v]} - D_{JK}K^2)(J+1) - 4D_J(J+1)^3 \qquad (5.85)$$

This expression applies to both prolate and oblate tops. Since B depends on the vibrational quantum numbers, we will get a separate pure-rotation spectrum for each set of vibrational quantum numbers; we will assume that only vibration–rotation levels with $v_1 = v_2 = \cdots = 0$ are significantly populated. (This is often not true for polyatomic molecules, which may have one or more low-frequency vibrational modes.) The value of B for all v_i equal to zero is called B_0. If centrifugal distortion is neglected, we get a series of equally spaced lines at frequencies $2B_0, 4B_0, 6B_0, \ldots$, just as for $^1\Sigma$ diatomic and linear polyatomic molecules. Inclusion of centrifugal distortion has two effects. The D_J term shifts the lines to low frequency, this shift increasing with increasing J. The D_{JK} term splits each line into several components, each corresponding to a different value of $|K|$; for the rotational level with quantum number J, there are $J+1$ possible values of $|K|$, and we get $J+1$ closely spaced lines. For reasonably heavy molecules, D_{JK} is usually so small that this splitting is not resolved for low J lines. As a typical case, the following constants were deduced[7] from the microwave spectrum of the symmetric top $CF_3C{\equiv}CH$:

$$B_0 = 2877.95 \text{ MHz}, \qquad D_J = 0.0002_4 \text{ MHz}, \qquad D_{JK} = 0.0063 \text{ MHz} \qquad (5.86)$$

For asymmetric tops with a dipole moment, the J and M selection rules are

$$\Delta J = 0, \pm 1 \qquad \Delta M = 0, \pm 1 \qquad (5.87)$$

For the asymmetric top, K is not a good quantum number. Asymmetric-top $\Delta J = 0$ transitions correspond to lines of nonzero frequency, in contrast to the symmetric top. Moreover, it is possible for an asymmetric-top level with quantum number $J-1$ to lie above a level with quantum number J (see Fig. 5.4); asymmetric-top microwave absorption lines with $\Delta J = -1$ are thus possible. Microwave transitions with $\Delta J = -1$, 0, and $+1$ are called P-, Q-, and R-branch lines, respectively, although these lines are interspersed among one another and do not really form branches as in the infrared spectrum. The dipole moment of an asymmetric top can be resolved into components along the three principal axes:

$$\mathbf{d} = d_a\mathbf{e}_a + d_b\mathbf{e}_b + d_c\mathbf{e}_c \qquad (5.88)$$

[7] W. E. Anderson, R. Trambarulo, J. Sheridan, and W. Gordy, *Phys. Rev.*, **82**, 58 (1951).

Transitions due to a nonvanishing dipole-moment component d_a are called a-type transitions. The selection rules for K_{pr} and K_{ob} for the three kinds of asymmetric-top microwave transitions are

$$\begin{aligned} &a \text{ type:} \quad \Delta K_{pr}=0, \pm 2, \pm 4, \ldots \qquad \Delta K_{ob}=\pm 1, \pm 3, \pm 5, \ldots \\ &b \text{ type:} \quad \Delta K_{pr}=\pm 1, \pm 3, \pm 5, \ldots \qquad \Delta K_{ob}=\pm 1, \pm 3, \pm 5, \ldots \\ &c \text{ type:} \quad \Delta K_{pr}=\pm 1, \pm 3, \pm 5, \ldots \qquad \Delta K_{ob}=0, \pm 2, \pm 4, \ldots \end{aligned} \tag{5.89}$$

In general, an asymmetric top can have nonvanishing components of **d** along all three axes, and all three types of transitions are possible; however, if there is a C_2 axis, then the dipole moment lies along this axis, which is also a principal axis, and we get only one type of transition. For all three types of transitions, the selection rules (5.87) hold. For asymmetric tops that are only slightly asymmetric, the $\Delta K_{pr}=0, \pm 1$ transitions will be most intense if B is near C, whereas $\Delta K_{ob}=0, \pm 1$ transitions will be most intense if B is near A. The patterns of asymmetric-top energy levels and allowed transitions are complex; in the microwave spectrum, P-, Q-, and R-branch, a-, b-, and c-type lines with high and low values of J are scattered among one another, with no clear-cut pattern.

At room temperature, many rotational levels of a molecule are well-populated because of the small energy differences between rotational levels. Thus, in contrast to electronic and vibrational absorption spectra, most of the lines in a pure-rotational absorption spectrum do not involve the ground state.

5.8 MICROWAVE SPECTROSCOPY

Microwave spectroscopy began in 1934 when Cleeton and Williams observed absorption at microwave frequencies for NH_3 contained in a rubber bag. The second molecule to have its microwave absorption observed was H_2O, the "work" being done by the U. S. Navy; during World War II, the Navy was developing radar and observed greater attenuation of signals than expected. Eventually it was realized that the radar frequency being used was one for which atmospheric water vapor absorbed. The development of radar provided the technology needed for microwave spectroscopy, and the field flourished after the war.

Microwave spectroscopy covers the frequency range from about 3 GHz to 300 GHz. The techniques used in the infrared, visible, and ultraviolet regions resemble one another closely, but the techniques of microwave spectroscopy differ considerably from those of optical spectroscopy. The source in microwave work is usually a *klystron*, which is an electronic tube that generates microwave radiation. Unlike the (nonlaser) sources in optical spectroscopy, a klystron provides nearly monochromatic radiation; hence no dispersing element is needed in the spectrometer. The frequency of the klystron's output can be varied by mechanically changing the size of a hollow resonant cavity in the tube, using a tuning screw. The microwave radiation from the klystron is propagated down a hollow metal pipe (usually of rectangular cross section) called a *waveguide*. The sample is a gas and is contained in a portion of the waveguide called the absorption cell; this cell is a few feet in length and is sealed off from the source and the detector by mica windows, which are essentially transparent to microwave radiation. A hole is drilled in the absorption cell and a glass-to-metal seal is made, which allows connection to a vacuum line to provide for introduction of the sample. The detector is a crystal diode consisting of a thin metal wire ("cat whisker") in contact with a piece of semiconducting silicon. The detector is mounted in the side of the waveguide and is in contact with a metal rod that extends across the waveguide and acts as an antenna for incident microwave radiation. The contact resistance in the detector is greater in one direction than the other, so that the diode rectifies the incident alternating-current microwave radiation. Accurate frequency measurements are made using a frequency standard; this consists of an electronic circuit containing a quartz-crystal oscillator whose frequency is adjusted (say to 1 MHz) by comparison with a standard frequency broadcast by radio station WWV of the U.S. National Bureau of Standards. The quartz-crystal frequency is first multiplied up to microwave frequencies by generating harmonics of it, using vacuum-tube and silicon-crystal multipliers, and then mixed in the waveguide with the klystron-generated frequency; this provides frequency markers of high accuracy. Microwave lines are routinely measured to 0.1 MHz and with more care can be measured to 0.01 MHz; this is high accuracy indeed, since a typical microwave frequency is 25,000 MHz.

Microwave absorption lines are weak compared to absorption lines in other frequency regions. To increase the sensitivity of the spectrometer, *Stark modulation* is widely used. In this process, introduced by Hughes and Wilson in 1947, a metal plate is inserted in the middle of the waveguide absorption cell; the plate runs the length of the cell and is insulated from the walls by strips of Teflon. A square-wave voltage (which is zero for a half cycle and a constant nonzero value for the other half cycle) of say 100 kHz frequency is applied to this plate. During the half cycle with nonzero voltage, the lines are shifted and split as compared to the unperturbed frequencies (Stark effect), thus giving rise to a 100 kHz ac signal at the detector crystal whenever an absorption line or one of its Stark components occurs. This ac signal is then amplified. (The principle of modulation is used in AM and FM radio transmission and detection.) Most microwave-spectroscopy research has been done using laboratory-built spectrometers; a commercial microwave spectrometer became available in 1964.

The sample must be in the gas phase. Microwave absorption lines are considerably broadened by molecular collisions at intermediate and high pressures; at atmospheric pressure, microwave absorption lines are tens of thousands of megacycles wide. Hence the gas is kept at low pressure, typically 0.01 to 0.1 torr; under these conditions, line widths run about $\frac{1}{2}$ MHz. The compound studied need not be a gas at room temperature, but it must have sufficient vapor pressure to give detectable absorption. To study involatile compounds such as the alkali halides, the waveguide must be heated to 500–1000°C; high-temperature microwave spectroscopy presents great experimental difficulties, but it has been used to study most of the alkali halides.

Although a permanent electric dipole moment is required for electric-dipole pure-rotation transitions, the microwave spectrum of O_2 is readily observed. The observed transitions are of magnetic-dipole type, due to the magnetic dipole moment associated with the triplet ground electronic state. Magnetic-dipole transitions are weaker than electric-dipole transitions, but for O_2 the microwave magnetic-dipole transitions are strong enough to be observed. The spectrometer used to study O_2 uses Zeeman rather than Stark modulation.

Aside from the possession of a permanent dipole moment and sufficient volatility, a molecule must be reasonably small for its microwave spectrum to be profitably studied. Large molecules have many low-frequency vibrational modes; these modes will be appreciably populated at room temperature, giving many strong pure-rotation transitions between levels with nonzero vibrational quantum numbers. The microwave spectrum of a large molecule will thus have so many lines that assignment of the lines will be virtually impossible.

The primary significance of microwave spectroscopy for chemistry is in determination of molecular structure. Assignment of microwave spectral lines to transitions between specific rotational levels allows determination of the rotational constants A_0, B_0, and C_0, and the corresponding moments of inertia. The moments of inertia are dependent on the molecular bond distances and bond angles.

For a symmetric top, the selection rules are such that we can determine only B_0 [see (5.85)]. Knowledge of $I_b{}^0$, the moment of inertia about a principal axis perpendicular to the symmetry axis, is not sufficient to determine the molecular structure, except for a diatomic molecule. To get added information, the microwave spectra of isotopically substituted species are examined. Isotopic substitution will cause a negligible change in the equilibrium bond distances and angles (the nuclear masses do not appear in the electronic Schrödinger equation). As an example, consider the symmetric top PCl_3; its structure is specified by two parameters: the P—Cl bond length and the ClPCl bond angle. A normal sample of PCl_3 is a mixture of $PCl_3{}^{35}$, $PCl_2{}^{35}Cl^{37}$, $PCl^{35}Cl_2{}^{37}$, and $PCl_3{}^{37}$. (The isotopic abundances are $P^{31}=100\%$, $Cl^{35}=75\%$, $Cl^{37}=25\%$.) The first and last species listed are symmetric tops, whereas the other two are asymmetric tops. The microwave spectrum yields[8] $B_0=2617.1$ MHz for $PCl_3{}^{35}$ and $B_0=2487.5$ MHz for $PCl_3{}^{37}$. (Since microwave lines are quite narrow at low pressures and since the B_0 values differ substantially, there is no need to separate the isotopic species when studying the microwave spectrum.) These two pieces of data allow a structural determination, the result being a P—Cl bond distance of 2.04 Å, and a bond angle of 100°. For molecules where the natural abundance of isotopically substituted species is very low, isotopically enriched samples can be prepared and studied.

The principal source of error in microwave structural determinations arises from the zero-point vibrations. Although R_e values remain unchanged on isotopic substitution, the B_0 values are not directly related to R_e values. The rotational constant B_0 is given by (5.72) with all v_i equal to zero. I_b^0 is calculated from B_0 as [see (5.64)]

$$I_b^0=\frac{h}{8\pi^2 B_0} \tag{5.90}$$

Since

$$\left\langle \frac{1}{\Sigma_\alpha m_\alpha b_\alpha^2} \right\rangle^{-1} \neq \left\langle \sum_\alpha m_\alpha b_\alpha^2 \right\rangle \tag{5.91}$$

(the brackets denote an average over the zero-point vibrations, and b_α is the distance of nucleus α from the b principal axis) an I_b^0 value does not really represent an average moment of inertia for the zero-point vibrations. Also, the magnitudes of zero-point vibrations change on isotopic substitution. Hence changes in B_0 are not directly related to either R_0 or R_e values (where these quantities are zero-point-averaged bond distances and equilibrium bond distances, respectively). This leads to discrepancies in structural determination. For example, B_0 values for four different pairs of isotopically substituted species of OCS give CO distances varying from 1.155 to 1.165 Å. One way to overcome this difficulty is to get the vibration–rotation interaction constants α_i for all the vibrational modes, and then calculate B_e from (5.72). This procedure is not practical (except

[8] P. Kisliuk and C. H. Townes, *J. Chem. Phys.*, **18**, 1109 (1950).

for diatomic and triatomic molecules), since it is only possible to observe pure-rotation transitions for excited vibrational modes of low frequency (which are well-populated at room temperature). The most practical way to get consistent bond distances and angles from microwave spectroscopic data is by the method proposed by Costain.[9] In this approach, the position of each different kind of atom is determined by isotopic substitution at that atom, using equations due to Kraitchman,[10] which relate the coordinates of an atom in the principal-axis system to the observed changes in moments of inertia on isotopic substitution of that atom. This method requires more work; for example, for PCl_3, as well as investigating PCl_3^{35} and PCl_3^{37}, one would also have to investigate a species with P isotopically substituted. The bond distances obtained by Costain's method are called R_s values (where s is for substitution), and are believed to lie between the R_0 and R_e values.

Microwave spectroscopy often gives more definite and precise information on the structure of polyatomic molecules than vibration–rotation and electronic spectra. For example, consider the simplest oxime: formaldoxime, $CH_2{=}NOH$. There are two likely structural configurations for this molecule; see Fig. 5.5. (The resonance structure $\overset{-}{C}H_2{-}N{=}\overset{+}{O}H$ favors planarity.) Two infrared studies of this molecule led to conflicting conclusions; the earlier study tended to favor configuration II, whereas the later study strongly favored configuration I. A microwave investigation[11] of $CH_2{=}NOH$ (which is an asymmetric top) gave I_a^0, I_b^0, and I_c^0 values for the molecule; however, this information is not sufficient to decide between I and II, since by adjusting bond distances and angles one can fit the observed moments of inertia with either configuration. To decide unambiguously between I and II, the isotopically substituted species $CH_2{=}NOD$ was prepared and studied. The hydrogen atom is relatively

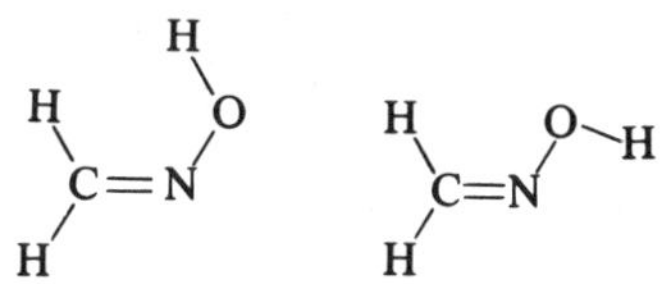

I II

Fig. 5.5 Two possible configurations for formaldoxime.

[9] C. C. Costain, *J. Chem. Phys.*, **29**, 864 (1958).

[10] J. Kraitchman, *Am. J. Phys.*, **21**, 17 (1953).

[11] I. N. Levine, *J. Mol. Spectry.*, **8**, 276 (1962); *J. Chem. Phys.*, **38**, 2326 (1963).

light, so that I and II will not differ much in their moments of inertia, but the *change* in moments of inertia on going from CH_2=NOH to CH_2=NOD would be quite different for I and II. The Kraitchman equations allowed calculation of the absolute values of the hydrogen-atom principal-axis coordinates; the result was that II is definitely the correct configuration. The complete R_s structure of formaldoxime was determined by investigating six species besides the normal species, with each atom in turn isotopically substituted. The resulting structure has the interesting feature that the *cis* HCN angle is 6° greater than the *trans* HCN angle: $121\frac{1}{2}°$ versus $115\frac{1}{2}°$.

Determination of the moments of inertia of a molecule generally allows one to decide whether the molecule is planar. If a_α, b_α, and c_α are the equilibrium principal-axis coordinates of nucleus α, then the equilibrium principal moments of inertia are

$$
\begin{aligned}
I_a^e &= \sum_\alpha m_\alpha (b_\alpha^2 + c_\alpha^2) \\
I_b^e &= \sum_\alpha m_\alpha (a_\alpha^2 + c_\alpha^2) \\
I_c^e &= \sum_\alpha m_\alpha (a_\alpha^2 + b_\alpha^2)
\end{aligned}
\tag{5.92}
$$

For a nonplanar molecule, these equations yield the inequality

$$I_c^e - I_a^e - I_b^e < 0 \tag{5.93}$$

For a planar molecule, one principal axis must be perpendicular to the molecular plane, and clearly this must be the axis of greatest moment of inertia, the c axis. Thus $c_\alpha = 0$ for all nuclei in a planar molecule. It follows that

$$I_c^e - I_a^e - I_b^e = 0 \tag{5.94}$$

for a planar molecule. The effective moments of inertia I_a^0, I_b^0, I_c^0, which are determined from the rotational constants A_0, B_0, C_0, are not true average moments of inertia [Eq. (5.91)]. Hence (5.94) does not hold for them; instead, it generally turns out that

$$\Delta \equiv I_c^0 - I_a^0 - I_b^0 > 0 \tag{5.95}$$

for a planar molecule. The quantity on the left of this inequality is the *inertial defect* Δ. A positive inertial defect is strong evidence for planarity. Inertial defects for planar molecules run about 0.02 to 0.20 amu Å^2. (For formaldoxime, the inertial defect is 0.054 amu Å^2.) It is possible for a slightly nonplanar molecule to have a small positive inertial defect: an example is formamide, H_2N—CHO, for which $\Delta = 0.01$ amu Å^2, but which a careful structural determination[12] shows to be nonplanar. The planar molecule phenol, C_6H_5OH, has a slightly negative Δ of -0.032 amu Å^2, due to the out-of-plane OH torsional vibration.

A second use of microwave spectroscopy is the measurement of dipole moments. These are obtained by measuring the frequency shifts of lines in the applied electric field of a Stark-modulated spectrometer. This method of dipole-moment determination is superior to the older method of measuring dielectric constants. For example, impurities in the sample will not affect the dipole moment as measured by microwave spectroscopy. The dipole moment of a substance present to the extent of a few percent can be accurately measured if its microwave spectral lines can be assigned. The components of **d** can be determined, thus giving its orientation in the molecule, in addition to its magnitude.

Although chemists tend to think of saturated hydrocarbons as nonpolar, this is not quite accurate. The C—H bond distance in a CCH_3 group may depart slightly from that in a CH group, and the CCH_3 bond angles may differ slightly from the tetrahedral angle. Hence the contributions of CH and CCH_3 groups to the dipole moment need not be exactly equal, and many saturated hydrocarbons have small dipole moments. The microwave spectra of propane, $(CH_3)_2CH_2$, and isobutane, $(CH_3)_3CH$, have been studied; these molecules have dipole moments of 0.083 D and 0.132 D, respectively. [D. R. Lide, *J. Chem. Phys.*, **33**, 1514, 1519, 1879 (1960).] The isotopically substituted molecules CH_3D and CH_3CD_3 have small dipole moments (see the discussion on HD in Section 4.11), and their pure-rotational spectra have been observed using microwave, molecular beam, and far-infrared techniques. The observed dipole moments are 0.0057 D and 0.0108 D, respectively. [I. Ozier, W. Ho, and G. Birnbaum, *J. Chem. Phys.*, **51**, 4873 (1969); E. Hirota and C. Matsumura, *ibid.*, **55**, 981 (1971).] Incredibly enough, the CH_4 molecule in its ground vibrational and electronic state has a tiny effective dipole moment due to centrifugal-distortion effects. High-J pure-rotational transitions of CH_4 have been observed in the far infrared at a gas pressure of 1 atm and an absorption path length of 139 m. [A. Rosenberg, I. Ozier, and A. K. Kudian, *J. Chem. Phys.*, **57**, 568 (1972).] The effective dipole moment is 5.4×10^{-6} D.

Another application of microwave spectroscopy is in the determination of vibrational frequencies. It might seem unlikely that study of the pure-

[12] C. C. Costain and J. M. Dowling, *J. Chem. Phys.*, **32**, 158 (1960).

rotation spectrum can give vibrational frequencies, but the dependence of the rotational constants on the vibrational quantum numbers makes this possible. This is done by measuring the ratio of intensities of the same pure-rotational transition for the ground vibrational state (all v_i equal to zero) and the vibrational state with one v_i equal to one, and all others zero. This ratio is equal to $\exp(h\nu_i/kT)$, where ν_i is the frequency of the excited mode. (This assumes that the nuclear statistical weights of corresponding rotational levels of the two vibrational levels are the same; see Section 6.10.) The method is applicable only to low-frequency vibrations, since only for these is there appreciable population of an excited vibrational state. An example is ketene, $CH_2{=}C{=}O$, for which microwave spectroscopy[13] found a vibrational mode with wave number 443 cm^{-1}. This low-frequency mode had not been found in previous infrared work.

A significant application of microwave spectroscopy is in the determination of barriers to internal rotation of one part of a molecule relative to another. Internal rotation is a vibrational motion, but has effects observable in the pure-rotation spectrum. If the barrier to internal rotation is very high, then the internal torsion is just like any other vibrational mode, and the rotational constants are affected in the usual way: $B_v = B_e - \alpha(v+\frac{1}{2})$. (An example is torsion about a C=C bond.) Measurement of the torsional frequency (by infrared spectroscopy, or by microwave spectroscopy as described in the previous paragraph) can yield an estimate of the barrier to internal rotation (see Problem 5.10).

In the limit of very high barrier, the internal torsional vibration of a group of identical atoms has a degeneracy associated with it, corresponding to different locations of these atoms in the potential well. For example, for CH_3—OH and CH_3—CH_3, the potential function for torsional vibration has three minima; in the limit of infinite rotational barrier, there is a threefold degeneracy of each torsional vibrational level, corresponding to vibrations about equilibrium positions differing from one another by a 120° rotation of one part of the molecule relative to the other. (This degeneracy is not included in the degeneracy of vibrational modes to be discussed in Section 6.4.) When the barrier is of finite height (as is actually the case in methanol and ethane), there is the possibility of tunneling through the barrier. The correct zeroth-order wave functions are linear combinations of the three vibrational functions that correspond to torsional vibration about the three equivalent equilibrium configurations; labeling these three configurations as I, II, and III, one finds the correct combinations to be

[13] A. P. Cox and A. S. Esbitt, *J. Chem. Phys.*, **38**, 1636 (1963).

$$3^{-1/2}(\psi_{I,v}+\psi_{II,v}+\psi_{III,v}) \tag{5.96}$$

$$3^{-1/2}(\psi_{I,v}+\omega\psi_{II,v}+\omega^2\psi_{III,v}) \tag{5.97}$$

$$3^{-1/2}(\psi_{I,v}+\omega^2\psi_{II,v}+\omega\psi_{III,v}) \tag{5.98}$$

where $\omega=\exp(2\pi i/3)$, and v is the torsional-vibration quantum number. (See Problem 5.13.) The energies of (5.97) and (5.98) are equal to each other (one function is simply the complex conjugate of the other) and differ from the energy of (5.96). Thus each threefold-degenerate torsional level is split into one nondegenerate sublevel and one doubly degenerate sublevel, as indicated for some of the levels in Fig. 5.6.

Let us consider the $v=0$ torsional level, since it will give the most intense transitions. Each of its two torsional sublevels has its own set of associated rotational levels, and the separation between a pair of rotational levels of the degenerate torsional sublevel is different from that between the corresponding pair of rotational levels of the nondegenerate torsional sublevel. Roughly speaking, the effective rotational constants differ for the two torsional sublevels. Each pure-rotation transition is then split into two lines, each line corresponding to a transition between rotational levels of one of the two torsional sublevels. The magnitude of the splitting allows the calculation of the barrier to internal rotation.

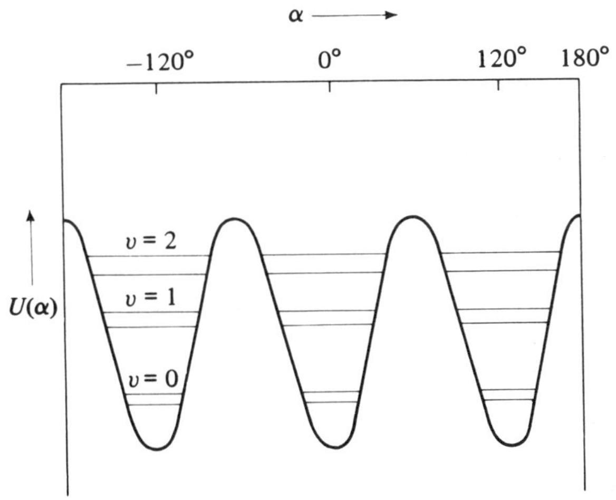

Fig. 5.6 Variation of the potential energy $U(\alpha)$ for torsional vibration; α is the angle of internal rotation in a molecule like ethane. The energy levels above the potential maxima have been omitted. For $\alpha=0°$, 120°, and $-120°$, we have the staggered form.

The most famous rotational barrier is that in ethane, but because the molecule is nonpolar its barrier is obtained from thermodynamic or infrared data, rather than from microwave spectroscopy. Microwave spectroscopy has provided barrier heights for a few dozen molecules. For molecules with three equivalent potential minima in the internal-rotation potential-energy function, the barriers usually range from 1 to 4 kcal/mole, except for very bulky substituents, where the barrier is higher. Interestingly, when the potential function has sixfold symmetry, the barrier is extremely low; for example, CH_3BF_2 has a barrier of 14 cal/mole.[14]

Besides internal rotation, another case where a molecule has more than one equivalent equilibrium configuration is NH_3; the nitrogen atom can be above the plane of the three hydrogens, or it can be below this plane. These two equilibrium positions (which correspond to inversion of the molecule) are separated by a potential maximum corresponding to the nitrogen's being in the plane of the three hydrogen atoms. For NH_3, the barrier to inversion is 5.8 kcal/mole. If $\psi_{\mathrm{I},v}$ and $\psi_{\mathrm{II},v}$ are the wave functions corresponding to vibration on each side of the potential barrier, then the correct linear combinations are

$$2^{-1/2}(\psi_{\mathrm{I},v} \pm \psi_{\mathrm{II},v}) \tag{5.99}$$

These two functions have different energies. If the barrier to inversion were infinitely high, then we would have doubly degenerate vibrational levels corresponding to vibration of the N atom on either side of the plane of the hydrogens. Since the barrier is of finite height, we get a splitting of each vibrational level into two closely spaced vibrational levels (called inversion levels), with the wave functions (5.99). The separation between these two levels for the ground vibrational state corresponds to a frequency of 24 GHz, which is a microwave frequency. This is an unusual situation in that transitions between vibrational levels usually correspond to infrared, rather than microwave, frequencies. The interaction between rotation and vibration gives a rotational fine structure to this transition, and we get a whole series of lines centered about 24 GHz. The rotational constant B_0 for NH_3 is 298 GHz, so that pure-rotation transitions occur at much higher frequencies than the transition between inversion levels.[15]

[14] For further discussion of rotational barriers, see *Sugden and Kenney*, Chapter 8; C. C. Lin and J. Swalen, *Rev. Mod. Phys.*, **31**, 841 (1959); E. B. Wilson, Jr., *Adv. Chem. Phys.*, **2**, 367 (1959); S. I. Miller, *J. Chem. Educ.*, **41**, 421 (1964); L. C. Allen, *Chem. Phys. Letters*, **2**, 597 (1968).

[15] For a discussion of inversion in NH_3, see G. W. Wheland, *Resonance in Organic Chemistry*, Wiley, New York, 1955, pp. 611–612.

Other molecules (e.g., AsH_3) have two equivalent equilibrium positions like NH_3, but the barrier between them is usually so high that the double degeneracy is split by a negligible amount.

SCF MO calculations of rotational barriers are generally quite successful. For the NH_3 inversion barrier, early SCF calculations yielded poor results. This was due to the inadequacy of the basis sets used. A calculation that included sufficient basis functions to reach near Hartree–Fock ψ's gave a barrier of 5.1 kcal/mole, in error by only 0.7 kcal/mole. [A. Rauk, L. C. Allen, and E. Clementi, *J. Chem. Phys.*, **52**, 4133 (1970).]

Microwave spectroscopy has been applied to the study of ring-puckering motions in small-ring compounds, and to the study of conformational isomers. See V. W. Laurie, *Accounts Chem. Res.*, **3**, 331 (1970).

There is another common cause of line splittings in microwave spectroscopy. Nuclei with spin $I \geqslant 1$ have an electric quadrupole moment. An electric quadrupole moment arises from a nonspherically symmetric charge distribution; for example, a uniformly charged ellipsoid of revolution has no electric dipole moment, but does have an electric quadrupole moment. The nuclear electric quadrupole moment is[16]

$$\int \rho_n(x,y,z)(3z^2 - r^2)\,dx\,dy\,dz \tag{5.100}$$

where ρ_n is the density of nuclear charge at point (x,y,z) of the nucleus, and where the z axis is taken parallel to the nuclear spin vector $\mathbf{I}$; the integral goes over the nuclear volume. The symbol for the nuclear quadrupole moment is eQ (where e is the proton's charge), and from (5.100) its dimensions are charge$\times$length^2. Tables usually list Q. Some values of Q follow:

Nucleus	H^2	B^{11}	N^{14}	O^{17}	Cl^{35}	Cl^{37}	Br^{79}	Br^{81}	I^{127}
Q (in units of 10^{-24} cm^2)	0.0028	0.036	0.016	-0.03	0.08	-0.06	0.33	0.28	-0.7

(5.101)

The interaction between the nuclear electric quadrupole moment and the electrons of the molecule couples the nuclear spin $\mathbf{I}$ to the rotational angular momentum $\mathbf{J}$, giving a resultant total angular momentum $\mathbf{F}$, of magnitude $\sqrt{F(F+1)}\,\hbar$. The quantum number F has the possible values

$$F = J + I,\ J + I - 1, \ldots, |J - I| \tag{5.102}$$

Each value of F corresponds to a different relative orientation of $\mathbf{J}$ and $\mathbf{I}$ and gives a slightly different energy. Thus each rotational level is split into $2I+1$ (or $2J+1$ if $I > J$) components, and the microwave line frequencies

[16] For the derivation, see *Townes and Schawlow*, pp. 130–135.

are also split. The splittings range from tenths of megacycles or less to a hundred or so megacycles for I^{127} compounds. If more than one nucleus has a quadrupole moment, we must form **F** as the resultant of **J** and more than one nuclear spin, giving further splitting. (Nuclei with $I=0$ or $\frac{1}{2}$ can be ignored in finding F, since these are not coupled to the rotational angular momentum.) The splittings due to nuclear quadrupole moments furnish information on chemical bonding, since they depend on the electrostatic interaction between the nucleus and the electrons.

The energy of the nuclear electric quadrupole in an *inhomogeneous* electric field varies with the orientation of the quadrupole moment. The electric field **E** at a particular nucleus in the molecule is determined by the electronic wave function and the electric fields produced by the other nuclei. There is no reason for **E** to be constant over the region around any nucleus. Rather, the components of **E** will vary, and we will have nonzero values of $\partial \mathcal{E}_x / \partial x$, $\partial \mathcal{E}_x / \partial y$, etc. The *electric potential* $V(x,y,z)$ at a point (x,y,z) in space can be defined such that $\mathbf{E} = -\mathbf{grad}\, V$, or [*Halliday and Resnick*, Section 29-7]

$$\mathcal{E}_x = -\partial V / \partial x, \qquad \mathcal{E}_y = -\partial V / \partial y, \qquad \mathcal{E}_z = -\partial V / \partial z \qquad (5.103)$$

In these equations, the xyz system is an arbitrary molecule-fixed set of axes located at the nucleus in question. Do not confuse the electric potential with potential energy (for which the same symbol is used). The electric field at a point is the force per unit charge at the point; therefore [Equation (1.7)], the electric potential is the potential energy per unit charge. At any point in the molecule, we can define a square matrix **q**:

$$\mathbf{q} = \begin{pmatrix} q_{xx} & q_{xy} & q_{xz} \\ q_{yx} & q_{yy} & q_{yz} \\ q_{zx} & q_{zy} & q_{zz} \end{pmatrix} \qquad (5.104)$$

$$q_{xx} \equiv \partial^2 V / \partial x^2, \qquad q_{xy} \equiv \partial^2 V / \partial x \partial y, \qquad \text{etc.} \qquad (5.105)$$

The **q** matrix is the negative of the electric-field gradient. Like the inertial tensor and the polarizability tensor, **q** is symmetric (since the order of partial differentiation is immaterial), and we can make an orthogonal transformation to a new set of axes α, β, γ such that **q** is diagonal, with diagonal elements $q_{\alpha\alpha}$, $q_{\beta\beta}$, $q_{\gamma\gamma}$. Note, however, that the origin for **q** is at the nucleus in question and the axes for which **q** is diagonal need bear no relation to the principal axes of inertia (unless the nucleus happens to lie on a symmetry element).

The interaction between a nuclear quadrupole moment eQ and the electric field gradient **q** at that nucleus gives a term in the Hamiltonian that involves $eQ\mathbf{q}$. (See *Wollrab*, Chapter 5.) By measuring the resulting splittings of pure-rotational lines, we can determine $eQq_{\alpha\alpha}$, $eQq_{\beta\beta}$, $eQq_{\gamma\gamma}$. These three quantities are called the *quadrupole coupling constants*, $\chi_{\alpha\alpha}$, $\chi_{\beta\beta}$, $\chi_{\gamma\gamma}$:

$$\chi_{\alpha\alpha} \equiv eQq_{\alpha\alpha}/h \quad \text{etc.} \tag{5.106}$$

Since the χ's are related to the molecular electronic charge distribution, they provide a good test of theoretically computed molecular wave functions. Hartree–Fock ψ's give generally good results for the quadrupole coupling constants. For example, a near-Hartree–Fock STO calculation on H_2O [T. H. Dunning, R. M. Pitzer, and S. Aung, *J. Chem. Phys.*, **57**, 5044 (1972)] gave the following values (in MHz) for O^{17} in H_2O^{17} (experimental values in parentheses): $\chi_{\alpha\alpha} = 10.8$ (10.2), $\chi_{\beta\beta} = -9.6$ (-8.9), $\chi_{\gamma\gamma} = -1.1$ (-1.3); similar good results are found for the deuterium quadrupole coupling constants in HDO. (Semiempirical EH and CNDO calculations of quadrupole coupling constants have given mixed results.)

For the rare molecule with nonzero total electron spin **S** in its ground electronic state, the interaction between the nuclear magnetic dipole moment (present in all nuclei with nonzero spin) and the electronic spin magnetic dipole moment causes the nuclear spin angular momenta to add to the rotational and electronic spin angular momenta, giving a hyperfine splitting of rotational lines. Examples are NO, NO_2, and ClO_2. (In considering Hund's coupling cases in Section 4.11, we omitted coupling the nuclear spin angular momentum to the other angular momenta.)

The use of microwave spectroscopy for chemical analysis has been suggested. A commercially available microwave spectrometer gives accurate intensity measurements, and applications to chemical analysis should increase.

The presence in interstellar space of many molecular species has been shown by observation of microwave spectral emission lines. Examples include H_2O, NH_3, H_2CO, HCN, CH_3OH, HCOOH, HCCCN, CO, $HCONH_2$, C_2H_5OH, and OH. See P. M. Solomon, *Physics Today*, March 1973, p. 32.

Compilations of microwave-spectroscopic results are available.[17]

For working problems, the following constant allows one to convert moments of inertia to rotational constants:

$$h/8\pi^2 = 505{,}379 \text{ MHz amu Å}^2 \tag{5.107}$$

[17] Landolt-Börnstein Tables, New Series, Group II, vols. 4 and 6: B. Starck, *Molecular Constants from Microwave Spectroscopy*, and J. Demaison *et al.*, *Molecular Constants*, Springer-Verlag, Berlin, 1967 and 1974.

PROBLEMS

5.1 Classify each of these molecules as a spherical, symmetric, or asymmetric top: (a) CH_4; (b) CH_3F; (c) CH_2F_2; (d) CH_3D; (e) SF_6; (f) SF_5Br; (g) *trans*-SF_4Br_2; (h) *cis*-SF_4Br_2; (i) HCN; (j) H_2S; (k) SF_4; (l) C_6H_6; (m) $C_6H_6Cr(CO)_3$; (n) XeF_4; (o) ClF_3; (p) spiropentane.

5.2 For which point groups must the molecule be a spherical top?

5.3 For which point groups must the molecule be a symmetric top?

5.4 For water, the equilibrium OH bond distance is 0.957 Å and the equilibrium bond angle is 104.5°. (a) Locate the principal axes and then calculate the equilibrium moments of inertia. (b) Calculate the rotational constants A_e, B_e, and C_e. (c) Calculate the frequencies corresponding to transitions between the $J=0$ level and each of the three $J=1$ levels; ignore the effects of zero-point vibrations. (d) Which of the transitions in part (c) are actually allowed for H_2O? (e) Give the frequencies of the allowed $J=1$ Q-branch transitions.

5.5 Verify (5.30) and (5.31).

5.6 Show that for a linear triatomic molecule

$$I_b = \frac{m_1 m_2 R_{12}^2 + m_1 m_3 R_{13}^2 + m_2 m_3 R_{23}^2}{m_1 + m_2 + m_3}$$

where the atoms are labeled 1, 2, and 3, and where R_{ij} is the distance between atoms i and j. *Hint*: use $(R_{13} - R_{12} - R_{23})^2 = 0$.

5.7 The lowest frequency microwave transitions of $HC^{12}N^{14}$ and $DC^{12}N^{14}$ occur at 88,631 and 72,415 MHz, respectively. (These are for the ground vibrational state.) Use the formula of Problem 5.6 to calculate the bond distances in HCN. (Ignore zero-point vibrations.)

5.8 (a) Show from (5.80) that grouping odd K and even K values separately when writing down the asymmetric-top secular determinant causes the determinant to be in block-diagonal form, one block having matrix elements with odd K' and K'', the other having matrix elements with even K' and K''. (b) What does the result of part (a) imply about which values of K occur in the expansion (5.77) for a given asymmetric-top energy level?

5.9 Solve the secular equation for the $J=2$ asymmetric-top levels. (Be sure to group even K values together.)

5.10 The potential function for internal rotation of a group of identical atoms is usually assumed to have the form $U(\alpha) = \frac{1}{2} U_0(1 - \cos n\alpha)$, where n is the number of minima and U_0 is the barrier height. For a torsional vibration, the vibration frequency is given by an expression analogous to

(4.27), except that k is replaced by the torque constant (defined as $\partial^2 U/\partial\alpha^2|_{\alpha=0}$) and μ is replaced by the reduced moment of inertia (defined as $I_1 I_2/(I_1+I_2)$, where I_1 and I_2 are the moments of inertia of the two halves of the molecule about the axis of torsion). If the barrier is very high, it is a good approximation to use the harmonic-oscillator expression for the energy of the lowest torsional levels. The wave number for torsional vibration about the C=C bond in $CH_2{=}CH_2$ is 1023 cm^{-1}. The structure of ethylene is $R_{CH}=1.09$ Å, $R_{CC}=1.33$ Å, $\angle HCC=122°$. Estimate the rotational barrier in ethylene.

5.11 A triatomic molecule has the formula A_2B. Its microwave spectrum shows strong lines at 15, 30, 45, ... kMHz, and no other lines. Which of the following structures is (are) compatible with this spectrum: (a) linear AAB; (b) linear ABA; (c) bent AAB; (d) bent ABA?

5.12 Use the generalized Hellmann–Feynman theorem (*Levine*, Section 14.3) to evaluate $\langle P_a^2\rangle$, $\langle P_b^2\rangle$, and $\langle P_c^2\rangle$ for each of the $J=1$ asymmetric-top states.

5.13 Show that if the overlap between torsional-vibration wave functions corresponding to oscillation about different equilibrium configurations is neglected, the perturbation-theory secular equation (1.207) for internal rotation in ethane has the same form as the secular equation for the Hückel MOs of the cyclopropenyl system, thereby justifying (5.96)–(5.98). Write down an expression (in terms of the Hamiltonian and the wave functions) for the energy splitting between sublevels of each torsional level.

5.14 (a) From the symmetric-top Schrödinger equation, find the ordinary differential equation satisfied by the function $G(\theta)$ in (5.68). (b) From the spherical-top Schrödinger equation, find the ordinary differential equation satisfied by the function $H(\theta)$ in (5.60). Verify that the differential equations for G and for H are identical, thereby showing the spherical-top and symmetric-top eigenfunctions to be identical.

5.15 Show that the following relations hold for the angles in Fig. 5.2:

$$\cos(aOZ) = -\sin\theta\cos\chi, \qquad \cos(bOZ) = \sin\theta\sin\chi$$

$$\cos(XOc) = \sin\theta\cos\varphi, \qquad \cos(YOc) = \sin\theta\sin\varphi$$

[The first two relations were used in the derivation of (5.29).]

5.16 Derive (5.36) and (5.37). (Use the second pair of relations in Problem 5.15.)

5.17 Typically, a Stark-modulated microwave spectrometer will readily resolve lines separated by at least 0.3 MHz. For $CF_3C{\equiv}CH$, what is the lowest value of J for which the microwave absorption spectrum will show more than one line? Calculate the $J+1$ theoretical frequencies for this value of J. Use the data of (5.86).

5.18 For a linear polyatomic molecule, there is no rotational angular momentum about the internuclear axis; therefore, the derivatives with respect to chi are absent from (5.40). What are the rotational wave functions for a linear polyatomic molecule?

5.19 Show that a planar symmetric top must be oblate.

5.20 Calculate the principal moments of inertia of HNO. The bond angle is 109° and the bond distances are $R(\text{NO}) = 1.21$ Å, $R(\text{HN}) = 1.06$ Å. Now locate the principal axes of inertia.

5.21 For each of the following, state whether the principal axes of the inertial tensor have the same orientation in the molecule as the principal axes of the polarizability tensor. (a) H_2O; (b) HDO; (c) D_2O.

5.22 What is the orientation of the principal axes of the electric-field-gradient tensor $\mathbf{q}$ at O^{17} in H_2O^{17}?

5.23 Let a molecule have electronic wave function ψ. Write down an expression for the electric field at nucleus a due to the electrons and the other nuclei.

5.24 Write a computer program that finds the principal moments and principal axes of inertia for a molecule. Do not use matrix diagonalization; instead, solve the secular equation by using the formula for the roots of a cubic equation. The input to the program is the set of atomic masses and coordinates in an arbitrary system with axes not necessarily at the center of mass.

CHAPTER 6

VIBRATION OF POLYATOMIC MOLECULES

6.1 INTRODUCTION

In Section 5.1, we noted that to a good approximation the nuclear motion of a polyatomic molecule can be separated into translational, vibrational, and rotational motions. If the molecule has N nuclei, then the nuclear wave function is a function of $3N$ coordinates. The translational wave function depends on the three coordinates of the molecular center of mass in a space-fixed coordinate system. For a nonlinear molecule, the rotational wave function depends on the three Eulerian angles θ, φ, and χ (Section 5.3), which describe the orientation of the principal axes a, b, and c with respect to a nonrotating set of axes with origin at the center of mass. For a linear molecule, the rotational quantum number K must be zero, and the wave function (5.68) is a function of θ and φ only; only two angles are needed to specify the orientation of a linear molecule. Thus the vibrational wave function will depend on $3N-5$ or $3N-6$ coordinates, according to whether the molecule is linear or nonlinear; we say there are $3N-5$ or $3N-6$ vibrational degrees of freedom.

Before considering the quantum-mechanical vibrational wave functions and energies, we must find the classical-mechanical Hamiltonian for vibration. (Quantum mechanics is peculiar in that it depends on classical mechanics to formulate the Hamiltonian, and yet classical mechanics is only a limiting case of the more general theory quantum mechanics.)

6.2 CLASSICAL MECHANICS OF VIBRATION

Our model for molecular vibration is a set of N point masses (the nuclei), each of which vibrates about an equilibrium position (which is determined by solving the electronic Schrödinger equation). The vibrational motion is described using the principal axes a, b, and c, which translate and rotate with the molecule. Let a_α, b_α, and c_α be the Cartesian coordinates of nucleus α in the principal-axis system, and let $a_{\alpha,e}$, $b_{\alpha,e}$, and $c_{\alpha,e}$ be the equilibrium values of these coordinates. The $3N$ Cartesian displacement coordinates, which measure the displacement of each nucleus from equilibrium, are defined as

$$x_\alpha = a_\alpha - a_{\alpha,e}, \qquad y_\alpha = b_\alpha - b_{\alpha,e}, \qquad z_\alpha = c_\alpha - c_{\alpha,e} \tag{6.1}$$

Since there are only $3N-6$ (or $3N-5$) vibrational degrees of freedom, these coordinates are not all independent. They are connected by six relations, which specify that the *abc* axes translate and rotate with the molecule.

The classical kinetic energy of vibration about the equilibrium positions is

$$T = \frac{1}{2}\sum_{\alpha=1}^{N} m_\alpha \left[\left(\frac{dx_\alpha}{dt}\right)^2 + \left(\frac{dy_\alpha}{dt}\right)^2 + \left(\frac{dz_\alpha}{dt}\right)^2\right] \tag{6.2}$$

To simplify this equation, we define the *mass-weighted Cartesian displacement coordinates* $q_1,\ldots,q_{3N}$:

$$\begin{aligned} &q_1 = m_1^{1/2}x_1, \qquad q_2 = m_1^{1/2}y_1, \qquad q_3 = m_1^{1/2}z_1 \\ &q_4 = m_2^{1/2}x_2, \qquad \ldots \qquad q_{3N} = m_N^{1/2}z_N \end{aligned} \tag{6.3}$$

The kinetic energy becomes

$$T = \frac{1}{2}\sum_{i=1}^{3N}\left(\frac{dq_i}{dt}\right)^2 = \frac{1}{2}\sum_{i=1}^{3N}\dot{q}_i^2 \tag{6.4}$$

where the overhead dot indicates differentiation with respect to time. Let $\dot{\mathbf{q}}$ be defined as the column vector with elements $\dot{q}_1,\ldots,\dot{q}_{3N}$. In Chapter 2, we used a tilde to indicate the transpose of a matrix. An equally common notation is to use a prime to indicate the transpose: $\mathbf{A}' \equiv \tilde{\mathbf{A}}$. For typographical convenience, *we shall denote all matrix transposes by primes in this section*. $\dot{\mathbf{q}}'$ is a row vector, and we can write (6.4) in matrix notation as follows [see (2.6) and (2.26)]:

$$T = \tfrac{1}{2}\dot{\mathbf{q}}'\dot{\mathbf{q}} \tag{6.5}$$

The potential energy of vibration is some function of the coordinates

$x_1,\ldots,z_N$; hence it is a function of the mass-weighted coordinates $q_1,\ldots,$ q_{3N}. For a molecule, the vibrational potential energy V is given by the function U, the sum of the purely electronic energy and the nuclear repulsions:

$$V = U(q_1,\ldots,q_{3N}) \tag{6.6}$$

We expand the potential energy in a Taylor series about the equilibrium positions, which correspond to $q_1 = q_2 = \cdots = q_{3N} = 0$. The potential energy is a function of several variables, and the Taylor-series expansion is (*Taylor and Mann*, pp. 222–223)

$$U = U_e + \sum_{i=1}^{3N}\left(\frac{\partial U}{\partial q_i}\right)_e q_i + \frac{1}{2}\sum_{i=1}^{3N}\sum_{k=1}^{3N}\left(\frac{\partial^2 U}{\partial q_i \partial q_k}\right)_e q_i q_k$$

$$+\frac{1}{6}\sum_{i=1}^{3N}\sum_{j=1}^{3N}\sum_{k=1}^{3N}\left(\frac{\partial^3 U}{\partial q_i \partial q_j \partial q_k}\right)_e q_i q_j q_k + \cdots \tag{6.7}$$

[Note that (6.7) reduces to (4.12) if U is a function of only one coordinate.] U_e is the equilibrium electronic energy, the value of the electronic energy including nuclear repulsion for the equilibrium nuclear configuration. The subscript e on the partial derivatives indicates that they are evaluated at the equilibrium nuclear configuration. For equilibrium, the potential energy is a minimum; hence

$$\left(\frac{\partial U}{\partial q_i}\right)_e = 0, \qquad i = 1,\ldots,3N \tag{6.8}$$

We shall neglect terms higher than quadratic in (6.7); this is a good approximation if the vibrations are small. Equation (6.7) becomes

$$U = U_e + \frac{1}{2}\sum_{i=1}^{3N}\sum_{k=1}^{3N} u_{ik} q_i q_k \tag{6.9}$$

$$u_{ik} \equiv \left(\frac{\partial^2 U}{\partial q_i \partial q_k}\right)_e \tag{6.10}$$

Comparison with (4.14)–(4.17) and (6.3) shows the u_{ik}'s to be $(m_i m_k)^{-1/2}$ times the force constants. In matrix notation (6.9) reads [compare (2.44) and (2.45)]

$$U = U_e + \tfrac{1}{2}\mathbf{q}'\mathbf{U}\mathbf{q} \tag{6.11}$$

where the matrix **U** has elements u_{ik} and the column vector **q** has elements $q_1,\ldots,q_{3N}$.

We now solve for the classical-mechanical vibrational motion for the potential energy (6.9). Newton's second law $F_x = ma_x$ and the definition (1.7) of potential energy give

$$F_{x,\alpha} = -\partial V/\partial x_\alpha = m_\alpha d^2 x_\alpha / dt^2 \tag{6.12}$$

where $F_{x,\alpha}$ is the x component of force acting on the mass m_α. The mass-weighted coordinate q_j is defined by (6.3) as $q_j = m_\alpha^{1/2} x_\alpha$; hence

$$\frac{\partial V}{\partial x_\alpha} = \frac{\partial V}{\partial q_j}\frac{\partial q_j}{\partial x_\alpha} = m_\alpha^{1/2}\frac{\partial V}{\partial q_j} \tag{6.13}$$

$$\frac{d^2 x_\alpha}{dt^2} = \frac{d^2}{dt^2}\left(\frac{q_j}{m_\alpha^{1/2}}\right) = \frac{1}{m_\alpha^{1/2}}\frac{d^2 q_j}{dt^2} \tag{6.14}$$

Substitution of (6.13) and (6.14) into (6.12) gives

$$\frac{d^2 q_j}{dt^2} + \frac{\partial V}{\partial q_j} = 0, \qquad j = 1,\ldots,3N \tag{6.15}$$

V is given by (6.9) and contains a double sum over the q's. Hence $\partial V/\partial q_j$ will contain a single sum over the q's. Thus *each* differential equation in (6.15) involves *all* the q_j's, which complicates the solution process. To get a simpler set of differential equations, we will carry out a change of variables. The change of variables we want is one that reduces the double sum in (6.9) to a single sum of squares. The double sum is a quadratic form and a change of variables using the square matrix of eigenvectors of **U** reduces it to a sum of squares (Section 2.2). Let **L** be the matrix of eigenvectors of **U**. The matrix **U** is real and symmetric ($u_{ij} = u_{ji}$ because the order of partial differentiation is immaterial for a well-behaved function); hence **L** can be chosen as orthogonal: $\mathbf{LL}' = \mathbf{I}$. We have [Eq. (2.46)]

$$\mathbf{UL} = \mathbf{L\Lambda} \qquad \text{or} \qquad \mathbf{L'UL} = \mathbf{\Lambda} \tag{6.16}$$

where $\mathbf{\Lambda}$ is the diagonal eigenvalue matrix. The eigenvalues λ_m of **U** can be found from the *vibrational secular equation* [compare (2.38)]:

$$\det(u_{jk} - \delta_{jk}\lambda_m) = 0 \tag{6.17}$$

After (6.17) is solved, we can find the orthonormal eigenvectors $\mathbf{L}^{(m)}$ of **U** by solving the linear simultaneous equations that correspond to (2.36) or (2.37):

$$(\mathbf{U}-\lambda_m\mathbf{I})\mathbf{L}^{(m)}=\mathbf{0}$$

$$\sum_{k=1}^{3N}(u_{jk}-\delta_{jk}\lambda_m)l_{km}=0, \qquad j=1,\ldots,3N \tag{6.18}$$

The eigenvectors $\mathbf{L}^{(m)}$ are the columns of $\mathbf{L}$.

We now define the *normal coordinates* Q_i as the following linear combinations of the mass-weighted Cartesian displacement coordinates [compare (2.47) and (2.48)]:

$$Q_i=\sum_{k=1}^{3N}l_{ki}q_k, \qquad i=1,\ldots,3N \tag{6.19}$$

$$\mathbf{Q}=\mathbf{L}'\mathbf{q} \qquad \text{or} \qquad \mathbf{q}=\mathbf{L}\mathbf{Q} \tag{6.20}$$

$$q_i=\sum_{k=1}^{3N}l_{ik}Q_k, \qquad i=1,\ldots,3N \tag{6.21}$$

where $\mathbf{Q}$ is the column vector with elements $Q_1,\ldots,Q_{3N}$. Substitution into (6.11) gives

$$U-U_e=\tfrac{1}{2}\mathbf{q}'\mathbf{U}\mathbf{q}=\tfrac{1}{2}(\mathbf{LQ})'\mathbf{U}(\mathbf{LQ})=\tfrac{1}{2}\mathbf{Q}'\mathbf{L}'\mathbf{U}\mathbf{L}\mathbf{Q}=\tfrac{1}{2}\mathbf{Q}'\boldsymbol{\Lambda}\mathbf{Q} \tag{6.22}$$

$$U=U_e+\frac{1}{2}\sum_{k=1}^{3N-6}\lambda_k Q_k^2 \tag{6.23}$$

[Compare the two equations following (2.48).] The reason for taking the sum through $3N-6$, rather than $3N$, will be discussed later.

The effect of this change of variables on T in (6.5) is

$$T=\tfrac{1}{2}\dot{\mathbf{q}}'\dot{\mathbf{q}}=\tfrac{1}{2}(\mathbf{L}\dot{\mathbf{Q}})'\mathbf{L}\dot{\mathbf{Q}}=\tfrac{1}{2}\dot{\mathbf{Q}}'\mathbf{L}'\mathbf{L}\dot{\mathbf{Q}}=\tfrac{1}{2}\dot{\mathbf{Q}}'\dot{\mathbf{Q}}$$

$$T=\frac{1}{2}\sum_{k=1}^{3N-6}\dot{Q}_k^2 \tag{6.24}$$

Thus when expressed in normal coordinates, the potential energy becomes a sum of terms, each involving only one coordinate, and the kinetic energy stays as such a sum. (Hence the Hamiltonian is separable in normal coordinates.)

We now express the equations of motion (6.15) using the new coordinates Q_k. Using (6.19) with indices i and k interchanged and (6.15), we have

$$\frac{d^2Q_k}{dt^2} = \sum_{i=1}^{3N} l_{ik}\frac{d^2q_i}{dt^2} = -\sum_{i=1}^{3N} l_{ik}\frac{\partial V}{\partial q_i} \tag{6.25}$$

However, the chain rule [Equation (1.112)] and (6.21) give

$$\frac{\partial V}{\partial Q_k} = \sum_{i=1}^{3N} \frac{\partial V}{\partial q_i}\frac{\partial q_i}{\partial Q_k} = \sum_{i=1}^{3N} \frac{\partial V}{\partial q_i} l_{ik} \tag{6.26}$$

which is the sum that appears on the right of (6.25). Hence (6.25) becomes

$$\frac{d^2Q_k}{dt^2} + \frac{\partial V}{\partial Q_k} = 0, \qquad k = 1,\ldots,3N \tag{6.27}$$

$$\frac{d^2Q_k}{dt^2} + \lambda_k Q_k = 0, \qquad k = 1,\ldots,3N \tag{6.28}$$

where (6.6) and (6.23) were used to find $\partial V/\partial Q_k$. These equations of motion can be solved at once to give [compare (1.97) and (1.100)]

$$Q_k = B_k \sin(\lambda_k^{1/2}t + b_k), \qquad k = 1,\ldots,3N \tag{6.29}$$

where B_k and b_k are constants. Using (6.21) and (6.29), we find for the mass-weighted Cartesian coordinates

$$q_i = \sum_{k=1}^{3N} A_{ik}\sin(\lambda_k^{1/2}t + b_k), \qquad i = 1,\ldots,3N \tag{6.30}$$

$$A_{ik} \equiv l_{ik}B_k \tag{6.31}$$

Let us look at the physical nature of the solutions (6.30). We first consider the special case with all arbitrary constants $B_k = 0$, except $B_m \neq 0$. For this case, (6.30) becomes

$$q_i = A_{im}\sin(\lambda_m^{1/2}t + b_m), \qquad i = 1,\ldots,3N \tag{6.32}$$

In this case, the coordinates of each atom vibrate in phase with one another with the same frequency ν_m; when t in (6.32) has increased by $2\pi/\lambda_m^{1/2}$, the sine function has gone through one cycle; thus the period of the vibration is $2\pi/\lambda_m^{1/2}$, and its frequency ν_m is

$$\nu_m = \lambda_m^{1/2}/2\pi \tag{6.33}$$

Each such vibration (6.32) is called a *normal mode* of vibration. For each normal mode, the vibrational amplitude A_{im} of each atomic coordinate is constant, but the amplitudes for different coordinates are, in general, different. The nature of the normal modes depends on the molecular geometry, the nuclear masses, and the values of the force constants u_{jk}. The eigenvalues λ_m of $\mathbf{U}$ determine the vibrational frequencies; the eigenvectors of $\mathbf{U}$ determine the relative amplitudes of the vibrations of the q_i's in each normal mode, since $A_{jm}/A_{im} = l_{jm}/l_{im}$. For H_2O there are $9-6=3$ normal modes, and the solution of (6.17) and (6.18) yields the vibrational modes shown in Fig. 6.1. For some molecules, two or more normal modes have the same vibrational frequency (corresponding to two or more equal roots of the secular equation); such modes are called *degenerate*. For example, a linear triatomic molecule has four normal modes, two of which have the same frequency. See Fig. 6.2. The general classical-mechanical solution (6.30) is an arbitrary superposition of the normal modes.

Let us summarize the results. Using the principal-axis system, we defined the nuclear mass-weighted Cartesian coordinates q_i by (6.1) and (6.3):

$$q_1 = m_1^{1/2}(a_1 - a_{1,e}) \qquad \text{etc.} \tag{6.34}$$

In terms of these coordinates, the vibrational kinetic and potential energies are given by (6.4) and (6.9). The vibrational motion is a superposition of normal modes, according to (6.30). The frequencies of the normal modes are (6.33), where the λ_m are the roots of the secular equation (6.17). After the secular equation has been solved, the relative amplitudes of vibration l_{km} for the mth normal mode are found from (6.18). The normal coordinates Q_k are related to the coordinates q_i by (6.19)–(6.21). When expressed in normal coordinates, the vibrational kinetic and potential energies have the simple forms (6.24) and (6.23).

We have taken the sums in (6.23) and (6.24) to include only $3N-6$ (rather than $3N$) normal coordinates. The reasons for this are as follows. We know that there are only $3N-6$ (if the molecule is linear, read 5 for 6 in this and each subsequent expression) vibrational degrees of freedom. We didn't take the trouble to use the six relations between the q's to eliminate six of them, but treated all $3N$ q's as if they were independent. It can be shown that the consequence of this is that six of the roots λ_k will be zero; these six vanishing roots correspond to translational and rotational degrees of freedom. We adopt the convention that the zero roots will be

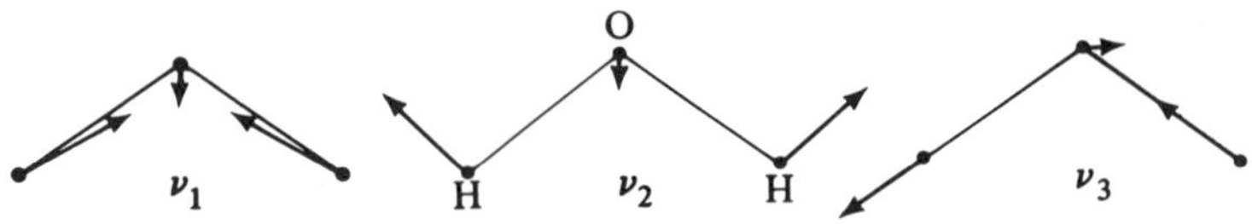

Fig. 6.1 Normal modes of H_2O. For ν_3, the vibrations of the hydrogen atoms are along the OH bonds; this is a consequence of the requirement that the vibration give rise to no rotational angular momentum (see *Herzberg, Infrared and Raman Spectra*, p. 133). For ν_1, the vibrations of the hydrogen atoms are not coincident with the OH bonds. For clarity, the magnitudes of the oxygen vibrations relative to those of hydrogen have been exaggerated.

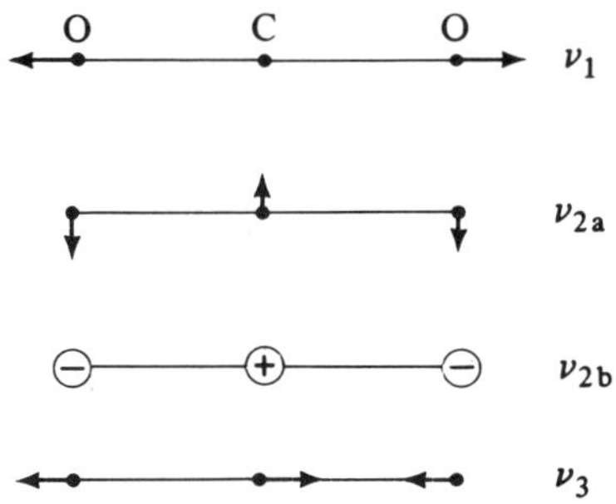

Fig. 6.2 The normal modes of CO_2. Two of the modes are degenerate; these correspond to bond bending in the plane of the paper and bond bending perpendicular to the plane of the paper.

numbered $\lambda_{3N-5}, \ldots, \lambda_{3N}$. Hence the sum in (6.23) need extend only through Q_{3N-6}. The *abc* coordinate system rotates and translates with the molecule, so that the kinetic-energy expression (6.4) includes only vibrational kinetic energy, and not translational or rotational kinetic energy. The expression (6.24) is equal to (6.4) and therefore (6.24) includes no translational or rotational kinetic energy; this means that $\dot{Q}_{3N-5}, \ldots, \dot{Q}_{3N}$, whose squares represent rotational and translational kinetic energy, must vanish. Hence the sum in (6.24) need extend only through $\dot{Q}_{3N-6}$.

As a simple example, we shall outline the normal-mode treatment of a diatomic molecule, leaving the reader to fill in the details (Problem 6.5). Let the nuclei be numbered 1 and 2. The *a* principal axis passes through the two nuclei; the *b* and *c* principal axes are perpendicular to the internuclear axis. The origin of the *abc* axes is at the center of mass. The equilibrium coordinates of the nuclei are $(a_{1,e}, 0, 0)$ and $(a_{2,e}, 0, 0)$. The *b* and *c* coordinates of each nucleus are always zero, since the axes translate and rotate with the molecule. From (6.3) and (6.1), the mass-weighted

Cartesian displacement coordinates are

$$q_1 = m_1^{1/2}(a_1 - a_{1,e}), \qquad q_2 = 0, \qquad q_3 = 0$$

$$q_4 = m_2^{1/2}(a_2 - a_{2,e}), \qquad q_5 = 0, \qquad q_6 = 0$$

From (6.9), the vibrational potential energy is

$$U = U_e + \tfrac{1}{2}(u_{11}q_1^2 + 2u_{14}q_1q_4 + u_{44}q_4^2)$$

From (4.17), we have $U = U_e + \frac{1}{2}k_e(R - R_e)^2$. Using $R = a_2 - a_1$, we find

$$U = U_e + \tfrac{1}{2}k_e\left[q_4^2 m_2^{-1} - 2q_1q_4(m_1m_2)^{-1/2} + q_1^2 m_1^{-1}\right]$$

$$u_{11} = k_e/m_1, \qquad u_{44} = k_e/m_2, \qquad u_{14} = u_{41} = -k_e/(m_1m_2)^{1/2}$$

with all other u_{ij}'s equal to zero. The secular equation (6.17) is

$$\begin{vmatrix} u_{11}-\lambda & 0 & 0 & u_{14} & 0 & 0 \\ 0 & -\lambda & 0 & 0 & 0 & 0 \\ 0 & 0 & -\lambda & 0 & 0 & 0 \\ u_{41} & 0 & 0 & u_{44}-\lambda & 0 & 0 \\ 0 & 0 & 0 & 0 & -\lambda & 0 \\ 0 & 0 & 0 & 0 & 0 & -\lambda \end{vmatrix} = 0$$

Interchange of rows 2 and 4 followed by interchange of columns 2 and 4 puts the secular determinant in block-diagonal form. The roots are found to be

$$\lambda_1 = k_e/\mu, \qquad \lambda_2 = \lambda_3 = \lambda_4 = \lambda_5 = \lambda_6 = 0$$

where the reduced mass is given by (4.5). The five zero roots correspond to translation and rotation. From (6.33), the vibrational frequency is $\lambda_1^{1/2}/2\pi = (k_e/\mu)^{1/2}/2\pi$, in agreement with (4.27). Substitution of λ_1 into (6.18) yields for the single normal mode of vibration

$$l_{11}/l_{41} = -(m_2/m_1)^{1/2}, \qquad l_{21} = l_{31} = l_{51} = l_{61} = 0$$

Note that l_{11}/l_{41} could have been derived from the condition that the origin remain at the center of mass. Normalization of $\mathbf{L}^{(1)}$ yields for the normal vibrational coordinate

$$Q_1 = -(\mu/m_1)^{1/2} q_1 + (\mu/m_2)^{1/2} q_4 = \mu^{1/2}(R - R_e)$$

Before the advent of computers, several procedures had been developed to simplify the very tedious hand calculation of normal coordinates and vibrational frequencies. These included the elimination of the six rotational and translational degrees of freedom to reduce the order of the secular determinant by six; the use of symmetry coordinates (group theory) to put the vibrational secular determinant in block-diagonal form; the setting up of the vibrational secular determinant using certain matrices, called the **F** and **G** matrices. (See *Wilson, Decius, and Cross* for descriptions of these methods.) When computer programs to solve the vibrational secular equation were originally developed, people wrote the programs using the above-mentioned procedures that were designed to simplify hand calculations. However, this gave programs that were large and unwieldy to use. In 1971 Gwinn pointed out that the simplifying procedures useful for hand calculations were not really needed for computer calculations, and by eliminating them from the computer program, he obtained a short, simple, easy-to-use program that is rapid and accurate. Gwinn's program sets up the **U** matrix using mass-weighted Cartesian coordinates and then diagonalizes **U** by an orthogonal similarity transformation to give the normal coordinates and vibrational frequencies. His program is available from the Quantum Chemistry Program Exchange (Section 1.21). As well as giving an improved computer program, elimination of the "simplifying" procedures makes life easier for students, who no longer have to learn the theory of **F** and **G** matrices. [See W. D. Gwinn, *J. Chem. Phys.*, **55**, 477 (1971).]

In the preceding discussion, the secular equation was set up in terms of the u_{ij}'s and then solved for the vibrational frequencies; the u_{ij}'s were assumed known from the solution of the molecular electronic Schrödinger equation. In practice, this is not true for most polyatomic molecules. What is usually done is the reverse of the previous procedure; namely, the molecular vibration frequencies are determined from the infrared and Raman spectra and these frequencies are used to calculate the force constants. Unless isotopically substituted species are studied, the number of pieces of experimental data (namely, the $3N-6$ normal-mode frequencies) is less than the number of (mass-weighted) force constants u_{ij} in (6.9). Hence various approximate forms of U that contain fewer parameters than (6.9) are often used in analyzing molecular force fields. One approach is the *valence-force* approximation; here, one assumes that the only terms in U are those involving changes in bond lengths and bond angles. For example, the valence-force approximation for a pyramidal XY_3 molecule like NH_3 is

$$U = U_e + \tfrac{1}{2}k\left[(R_1 - R_{1,e})^2 + (R_2 - R_{2,e})^2 + (R_3 - R_{3,e})^2\right]$$

$$+ \tfrac{1}{2}k_\alpha\left[(\alpha_1 - \alpha_{1,e})^2 + (\alpha_2 - \alpha_{2,e})^2 + (\alpha_3 - \alpha_{3,e})^2\right]$$

where the R's are the three X—Y bond lengths and the α's are the three YXY bond angles; the subscript e denotes the equilibrium values. The constant k_α is called a *torque constant*. By expressing the R's and α's in terms of Cartesian coordinates, we can relate k and k_α to the u_{ij}'s of (6.9). The valence-force approximation works reasonably well, but is not highly accurate. Using it, one can calculate force and torque constants for bond stretching and bending in molecules whose vibrational frequencies have been measured. It is found that the force constant for a given bond remains roughly constant from one molecule to another. Typical values are 4.7–5.1 mdyn/Å for C—H bond stretching and 0.5×10^{-11} ergs/rad^2 for HCH bond bending in a methyl group. Bond-stretching constants range from 1 to 20 mdyn/Å; bond-bending constants range from $0.1 R_1 R_2$ to $2 R_1 R_2$ mdyn/Å, where R_1 and R_2 are the equilibrium bond lengths defining the angle.

An improvement on the valence-force function is the *Urey–Bradley* potential function, which adds terms to the valence-force function to represent the van der Waals forces between nonbonded atoms.

The theoretical calculation of polyatomic-molecule force constants and vibrational frequencies requires the calculation of the electronic energy U for many points in the region of the equilibrium geometry, so as to generate the U potential-energy surface near equilibrium and thereby evaluate the partial derivatives (6.10). It is found that Hartree–Fock wave functions yield only mediocre force constants. For example, a near-Hartree–Fock calculation on water [T. H. Dunning, R. M. Pitzer, and S. Aung, *J. Chem. Phys.*, **57**, 5044 (1972)] gave force constants with errors of 10–23 percent. An accurate force-constant calculation requires CI. For *diatomic* molecules, the optimized-valence-configuration CI method (Section 1.20) yields accurate $U(R)$ curves and spectroscopic constants. For example, for F_2 this method gives the following values in cm^{-1} (experimental values in parentheses): $\tilde{\nu}_e = 942$ (932); $\tilde{\alpha}_e = 0.016$ (0.015); $\tilde{B}_e = 0.88$ (0.89). [G. Das and A. C. Wahl, *J. Chem. Phys.*, **56**, 3532 (1972).] For comparison, the Hartree–Fock $\tilde{\nu}_e$ is 1257 cm^{-1} for F_2. As to semiempirical methods, force constants calculated by CNDO and INDO generally have very large errors, often exceeding 100 percent. However, MINDO/2 force constants are not bad, being in error by 10–25 percent, typically.

6.3 SYMMETRY AND NORMAL VIBRATIONS

A normal mode of vibration is depicted by displacement vectors showing the direction and relative amplitude of vibration of each nucleus; examples are Figs. 6.1 and 6.2. We now consider the effect of a symmetry operation on the configuration of displacement vectors of a normal mode.

If a symmetry operation $\hat{R}$ is performed on the molecule, we get a nuclear configuration that is indistinguishable from the original one; hence the potential-energy function U remains unchanged when the corresponding operator $\hat{O}_R$ (Section 1.18) is applied to U. The function U is given by (6.23), and from $\hat{O}_R U = U$, we get

$$\sum_{k=1}^{3N-6} \lambda_k \left(\hat{O}_R Q_k\right)^2 = \sum_{k=1}^{3N-6} \lambda_k Q_k^2 \tag{6.35}$$

If no two λ_k's are equal, then the only way (6.35) can be satisfied is if

$$\left(\hat{O}_R Q_k\right)^2 = Q_k^2$$

$$\hat{O}_R Q_k = \pm Q_k, \qquad k = 1, \ldots, 3N-6 \tag{6.36}$$

Thus if there are no degenerate vibrations, each normal coordinate is either unchanged or multiplied by -1 upon application of a symmetry operation. Each Q_k is a linear combination of the mass-weighted Cartesian displacement coordinates of the nuclei. If Q_k is multiplied by -1, each Cartesian displacement coordinate is multiplied by -1, which reverses the directions of all the displacement vectors. If Q_k is unchanged by a symmetry operation, then the symmetry operation sends the displacement vectors to a configuration indistinguishable from the original one. (The displacement vectors are defined relative to molecule-fixed axes, which in turn are defined relative to the nuclear positions. The effect of a symmetry operation $\hat{R}$ on a normal coordinate is found by applying $\hat{R}$ to the displacement vectors, keeping the nuclei fixed in their equilibrium positions.)

If two or more of the λ_k's are equal, then (6.36) does not necessarily hold for the Q_k's corresponding to the degenerate modes. Thus let $\lambda_j = \lambda_k$; suppose

$$\hat{O}_R Q_j = aQ_j + bQ_k, \qquad \hat{O}_R Q_k = cQ_j + dQ_k$$

where the constants a, b, c, and d satisfy

$$b^2 + d^2 = a^2 + c^2 = 1, \qquad ab + cd = 0 \tag{6.37}$$

We then have

$$\lambda_j(\hat{O}_R Q_j)^2 + \lambda_j(\hat{O}_R Q_k)^2 = \lambda_j Q_j^2 + \lambda_j Q_k^2$$

so that (6.35) is satisfied. [Equations (6.37) can be satisfied if $a = \pm\cos\alpha$, $b = \sin\alpha$, $c = \mp\sin\alpha$, $d = \cos\alpha$, where α is some angle.] The result is readily extended to degeneracies higher than twofold. The general result is that a symmetry operation converts a normal coordinate of a degenerate mode into a linear combination of the normal coordinates that correspond to the degenerate frequency.

For each nondegenerate mode, we must have

$$\hat{O}_R Q_k = \begin{cases} Q_k & (6.38) \\ \text{or} \\ -Q_k & (6.39) \end{cases}$$

We say that the kth mode is symmetric or antisymmetric with respect to the symmetry operation $\hat{R}$, according to whether (6.38) or (6.39) holds.

The symmetry requirements severely restrict the possible forms of the normal vibrations. Thus for a nondegenerate mode, nuclei lying in a molecular symmetry plane can only vibrate in this plane or perpendicular to it, while nuclei lying on an axis of symmetry can only vibrate along the axis or perpendicular to it. Symmetry can frequently be used to deduce the qualitative appearance of the normal modes, without solving the vibrational secular equation.[1]

Consider the symmetry behavior of the normal modes of H_2O and CO_2. For water, the symmetry operations are $\hat{E}$, $\hat{C}_2(b)$, $\hat{\sigma}_v(ab)$, and $\hat{\sigma}_v(bc)$, where the a and b principal axes lie in the molecular plane; the bond angle in water is relatively obtuse, so that the C_2 axis is the axis of intermediate moment of inertia. Each of these operations converts the displacement vectors of ν_1 in Fig. 6.1 to an indistinguishable configuration; the same is true for ν_2. Normal modes that are symmetric with respect to all the molecular symmetry operations are called *totally symmetric*. For H_2O, the modes ν_1 and ν_2 are totally symmetric, but ν_3 is not. Figure 6.3 shows that ν_3 is antisymmetric with respect to $\hat{C}_2(b)$. We can tabulate the behavior of the H_2O normal modes with respect to the molecular symmetry operations:

	$\hat{E}$	$\hat{C}_2(b)$	$\hat{\sigma}_v(bc)$	$\hat{\sigma}_v(ab)$
ν_1, ν_2:	+1	+1	+1	+1
ν_3:	+1	−1	−1	+1

[1]See *Herzberg, Infrared and Raman Spectra*, pp. 87–94, 131–140.

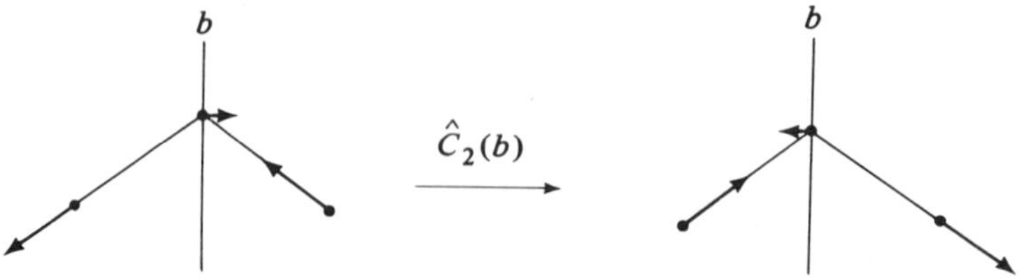

Fig. 6.3 The effect of $\hat{C}_2(b)$ on the ν_3 normal mode of H_2O.

Normal modes that have the same behavior with respect to the set of symmetry operations of a molecule are said to belong to the same *symmetry species*. An example is ν_1 and ν_2 for H_2O. (Recall that we classified molecular orbitals and molecular electronic wave functions according to their symmetry species. Further discussion of symmetry species will be found in Chapter 9.)

For CO_2, the modes ν_1 and ν_3 behave as follows:

	$\hat{E}$	$\hat{C}_\infty^\varphi$	$\hat{\sigma}_h$	$\infty\hat{C}_2$	$\infty\hat{\sigma}_v$	$\hat{S}_\infty^\varphi$	$\hat{i}$
ν_1:	+1	+1	+1	+1	+1	+1	+1
ν_3:	+1	+1	−1	−1	+1	−1	−1

where $\hat{C}_\infty^\varphi$ and $\hat{S}_\infty^\varphi$ are proper and improper rotations about the internuclear axis by an angle φ; there are an infinite number of C_2 axes and σ_v planes. The behavior of the two degenerate modes of frequency ν_2 is not so simply tabulated. A 90° rotation about the C_∞ axis converts each mode to the other, whereas a rotation about C_∞ by an angle not an integral multiple of 90° converts each degenerate mode to a linear combination of the two degenerate modes.

In the standard convention[2] for numbering the frequencies of normal modes, the totally symmetric frequencies (which are always nondegenerate—see Chapter 9) are numbered $\nu_1, \nu_2, \ldots, \nu_f$ in order of decreasing frequency, where f is the number of totally symmetric modes. After that, the remaining frequencies are numbered $\nu_{f+1}, \nu_{f+2}, \ldots,$ with nondegenerate frequencies being numbered before degenerate frequencies, using rules that we will not worry about; degenerate modes of the same frequency are given the same number. As with most rules, there is an exception: the doubly degenerate bending frequency of linear triatomic molecules is always called ν_2.

[2]R. S. Mulliken, *J. Chem. Phys.*, **23**, 1997 (1955).

Up to this point the treatment of molecular vibrations has been purely classical. Quantum mechanics does not allow the specification of the exact path taken by a particle; hence the picture of each nucleus executing the appropriate simple harmonic motion for a given normal mode should not be taken literally. On the other hand, nuclei are relatively heavy (compared to electrons), so that the classical picture of the motion is not totally lacking in validity.

6.4 QUANTUM MECHANICS OF VIBRATION

Adding the kinetic and potential energies (6.24) and (6.23) and replacing classical quantities by operators, we have as the (approximate) quantum-mechanical vibrational Hamiltonian for a polyatomic molecule

$$\hat{H} = \frac{1}{2}\sum_{k=1}^{3N-6} \hat{\dot{Q}}_k^2 + \frac{1}{2}\sum_{k=1}^{3N-6} \lambda_k \hat{Q}_k^2 + U_e \tag{6.40}$$

We need expressions for the operators $\hat{Q}_k$ and $\hat{\dot{Q}}_k$.

The operator $\hat{q}_i$ is simply multiplication by q_i; since Q_k is a linear combination of the q_i's [Eq. (6.19)], the operator $\hat{Q}_k$ is simply multiplication by Q_k.

Differentiating (6.19) with i and k interchanged with respect to time, and then replacing physical quantities by operators, we get

$$\hat{\dot{Q}}_k = \sum_{i=1}^{3N} l_{ik} \hat{\dot{q}}_i \tag{6.41}$$

We know the operator for the x component of momentum. Since $p_x = m\dot{x}$, we have

$$\hat{\dot{x}} = \frac{\hbar}{mi}\frac{\partial}{\partial x}$$

From (6.3), the operator for $\dot{q}_1$ is then

$$\hat{\dot{q}}_1 = m_1^{1/2}\hat{\dot{x}}_1 = \frac{\hbar}{im_1^{1/2}}\frac{\partial}{\partial x_1}$$

Now

$$\frac{\partial}{\partial q_1} = \frac{\partial x_1}{\partial q_1}\frac{\partial}{\partial x_1} = \frac{1}{m_1^{1/2}}\frac{\partial}{\partial x_1}$$

$$\hat{\dot{q}}_1 = \frac{\hbar}{i}\frac{\partial}{\partial q_1}$$

with similar equations for the other $\hat{q}_i$'s. Equation (6.41) becomes

$$\hat{Q}_k = \frac{\hbar}{i} \sum_{i=1}^{3N} l_{ik} \frac{\partial}{\partial q_i} \tag{6.42}$$

Equation (6.26) with V deleted tells us that the sum on the right of (6.42) is simply the operator $\partial/\partial Q_k$. Hence (6.42) becomes

$$\hat{Q}_k = \frac{\hbar}{i} \frac{\partial}{\partial Q_k} \tag{6.43}$$

The Hamiltonian (6.40) becomes

$$\hat{H} = -\frac{\hbar^2}{2} \sum_{k=1}^{3N-6} \frac{\partial^2}{\partial Q_k^2} + \frac{1}{2} \sum_{k=1}^{3N-6} \lambda_k Q_k^2 + U_e \tag{6.44}$$

The equilibrium electronic energy is a constant, and we shall drop it from the Hamiltonian; this does not affect the eigenfunctions of (6.44) and simply decreases the eigenvalues by U_e (Problem 1.6). We write the vibrational Hamiltonian as

$$\hat{H}_{\text{vib}} = \sum_{k=1}^{3N-6} \hat{H}_k \tag{6.45}$$

$$\hat{H}_k = -\frac{\hbar^2}{2} \frac{\partial^2}{\partial Q_k^2} + \tfrac{1}{2}\lambda_k Q_k^2 \tag{6.46}$$

The vibrational Schrödinger equation is

$$\hat{H}_{\text{vib}} \psi_{\text{vib}} = E_{\text{vib}} \psi_{\text{vib}} \tag{6.47}$$

Since the vibrational Hamiltonian is a sum of terms, each of which involves only one coordinate, the separation-of-variables theorem of Section 1.2 shows that

$$\psi_{\text{vib}} = \prod_{k=1}^{3N-6} \psi_k(Q_k) \tag{6.48}$$

$$E_{\text{vib}} = \sum_{k=1}^{3N-6} E_k \tag{6.49}$$

where

$$\hat{H}_k \psi_k(Q_k) = E_k \psi_k(Q_k) \tag{6.50}$$

The $3N-6$ one-dimensional Schrödinger equations (6.50) are easily solved. The one-dimensional harmonic-oscillator Hamiltonian is

$$-\frac{\hbar^2}{2m}\frac{d^2}{dx^2} + \tfrac{1}{2}kx^2$$

We see that $\hat{H}_k$ in (6.46) is a harmonic-oscillator Hamiltonian with Q_k corresponding to x, with λ_k corresponding to the force constant k, and with $m = 1$. (The normal coordinate Q_k is a combination of mass-weighted coordinates, which is why m is absent from $\hat{H}_k$.) Although the boundary conditions for (6.50) differ from harmonic-oscillator boundary conditions, the difference is unimportant (Section 4.1). From (1.133)–(1.135) and (1.140), we have

$$\psi_k(Q_k) = \frac{1}{(2^{v_k} v_k!)^{1/2}} \left(\frac{\alpha_k}{\pi}\right)^{1/4} e^{-\alpha_k Q_k^2/2} H_{v_k}(\alpha_k^{1/2} Q_k) \tag{6.51}$$

$$E_k = (v_k + \tfrac{1}{2})h\nu_k, \qquad v_k = 0, 1, 2, \ldots \tag{6.52}$$

$$\alpha_k = 2\pi\nu_k/\hbar = \lambda_k^{1/2}/\hbar \tag{6.53}$$

where v_k is the vibrational quantum number of the kth normal mode. The total vibrational energy is the sum of the energies associated with each normal mode:

$$E_{\text{vib}} = \sum_{k=1}^{3N-6} (v_k + \tfrac{1}{2})h\nu_k \tag{6.54}$$

Since each ψ_k is normalized, the vibrational wave function is normalized:

$$\int_{-\infty}^{\infty} \cdots \int_{-\infty}^{\infty} |\psi_1(Q_1)|^2 \cdots |\psi_{3N-6}(Q_{3N-6})|^2 \, dQ_1 \cdots dQ_{3N-6} = 1 \tag{6.55}$$

The ground vibrational level has the zero-point energy

$$\frac{1}{2}\sum_{k=1}^{3N-6} h\nu_k$$

Because of the large number of vibrational modes, the zero-point energy of a polyatomic molecule can be substantial.

Figure 6.4 shows the vibrational energy levels for a molecule with three normal modes. The vibrational quantum numbers of each mode are given in parentheses: $(v_1 v_2 \cdots v_{3N-6})$. Levels with one v_k equal to 1 and all other vibrational quantum numbers equal to zero are called *fundamental levels*. In Fig. 6.4, these are the levels (100), (010), and (001). Levels with one v_k greater than 1 and all others zero are *overtone levels*. Levels with more than one v_k nonzero are *combination levels*. The frequencies of most diatomic molecules are high enough to make the population of excited vibrational levels negligible at room temperature; polyatomic molecules with, say, five or more atoms generally have one or more low-lying vibrational levels, which are appreciably populated at room temperature.[3]

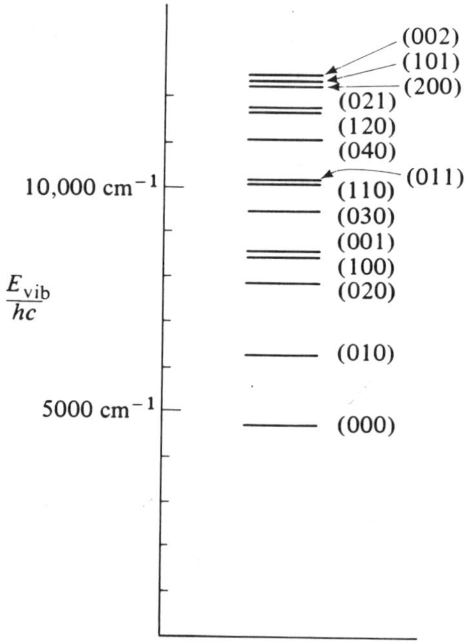

Fig. 6.4 Vibrational energy levels of a molecule with $\nu_1 = 3800$ cm^{-1}, $\nu_2 = 1600$ cm^{-1}, and $\nu_3 = 3900$ cm^{-1}. (These are roughly the values for H_2O.) The energy is calculated from (6.54), and does not include anharmonicity corrections.

Consider the degeneracies of the vibrational levels in the harmonic-oscillator approximation. For the ground level there is only one possible set of vibrational quantum numbers: $(00 \cdots 0)$; hence the ground vibrational level is always nondegenerate. If none of the normal modes are

[3]For a tabulation of vibration frequencies of polyatomic molecules, see *Herzberg, Electronic Spectra of Polyatomic Molecules*, pp. 580–668.

degenerate, then each different set of vibrational quantum numbers gives a different value to (6.54), and each vibrational level is nondegenerate. [This conclusion ignores the possibility of accidental degeneracy; e.g., if it happens that ν_1 is exactly three times ν_2, then there is degeneracy—the states $(10v_3 \cdots v_{3N-6})$ and $(03v_3 \cdots v_{3N-6})$ have the same energy, as do other pairs of states. The probability that the ratio of two frequencies will be exactly equal to the ratio of two integers is essentially zero; however, near degeneracies are very common.] If at least one vibrational mode is degenerate, then there is degeneracy in the energy levels. For example, for CO_2 (or any linear triatomic molecule) the modes ν_{2a} and ν_{2b} are degenerate. The states $(v_1 01 v_3)$ and $(v_1 10 v_3)$ belong to a doubly degenerate energy level; the states $(v_1 20 v_3)$, $(v_1 11 v_3)$, and $(v_1 02 v_3)$ belong to a triply degenerate level; and so on. We have written the quantum numbers of the CO_2 vibrational states as $(v_1 v_{2a} v_{2b} v_3)$. If we specify the vibrational level, rather than the state, we write the quantum numbers as $(v_1 v_2 v_3)$, where $v_2 \equiv v_{2a} + v_{2b}$; for example, the triply degenerate level mentioned above is the $(v_1 2 v_3)$ level. We emphasize that the discussion has been based on the harmonic-oscillator approximation; anharmonicity and other perturbations will split the degenerate vibrational levels (see Section 6.8).

In Section 6.3 we examined the behavior of the normal coordinates with respect to the molecular symmetry operations $\hat{R}$. We now do the same for the vibrational wave functions. The exponential factor in ψ_{vib} is [(6.48), (6.51), and (6.53)]

$$\exp\left[-\sum_{k=1}^{3N-6} \frac{\pi \nu_k Q_k^2}{\hbar}\right]$$

For nondegenerate modes, we saw that $\hat{O}_R Q_k^2 = Q_k^2$; for doubly degenerate modes, we saw that $\hat{O}_R(Q_j^2 + Q_k^2) = (Q_j^2 + Q_k^2)$; and so on. Hence $\hat{O}_R$ has no effect on the exponential factor in the vibrational wave function. The symmetry behavior of ψ_{vib} thus depends on the Hermite-polynomial factor:

$$\prod_{k=1}^{3N-6} H_{v_k}(\alpha_k^{1/2} Q_k)$$

For the ground state, each of the Hermite polynomials is a constant and the ground-state ψ_{vib} is unchanged (totally symmetric) under all symmetry operations.

Recall (Section 1.5) that $v = 0, 2, 4, \ldots$ oscillator states have a polynomial factor that contains only even powers of the coordinate, whereas the

$v=1,3,5,\ldots$ states have a polynomial factor with only odd powers. Hence

$$H_{v_k}(-\alpha_k^{1/2}Q_k)=(-1)^{v_k}H_{v_k}(\alpha_k^{1/2}Q_k)$$

Consider an overtone level of a nondegenerate vibration ν_k. We know that $\hat{O}_R Q_k = \pm Q_k$. If v_k is even, then $\hat{O}_R H_{v_k} = H_{v_k}$ whether or not Q_k is changed in sign by $\hat{O}_R$. Hence overtone levels of a nondegenerate mode are totally symmetric with respect to all symmetry operations when the excited vibrational quantum number is even. If v_k is odd, then $\hat{O}_R H_{v_k} = -H_{v_k}$ if $\hat{O}_R Q_k = -Q_k$, and $\hat{O}_R H_{v_k} = H_{v_k}$ if $\hat{O}_R Q_k = Q_k$; that is, an overtone level of a nondegenerate vibration ν_k belongs to the same symmetry species as Q_k when v_k is odd.

For a combination level with only nondegenerate vibrations excited, we determine the behavior of each H_{v_k} with respect to a given $\hat{R}$ (as in the last paragraph) and then multiply the $+1$'s and -1's to get the behavior of ψ_{vib}.

For states with degenerate vibrations excited, the degenerate Q_k's are transformed into linear combinations of one another, and the symmetry of the wave function is best determined using group theory; see Chapter 9.

As an example, we take H_2O. In Section 6.3 we tabulated the symmetry behavior of the normal modes. The totally symmetric species (to which ν_1 and ν_2 belong) is A_1 and the symmetry species to which ν_3 belongs is designated B_2; see (1.283). From the preceding discussion, we tabulate the symmetry behavior of a few vibrational states:

state	000	100	010	001	200	020	002	011	101	110
symmetry species	A_1	A_1	A_1	B_2	A_1	A_1	A_1	B_2	B_2	A_1

The vibrational levels (6.54) are only approximate, because we neglected the anharmonic cubic, quartic, ... terms in the potential energy (6.7). For a diatomic molecule, anharmonicity adds the correction $-h\nu_e x_e(v+\frac{1}{2})^2$ to the vibrational levels [Eq. (4.67)]. For a polyatomic molecule, the anharmonicity correction modifies (6.54) to

$$E_{\text{vib}}=\sum_{k=1}^{3N-6}(v_k+\tfrac{1}{2})h\nu_{k,e}+\sum_{i=1}^{3N-6}\sum_{k\geqslant i}hx_{ik}(v_i+\tfrac{1}{2})(v_k+\tfrac{1}{2})+hG_0 \tag{6.56}$$

The subscript e has been added to emphasize that the vibrational frequencies correspond to the equilibrium positions [(6.23) and (6.33)]:

$$\nu_{k,e}=\frac{1}{2\pi}\left[\left(\frac{\partial^2 U}{\partial Q_k^2}\right)_e\right]^{1/2} \tag{6.57}$$

G_0 is a constant that arises partly from anharmonicity [like Y_{00} in (4.67)] and partly from vibration–rotation interaction. It is usually omitted since it is small, but for light molecules (e.g., H_2O), it should be included.[4] The convention used in defining the anharmonicity correction for a polyatomic molecule differs from that used for diatomic molecules; x_{ik} and x_e differ in sign and differ by a frequency factor. Most of the experimentally determined x_{ik}'s are negative, causing the higher vibrational levels to be lower in energy than the corresponding harmonic-oscillator levels. (Recall that x_e is generally positive.) As an example, analysis of the infrared spectrum of H_2O gives the vibrational constants (in cm^{-1}) as[4,5]

$$\tilde{\nu}_{1,e}=3832.2 \qquad \tilde{\nu}_{2,e}=1648.5 \qquad \tilde{\nu}_{3,e}=3942.5$$

$$\tilde{x}_{11}=-42.6 \qquad \tilde{x}_{22}=-16.8 \qquad \tilde{x}_{33}=-47.6$$

$$\tilde{x}_{12}=-15.9 \qquad \tilde{x}_{13}=-165.8 \qquad \tilde{x}_{23}=-20.3$$

$$\tilde{G}_0=1.6 \tag{6.58}$$

where the tilde denotes division by c. The zero-point energy is 4636 cm^{-1}. From Fig 6.1, we see that ν_1 and ν_3 both are O—H bond-stretching vibrations, so that the similarity of their frequencies is reasonable. Bond-bending frequencies are generally lower than the corresponding bond-stretching frequencies, so that ν_2 is less than ν_1 and ν_3.

Higher-order anharmonicity corrections analogous to (4.79) can also be included in E_{vib}, but usually the data available are not sufficient to allow their evaluation.

To get the total energy (excluding translational energy) of a polyatomic molecule in a given state of nuclear and electronic motion, we add the equilibrium electronic energy U_e and the rotational energy to (6.56):

$$E=U_e+E_{\text{vib}}+E_{\text{rot}} \tag{6.59}$$

where E_{rot} was discussed in Chapter 5.

6.5 SELECTION RULES FOR VIBRATION–ROTATION AND PURE-ROTATION TRANSITIONS

We now consider the electric-dipole selection rules for radiative transitions between energy levels of the same electronic state of a polyatomic molecule. The electric-dipole transition moment is (4.91), which becomes

[4] M. Wolfsberg, A. A. Massa, and J. W. Pyper, *J. Chem. Phys.*, **53**, 3138 (1970).
[5] W. S. Benedict, N. Gailar, and E. K. Plyler, *J. Chem. Phys.*, **24**, 1139 (1956).

[see (4.92) and the discussion following it]

$$\langle\psi|\hat{\mathbf{d}}|\psi'\rangle=\int\psi_N^*\mathbf{d}\psi_N'\,d\tau_N \tag{6.60}$$

The permanent electric dipole moment **d** is

$$\mathbf{d}=\mathbf{e}_X d_X+\mathbf{e}_Y d_Y+\mathbf{e}_Z d_Z=\mathbf{e}_a d_a+\mathbf{e}_b d_b+\mathbf{e}_c d_c \tag{6.61}$$

where X, Y, and Z form a set of nonrotating axes with origin at the molecular center of mass; that is, the XYZ axes are space-fixed, except for translating with the molecule. The axes *abc* are the molecule-fixed principal axes of inertia. The **e**'s are unit vectors along the axes. We have

$$d_Z=\mathbf{e}_Z\cdot\mathbf{d}=(\mathbf{e}_Z\cdot\mathbf{e}_a)d_a+(\mathbf{e}_Z\cdot\mathbf{e}_b)d_b+(\mathbf{e}_Z\cdot\mathbf{e}_c)d_c \tag{6.62}$$

$$d_Z=d_a\cos(ZOa)+d_b\cos(ZOb)+d_c\cos(ZOc) \tag{6.63}$$

where the angles are those between the Z axis and the a, b, and c axes (Fig. 5.2). We have similar equations for d_X and d_Y. The cosines in (6.63) are functions of the Eulerian angles θ, φ, and χ [see (5.29)]:

$$\cos(ZOa)=-\sin\theta\cos\chi,\qquad \cos(ZOb)=\sin\theta\sin\chi$$
$$\cos(ZOc)=\cos\theta \tag{6.64}$$

The principal-axis dipole-moment components d_a, d_b, and d_c depend on the configuration of the nuclei relative to one another, but are independent of the spatial orientation of the molecule; they thus depend on the vibrational coordinates, but not on the Eulerian angles. We have

$$\int\psi_N^* d_Z\psi_N'\,d\tau_N=\int\psi_{\text{vib}}^* d_a\psi_{\text{vib}}'\,d\tau_{\text{vib}}\int\cos(ZOa)\psi_{\text{rot}}^*\psi_{\text{rot}}'\,d\tau_{\text{rot}}$$
$$+(a\to b)+(a\to c) \tag{6.65}$$

where the second and third terms are obtained from the first by replacing a by b or c. Next we expand d_a, d_b, and d_c in Taylor series about the equilibrium nuclear configuration (all $Q_k=0$):

$$d_a(Q_1,\ldots,Q_{3N-6})=d_{a,e}+\sum_{k=1}^{3N-6}\left(\frac{\partial d_a}{\partial Q_k}\right)_e Q_k$$
$$+\frac{1}{2}\sum_{j=1}^{3N-6}\sum_{k=1}^{3N-6}\left(\frac{\partial^2 d_a}{\partial Q_j\partial Q_k}\right)_e Q_jQ_k+\cdots \tag{6.66}$$

with similar equations for d_b and d_c. The quantity $d_{a,e}$ is the a component of the dipole moment for the equilibrium nuclear configuration. The quadratic terms and higher terms in (6.66) are believed to be small, and we shall neglect them. Equation (6.65) becomes

$$\int \psi_N^* d_Z \psi_N' \, d\tau_N = \int \psi_{\text{vib}}^* \psi_{\text{vib}}' \, d\tau_{\text{vib}} [d_{a,e} I_{ZOa} + d_{b,e} I_{ZOb} + d_{c,e} I_{ZOc}]$$

$$+ \sum_{k=1}^{3N-6} \int \psi_{\text{vib}}^* Q_k \psi_{\text{vib}}' \, d\tau_{\text{vib}} \left[\left(\frac{\partial d_a}{\partial Q_k} \right)_e I_{ZOa} \right.$$

$$\left. + \left(\frac{\partial d_b}{\partial Q_k} \right)_e I_{ZOb} + \left(\frac{\partial d_c}{\partial Q_k} \right)_e I_{ZOc} \right] \tag{6.67}$$

where

$$I_{ZOa} \equiv \int \cos(ZOa) \psi_{\text{rot}}^* \psi_{\text{rot}}' \, d\tau_{\text{rot}} \tag{6.68}$$

with similar definitions for I_{ZOb} and I_{ZOc}. We have equations similar to (6.67) for the matrix elements of d_X and d_Y.

The vibrational wave function is a product of harmonic-oscillator functions, one for each normal coordinate; hence

$$\int \psi_{\text{vib}}^* \psi_{\text{vib}}' \, d\tau_{\text{vib}} = \prod_{k=1}^{3N-6} \int \psi_k^*(Q_k) \psi_k'(Q_k) \, dQ_k \tag{6.69}$$

where the ψ_k are given by (6.51). The ψ_k are orthonormal, so that the integral (6.69) vanishes unless

$$v_k' = v_k, \qquad k = 1, 2, \ldots, 3N-6 \tag{6.70}$$

Thus the first term on the right of (6.67) corresponds to a transition with no change in vibrational quantum numbers, that is, a pure-rotation transition. The multiplier in (6.67) of the integral (6.69) vanishes if the molecule has no equilibrium dipole moment. Hence a molecule must have a permanent equilibrium dipole moment to exhibit pure-rotation transitions, as was stated in Chapter 5.

The selection rules for pure-rotation transitions are found by evaluation of the nine integrals $I_{XOa}, \ldots, I_{ZOc}$; these involve the nine *direction cosines*[6] $\cos(XOa), \ldots, \cos(ZOc)$. The volume element in Eulerian angles can be shown to be

$$d\tau_{\text{rot}} = \sin\theta \, d\theta \, d\varphi \, d\chi \tag{6.71}$$

[6]For a list of all nine, see *Wilson, Decius, and Cross*, p. 286.

For a symmetric top, symmetry requires the dipole moment to lie along the symmetry axis, so that two of the three principal-axis components of **d** must vanish. In deriving the symmetric-top wave functions in Section 5.5, we assumed that the c axis was the symmetry axis; hence to use the eigenfunctions (5.68) to find the selection rules, we must take $d_a = d_b = 0$, $d_c \neq 0$. For a symmetric top, we thus must evaluate only the three integrals I_{XOc}, I_{YOc}, and I_{ZOc}. The three relevant direction cosines are given in (6.64) and Problem 5.15; they are independent of χ. Since the integral

$$\int_0^{2\pi} [e^{iK\chi}]^* e^{iK'\chi}\, d\chi \tag{6.72}$$

vanishes unless $K = K'$, we have the symmetric-top selection rule $\Delta K = 0$ [Equation (5.84)]. Derivation of the M selection rule is left as an exercise; derivation of the J selection rule is omitted.

The asymmetric-top eigenfunctions are linear combinations of symmetric-top eigenfunctions, which allows the asymmetric-top selection rules to be found. Since the dipole moment can have an arbitrary orientation, we get the three kinds of selection rules in (5.89); details are omitted.

Now consider the second term on the right side of (6.67). The $3N-6$ integrals in the sum each have the form

$$\int \psi_{\text{vib}}^* Q_k \psi_{\text{vib}}' \, d\tau_{\text{vib}} = \int \psi_k^*(Q_k) Q_k \psi_k'(Q_k)\, dQ_k \prod_{j \neq k} \int \psi_j^*(Q_j) \psi_j'(Q_j)\, dQ_j \tag{6.73}$$

The $3N-5$ integrals in the product on the right vanish unless $v_j' = v_j$, because of orthogonality; the first integral on the right vanishes unless $v_k' = v_k \pm 1$ [Eq. (4.45)]. Thus the second term on the right of (6.67) gives rise to infrared transitions in which one vibrational quantum number changes by one unit, while all other vibrational quantum numbers are unchanged:

$$\left.\begin{aligned} \Delta v_k &= \pm 1 \\ \Delta v_j &= 0, \qquad j \neq k \end{aligned}\right\} k = 1, \ldots, 3N-6 \tag{6.74}$$

The factor in (6.67) that multiplies the integral (6.73) contains the derivatives of the dipole-moment components with respect to the normal coordinate Q_k, evaluated at the equilibrium configuration. We conclude that a radiative infrared transition in which the vibrational quantum number of the kth normal mode changes by one is forbidden unless the kth mode has a change in dipole moment associated with it. The value of the equilibrium dipole moment $\mathbf{d}_e$ is irrelevant for infrared transitions of a polyatomic molecule.

Actually, many other infrared transitions occur besides those allowed by the selection rule (6.74). The neglected terms in the expansion (6.66) will give transitions with a change of 2 or more in a given vibrational quantum number and transitions in which more than one vibrational quantum number changes; moreover, anharmonicity corrections to the vibrational wave function will add to the probability of such transitions. Generally, the transitions (6.74) are the strongest.

The rotational fine structures of infrared vibration–rotation transitions are determined by the same integrals $I_{XOa}, \ldots, I_{ZOc}$ as determine pure-rotation transitions. Consider first symmetric tops. As a consequence of symmetry, it can be shown that any allowed vibrational transition of a symmetric top changes either the component of **d** along the symmetry axis or a component of **d** perpendicular to the symmetry axis. These two kinds of transitions are called *parallel* (||) and *perpendicular* ($\perp$), respectively. Consider first a parallel transition, which has

$$\left(\frac{\partial d_a}{\partial Q_k}\right)_e = \left(\frac{\partial d_b}{\partial Q_k}\right)_e = 0, \qquad \left(\frac{\partial d_c}{\partial Q_k}\right)_e \neq 0 \tag{6.75}$$

where the c axis is assumed to be the symmetry axis. We see from (6.67) that the relevant direction-cosine integrals for the rotational selection rules of a parallel vibration–rotation transition are the same three integrals that determine the pure-rotation selection rules of a symmetric top; hence from (5.84) we have

$$\left.\begin{array}{ll} \Delta J = 0, \pm 1 & \Delta K = 0 \\ \Delta J = 0 \quad \text{forbidden if} & K = 0 \end{array}\right\} \begin{array}{l}\text{symmetric-top} \\ \parallel \text{ band}\end{array} \tag{6.76}$$

For a perpendicular transition, evaluation of the relevant direction-cosine integrals gives

$$\Delta J = 0, \pm 1 \qquad \Delta K = \pm 1\}\text{symmetric-top } \perp \text{ band} \tag{6.77}$$

A spherical top is a special case of a symmetric top; the rotational energy depends only on J, and the J selection rule for the vibration–rotation transitions is the same as for symmetric tops:

$$\Delta J = 0, \pm 1 \qquad \text{spherical top} \tag{6.78}$$

For an asymmetric top, the rotational quantum numbers undergo the same possible changes in vibration–rotation transitions as in pure-rotation transitions; however, it is the direction of the change in the dipole moment, rather than the direction of the dipole moment, that is relevant. The selection rules are

$$\Delta J = 0, \pm 1, \text{etc.} \qquad \text{asymmetric top} \tag{6.79}$$

where *etc*. indicates the possible changes in K_{ob} and K_{pr} [Eq. (5.89)].

For completeness, we note that the restriction

$$J = 0 \leftarrow\!|\!\rightarrow J = 0 \tag{6.80}$$

always holds.

6.6 INFRARED SPECTROSCOPY

In 1800 Sir William Herschel found that a thermometer placed in the region beyond the red end of the solar spectrum (obtained by dispersing solar radiation with a prism) was heated even more than when placed in the visible portion; Herschel had discovered infrared radiation. Around 1900 infrared (IR) absorption investigations of molecules began.

The traditional source in IR absorption spectroscopy is a glowing rod or wire heated by the passage of an electric current; the hot body emits radiation over a continuous frequency range. The radiation is dispersed using a prism; NaCl, which is transparent over much of the IR region, is commonly used for IR prisms and windows. The sample may be a solid, liquid, or gas. Various detectors are used; the most common are thermocouples, photoconductive materials such as PbS, bolometers (which are temperature-dependent resistors), and the Golay cell (which uses the thermal expansion of a gas contained in a chamber).

Prior to 1940, the applications of IR spectroscopy were limited to research in molecular structure using spectrometers built by research workers. These spectrometers were difficult and tedious to use; the spectrum had to be measured and plotted point by point, and the spectrometers were extremely sensitive to disturbances such as small fluctuations in room temperature. Interest in the military applications of IR radiation led to the development of improved IR detectors during World War II; this made possible the construction of commercial automatic-recording IR spectrometers. The resulting simplicity and rapidity of operation made IR spectroscopy a standard tool in research and industrial work.

The operation of a commercial recording IR spectrometer (Fig. 6.5) is as follows. The source radiation is reflected from two mirrors to give two identical beams; one beam passes through the sample cell, the other through an empty reference cell (if the sample is in solution, the reference cell is filled with the solvent). A chopper system then causes alternately the sample beam and the reference beam to fall on the dispersing prism. If there is no sample absorption, the sample and reference beams reaching the detector are equal in intensity, and the ac amplifier connected to the detector gives no output. When there is sample absorption, the reference and sample signals reaching the detector are unequal, and the ac amplifier (which is tuned to the chopper frequency) gives an output proportional to the amount of absorption. The amplifier output drives the recorder pen and also drives an attenuating wedge into the reference beam. The wedge and pen are driven until the reference signal is equal to the sample signal, which makes the amplifier output zero. The frequency of the radiation falling on the detector is varied by a slowly rotating mirror between the prism and detector; the mechanism rotating this mirror also drives the drum holding the chart paper on which the spectrum is recorded as a function of frequency. Use of the reference cell eliminates errors due to solvent and atmospheric absorption and variations in source output and detector sensitivity. To give higher resolution, a diffraction grating is used in addition to the prism.

IR wavelengths are traditionally divided into three regions. The portion that adjoins the visible region is the *near infrared*: $\lambda = 0.8$–$2.5\ \mu m$; the *mid infrared* extends from 2.5 to 50 μm; the *far infrared* extends from 50 to 1000 μm. Most commercial IR spectrometers operate in the mid IR. Pure rotational transitions of light molecules, and low-frequency vibrational transitions of heavy molecules occur in the far IR.

The use of CW tunable semiconductor lasers as a source in IR spectroscopy research makes possible a very great increase in resolving power over traditional IR grating spectrometers. IR studies with laser sources have been done on several gases (e.g., H_2O, NH_3, SF_6, NO). The laser line width is typically 1/100th the width of the Doppler-broadened absorption lines of the gases, so the fine details of the IR line shapes are

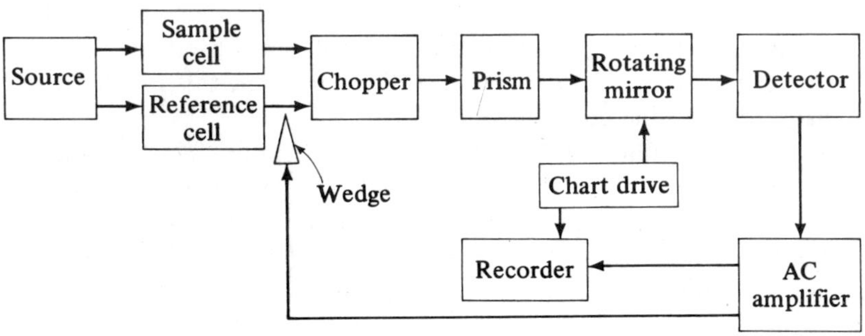

Fig. 6.5 Block diagram of a double-beam IR spectrometer.

easily obtained. [See K. W. Nill, *Bull. Am. Phys. Soc.*, **18**, 11 (1973).]

The infrared spectrum of a polyatomic molecule consists of a series of bands, which may show a rotational fine structure under high resolution. We first consider the positions of the band origins, ignoring the rotational fine structure. The strongest bands are generally those corresponding to the selection rule (6.74). Since the ground vibrational level is generally the most populated level, the most intense absorption transitions are those from the $(00\cdots 0)$ ground level to the fundamental levels $(10\cdots 0)$, $(01\cdots 0),\ldots,$ $(00\cdots 1)$; however, only for those vibrational modes that involve a change in dipole moment do we observe a fundamental IR band. If $(\partial \mathbf{d}/\partial Q_k)$ happens to be small but nonzero, then the corresponding fundamental band will be weak, perhaps weaker than some of the non-fundamental bands. The absorption frequencies corresponding to the transitions from the ground level to the fundamental levels are called the *fundamental frequencies*; because of anharmonicity, the fundamental frequencies differ somewhat from the equilibrium vibrational frequencies $\nu_{1,e}, \nu_{2,e}, \ldots$. Since the available data are usually not sufficient to allow determination of all the anharmonicity constants, the fundamental frequencies, rather than the equilibrium frequencies, are usually used in a force-constant analysis; this involves an approximation. In the literature, the wave numbers of the equilibrium frequencies (also called the *zero-order frequencies*) are denoted by $\omega_1, \omega_2, \ldots,$ whereas the wave numbers of the fundamental frequencies are denoted by $\nu_1, \nu_2, \ldots$.

For H_2O, we see from Fig. 6.1 that each normal mode causes a change in dipole moment. Hence three fundamental bands are observed; these occur at 1595, 3657, and 3756 cm^{-1}. The band-origin vacuum wave numbers of vibrational bands of water vapor are listed in Table 6.1. Because the fundamental vibrational frequencies are all rather high, the population of excited vibrational levels is negligible at room temperature, and all transitions originate from the ground vibrational level. In addition to the three strong fundamental bands, many weaker bands corresponding to transitions to overtone and combination levels are observed. Some of these bands occur in the visible portion of the spectrum; this is often cited as the reason for the blue color of water.

For CO_2, it is clear that ν_1 leaves the dipole moment unchanged, whereas ν_2 and ν_3 give a change in dipole moment. We say that ν_1 is *infrared inactive*, whereas ν_2 and ν_3 are *infrared active*. Table 6.2 lists some of the IR bands of gaseous CO_2. The superscript on v_2 will be explained in Section 6.8. Since ν_2 is rather low, some of the bands are hot bands, that is, bands for which the initial level is not (000).

Table 6.1 Infrared Bands of H_2O Vapor[a]

σ_0 (cm^{-1})	Upper State $v_1v_2v_3$	Lower State $v_1v_2v_3$	σ_0 (cm^{-1})	Upper State $v_1v_2v_3$	Lower State $v_1v_2v_3$
1595	010	000	12151	211	000
3152	020	000	12565	013	000
3657	100	000	13831	301	000
3756	001	000	14319	103	000
5331	011	000	15348	311	000
6872	021	000	15832	113	000
7252	101	000	16822	321	000
8807	111	000	16899	401	000
10613	201	000	17495	203	000
11032	003	000			

[a]Data from references cited in W. S. Benedict, N. Gailar, and E. K. Plyler, *J. Chem. Phys.*, **24**, 1139 (1956).

The appearance of the IR spectrum of a compound depends somewhat on the sample's phase. Under high resolution, gas-phase IR bands consist of closely spaced lines—the rotational fine structure; however, IR bands of liquids and solids very rarely show rotational fine structure. In most solids, the molecules are held in fixed lattice positions and are not free to rotate. In liquids, the high rate of intermolecular collisions and the substantial intermolecular interactions cause random shifts in the rotational energies, thereby broadening the rotational lines of a band sufficiently to merge them into one another, and eliminate the rotational fine structure. (Broadening of fine structure lines is also observed in gas-phase spectra when the pressure is increased.)

There are a few exceptions to the statements of the previous paragraph. The vibrational Raman spectrum of liquid H_2 shows rotational fine structure; for H_2, the rotational levels are widely spaced and intermolecular forces are reasonably small. Certain solids when heated undergo a transition to a solid state in which molecular rotation in the crystal is possible. Solid H_2 undergoes such a transition, as shown by the heat-capacity curve; see *Davidson*, Section 16-9.

Vibrational energies are much greater than rotational energies, and the relative changes in vibrational frequencies on going from the gas phase to

Table 6.2 Infrared Bands[a] of Gaseous CO_2

	Upper State			Lower State				Upper State			Lower State		
σ_0 (cm^{-1})	v_1	v_2^l	v_3	v_1	v_2^l	v_3	σ_0 (cm^{-1})	v_1	v_2^l	v_3	v_1	v_2^l	v_3
667	0	1^1	0	0	0^0	0	618	0	2^0	0	0	1^1	0
1932	0	3^1	0	0	0^0	0	668	0	2^2	0	0	1^1	0
2077	1	1^1	0	0	0^0	0	721	1	0^0	0	0	1^1	0
2349	0	0^0	1	0	0^0	0	1881	0	4^0	0	0	1^1	0
3613	0	2^0	1	0	0^0	0	2093	1	2^2	0	0	1^1	0
3715	1	0^0	1	0	0^0	0	2130	2	0^0	0	0	1^1	0
4854	0	4^0	1	0	0^0	0	597	0	3^1	0	0	2^2	0
4978	1	2^0	1	0	0^0	0	647	0	3^1	0	0	2^0	0
5100	2	0^0	1	0	0^0	0	742	1	1^1	0	0	2^2	0
6076	0	6^0	1	0	0^0	0	961	0	0^0	1	1	0^0	0
6228	1	4^0	1	0	0^0	0	1064	0	0^0	1	0	2^0	0
6348	2	2^0	1	0	0^0	0							
6503	3	0^0	1	0	0^0	0							
6972	0	0^0	3	0	0^0	0							

[a]Data from C. P. Courtoy, *Ann. Soc. Sci. Bruxelles, Ser. I*, **73**, 5 (1959). See the discussion near the end of Section 6.8.

the liquid or solid phases are usually small. Usually there is a small decrease in vibrational frequencies in the liquid and solid states as compared to the gas state. The changes are greater for polar molecules than for nonpolar ones. For example, the fundamental vibration frequency of HCl is 2886, 2785, and 2768 cm^{-1} for gas, liquid, and solid phases, respectively; the ν_1 fundamental frequency of CH_4 is 2914, 2909, and 2906 cm^{-1} for the gas, liquid, and solid phases, respectively. The shapes and intensities of vibrational bands are also somewhat dependent on the phase.

In addition to frequency shifts, new vibrational bands occasionally appear in the liquid state as compared to the gas phase. An example is formic acid, HCOOH. At high temperatures, the vapor shows a characteristic O—H stretching vibrational band at 3750 cm^{-1}; as the temperature of the gas is lowered, this band disappears and a new band at 3080 cm^{-1} appears. In the liquid state, only the 3080 cm^{-1} band is evident. The new band is due to O—H vibration in the hydrogen-bonded dimer. Hydrogen bonding also affects the IR spectra of liquid CH_3OH and H_2O.

Frequently the IR spectrum of a compound is observed in solution. The appearance of the spectrum depends somewhat on the solvent used, due to varying solute–solvent interactions.

We now consider the rotational fine structure of gas-phase IR bands, beginning with linear molecules. For nearly all known linear polyatomic molecules, the ground electronic state is a $^1\Sigma$ state and we do not have to worry about the interaction of rotational and electronic angular momenta. A linear molecule in a $^1\Sigma$ electronic state is a symmetric top with $K=0$; the selection rules are [(6.76) and (6.77)]

$$\Delta J = \pm 1 \qquad \| \text{ band of linear molecule}$$
$$\Delta J = 0, \pm 1 \qquad \perp \text{ band of linear molecule} \tag{6.81}$$

A parallel band of a linear molecule has no Q-branch lines. (The single vibration of a diatomic molecule is a parallel mode.) See Fig. 6.6. In CO_2, we see from Fig. 6.2 that ν_3 changes the dipole-moment component along the symmetry axis; hence the ν_3 fundamental band is a parallel band; in contrast, the ν_2 fundamental is a perpendicular band.

From the selection rules given in Section 6.5, we see that nonlinear-symmetric-top and asymmetric-top bands show Q branches, as well as P and R branches. Detailed consideration of the band shapes is omitted.[7]

Analysis of the rotational fine structure of IR bands yields the moments of inertia I_a^0, I_b^0, and I_c^0. From these, the molecular structure can be fitted. (It may be necessary to assign spectra of isotopically substituted species in order to have sufficient data for a structural determination.) Such structures are subject to the usual errors due to zero-point vibrations. Values of moments of inertia determined from IR work are less accurate than those obtained from microwave work. However, the pure-rotation spectra of many polyatomic molecules cannot be observed because the molecules have no permanent electric dipole moment; in contrast, all polyatomic molecules have IR-active vibration–rotation bands, from which the rotational constants and structure can be determined. For example, the structure of the nonpolar molecule ethylene, $CH_2{=}CH_2$, was determined from IR study of the normal species and of $CD_2{=}CD_2$ to be[8]

$$R_{\text{C-H}} = 1.09 \text{ Å}, \qquad R_{\text{C=C}} = 1.34 \text{ Å}, \qquad \angle\text{HCC} = 121°$$

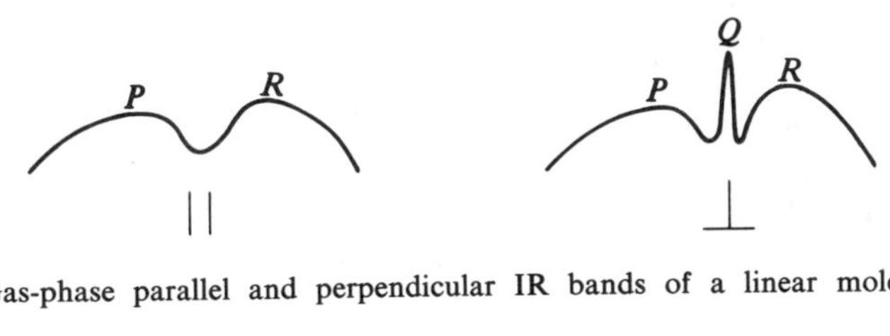

Fig. 6.6 Gas-phase parallel and perpendicular IR bands of a linear molecule for low resolution.

[7]See *Herzberg, Infrared and Raman Spectra*, pp. 414–440, 468–489.

[8]H. C. Allen and E. K. Plyler, *J. Am. Chem. Soc.*, **80**, 2673 (1958).

The application of IR spectroscopy most familiar to organic chemists is its qualitative use for determining the functional groups in an unknown mixture or compound. The IR bands of most common functional groups vary but little from molecule to molecule. For example, most simple ketones show an IR band lying in the range 1710–1720 cm^{-1}, which is assigned as a C=O stretching fundamental vibration. The corresponding band in aldehydes occurs in the range 1715–1735 cm^{-1}; in organic acids the C=O stretch is usually found in the range 1750–1770 cm^{-1}. (These frequencies are for dilute solutions using a nonpolar solvent.) Figure 6.7 shows the vibrational frequencies of common functional groups. In addition to frequencies associated with vibrations of particular functional groups, there are frequencies which involve vibration of the whole molecular skeleton; these frequencies are peculiar to the individual molecule. For example, C—C skeletal stretching frequencies appear in the wide range 800–1200 cm^{-1} ($8\frac{1}{2}$–$12\frac{1}{2}$ μm); their locations depend on the individual compound. The overall IR spectrum is a highly characteristic property of a compound and has been compared to a fingerprint. The frequency region 800–1400 cm^{-1}, where the skeletal vibrations occur, is called the "fingerprint region."

At first sight, the existence of characteristic group frequencies may seem hard to reconcile with the concept of normal vibrations. For a normal mode, all (or nearly all) the atoms of a molecule are vibrating. However, for certain normal modes, the vibrational amplitudes of most of the atoms may be quite small, and the vibrational mode can be considered to be essentially localized within a chemical group of two or three atoms; the valence force constants (Section 6.2) of a given chemical group are approximately constant from molecule to molecule, leading to the characteristic group frequencies.

For example, consider the O—H group. The mass of the hydrogen atom

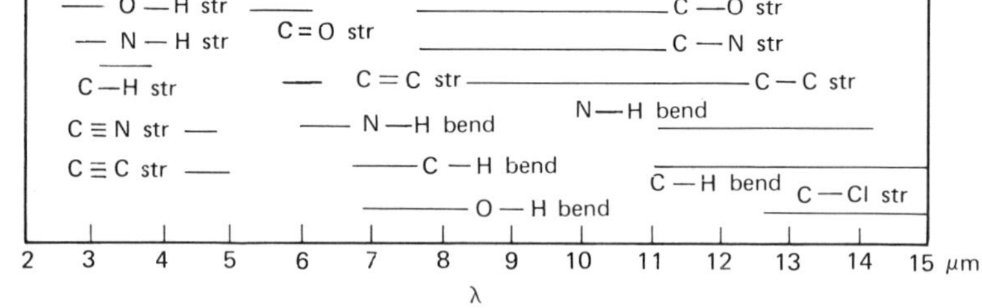

Fig. 6.7 Group frequencies for IR absorption. The scale is linear in wavelength, not frequency.

is so small compared to other nuclei that in treating the vibrations of the hydrogen atom, we can consider it to be vibrating against an infinitely heavy mass. The O—H bond-stretching and bond-bending frequencies will then be essentially independent of the nature of the rest of the molecule. Thus for the O—H stretching frequency, we have

$$\nu(\text{O—H str}) = \frac{1}{2\pi}\left[\frac{k(\text{O—H str})}{\mu}\right]^{1/2}$$

The reduced mass is

$$\mu = \frac{m_H m_R}{m_H + m_R} \approx \frac{m_H m_R}{m_R} = m_H$$

where the mass of the rest of the molecule m_R is much greater than m_H. Hence

$$\nu(\text{O—H str}) \approx \frac{1}{2\pi}\left[\frac{k(\text{O—H str})}{m_H}\right]^{1/2} \qquad (6.82)$$

The typical wave number for an O—H stretch is 3600 cm^{-1}; this corresponds to a valence force constant $k(\text{O—H str}) = 7.6 \times 10^5$ dynes/cm. The frequency of the O—H bending vibration is treated similarly. If there is more than one O—H bond in the molecule, the hydrogen atoms will all undergo either bond stretching or bond bending simultaneously, giving rise to several normal-mode frequencies; however, those normal modes involving O—H stretching will have closely similar frequencies, as will the modes involving O—H bond bending. An example is H_2O, where ν_1 and ν_3 are essentially O—H bond stretches of slightly different fundamental frequencies (3657 and 3756 cm^{-1}), whereas ν_2 is essentially an O—H bond bending of fundamental frequency 1595 cm^{-1}. The situation is roughly similar to the perturbation treatment of the helium $1s2s$ configuration; we have two O—H bonds, and ignoring interactions between them, we would have two zeroth-order wave functions, one for each O—H stretching vibration. However, the symmetry of the molecule requires that the correct zeroth-order wave functions be formed as the symmetric and antisymmetric linear combinations, leading to symmetric (ν_1) and antisymmetric (ν_3) O—H stretching vibrations of slightly different frequencies.

Exactly the same considerations as for O—H apply to the C—H and N—H groups. In an organic molecule with —CH_2— and —CH_3 groups, an IR spectrometer of good resolution shows four closely spaced C—H stretching vibrations in the region near 3000 cm^{-1}. Two of these are the symmetric and antisymmetric C—H stretching vibrations of the $—CH_2—$ group, similar to the corresponding vibrations in H_2O; the other two are symmetric and asymmetric C—H stretching vibrations of the $—CH_3$ group. Several C—H bending bands are observed near 1400 cm^{-1}; these are described by such picturesque terms as wagging, scissoring, twisting, and rocking.

We still have to explain why a group like >C=O, which does not involve a light atom, has a characteristic vibration frequency. The explanation was provided by Bartholomé and Teller in 1932. They considered the normal modes of a model system consisting of points of similar mass connected by springs, each spring being of a very different stiffness from every other spring. Consider two of the masses a and b that are connected by a spring with force constant k_{ab}. If all other masses were absent, we would have a stretching vibration of frequency ν_{ab} which is characteristic of m_a and m_b and k_{ab}. Study of the model system showed that among the normal modes of the system is one for which the vibrational amplitudes of all atoms except a and b are small; the frequency of this mode is close to ν_{ab}. Thus when the bond force constant of a group is quite different from that of other bonds in the molecule, a characteristic group frequency will be observed. For example, in a ketone or aldehyde, the CO double bond has a force constant considerably greater than the single-bond force constants, and we get a characteristic C=O bond-stretching frequency. Also, if the bond force constants in a molecule are all similar in magnitude, but one group involves a large difference in the masses, there will be a characteristic group frequency. Examples are C—Cl and C—Br bonds in organic molecules; another example is the C—H bond discussed previously. As long as the characteristic group frequencies are substantially different from those of the rest of the molecule, the group frequencies are preserved in the molecule.

When there are several bonds of similar strength between atoms of similar mass, we get several skeletal vibrations occurring over a wide frequency range; an example is the C—C skeletal vibrations of organic molecules.

IR spectroscopy is useful in biochemical research and has aided in the structural determinations of amino acids, nucleic acids, and polypeptides; the IR spectra of viruses and bacteria have been studied.[9] The controversial cancer drug Krebiozen was identified by its IR spectrum as the amino acid derivative creatine, which is present in humans in large quantities.[10]

[9]F. S. Parker, *Applications of Infrared Spectroscopy in Biochemistry, Biology, and Medicine*, Plenum, 1971.

[10]*Science*, **141**, 1021 (1963); *The New York Times*, Sept. 7, 1963, p. 21; Sept. 8, 1963, p. 84.

The *matrix isolation* technique allows one to study the IR spectra of free radicals and other reactive species. Here, one cools a mixture of a compound of interest and a gas such as Ar or N_2 to a temperature of a few degrees Kelvin. Photolysis of the frozen sample produces free radicals trapped in an inert matrix; their IR spectra can then be observed. Species studied include CH_3, CF_2, HCO, SiF_2. [See L. Andrews, *Ann. Rev. Phys. Chem.*, **22**, 109 (1971).]

Simple HMO theory predicts cyclobutadiene (C_4H_4) to have a triplet ground state (with one electron in each of two degenerate pi MOs). After years of fruitless efforts, C_4H_4 was finally detected; it is a highly reactive intermediate with only a short lifetime at room temperature. Its matrix-isolation IR spectrum [O. L. Chapman, C. L. McIntosh, and J. Pacansky, *J. Am. Chem. Soc.*, **95**, 614 (1973)] indicates a square-planar ground-state geometry, in agreement with an *ab initio* calculation [A. Krantz, C. Y. Lin, and M. D. Newton, *ibid.*, **95**, 2744 (1973)] which predicts a triplet ground state with a square-planar geometry. However, the absence of an electron spin resonance spectrum, together with a MINDO calculation that predicts a singlet ground state with rectangular geometry, has put the nature of the C_4H_4 ground state in doubt [S. Masamune, N. Nakamura, M. Suda, and H. Ona, *ibid.*, **95**, 8481 (1973)].

Several applications of IR spectroscopy to astrophysics have been made. Small amounts of methane in the earth's atmosphere have been detected by the observation of weak IR absorption lines in solar radiation that has passed through the earth's atmosphere. Intense IR absorption bands of CH_4 have been found in the spectra of the atmospheres of Jupiter, Saturn, Uranus, and Neptune. Bands of ammonia have been observed for Jupiter and Saturn; bands of CO_2 have been observed in the Venusian spectrum; and bands of H_2O have been observed in the Martian spectrum.

Because of the many normal modes and the presence of overtone and combination bands, hot bands, and impurity bands, which may overlap one another, the IR spectra of medium-sized and large molecules are complex and may be difficult to assign. Incorrect IR vibrational assignments have, unfortunately, been quite common.

IR spectra of over 100,000 compounds have been cataloged.

6.7 RAMAN SPECTROSCOPY

Raman spectroscopy (Section 4.10) aids the study of the vibrations of polyatomic molecules. For a vibration to be Raman active, it must give a change in the molecular polarizability. For many molecules with some symmetry, one or more of the normal modes correspond to no change in dipole moment, so that study of the IR spectrum may not give all the normal-mode frequencies. These IR-inactive modes may well be Raman active, thus allowing the determination of their frequencies. For example, the tetrahedral molecule methane has one nondegenerate vibration (ν_1), one doubly degenerate vibration (ν_2), and two triply degenerate vibrations (ν_3 and ν_4); of these, only ν_3 and ν_4 are IR active, but all four are Raman active. Knowledge of all the fundamental frequencies is needed for a force-constant analysis and for calculation of thermodynamic properties from the partition function.

The rule of mutual exclusion states that for a molecule with a center of symmetry, a given vibrational transition cannot appear in both the IR and Raman spectra. (For the proof, see Chapter 9.) Some fundamentals may be both IR and Raman inactive; their frequencies can often be determined from IR or Raman combination bands.

The pure-rotational Raman spectrum of a polyatomic molecule provides information on the moments of inertia, hence allowing a structural determination. For a molecule to exhibit a pure-rotational Raman spectrum, the polarizability must be anisotropic; that is, the polarizability ellipsoid must not be a sphere. As noted in Section 5.2, a spherical top has a spherical polarizability ellipsoid, and so gives no pure-rotational Raman spectrum. Symmetric and asymmetric tops have asymmetric polarizabilities. The structures of several nonpolar molecules (which cannot be studied by microwave spectroscopy) have been determined from their pure-rotational Raman spectra; these include F_2, C_2H_4, and C_6H_6.

Observation of the Raman effect presents difficulties in that the lines are quite weak; only $\frac{1}{1000}$th of the scattered radiation is Raman scattered, the rest being Rayleigh scattered with no change in frequency. However, the availability of laser light sources sparked a renaissance of Raman spectroscopy in the late 1960s. The laser provides an intense, monochromatic light source ideally suited for Raman work; also, the coherence of laser light makes it easy to focus and collimate the radiation. Unlike the infrared, where separate instruments are required for mid and far IR regions, a single Raman spectrometer covers the full range of molecular vibrational frequencies. Several commercial laser Raman spectrometers are available; they use a CW gas laser. Like IR spectroscopy, Raman spectroscopy can be used for compound identification (fingerprinting). Characteristic group vibrational frequencies occur in Raman spectra, as in IR spectra. In molecules with little symmetry, most vibrational modes appear in both the IR and the Raman spectrum, but usually with quite different intensities, since the change in polarizability for a vibration differs from the change in dipole moment for the vibration. Thus Raman information complements IR information. The strongest bands in the vibrational Raman spectrum are the allowed $\Delta v = 1$ fundamentals. Raman spectroscopy is valuable for studying water-soluble biologically active compounds; water is a poor Raman scatterer, but strongly absorbs IR radiation, so water is not a good IR solvent. Laser Raman spectroscopy of polymers and crystals is a growing field. For further information, see M. C. Tobin, *Laser Raman Spectroscopy*, Wiley-Interscience, New York, 1971; S. K. Freeman, *Applications of Laser Raman Spectroscopy*, Wiley-Interscience, 1973.

6.8 DEGENERATE AND NEAR-DEGENERATE VIBRATIONAL LEVELS

There are several important effects associated with degenerate vibrational levels. We begin by considering the doubly degenerate ν_{2a} and ν_{2b} vibrational modes of a linear triatomic molecule. In these modes, the three atoms each vibrate in a plane perpendicular to the molecular axis, the vibrations being either in the xz or the yz plane, where the z axis is the molecular axis (Fig. 6.2). Classically, if both these modes are excited, the vibrations may give rise to vibrational angular momentum about the internuclear axis; for example, if ν_{2a} and ν_{2b} are of equal amplitude and differ in phase by 90°, then the resultant motion of each nucleus is a circle about the molecular axis, as shown in Fig. 6.8 (see Problem 6.18).

Quantum mechanically, there is also the possibility of nuclear vibrational angular momentum for these degenerate modes. We denote the normal coordinates for the doubly degenerate vibrations by Q_x and Q_y. From Fig. 6.2, we have

$$Q_x = ax_1 - bx_2 + cx_3, \qquad Q_y = ay_1 - by_2 + cy_3 \tag{6.83}$$

where the nuclei are numbered 1, 2, 3, from left to right, and where (x_1, y_1, z_1) are the Cartesian displacement coordinates of nucleus 1. The quantities a, b, and c are positive constants; for a $\mathscr{D}_{\infty h}$ molecule, $a = c$. The factor in the vibrational wave function due to the two degenerate modes is (except for a normalization constant)

$$H_{v_{2a}}(\alpha_2^{1/2}Q_x)H_{v_{2b}}(\alpha_2^{1/2}Q_y)e^{-\alpha_2(Q_x^2+Q_y^2)/2} \tag{6.84}$$

where v_{2a} and v_{2b} are the quantum numbers of each degenerate mode.

Consider a degenerate vibrational energy level with $v_{2a} + v_{2b} = 2$; the level is triply degenerate, corresponding to the wave functions

$$f \equiv \psi(v_1 20 v_3), \qquad g \equiv \psi(v_1 11 v_3), \qquad h = \psi(v_1 02 v_3) \tag{6.85}$$

From (1.137), the polynomial factors in these functions involve

$$f\colon\ 4\alpha_2 Q_x^2 - 2, \qquad g\colon\ 4\alpha_2 Q_x Q_y, \qquad h\colon\ 4\alpha_2 Q_y^2 - 2 \tag{6.86}$$

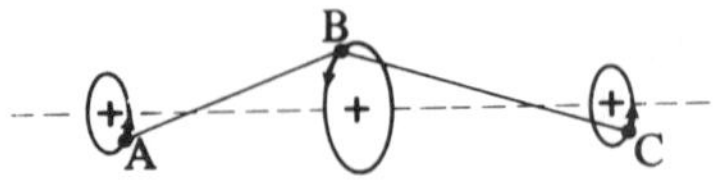

Fig. 6.8 Angular momentum arising from excitation of the doubly degenerate vibrational modes of a linear molecule. The plusses mark the equilibrium positions. Note how the motion resembles the rotation of a slightly bent molecule.

Any linear combination of the three functions (6.85) is an eigenfunction of the vibrational Hamiltonian (within the harmonic-oscillator approximation) and represents a possible state. There are three linearly independent linear combinations of special importance; these are

$$f-h+2ig, \qquad f+h, \qquad f-h-2ig \tag{6.87}$$

From (6.86), the polynomial factors of the functions (6.87) involve

$$4\alpha_2(Q_x^2-Q_y^2+2iQ_xQ_y), \qquad 4\alpha_2(Q_x^2+Q_y^2)-4$$

$$4\alpha_2(Q_x^2-Q_y^2-2iQ_xQ_y) \tag{6.88}$$

To see the significance of these functions, we define the *plane polar normal coordinates* ρ and φ (Fig 6.9) by

$$Q_x=\rho\cos\varphi, \qquad Q_y=\rho\sin\varphi \tag{6.89}$$

Let ρ_1 and φ_1 be the polar coordinates of nucleus 1 in the xy molecular plane; from (6.83) we have

$$Q_x=a\rho_1\cos\varphi_1-b\rho_2\cos\varphi_2+c\rho_3\cos\varphi_3 \tag{6.90}$$

However, in the normal mode Q_x, each nucleus has the same angle of rotation about the molecular axis: $\varphi_1=\varphi_2=\varphi_3\equiv\varphi'$. Hence in plane polar

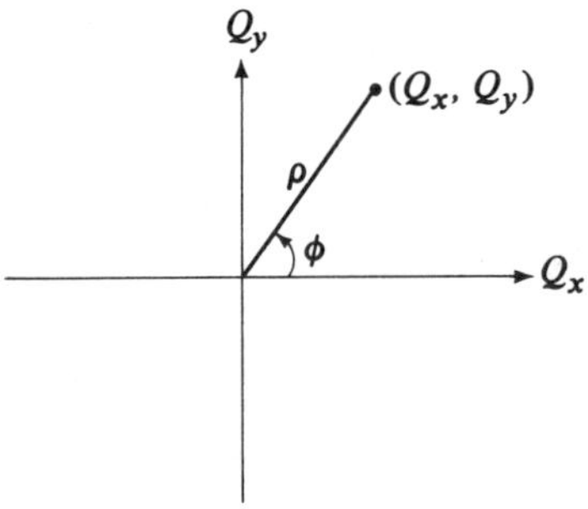

Fig. 6.9 Plane polar normal coordinates. The Q_xQ_y plane is not a plane in physical space, but rather a plane in the abstract normal-coordinate "space."

coordinates, (6.83) reads

$$
\begin{aligned}
Q_x &= \cos\varphi'(a\rho_1 - b\rho_2 + c\rho_3) \\
Q_y &= \sin\varphi'(a\rho_1 - b\rho_2 + c\rho_3)
\end{aligned}
\tag{6.91}
$$

Comparing these equations with (6.89), we have

$$\varphi = \varphi', \qquad \rho = a\rho_1 - b\rho_2 + c\rho_3 \tag{6.92}$$

Thus the polar normal coordinate φ is the angle of rotation of each nucleus about the molecular axis of the linear molecule. Substitution of (6.89) into (6.88) and use of trigonometric identities gives

$$
\begin{aligned}
f - h + 2ig&: \quad 4\alpha_2\rho^2 e^{2i\varphi} \\
f + h&: \quad 4\alpha_2\rho^2 - 4 \\
f - h - 2ig&: \quad 4\alpha_2\rho^2 e^{-2i\varphi}
\end{aligned}
\tag{6.93}
$$

The exponential factor in (6.84) becomes $e^{-\alpha_2\rho^2/2}$.

The operator for the z component of angular momentum is $\hat{L}_z = -i\hbar\partial/\partial\varphi$. The three functions (6.93) are eigenfunctions of $\hat{L}_z$ with eigenvalues $+2\hbar$, 0, and $-2\hbar$; since the normal coordinates Q_1 and Q_3 involve only the z coordinates of the nuclei and not the φ coordinate, the linear combinations (6.87) are also eigenfunctions of $\hat{L}_z$ with these eigenvalues. Thus the three states (6.87) are states in which the vibrational angular momentum of the nuclei about the molecular axis has a definite value. In the approximation of treating the vibrations as harmonic, there is no reason to prefer the functions (6.87) to other linear combinations. However, when the perturbation due to anharmonicity is considered, it turns out that the functions (6.87) are the correct zeroth-order vibrational wave functions; hence for the actual molecule, which is vibrating anharmonically, the functions (6.87) are a good approximation to the true vibrational wave functions, which is not true of the functions (6.85).

The functions (6.85) each correspond to a specific pair of the normal-mode vibrational quantum numbers v_{2a} and v_{2b}, whereas the correct zeroth-order functions (6.87) do not. Hence the quantum numbers v_{2a} and v_{2b} are not particularly physically significant. The quantum numbers of physical significance are $v_2 \equiv v_{2a} + v_{2b}$, which (together with v_1 and v_3) specifies which degenerate vibrational level we are dealing with, and the vibrational angular-momentum quantum number.

We have considered only one particular degenerate vibrational level. The general case is dealt with by solving the isotropic two-dimensional harmonic oscillator in plane polar coordinates (Problem 6.19). The result is that the factor in the correct zeroth-order vibrational wave function due to the doubly degenerate vibrational modes of a linear triatomic molecule is

$$e^{-\alpha_2\rho^2/2}F_{v_2}^{l}(\alpha_2^{1/2}\rho)e^{im_N\varphi} \tag{6.94}$$

$$m_N = v_2,\ v_2-2,\ v_2-4, \ldots,\ -v_2 \tag{6.95}$$

where the function F is a certain polynomial. The nuclear vibrational angular-momentum quantum number m_N ranges from v_2 to $-v_2$ in steps of 2. For the case we considered, the quantum numbers are $v_2=2$, $m_N=2,0,-2$. In the standard notation for polyatomic molecules, the absolute value of m_N is designated l: $l=|m_N|$;

$$l = v_2,\ v_2-2,\ v_2-4, \ldots,\ 1 \text{ or } 0 \tag{6.96}$$

The functions (6.94) are the linear combinations of (6.84) that correspond to definite values of the vibrational angular momentum; the absolute value of the vibrational angular momentum about the molecular axis is $l\hbar$.

Since we have abandoned the individual quantum numbers of the degenerate modes Q_x and Q_y, we will replace the summation over the $3N-6$ normal modes in energy expressions by a summation over the distinct vibrational frequencies. For a linear triatomic molecule there are four normal modes, but only three distinct vibrational frequencies. In the harmonic-oscillator approximation, the energy contribution of the doubly degenerate modes is

$$(v_{2a}+\tfrac{1}{2})h\nu_2+(v_{2b}+\tfrac{1}{2})h\nu_2=(v_2+1)h\nu_2 \tag{6.97}$$

For a vibrational frequency ν_i that is d_i-fold degenerate, the energy contribution is $(v_i+\frac{1}{2}d_i)h\nu_i$, with the possible values of d_i being $1,2,3,\ldots$. Equation (6.54) becomes

$$E_{\text{vib}} = \sum_i \left(v_i + \frac{d_i}{2}\right)h\nu_i \tag{6.98}$$

where the vibrational frequencies, rather than the normal modes, are being summed over.

Recall that linear molecules have $\Lambda\hbar$ as the absolute value of the axial component of electronic orbital angular momentum; the *electronic* wave functions are classified as $\Sigma,\Pi,\Delta,\Phi,\ldots$ according to whether Λ is $0,1,2,3,\ldots$. Similarly, *nuclear vibrational* wave functions are classified as $\Sigma,\Pi,\Delta,\Phi,\ldots$, according to whether l is $0,1,2,3,\ldots$. The context will usually indicate whether a given Greek-letter symbol refers to an electronic or a vibrational wave function; if necessary, the left superscript e or v can be added for clarity: ${}^e\Sigma, {}^e\Pi, \ldots, {}^v\Sigma, {}^v\Pi, \ldots$.

The formula (6.56), which includes anharmonicity corrections, is applicable only to molecules with no degenerate vibrational modes. Degenerate vibrational energy levels must be treated using degenerate perturbation theory. The result for the vibrational energy of a linear triatomic molecule is found to be

$$E_{\text{vib}} = \sum_{i=1}^{3}\left(v_i + \frac{d_i}{2}\right)h\nu_{i,e} + \sum_{i=1}^{3}\sum_{k\geqslant i} hx_{ik}\left(v_i + \frac{d_i}{2}\right)\left(v_k + \frac{d_k}{2}\right) + hg_{22}l^2 \quad (6.99)$$

where g_{22} and the x_{ik}'s are anharmonicity constants. Thus the anharmonicity splits degenerate vibrational states that have different values of l. The quantum number l is written as a superscript to v_2, so that the vibrational levels of a linear triatomic molecule are specified as $(v_1v_2^lv_3)$.

As an example, consider the three states of the (v_12v_3) level of a linear triatomic molecule. In the harmonic-oscillator approximation, these states have the same energy and the level is triply degenerate. Anharmonicity splits this degenerate level into the nondegenerate Σ level $(v_12^0v_3)$ and the doubly degenerate Δ level $(v_12^2v_3)$; the Δ level is doubly degenerate since $l=2$ corresponds to the two values $+2$ and -2 for m_N. The wave functions are (6.87). For CO_2, analysis of the vibrational levels (Table 6.2) yields the following constants[11] (all in cm^{-1}):

$$\begin{array}{lll} \tilde{\nu}_{1,e}=1354.0 & \tilde{\nu}_{2,e}=673.2 & \tilde{\nu}_{3,e}=2396.3 \\ \tilde{x}_{11}=-2.9 & \tilde{x}_{22}=+1.1 & \tilde{x}_{33}=-12.5 \\ \tilde{x}_{12}=-3.6 & \tilde{x}_{13}=-19.7 & \tilde{x}_{23}=-12.4 \\ \tilde{g}_{22}=-0.9 & & \end{array} \quad (6.100)$$

For linear molecules with several atoms, there is more than one doubly degenerate vibrational frequency. Thus for acetylene, the vibrations ν_1, ν_2, and ν_3 are nondegenerate, whereas ν_4 and ν_5 are each doubly degenerate (Fig 6.10). Each doubly degenerate vibration ν_i has associated with it the quantum numbers v_i and l_i; the individual l_i are combined to give a total vibrational angular-momentum quantum number l. (The situation is analogous to combining the λ_i of individual electrons to get the electronic orbital angular-momentum quantum number Λ.) For example, for a vibration in acetylene with $v_4=1$ and $v_5=1$, we must have $l_4=1$ and $l_5=1$; these values of l_4 and l_5 correspond to $m_{N4}=\pm 1$, $m_{N5}=\pm 1$. Algebraic addition of the m_N's gives the possibilities $+2$, -2, 0, and 0; the values ± 2 belong to a doubly degenerate vibrational level with $l=2$, that is, a ${}^{v}\Delta$ level. The values 0 and 0 correspond to two levels with $l=0$, that is, two ${}^{v}\Sigma$ levels. Recall that a $\pi\pi$ electronic configuration gives rise to ${}^{e}\Sigma^+$, ${}^{e}\Sigma^-$, and ${}^{e}\Delta$ terms (Section 1.19). Similarly, the two ${}^{v}\Sigma$ states can be classified as ${}^{v}\Sigma^+$

[11] I. Suzuki, *J. Mol. Spectry.*, **25**, 479 (1968).

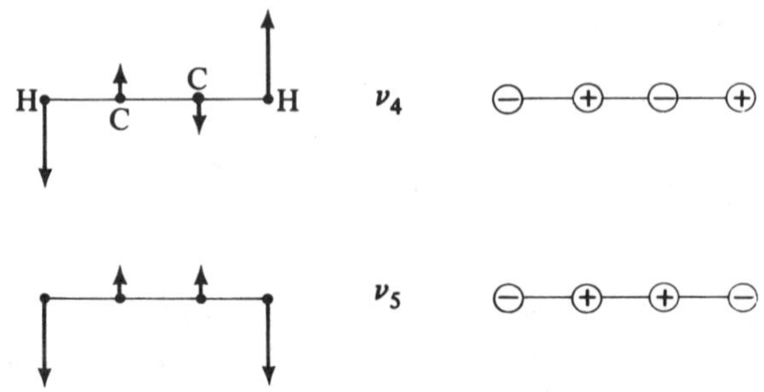

Fig. 6.10 Degenerate normal modes of acetylene. The relative displacements of the carbon atoms are exaggerated.

and ${}^{v}\Sigma^-$ states; these states correspond to vibrational wave functions that are multiplied by $+1$ and -1, respectively, when the vibrational normal coordinates are reflected in a plane containing the molecular axis. In the harmonic-oscillator approximation, the level with $v_4=1, v_5=1$ is fourfold degenerate; anharmonicity splits this level into a doubly degenerate ${}^{v}\Delta$ level and nondegenerate ${}^{v}\Sigma^+$ and ${}^{v}\Sigma^-$ levels.

Just as the electronic wave functions of $\mathcal{D}_{\infty h}$ molecules can be classified as g or u, so can the vibrational wave functions of such molecules. This classification of ψ_{vib} refers to its behavior on inversion of the normal coordinates with respect to molecule-fixed axes. The ground ψ_{vib} of a $\mathcal{D}_{\infty h}$ molecule is always a ${}^{v}\Sigma_g^+$ function, since the polynomial factor in the ground-state ψ_{vib} is a constant.

Spherical tops and nonlinear symmetric tops also have degenerate vibrational modes. For a linear molecule, there is no rotational angular momentum about the C_∞ symmetry axis, only vibrational angular momentum (for degenerate vibrational levels). For nonlinear molecules, there is rotational angular momentum, as well as the possibility of vibrational angular momentum about symmetry axes. This leads to complications in treating the degenerate vibrational levels, consideration of which we omit.[12]

Asymmetric tops have no degenerate vibrational modes (this will be shown in Chapter 9); hence we need not worry about vibrational angular momentum for them.

The presence of degenerate vibrational modes affects the rotational energies. In Section 4.11, we saw the effect of electronic angular momentum on the rotational energies of a diatomic molecule. In this section, we shall assume that there is no electronic angular momentum, only nuclear vibrational and rotational angular momentum. If both electronic and

[12]*Herzberg, Infrared and Raman Spectra*, pp. 401–406, 447–449.

vibrational angular momenta are present, their interaction with each other and with the rotational angular momentum must be considered.[13]

A linear molecule in a vibrational level with vibrational angular momentum quantum number l can be viewed as a symmetric top with $l\hbar$ replacing $|K|\hbar$ as the absolute value of the component of nuclear angular momentum about the symmetry axis. Replacing $|K|$ with l in (5.67), we have as the rotational energy of a linear molecule for a degenerate vibrational level

$$B_{[v]}h[J(J+1)-l^2]+A_{[v]}hl^2 \tag{6.101}$$

where $A_{[v]}$ is the effective moment of inertia of the nuclei about the molecular axis for the degenerate vibration. From (5.66), we have $J \geqslant l$. The quantity $(A_{[v]}-B_{[v]})l^2$ is usually considered as part of the vibrational energy of the linear molecule; for example, for a linear triatomic molecule, it is absorbed into the $g_{22}l^2$ term in (6.99). Hence the usual formula for the rotational energy of a linear molecule,

$$E_{\text{rot}}=hB_{[v]}J(J+1) \tag{6.102}$$

is applicable, except that the values $J=0,1,2,\ldots,l-1$ are excluded.

We omit consideration of the rotational energies of symmetric tops and spherical tops in degenerate vibrational levels.[14]

Recall that the doubly degenerate ${}^e\Pi, {}^e\Delta, \ldots$ electronic levels of a linear molecule are split into nondegenerate ${}^e\Pi^+, {}^e\Pi^-, {}^e\Delta^+, {}^e\Delta^-, \ldots$ electronic states by the interaction of electronic and rotational motions (Λ-type doubling). Similarly, the doubly degenerate ${}^v\Pi, {}^v\Delta, \ldots$ vibrational levels of a linear molecule are split into nondegenerate ${}^v\Pi^+, {}^v\Pi^-, {}^v\Delta^+, {}^v\Delta^-, \ldots$ vibrational states by the interaction of vibrational and rotational motions; this splitting is called *l-type doubling* As with Λ doubling, the l splittings are small. The correct zeroth-order wave functions for the two states of the l doublet are the linear combinations

$$2^{-1/2}[\psi(+l)+\psi(-l)] \quad \text{and} \quad 2^{-1/2}[\psi(+l)-\psi(-l)] \tag{6.103}$$

where $\psi(+l)$ is the vibrational wave function corresponding to vibrational angular momentum $+l\hbar$ along the molecular axis. As a result of the l-type doubling, pure-rotational transitions of a linear molecule in a degenerate vibrational level are split into two lines, since the effective moments of inertia are different for the two components of the l doublet.

[13] *Herzberg, Electronic Spectra of Polyatomic Molecules*, pp. 63–104.

[14] *Herzberg, Infrared and Raman Spectra*, pp. 403, 447.

So far we have considered interactions that lift the degeneracy of vibrational levels that are exactly degenerate in the harmonic-oscillator approximation. For two levels that are not exactly degenerate, but by accident are nearly degenerate in the harmonic-oscillator approximation, the formula (6.56) for the vibrational energy breaks down. The anharmonicity corrections in (6.56) are derived using the usual nondegenerate-perturbation-theory expansions of Section 1.10. However, when the energy difference between two unperturbed states is very small, the use of these expansions for these two states is not justified. When $E_n^{(0)} - E_m^{(0)}$ is very small, (1.200) tells us that the perturbed wave function for state n will have a large contribution from $\psi_m^{(0)}$ and vice versa. The proper procedure in the case of near degeneracy is to set up a secular equation analogous to the secular equation (1.207) of degenerate perturbation theory. The matrix elements in this secular equation are between the states that are near degenerate; most commonly, there are two near-degenerate states and the secular determinant is 2×2. Solution of the secular determinant then gives the first-order energy corrections for the near-degenerate states.

The large perturbations between near-degenerate vibrational levels produced by anharmonicity were first pointed out by Fermi, and such vibrational perturbations are called *Fermi resonance*. (The usual warning against the literal interpretation of the word "resonance" applies.)

Fermi resonance occurs in CO_2. We note from (6.100) that $2\nu_{2,e}$ (=1346 cm^{-1}) is very close to $\nu_{1,e}$ (=1354 cm^{-1}). Hence harmonic-oscillator levels of CO_2 for which $2v_1' + v_2' = 2v_1 + v_2$ (where the primed and unprimed quantum numbers refer to different vibrational levels) are quite close together and we have Fermi resonance. For example, (6.99) and (6.100) predict the levels ($10^0 0$), ($02^0 0$), and ($02^2 0$) to lie 1335, 1339, and $1335\frac{1}{2}$ cm^{-1}, respectively, above the ground vibrational level; the observed spectrum (Table 6.2) shows these levels actually lie 1388, 1285, and 1335 cm^{-1}, respectively, above the ground level (Problems 6.20 and 6.21). Clearly, the ($10^0 0$) and ($02^0 0$) levels have interacted with each other, thus shifting their energies. The ($02^2 0$) level is unaffected because the matrix element H_{ij}' is zero if states i and j have different values of l (Problem 6.22); Fermi resonance occurs only between states of the same symmetry. Fermi resonance between two levels increases the energy of the upper level and decreases the energy of the lower level; the levels "repel" each other.

In the 1960s, it was realized that the traditional labeling of CO_2 vibrational levels involved in Fermi resonance was in error. For example, the ($02^0 0$) and ($10^0 0$) labels had to be interchanged. [See the reference of footnote 11 and R. Oberly *et al.*, *J. Mol. Spectry.*, **25**, 138 (1968).] The labeling given in Table 6.2 is the traditional labeling, which is *incorrect* in many cases. The traditional labeling is being replaced by a notation which lists the two (or more) unperturbed harmonic-oscillator levels that make the main contributions to the level in question. Thus the levels traditionally labeled (10^00) and (02^00) are now called $(10^00, 02^00)_{\mathrm{I}}$ and $(10^00, 02^00)_{\mathrm{II}}$, respectively, where the I level lies higher. This notation is sometimes abbreviated to $(10^00)_{\mathrm{I}}$ and $(10^00)_{\mathrm{II}}$.

As well as shifting the expected positions of vibrational bands, Fermi resonance affects their intensities. For example, the wave function of the (02^00) level of CO_2 has a considerable contribution from the harmonic-oscillator function corresponding to the (10^00) level; since the latter is a fundamental level, the transition from the ground level (00^00) to the (02^00) level is much more intense than would be expected for an overtone band. This transition is Raman active, but IR inactive; the Raman spectrum of CO_2 shows two very strong bands at 1285 and 1388 cm^{-1}, corresponding to transitions to the (02^00) and (10^00) levels from the ground level.

For H_2O, the fundamental frequencies ν_1 and ν_3 are fairly close together and one might expect Fermi resonance between the (100) and (001) levels; however, these levels have different symmetry species (Section 6.3) and cannot interact. The levels (200) and (002) both have symmetry species A_1 and do interact. A little thought shows that only H_2O vibrational levels with both v_1 and v_3 less than 2 do not undergo Fermi resonance. (Details of the Fermi-resonance calculations for H_2O and CO_2 are given in *Herzberg, Infrared and Raman Spectra*, pp. 216–219.)

In benzene, the fundamental level corresponding to one of the C—H stretching vibrations lies very close to a combination level of two other fundamentals. Fermi resonance leads to two strong bands at 3099 and 3045 cm^{-1} for gas-phase C_6H_6, instead of one strong fundamental band and one weak combination band. It was once suggested that the presence of these two strong bands was "unambiguous proof" that benzene did not have $\mathcal{D}_{6h}$ symmetry, since only one strong band in the 3000 cm^{-1} region was to be expected for $\mathcal{D}_{6h}$ symmetry; the proponents of this view overlooked the possibility of Fermi resonance.

For molecules with several fundamental frequencies, there is a substantial probability for vibrational levels of the same symmetry to fall close together. Hence Fermi resonance is fairly common and adds to the difficulties of assigning IR spectra.

6.9 PARITY OF POLYATOMIC-MOLECULE WAVE FUNCTIONS

Inversion of the spatial coordinates of all particles leaves all interparticle distances unchanged and has no effect on the complete Hamiltonian of a polyatomic molecule. Since $\hat{\Pi}$ and $\hat{H}$ commute, the overall wave functions of any polyatomic molecule can be classified according to their parity. As with diatomic molecules (Section 4.7), polyatomic-molecule states are called positive or negative according to whether the wave function has the parity eigenvalue $+1$ or -1. Since inversion of all particle spatial coordinates leaves $\hat{H}_{\text{el}}$, $\hat{H}_{\text{vib}}$, and $\hat{H}_{\text{rot}}$ unchanged, the electronic, vibrational, and rotational wave functions can each be classified according to their parity. Recall that we must examine the parity of each function with respect to inversion of electronic and nuclear *space-fixed* coordinates.

Consider linear polyatomic molecules. The same reasoning as for diatomic molecules shows that ψ_{el} has parity eigenvalue $+1$ for ${}^e\Sigma^+, {}^e\Pi^+, \ldots$ electronic states and -1 for ${}^e\Sigma^-, {}^e\Pi^-, \ldots$ states. (If Λ-type doubling is negligible, then two electronic states of opposite parities coincide for $\Pi, \Delta, \ldots$ levels.) The vibrational wave functions have the factor $e^{\pm il\varphi}$ (where φ is a nuclear coordinate), which is analogous to the factor $e^{\pm i\Lambda\varphi}$ (where φ is an electronic coordinate) in the electronic wave functions; hence the parity eigenvalue of ψ_{vib} is $+1$ for ${}^v\Sigma^+, {}^v\Pi^+, \ldots$ vibrational wave functions and -1 for ${}^v\Sigma^-, {}^v\Pi^-, \ldots$ functions. (See also the discussion of planar molecules in the next paragraph.) The rotational wave functions of a linear polyatomic molecule are the same as for a diatomic molecule, namely, the spherical harmonics (Problem 5.18). Hence, as for diatomic molecules, the parity eigenvalue of ψ_{rot} is $(-1)^J$. The ground electronic state of a linear polyatomic molecule nearly always has the symmetry ${}^e\Sigma^+$. We showed in Section 6.4 that the polynomial factor in the ground vibrational wave function is a constant; for linear molecules, this corresponds to ${}^v\Sigma^+$ symmetry. Hence for the ground electronic and vibrational state of most linear polyatomic molecules, the rotational levels with even J are positive, whereas those with odd J are negative.

Next we consider planar molecules. The electronic wave function is expressed with respect to molecule-fixed axes, which we can take to be the *abc* principal axes of inertia. To achieve inversion of all particles with respect to space-fixed axes, we first rotate all the electrons and nuclei by 180° about the c axis (which is perpendicular to the molecular plane); we then reflect all the electrons in the molecular *ab* plane. The net effect of these two transformations is the desired space-fixed inversion of all particles. (Compare the corresponding discussion for diatomic molecules in Section 4.7.) The first step rotates the electrons and nuclei together and therefore has no effect on the molecule-fixed coordinates of either the electrons or the nuclei. (The *abc* axes rotate with the nuclei.) Thus the first step has no effect on ψ_{el}. The second step is a reflection of electronic spatial coordinates in the molecular plane; this is a symmetry plane and the corresponding operator $\hat{O}_\sigma$ has the possible eigenvalues $+1$ and -1 (since its square is the unit operator). The electronic wave functions of a planar molecule can thus be classified as having eigenvalue $+1$ or -1 with respect to $\hat{O}_\sigma$, and this eigenvalue is the desired parity eigenvalue of ψ_{el}. The ground electronic state of most molecules is symmetric (eigenvalue $+1$) with respect to all symmetry operations; hence the ground-state ψ_{el} is usually of even parity for planar molecules. Now consider ψ_{vib} for planar molecules. The vibrational wave function is a function of the normal coordinates Q_k, which are defined with respect to molecule-fixed axes. To achieve inversion with respect to space-fixed coordinates of the displacement vectors representing a given vibration, we first rotate the nuclei (in their equilibrium positions) and the displacement vectors by 180° about the c axis; we then reflect all displacement vectors in the molecular ab plane. The first operation moves the displacement vectors together with the molecule-fixed axes and hence (just as for the corresponding operation in the discussion of ψ_{el}) leaves the Q_k and ψ_{vib} unaffected. The second step is a reflection of the Q_k in the molecular symmetry plane; we saw in Section 6.4 how to determine the behavior of the vibrational wave functions with respect to a symmetry operation. The parity of ψ_{vib} is the same as its eigenvalue for reflection in the molecular plane. For a molecule with only in-plane vibrational modes (e.g., H_2O), ψ_{vib} always has parity eigenvalue $+1$. For BF_3, all the normal modes but one are in-plane vibrations (Fig. 6.11); the normal coordinate for the out-of-plane mode is antisymmetric with respect to reflection in the molecular plane. The vibrational eigenfunctions of BF_3 thus have parity eigenvalue $(-1)^{v_k}$, where v_k is the vibrational quantum number of the out-of-plane mode. Now consider ψ_{rot} for a planar molecule. Planar molecules can be symmetric or asymmetric tops, but not spherical tops, since the presence of two or more noncoincident C_3 or higher axes rules out planarity. We first consider planar symmetric tops. The rotational wave function depends on the Eulerian angles θ, φ, and χ; the angles θ and φ give the orientation of the positive direction of the molecular symmetry axis with respect to space-fixed coordinates, while χ is the angle of rotation about the symmetry axis. The positive direction of the symmetry axis is defined by numbering the successive identical nuclei as $1, 2, 3, \ldots$ and taking the positive direction as the direction the unbent thumb of the right hand points to when the fingers of the right hand push nucleus 1 toward nucleus 2. The identical nuclei of a planar symmetric top will be located at points on the ab circle in Fig. 5.2. Inversion of the nuclear space-fixed coordinates through point O sends each nucleus to a point on the opposite side of this circle, and preserves the relative orientation of nuclei $1, 2, 3, \ldots$; therefore, the positive direction of the symmetry axis is unchanged upon inversion. Hence θ and φ are unchanged by inversion. However, χ is increased by π since inversion of the nuclei is equivalent to rotation of the nuclei of the planar molecule by 180° about the symmetry axis, thereby rotating the b axis (and the a axis) by 180° (since these axes are defined with respect to the nuclear locations). The angle χ enters the symmetric-top wave function as the factor $e^{iK\chi}$. Therefore inversion multiplies ψ_{rot} by $e^{iK\pi} = (-1)^K$; the parity eigenvalue of ψ_{rot} is $+1$ for even K and -1 for odd K, for planar symmetric tops. Now consider planar asymmetric tops. We saw (Problem 5.8) that the symmetric-top wave functions that occur in the expansion of a given asymmetric-top wave function all have either even values of K or odd values of K. The parity of a planar asymmetric-top wave function is thus the same as the parity of each of the planar symmetric-top wave functions of which it is a linear combination. A planar symmetric top must be an oblate top (Problem 5.19). The asymmetric-top levels are designated as $J_{K_{pr}K_{ob}}$. If the asymmetric-top level correlates with an oblate-top level with K even, then it must be a linear combination of oblate-top functions with even K values; similarly for correlation with odd K_{ob}. Hence the asymmetric-top wave functions have parity eigenvalues $+1$ or -1 according to whether K_{ob} is even or odd, respectively. Having determined the parity eigenvalues of ψ_{el}, ψ_{vib}, and ψ_{rot}, we multiply these eigenvalues together to get the parity of ψ for planar molecules.

Finally, we come to nonplanar molecules. Here, the parity of ψ is usually unimportant for the following reasons. Inversion of a nonplanar molecule changes the nuclear configuration to a configuration physically equivalent to, but distinguishable from, the original one. We call these two configurations the *right form* and the *left form*. One cannot be obtained from the other by rotation, but only by inversion. Almost always, the potential-energy barrier to inversion is extremely high and for all practical purposes can be considered as infinite. Let the molecular wave functions corresponding to the right and left forms be ψ_{I} and ψ_{II}. In the limit of infinite inversion barrier, these functions are degenerate; the linear combinations

$$2^{-1/2}(\psi_{I}+\psi_{II}) \quad \text{and} \quad 2^{-1/2}(\psi_{I}-\psi_{II}) \tag{6.104}$$

have parity eigenvalues $+1$ and -1, respectively. However, because the barrier is essentially infinite, the positive and negative wave functions in (6.104) have essentially the same energy. Thus every energy level of a nonplanar molecule has a twofold inversion degeneracy, with positive and negative states coinciding in energy. The coincident states have the same quantum numbers except for the parity eigenvalue. Thus the usual parity selection rule has no effect on the spectrum, and there is no point to considering the parities of the states. In certain cases, however, the parity of states of nonplanar molecules is significant. For NH_3 and certain of its derivatives, the inversion barrier is fairly low; this causes the energies of the two wave functions (6.104) to split (Section 5.8). When this inversion splitting occurs, the positive and negative states no longer have the same energy and the parity of the states becomes important.

In discussing the parity of the molecular wave functions, we have assumed that ψ can be factored into electronic, vibrational, and rotational factors:

$$\psi=\psi_{el}\psi_{vib}\psi_{rot} \tag{6.105}$$

Of course, this is an approximation, and (6.105) is not an exact eigenfunction of the molecular Hamiltonian $\hat{H}$, but is an eigenfunction of some operator which we shall call $\hat{H}^0$. (We shall not specify $\hat{H}^0$ explicitly.) A better approximation to the true molecular wave function is obtained by applying perturbation theory, taking the perturbation Hamiltonian $\hat{H}'$ as

$\hat{H}-\hat{H}^0$. The perturbation $\hat{H}'$ contains interaction terms between the electronic, vibrational, and rotational motions, and adds a correction term to the simple product wave function:

$$\psi = \psi_{\text{el}}\psi_{\text{vib}}\psi_{\text{rot}} + \psi'_{\text{evr}}$$

Even though (6.105) is only an approximation, its symmetry properties (e.g., its parity) are the same as those of the true molecular wave function, since the symmetry properties of the wave function follow rigorously from the symmetry of the true Hamiltonian $\hat{H}$; the perturbation $\hat{H}'$ cannot change the overall symmetry of ψ.

In linear molecules, the electronic–rotation interaction terms in $\hat{H}'$ cause the Λ-type doubling of electronic states, whereas the vibration–rotation interaction terms in $\hat{H}'$ cause the l-type doubling of vibrational states. In addition, the perturbation $\hat{H}'$ can cause interactions between vibration–rotation levels of different electronic states. If it happens that two vibration–rotation levels of different electronic states of a molecule have nearly the same energy [in the approximation of neglecting $\hat{H}'$ and taking ψ as (6.105)], then $\hat{H}'$ may lead to substantial mixing of the zeroth-order wave functions of the two levels and substantial shifts in their energies. The resulting shifts in the spectral lines involving transitions to or from one of these levels are called *spectroscopic perturbations*.

The complete wave function of a molecule is called the *rovibronic* wave function. In the simplest approximation, the rovibronic function is a product of rotational, vibrational, and electronic functions. For certain applications, the rotational motion is first neglected, and the vibrational and electronic motions are treated together. The rotational motion is then taken into account. The wave function for electronic and vibrational motion is called the *vibronic* wave function. Just as we separately classified the electronic and vibrational wave functions according to their symmetries, we can do the same for the vibronic functions. In the simplest approximation, the vibronic wave function is a product of electronic and vibrational wave functions, and we can thus readily determine its symmetry. For example, if the electronic state is an ${}^{e}\Sigma^-$ state and the vibrational state is a ${}^{v}\Sigma^-$ state, then the vibronic wave function is designated ${}^{ev}\Sigma^+$.

6.10 NUCLEAR SPIN AND THE PAULI PRINCIPLE

As for diatomic molecules (Section 4.8), the wave function of a polyatomic molecule should include the nuclear spin coordinates, as well as their spatial coordinates. If all the nuclei are different, then the wave function simply contains the factor

$$|M_{I_A}(\mathrm{A})\rangle|M_{I_B}(\mathrm{B})\rangle\cdots|M_{I_G}(\mathrm{G})\rangle$$

where $I_A, I_B, \ldots, I_G$ are the spins of nuclei A, B, ..., G. The nuclear statistical weight of each molecular energy level is

$$(2I_A+1)(2I_B+1)\cdots(2I_G+1)$$

Since the statistical weights are the same for all levels, the weights can usually be omitted; see *Davidson*, Section 9-3. Since all the nuclei are different, the Pauli principle is irrelevant.

If some of the nuclei are identical, the Pauli principle may be relevant. We first consider molecules with identical nuclei that are located in nonequivalent positions in the molecule. An example is the linear molecule NNO. The N^{14} nucleus is a boson, and it might be thought that the Pauli principle required the complete wave function of $N_2^{14}O$ to be symmetric with respect to interchange of the space and spin coordinates of the two N^{14} nuclei. However, the potential barrier to interchange of the two nitrogen nuclei in N_2O is extremely high, and the two nitrogen atoms can be regarded as distinguishable by the fact that one lies adjacent to the oxygen nucleus and the other does not. Hence there is no need to satisfy the Pauli principle in this case. [It might be argued that since the barrier to interchange of the two N nuclei is not infinite, there is a small nonzero probability for this interchange to occur, and therefore the molecular wave function should be symmetrized with respect to this interchange. Strictly speaking, this viewpoint is correct, but it leads to the same degeneracy as obtained when the nitrogen nuclei are considered as distinguishable: The molecular wave function is based on a given framework of nuclei, and interchange of the two N nuclei in NNO changes this framework. To construct a wave function symmetric with respect to interchange of the N nuclei, we must take the function $(\psi_A+\psi_B)/\sqrt{2}$ where ψ_A is a wave function based on the framework N_1N_2O and ψ_B is a wave function with the same quantum numbers as ψ_A, but based on the framework N_2N_1O. However, this symmetrization of the wave function with respect to interchange of the nitrogen nuclei introduces no new degeneracy, because there is only one linear combination of the wave functions based on different frameworks that is symmetric with respect to interchange of the nitrogen nuclei, namely, $(\psi_A+\psi_B)/\sqrt{2}$. In the general case of several identical nuclei, we should make ψ have the proper symmetry with respect to all possible permutations of identical nuclei. However, if we construct a wave function that has the proper symmetry only with respect to those permutations that don't change the molecular framework (these are the interchanges that can be brought about by proper rotations of the framework), the statistical weights will be the same as for the wave function that has the proper symmetry with respect to all possible nuclear permutations. This is so because there is only one linear combination of wave functions based on different frameworks that has the proper symmetry with respect to all permutations of identical nuclei.[15] Hence we need only worry about satisfying the Pauli principle for those interchanges of identical nuclei that can be brought about by proper rotations of the nuclear framework.]

In molecules like CO_2 or $CH_2{=}CH_2$, the identical nuclei are in chemically equivalent positions and can be exchanged by rotations of the molecule; for such molecules, the Pauli principle must be satisfied. (In a molecule like CH_3OH, for any given conformation the methyl hydrogens are distinguishable from one another by their positions relative to the hydroxyl hydrogen; however, the barrier to internal rotation is low, and tunneling from one equivalent conformation to another readily occurs. Hence the wave functions must be antisymmetrized with respect to exchange of methyl hydrogens. For a molecule like $CFBr{=}CH_2$, the barrier to rotation about the double bond is high, and the two hydrogen nuclei can be treated as distinguishable.)

We begin by considering linear polyatomic molecules. For $\mathcal{C}_{\infty v}$ molecules, rotation cannot interchange identical indistinguishable nuclei, and the nuclear statistical weight of each molecular rotational level is the same. (For a molecule like $HC^{12}{\equiv}C^{13}H$, the two hydrogen atoms can be distinguished by which carbon atom each is bonded to.)

For $\mathcal{D}_{\infty h}$ molecules, a $\hat{C}_2$ rotation of the nuclear framework about an axis perpendicular to the molecular axis at its midpoint will interchange one or more pairs of identical indistinguishable nuclei. The behavior of the complete molecular wave function ψ under this rotation depends on the spins of the nuclei. Interchange of a pair of fermions multiplies ψ by -1, whereas interchange of a pair of bosons multiplies ψ by $+1$. A $\mathcal{D}_{\infty h}$ molecule has the general formula $R(AB\cdots G)_2$ or $(AB\cdots G)_2$. If the nuclei $A, B, \ldots, G$ contain an odd number of fermions, then the $\hat{C}_2$ rotation of the nuclei multiplies ψ by -1; otherwise ψ is unchanged by the $\hat{C}_2$ rotation.

The overall wave function of a $\mathcal{D}_{\infty h}$ molecule can be written as

$$\psi = \psi_{el}\psi_{vib}\psi_{rot}\psi_{ns}$$

We first consider the case where there is no central atom R. The nuclear spin function ψ_{ns} is then the product of the spin functions $\varphi_A, \varphi_B, \ldots, \varphi_G$,

[15]The proof of this statement is given in E. B. Wilson, Jr., *J. Chem. Phys.*, **3**, 276 (1935).

one for each pair of identical nuclei. Each of $\varphi_A, \ldots, \varphi_G$ can be either symmetric or antisymmetric with respect to exchange of nuclear coordinates. The number of possible symmetric and antisymmetric ψ_{ns} is found by an extension of the method used in Section 4.8 for diatomic molecules. Since there are $2I+1$ possible nuclear spin functions for a nucleus of spin I, the total number of possible ψ_{ns} for the molecule $(AB\cdots G)_2$ is

$$(2I_A+1)^2(2I_B+1)^2\cdots(2I_G+1)^2 \tag{6.106}$$

There are $2I_A+1$ symmetric functions φ_A of the form (4.127). There are thus

$$(2I_A+1)(2I_B+1)\cdots(2I_G+1) \tag{6.107}$$

functions ψ_{ns} that are products of spin functions of the form (4.127); these ψ_{ns} are symmetric with respect to simultaneous interchange of all identical nuclei. There remain

$$\left[(2I_A+1)^2\cdots(2I_G+1)^2\right]-\left[(2I_A+1)\cdots(2I_G+1)\right] \tag{6.108}$$

spin functions ψ_{ns}; of these, half are symmetric and half antisymmetric with respect to simultaneous interchange of all identical nuclei, as we shall now prove. First consider φ_A. Besides the $2I_A+1$ functions φ_A of the form (4.127), there are φ_A functions of the forms (4.128) and (4.129). The number of symmetric φ_A of the form (4.128) is the same as the number of antisymmetric φ_A of the form (4.129). We now multiply the φ_A of the forms (4.128) and (4.129) by the φ_B functions having each of these forms. We have the following possibilities:

form of φ_A	form of φ_B
(4.128)	(4.128)
(4.129)	(4.129)
(4.128)	(4.129)
(4.129)	(4.128)

Of these possibilities, the first two give functions $\varphi_A\varphi_B$ that are symmetric with respect to simultaneous interchange of the two A and two B nuclei, whereas the third and fourth give functions $\varphi_A\varphi_B$ that are antisymmetric; the number of symmetric and antisymmetric $\varphi_A\varphi_B$ clearly are equal. We next multiply the $\varphi_A\varphi_B$ by φ_C functions of the forms (4.128) and (4.129); and so on. Q.E.D. Taking half of (6.108) and adding it to (6.107), we have as the number of symmetric ψ_{ns}

$$\tfrac{1}{2}(2I_A+1)\cdots(2I_G+1)[(2I_A+1)\cdots(2I_G+1)+1] \qquad (6.109)$$

Taking half of (6.108), we have as the number of antisymmetric ψ_{ns}

$$\tfrac{1}{2}(2I_A+1)\cdots(2I_G+1)[(2I_A+1)\cdots(2I_G+1)-1] \qquad (6.110)$$

These are the nuclear statistical weights.

For $\mathfrak{D}_{\infty h}$ molecules of the form $R(AB\cdots G)_2$, the quantities (6.109) and (6.110) must each be multiplied by $2I_R+1$.

Now we consider the nuclear-interchange symmetry of the remaining factors in the wave function. For ψ_{rot}, the wave functions are spherical harmonics, just as for diatomic molecules; hence, as for diatomic molecules (Section 4.8), the eigenvalue of ψ_{rot} with respect to simultaneous interchange of the coordinates of all identical nuclei in a $\mathfrak{D}_{\infty h}$ molecule is $(-1)^J$ for $^1\Sigma$ electronic states. (For Σ electronic states that are not singlets, the eigenvalue is $(-1)^N$; see Section 4.11.)

The symmetry with respect to nuclear exchange of ψ_{vib} for linear polyatomic molecules is found by the same considerations we used for ψ_{el} of diatomic molecules. We first invert the space-fixed coordinates of all nuclei and all displacement vectors; we then reinvert the space-fixed coordinates of the displacement vectors. The net effect is the inversion of all nuclear space-fixed coordinates. For a $\mathfrak{D}_{\infty h}$ molecule, this inversion of nuclear coordinates interchanges all pairs of identical nuclei. We showed in Section 6.9 that inversion of the nuclei and the displacement vectors has eigenvalue $+1$ for ${}^v\Sigma^+, {}^v\Pi^+, \ldots$ vibrational states and eigenvalue -1 for ${}^v\Sigma^-, {}^v\Pi^-, \ldots$ vibrational states. The reinversion of the displacement vectors has eigenvalue $+1$ for g vibrational states and eigenvalue -1 for u vibrational states. Hence ψ_{vib} is symmetric with respect to nuclear exchange for ${}^v\Sigma_g^+, {}^v\Sigma_u^-, {}^v\Pi_g^+, {}^v\Pi_u^-, \ldots$ functions and antisymmetric for ${}^v\Sigma_g^-, {}^v\Sigma_u^+, \ldots$ functions. For homonuclear diatomic molecules, the single vibrational mode is totally symmetric and ψ_{vib} is always a ${}^v\Sigma_g^+$ function, which is symmetric with respect to nuclear exchange.

The symmetry of ψ_{el} for linear polyatomic molecules follows from the same arguments as for diatomic molecules; thus ψ_{el} is symmetric with respect to nuclear exchange for ${}^e\Sigma_g^+, {}^e\Sigma_u^-, \ldots$ functions and antisymmetric for ${}^e\Sigma_g^-, {}^e\Sigma_u^+, \ldots$ functions.

As with homonuclear diatomic molecules, the letters s and a are used to give the nuclear-exchange symmetry of ψ/ψ_{ns} for $\mathfrak{D}_{\infty h}$ polyatomic molecules.

Now consider some examples. For $C^{12}O_2^{16}$, the nuclei all have zero spin.

There is only one possible ψ_{ns}, and this function is symmetric with respect to nuclear exchange. The ground electronic state is $^1\Sigma_g^+$ and the ground vibrational state (as for all $\mathcal{D}_{\infty h}$ molecules) is $^v\Sigma_g^+$. Hence the symmetry eigenvalue for nuclear exchange is $(-1)^J$ for the ground vibronic state. However, O^{16} is a boson. Hence odd values of J cannot occur for the (00^00) vibrational level of the ground electronic state. The vibrational mode ν_1 is a Σ_g^+ mode and the odd J rotational levels of the (10^00) vibrational level are missing for the ground electronic state. The (00^01) vibrational level of the ground electronic state has even J rotational levels missing because ν_3 is a Σ_u^+ mode (Fig 6.2). Similar considerations hold for other vibrational levels. The fact that half the rotational levels are missing is verifiable from the rotational structure of CO_2 IR and Raman vibration–rotation bands and from the pure-rotation Raman spectrum.

For $C_2^{12}H_2^1$, the nuclear spin of C^{12} is zero and contributes a factor of 1 to the nuclear statistical weights. The statistical weights are therefore the same as in H_2^1. For the ground vibronic state, the even J levels are s and have nuclear statistical weight 1, corresponding to the one possible ψ_{ns}; the odd J levels are a and have nuclear statistical weight 3. The usual selection rule (4.138) holds for collisions as well as radiative transitions, and we have ortho and para acetylene. The two forms have not been separated.

For nonlinear molecules with identical nuclei that are interchangeable by rotation, the derivation of the nuclear statistical weights is not as simple as for linear molecules. The problem is most efficiently dealt with using group theory. We will not attempt a complete discussion, but will only point out some features of the results.[16]

For $\mathcal{C}_{2v}$ molecules where the $\hat{C}_2$ rotation interchanges two spin-$\frac{1}{2}$ nuclei (e.g., the normal isotopic species of H_2O and $H_2C{=}O$), the function ψ_{ns} involves the same spin functions as for H_2. Thus the a levels of H_2O have a nuclear statistical weight of 3, whereas the s levels have a nuclear weight of 1; we have ortho and para water. For molecules where the $\hat{C}_2$ rotation interchanges two spin-zero nuclei (e.g., SO_2^{16}), half the rotational levels are missing, just as for CO_2^{16}.

For $CH_2{=}CH_2$, there are four symmetry modifications, which might be called ss, sa, as, and aa. The letters refer to the symmetry of ψ/ψ_{ns} for nuclear exchanges brought about by rotations of the framework about two of the C_2 axes. There are three C_2 axes, but since $\hat{C}_2(z)=\hat{C}_2(x)\hat{C}_2(y)$, the symmetry behavior with respect to two of the $\hat{C}_2$ rotations determines the behavior with respect to the third $\hat{C}_2$ rotation.

For symmetric and spherical tops, there are three or more identical nuclei interchangeable by rotation, and we can no longer factor the

[16]For further discussion, see *Townes and Schawlow*, pp. 69–73, 102–105; *Herzberg, Infrared and Raman Spectra*, pp. 409–411, 451, 462–465; *King*, pp. 296–301, 309–312.

nuclear wave function into the product $\psi_{vib}\psi_{rot}\psi_{ns}$. (Recall that for atoms and molecules with three or more electrons, the electronic wave function cannot be written as the product of an electronic spatial and an electronic spin function.) Rather, the nuclear wave function involves a sum of terms, each of which is a product of a nuclear spatial and a nuclear spin factor. As an example of the results obtained, for a totally symmetric vibronic state of the symmetric top CH_3Cl, the $|K|=0,3,6,9,\ldots$ rotational levels have twice the statistical weight of the $|K|=1,2,4,5,7,8,\ldots$ levels; the levels with K a multiple of 3 do not combine with the levels for which K is not a multiple of three, and there are two modifications of CH_3Cl (which have not been separated). For symmetric tops where the identical nuclei that are interchanged by rotation of the framework have spin zero, certain of the rotational levels do not occur (just as for O_2^{16} and $C^{12}O_2^{16}$). For CH_4, there are three noncombining modifications (sometimes called meta, ortho, and para methane); these correspond to total nuclear spin quantum number $T=2, 1$, and 0.[17]

The nuclear statistical weights of the rotational levels affect the rotational partition function. The levels with different nuclear statistical weights must be summed over separately. However, for sufficiently high temperatures, it can be shown[18] to be a good approximation to evaluate the rotational partition function by giving each rotational level an average nuclear statistical weight that is equal to the total spin multiplicity [obtained as the product $(2I_1+1)(2I_2+1)\cdots(2I_N+1)$ over all N nuclei] divided by the *symmetry number*. The symmetry number is the number of different indistinguishable framework orientations obtainable by proper rotations of the nuclear framework. For example, the symmetry numbers of $\mathcal{C}_{\infty v}$ and $\mathcal{D}_{\infty h}$ molecules are 1 and 2, respectively. For C_6H_6, the symmetry number is 12: There are six different indistinguishable configurations obtainable by application of the rotations $\hat{E}, \hat{C}_6, \hat{C}_6^2, \ldots, \hat{C}_6^5$ to an initial configuration; application of rotations about each of the six in-plane C_2 axes to an initial configuration gives six more indistinguishable configurations. For CH_4, the symmetry number is 12: Application of $\hat{C}_3$ and $\hat{C}_3^2$ rotations about each of the four C_3 axes gives eight configurations indistinguishable from the original one; in addition, we have three configurations obtainable by rotating about each of the three C_2 axes, plus the original unrotated configuration.[19]

[17] For IR spectroscopic observations on the rates of interconversion of these forms of methane, see F. H. Frayer and G. E. Ewing, *J. Chem. Phys.*, **48**, 781 (1968).

[18] See *Herzberg, Infrared and Raman Spectra*, pp. 507–509.

[19] For nonlinear molecules, the symmetry number is simply the order of the rotational subgroup (these terms will be defined in Chapter 9).

PROBLEMS

6.1 How many normal modes of vibration does each of the following molecules have? (a) HCl; (b) HCN; (c) $PbCl_2$; (d) $HgCl_2$; (e) XeF_6.

6.2 Write down the three relations involving the $3N$ nuclear coordinates $x_1,\ldots,z_N$ [Eq. (6.1)] that specify that the *abc* system of axes translates with the molecule.

6.3 For the $v=0$ and $v=1$ vibrational levels of CO, calculate the maximum departure of each nucleus from its equilibrium position in the principal-axis coordinate system if it is assumed the nuclei move classically. Assume harmonic-oscillator energy levels.

6.4 The normal modes of BF_3 are sketched in Fig. 6.11. Which modes must be degenerate?

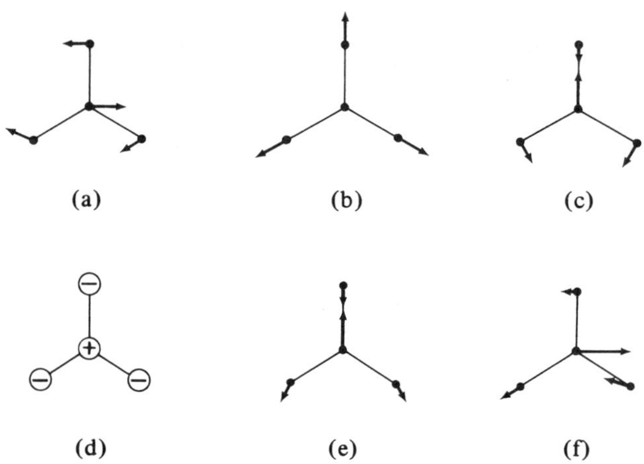

Fig. 6.11 The normal modes of BF_3. The relative magnitudes of the fluorines' motions are exaggerated.

6.5 Verify all the equations given in the normal-mode treatment of a diatomic molecule in Section 6.2.

6.6 Classify the normal modes of ethylene shown in Fig. 6.12 according to the $\mathcal{D}_{2h}$ symmetry species listed in Section 9.12. (See also Problem 1.12.)

6.7 Consider a symmetric system of three masses A—B—A confined to move in one dimension. Springs, each with force constant k, connect B to each A. (a) Set up and solve the vibrational secular equation to find the

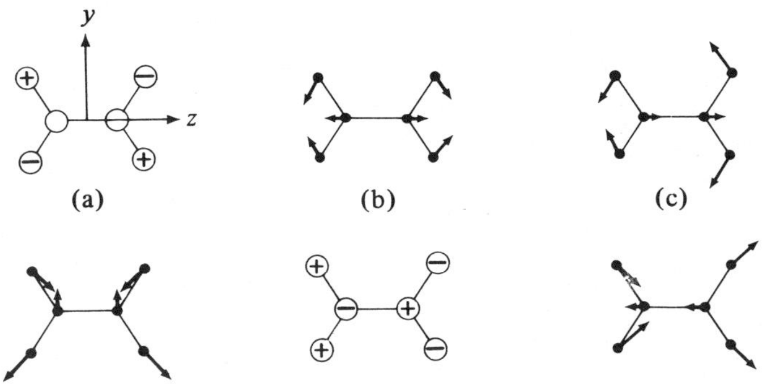

Fig. 6.12 Six of the 12 normal modes of C_2H_4. The x axis is perpendicular to the molecular plane.

normal-vibration frequencies and the normal coordinates. How many translational and how many genuine vibrational modes are found? (b) From the results of part (a), calculate the predicted ratio $\nu_{3,e}/\nu_{1,e}$ for CO_2; compare with (6.100). Explain the discrepancy. (c) Estimate the bond-stretching force constant in CO_2.

6.8 Look up the fundamental vibrational frequencies and calculate the approximate zero-point energy for each of the following: (a) PH_3; (b) C_6H_6; (c) C_2H_4. If it is assumed that all anharmonicity constants are negative, are the true zero-point energies less than or greater than those calculated in this problem?

6.9 Give the symmetry species of each of the following vibrational states of H_2O: (a) (111); (b) (012); (c) (021); (d) (130); (e) (541); (f) (031).

6.10 For C_2H_4, give the symmetry species for the vibrational state with vibrations (a) through (f) of Fig. 6.12 having the vibrational quantum numbers 2, 1, 3, 0, 1, 1, respectively, and all other modes having vibrational quantum number 0.

6.11 Which of the ethylene vibrational modes in Fig. 6.12 are IR active? Which mode corresponds to torsion about the double bond?

6.12 Use the vibrational constants (6.58) of water to calculate σ_0 for vibrational transitions from the (000) level to each of the following levels; compare the calculated values with the observed values in Table 6.1: (a) (001); (b) (111); (c) (211). Why is there a large discrepancy between calculated and observed values for (c)?

6.13 The typical O—H stretching frequency is 3600 cm^{-1}. What is the typical O—D stretching frequency?

6.14 A certain molecule XY_2 has strong IR bands at 400 and 1500 cm^{-1} and a strong Raman band at 650 cm^{-1}. All other IR and Raman bands are weak. Which structure is most probable? (a) linear XYX; (b) linear XXY; (c) bent XYX; (d) bent XXY.

6.15 The linear molecule AB_2 shows strong IR absorption bands at 600, 1300, and 2200 cm^{-1}. Is the order of the atoms ABA or AAB?

6.16 A certain triatomic molecule shows three strong IR bands, each having a simple rotational structure. One band has P, Q, and R branches, while the other two have only P and R branches. Is the molecule linear or bent?

6.17 Show that the M selection rule given in (5.84) follows from the discussion of Section 6.5.

6.18 Show mathematically that superposition of two vibrations at right angles to each other gives circular motion if the vibrations have equal amplitudes and differ in phase by 90°. What kind of motion is obtained if the phase difference is 90° but the vibrations are of unequal amplitudes?

6.19 Solve the Schrödinger equation for the two-dimensional isotropic harmonic oscillator using plane polar coordinates. First show that

$$\nabla^2 = \frac{\partial^2}{\partial\rho^2} + \frac{1}{\rho}\frac{\partial}{\partial\rho} + \frac{1}{\rho^2}\frac{\partial^2}{\partial\varphi^2}$$

Then separate variables, according to $\psi = G(\rho)\Phi(\varphi)$. To solve for $G(\rho)$, let

$$G(\rho) = e^{-\alpha\rho^2/2}F(\rho)$$

$$F(\rho) = \rho^s \sum_{n=0}^{\infty} a_n\rho^n$$

and find s. Assume the power series for F must be broken off to keep G well behaved for large ρ, and show that

$$E = (v+1)h\nu \qquad v = 0, 1, 2, \ldots$$

$$\Phi(\varphi) = (2\pi)^{-1/2}e^{im_N\varphi} \qquad m_N = v, v-2, \ldots, -v$$

6.20 Use the CO_2 constants (6.100) to calculate σ_0 for transitions from the (00^00) vibrational level to each of the following levels: (a) (10^00); (b) (02^00); (c) (02^20). Explain why the results do not always agree with the observed values in Table 6.2.

6.21 Use the data in Table 6.2 to find how far above the ground vibrational level each of the following CO_2 levels lies: (10^00); (02^00); (02^20); (20^00).

6.22 Prove that Fermi resonance cannot occur between two levels with different values of l.

6.23 Give the nuclear statistical weights for the s and a levels of the following molecules: (a) $H^1(C^{13})_4H^1$; (b) $H^2(C^{13})_4H^2$; (c) $C^{13}H^1_2{=}O^{16}$; (d) $H^2_2O^{16}$.

6.24 Give the symmetry number for molecules belonging to each of the following point groups: (a) $\mathcal{C}_1$; (b) $\mathcal{C}_s$; (c) $\mathcal{C}_i$; (d) $\mathcal{C}_n$; (e) $\mathcal{C}_{nh}$; (f) $\mathcal{C}_{nv}$; (g) $\mathcal{D}_n$; (h) $\mathcal{D}_{nh}$; (i) $\mathcal{D}_{nd}$; (j) $\mathcal{S}_{2n}$; (k) $\mathcal{T}_d$; (l) $\mathcal{O}_h$.

6.25 Stoicheff investigated the pure rotational Raman spectrum of CS_2. The first few lines could not be observed because of the width of the exciting line. The average values of the Stokes and anti-Stokes shifts for the first few observable lines (accurate to ± 0.02 cm^{-1}) are $\Delta\nu = 4.96$, 5.87, 6.76, 7.64, and 8.50 cm^{-1}. (a) Calculate the C═S bond length in carbon disulfide. (Assume centrifugal distortion is negligible. The rotational Raman selection rule for linear molecules in $^1\Sigma$ electronic states is $\Delta J = 0, \pm 2$.) (b) Is this an R_0 or R_e value? (c) Predict the shift for the $J = 0 \to 2$ transition.

6.26 The result of Problem 6.25 is inaccurate for a number of reasons. (a) Show that the procedure of averaging the differences of successive Raman shifts is poor in that it throws away all observations but two. (b) Because of the ± 0.02 cm^{-1} error in the shifts, it is desirable to use measured shifts for high J values, for which the percent error is smaller. However, centrifugal distortion must then be taken into account. Show that the Raman shifts $\Delta\nu$ for $J \to J+2$ equal

$$\left(4B_0 - 6\bar{D}\right)\left(J+\tfrac{3}{2}\right) - 8\bar{D}\left(J+\tfrac{3}{2}\right)^3$$

Hence plotting $\Delta\nu/(J+\frac{3}{2})$ against $(J+\frac{3}{2})^2$ gives a straight line. (c) For $J = 40, 50, 60, 70, 80, 90$, the observed CS_2 shifts are 18.11, 22.47, 26.83, 31.18, 35.53, 39.86 cm^{-1}. Evaluate B_0 and $\bar{D}$, and recalculate the bond length.

6.27 Explain how infrared spectroscopy might be used to find the temperature of the Jovian atmosphere.

6.28 The planar molecule 1,3-butadiene might have either the *s-cis* or the *s-trans* configuration in which the CH_2 groups lie either on the same side or on opposite sides of the carbon–carbon single bond, respectively. (a) It is observed that butadiene does not show a microwave spectrum. Is this evidence for either form? Explain. (b) Investigation of the rotational Raman spectrum of butadiene shows a simple spectrum with lines of nearly equal spacing. Although both possible forms are asymmetric tops, one form would have two moments of inertia nearly equal and its rotational spectrum would resemble the simple spectrum of a symmetric top. Which form does the rotational Raman spectrum favor? [This problem does not require any calculation; use (5.94).]

6.29 Spectroscopists use a shorthand notation to specify the IR absorption bands of a molecule. Examples of this notation for a molecule with three vibrational frequencies are

transition	notation
$(000) \rightarrow (010)$	ν_2
$(000) \rightarrow (200)$	$2\nu_1$
$(000) \rightarrow (012)$	$\nu_2 + 2\nu_3$
$(010) \rightarrow (020)$	$2\nu_2 - \nu_2$
$(010) \rightarrow (001)$	$\nu_3 - \nu_2$

If anharmonicity constants were negligible, the band-origin frequencies would be given by the shorthand notation. Give the shorthand designations for each of the following IR bands: (a) $(000) \rightarrow (202)$; (b) $(010) \rightarrow (011)$; (c) $(110) \rightarrow (300)$.

6.30 Some of the band origins (in cm^{-1}) and their relative intensities for the gas-phase IR spectrum of SO_2 are

σ_0	518	845	1151	1362	1876	2296	2500	2715	2808	3630
relative intensity	455	0.5	565	1000	6	5.5	20	0.2	0.8	0.8

Assign these bands to vibrational transitions; use the notation of Problem 6.29. (The lowest and highest frequency fundamentals are ν_2 and ν_3, respectively.)

6.31 Boron trifluoride has two nondegenerate and two twofold degenerate fundamental vibrational frequencies. (a) From Fig. 6.11, how many fundamental bands should appear in the IR spectrum of BF_3? (b)

The observed IR bands (in cm^{-1}) of $B^{10}F_3$ and $B^{11}F_3$ are

$B^{10}F_3$	482	718	1370	1505	1838	1985	2243	2385	3008	3263
$B^{11}F_3$	480	691	1370	1454	1838	1932	2243	2336	2903	3214
intensity	s	s	m	vs	w	w	w	w	w	w

where the abbreviations stand for very strong, strong, medium, and weak. The order of increasing vibration frequency of the fundamentals is $\nu_4 < \nu_2 < \nu_1 < \nu_3$. Assign the observed bands using the notation of Problem 6.29. (It might be thought that the 1370 cm^{-1} band is the overtone $2\nu_2$, but this can be ruled out; if this band were $2\nu_2$, it would not appear at the same frequency for the two isotopic species.) (c) The Raman spectrum of boron trifluoride shows a strong band that does not appear in the IR spectrum. This Raman band occurs at the same frequency for each isotopic species. Predict the frequency of this Raman band. Which normal mode of BF_3 is it due to?

6.32 Show that Fermi resonance between two levels increases the energy of the higher level and decreases the energy of the lower level.

6.33 The sum of the first through eighth ionization potentials of the oxygen atom is 2043.86 eV. $\Delta H^{\circ}_{f,298}$ for $H_2O(g)$ is -57.7979 kcal/mole. Use these data, data available in Chapters 4 and 6, and simple ideas of statistical mechanics to calculate the experimental electronic energy (including nuclear repulsion) of H_2O.

6.34 In deriving (6.24), we used $\dot{\mathbf{q}} = \mathbf{L}\dot{\mathbf{Q}}$. Show that this equation follows from (6.21).

CHAPTER 7

ELECTRONIC SPECTROSCOPY

7.1 ELECTRONIC SPECTROSCOPY

We now investigate transitions from one electronic state to another. The differences between valence electronic energies usually correspond to radiation in the visible and ultraviolet (UV) regions. Such spectra are studied most commonly in absorption, but frequently in emission. For electronic absorption spectroscopy, the traditional sources are a tungsten-wire lamp in the visible region and a gas-discharge lamp in the UV region. Gratings or prisms separate the frequencies. The detector is a photoelectric cell or photographic plate. In emission spectroscopy the molecules may be excited in a flame or a gas-discharge tube. Flame spectroscopy allows the emission spectra of free radicals (e.g., OH, CH, NH, NH_2) to be studied. Electronic absorption spectra of free radicals can be studied by using flash photolysis to decompose the appropriate parent compound, and then recording the absorption spectrum on a photographic plate.

The use of tunable lasers as sources in electronic absorption and emission spectroscopy has made possible a very considerable increase in resolution and precision. Electronic spectra are often difficult to analyze because of the many transitions involved. However, with a tunable laser source, one can tune the laser frequency to a specific absorption frequency of the molecule under study and thus populate a single excited electronic vibration–rotation energy level; the resulting fluorescence emission spectrum is then simple, and easy to analyze.

We shall approximate the molecular energy as the sum of electronic, vibrational, and rotational energies; the wave function will be approximated as the product of wave functions for these three kinds of motion. The (equilibrium) electronic energy of a given electronic state is defined as

U_e in (6.7), the value at the equilibrium geometry of the potential energy for nuclear motion. The convention is to denote quantities of the upper electronic state of a transition by single primes, and quantities of the lower electronic state by double primes. The symbols T_e, G, and F are used for the electronic, vibrational, and rotational energies in cm^{-1}:

$$T_e = U_e/hc, \qquad G = E_{\text{vib}}/hc, \qquad F = E_{\text{rot}}/hc \tag{7.1}$$

The wave number σ of an electronic transition is then

$$\sigma = T_e' + G' + F' - T_e'' - G'' - F'' \tag{7.2}$$

Usually

$$T_e' - T_e'' \gg |G' - G''| \gg |F' - F''| \tag{7.3}$$

Thus a transition between two given electronic states shows many bands, each such band corresponding to a different pair of initial and final vibrational states; under high resolution, each band shows many closely spaced lines, each such line corresponding to a different pair of initial and final rotational states. (The electronic spectra of molecules are called *band* spectra, whereas the electronic spectra of atoms are called *line* spectra.)

Consider the selection rules for electronic transitions. The electric dipole-moment operator $\hat{\mathbf{d}}$ can be separated into a sum over electrons and a sum over nuclei [Eq. (1.287)]:

$$\hat{\mathbf{d}} = -\sum_i e\mathbf{r}_i + \sum_\alpha Z_\alpha e\mathbf{r}_\alpha = \hat{\mathbf{d}}_{\text{el}} + \hat{\mathbf{d}}_N$$

The molecular wave function is (approximately) $\psi = \psi_{\text{el}}\psi_N$. The electric-dipole transition moment is

$$\langle\psi'|\hat{\mathbf{d}}|\psi''\rangle = \int\int \psi_{\text{el}}'^*\psi_N'^*(\hat{\mathbf{d}}_{\text{el}} + \hat{\mathbf{d}}_N)\psi_{\text{el}}''\psi_N''\,d\tau_{\text{el}}\,d\tau_N \tag{7.4}$$

$$\langle\psi'|\hat{\mathbf{d}}|\psi''\rangle = \int \psi_N'^*\psi_N''\left[\int \psi_{\text{el}}'^*\hat{\mathbf{d}}_{\text{el}}\psi_{\text{el}}''\,d\tau_{\text{el}}\right]d\tau_N$$

$$+\int \psi_N'^*\hat{\mathbf{d}}_N\psi_N''\left[\int \psi_{\text{el}}'^*\psi_{\text{el}}''\,d\tau_{\text{el}}\right]d\tau_N \tag{7.5}$$

[Since ψ_{el} depends parametrically on the nuclear coordinates, we cannot take the integrals over electronic coordinates outside the integrals over nuclear coordinates.]

The electronic wave functions ψ_{el}' and ψ_{el}'' are eigenfunctions of the Hermitian operator $\hat{H}_{\text{el}}$ with different eigenvalues; hence they are orthogonal, and (7.5) reduces to

$$\langle\psi'|\hat{\mathbf{d}}|\psi''\rangle = \int \psi_N'^* \psi_N'' \left[\int \psi_{\text{el}}'^* \hat{\mathbf{d}}_{\text{el}} \psi_{\text{el}}'' \, d\tau_{\text{el}} \right] d\tau_N \tag{7.6}$$

When the integral in brackets in (7.6) vanishes, the electronic transition is forbidden. For example, since ψ'_{el} and ψ''_{el} are both eigenfunctions of the Hermitian operator $\hat{S}^2$ (provided spin–orbit interaction is small), and since $\hat{S}^2$ commutes with $\hat{\mathbf{d}}_{\text{el}}$, we conclude [Equation (1.51)] that electronic transitions with a change in S are forbidden (just as in atoms):

$$\Delta S = 0 \tag{7.7}$$

Actually, violations of this selection rule are not unknown, even in relatively light molecules. Thus for CO, the Cameron system of visible and UV bands is due to a transition between the singlet ground state and an excited triplet electronic state.

7.2 ELECTRONIC SPECTRA OF DIATOMIC MOLECULES

We now specialize to diatomic molecules, although many of the ideas of this section will apply also to polyatomics. For diatomics, (7.2) becomes

$$\sigma = T_e' + G'(v') + F'(J') - T_e'' - G''(v'') - F''(J'')$$

where v' and J' are the vibrational and rotational quantum numbers of the level of the upper electronic state.

The electronic quantum numbers of diatomic molecules are discussed in Sections 1.19 and 4.11. The selection rule for Λ can be shown to be

$$\Delta\Lambda = 0, \pm 1 \tag{7.8}$$

This rule holds for both case (a) and case (b) coupling. Note the resemblance of (7.8) to the selection rule (3.77) for the atomic quantum number m. We classified Σ electronic states as Σ^+ or Σ^-, and the following selection rule holds:

$$\Sigma^+ \leftarrow\!|\!\rightarrow \Sigma^- \tag{7.9}$$

The proof is based on the fact that the integral over all space of an integrand that is an odd function of some of the Cartesian coordinates is zero (Section 1.8); see Problem 7.1.

For homonuclear diatomic molecules, the electronic wave functions have definite parity (g or u), and since $\hat{\mathbf{d}}_{\text{el}}$ is of odd parity, we must have a change in parity of ψ_{el} (corresponding to the Laporte rule in atoms):

$$g \leftarrow\mid\rightarrow g, \qquad u \leftarrow\mid\rightarrow u \tag{7.10}$$

For case (a) coupling, the following selection rules hold:

$$\Delta\Sigma = 0, \qquad \Delta\Omega = 0, \pm 1 \tag{7.11}$$

Recall that homonuclear diatomic molecules have no vibration–rotation or pure-rotation spectra due to the vanishing of the permanent electric dipole moment. For electronic transitions, the transition-moment integral (7.4) does not involve the dipole moment $\mathbf{d}$; hence electric-dipole electronic transitions are allowed for homonuclear diatomic molecules, subject to the above selection rules, of course. [The electric dipole moment $\mathbf{d}$ is given by (1.289), and should be distinguished from the electric dipole-moment operator $\hat{\mathbf{d}}$, which is given by (1.286).] Analysis of the vibrational and rotational structure of an electronic transition in a homonuclear diatomic molecule allows the determination of the vibrational and rotational constants of the electronic states involved, which is information that cannot be provided by IR or microwave spectroscopy. (Raman spectroscopy can also furnish information on the constants of the ground electronic state of a homonuclear diatomic molecule.)

Having considered the possible changes in electronic quantum numbers, we now discuss the allowed changes in v and J. To do so, we must deal with the complete integral (7.6). The bracketed integral in (7.6) is a function of the internuclear distance R; calling this integral $\mathbf{P}_{\text{el}}$, we have

$$\mathbf{P}_{\text{el}}(R) = \int \psi_{\text{el}}'^{*}(q_{\text{el}}; R)\,\hat{\mathbf{d}}_{\text{el}}\,\psi_{\text{el}}''(q_{\text{el}}; R)\, d\tau_{\text{el}} \tag{7.12}$$

where q_{el} symbolizes the electronic coordinates. The expression for the transition moment (7.6) is then

$$\langle \psi' | \hat{\mathbf{d}} | \psi'' \rangle = \int \psi_N'^{*} \psi_N'' \mathbf{P}_{\text{el}}(R)\, d\tau_N \tag{7.13}$$

If both electronic states are $^1\Sigma$ states, then

$$\psi_N = \psi_{\text{vib}}(R)\, Y_J^M(\theta_N, \varphi_N)\, \psi_{\text{ns}} \tag{7.14}$$

The function $\mathbf{P}_{\text{el}}$ is independent of nuclear spin, so that the summation over nuclear-spin coordinates in (7.13) is

$$\sum_{M_{I_a}} \sum_{M_{I_b}} \psi_{\text{ns}}'^{*} \psi_{\text{ns}}''$$

which gives the previously discussed strong selection rule (4.138) for transitions between rotational levels of a homonuclear diatomic. $\mathbf{P}_{el}$ is given by (7.12), and when ψ'_{el} and ψ''_{el} are both Σ states, we can prove (Problem 7.1b) that the vector $\mathbf{P}_{el}$ is directed along the molecular axis. Therefore, analogous to (4.94), we have

$$\mathbf{P}_{el}(R) = P_{el}(R)[\mathbf{i}\sin\theta_N\cos\varphi_N + \mathbf{j}\sin\theta_N\sin\varphi_N + \mathbf{k}\cos\theta_N] \tag{7.15}$$

where $P_{el}(R)$ is the magnitude of $\mathbf{P}_{el}(R)$, and θ_N and φ_N are angles with respect to space-fixed coordinates (Fig. 4.1). Substituting (7.15) and (7.14) into (7.13), we get integrals over θ_N and φ_N that are identical to those occurring in (4.97). We thus have the rotational selection rule [Eq. (4.99)]:

$$\Delta J = \pm 1 \qquad {}^1\Sigma - {}^1\Sigma \quad \text{transitions} \tag{7.16}$$

If one or both electronic states are not $^1\Sigma$ states, then it turns out that we usually can have $\Delta J = 0$ also, giving a Q branch, in addition to P and R branches:

$$\begin{matrix} \Delta J = 0, \pm 1 \\ J = 0 \leftarrow\!|\!\rightarrow J = 0 \end{matrix} \qquad \text{transitions other than } {}^1\Sigma - {}^1\Sigma \tag{7.17}$$

Note the resemblance of (7.17) to the selection rule (3.80) for atoms. The selection rule (7.17) is not quite true, in that $\Delta J = 0$ transitions are forbidden if both electronic states have $\Omega = 0$ [case (a) coupling].

The above selection rules for J are applicable for case (a) coupling. For case (b) coupling, we saw that for Σ states ($\Lambda = 0$), the quantum number N represents the rotational angular momentum; therefore, analogous to (7.16), we have for case (b)

$$\Delta N = \pm 1 \qquad \Sigma - \Sigma \text{ transitions} \tag{7.18}$$

If one or both electronic states are not Σ states, then we have for case (b)

$$\Delta N = 0, \pm 1 \qquad \text{transitions other than } \Sigma - \Sigma \tag{7.19}$$

The transition moment (7.13) is thus proportional to

$$\langle \psi' | \hat{\mathbf{d}} | \psi'' \rangle \propto \int \psi'^{*}_{vib} \psi''_{vib} P_{el}(R)\, dR \tag{7.20}$$

where the proportionality constant is found from the integral over θ_N and φ_N and the sum over nuclear spins. Now the condition for the validity of the Born–Oppenheimer approximation is that the electronic wave functions vary slowly with R; if ψ_{el} changed greatly for a small change in R, then the electrons would not be able to adjust fast enough to follow the nuclear motion. We therefore expect that the integral (7.12) over the electronic wave functions will also vary slowly with R; we are thus led to the *approximation* of replacing the function $P_{el}(R)$ by some constant $P_{el}(\bar{R})$, where $\bar{R}$ is some sort of average internuclear distance for the transition. The appropriate value of $\bar{R}$ is called the *R-centroid* for the transition. As well as depending on which two electronic states are involved, $\bar{R}$ also depends on which vibrational levels are involved in the transition; we write

$$P_{el}(R) \approx P_{el}(\bar{R}_{v'v''}) \tag{7.21}$$

With this approximation, (7.20) becomes

$$\langle\psi'|\hat{\mathbf{d}}|\psi''\rangle \propto P_{el}(\bar{R}_{v'v''}) \int \psi'^{*}_{\text{vib}}\psi''_{\text{vib}}\, dR \tag{7.22}$$

Now consider the selection rules for the vibrational quantum number. A too hasty glance at (7.22) might lead one to conclude that the orthogonality of the vibrational wave functions gives $\Delta v = 0$; this is wrong. The vibrational wave functions are found by solving (4.9) or its approximate form (4.23). The functions ψ'_{vib} and ψ''_{vib} are eigenfunctions of Hermitian operators; but, for different electronic states, ψ'_{vib} and ψ''_{vib} are eigenfunctions of *different* Hermitian operators, since the potential-energy function $U(R)$ is different for different electronic states. (Harmonic-oscillator wave functions with different values of v are only orthogonal if they correspond to the same force constant and equilibrium internuclear separation, both of which differ for different electronic states.) In general, there is no reason for the integral

$$\int \psi'^{*}_{\text{vib}}\psi''_{\text{vib}}\, dR \tag{7.23}$$

to vanish, and the selection rule for electronic transitions of a diatomic molecule is that

$$\Delta v = 0, \pm 1, \pm 2, \pm 3, \pm \cdots \tag{7.24}$$

However, the vibrational bands in an electronic transition vary in intensity, in proportion to the square of the integral (7.23). Thus the magnitude of the overlap between ψ'_{vib} and ψ''_{vib} determines the strength of the vibrational band.

Figure 7.1 shows the vibrational wave functions for some vibrational levels of two electronic states that have different values of R_e (as is usually the case). For $v = 0$, the function ψ_{vib} is largest at or near R_e and is small at the classical turning points of the vibration. For larger v, the vibrational wave functions are large near the classical turning points. (This is in accord

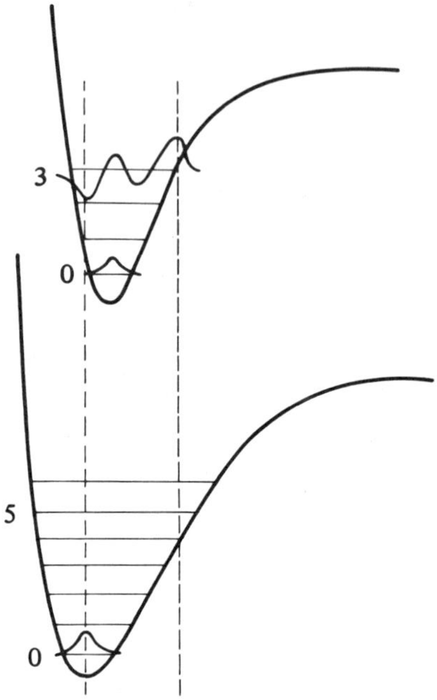

Fig. 7.1 Overlap between vibrational wave functions of two electronic states.

with the correspondence principle, since classically the oscillator spends more time near the turning points, where it is moving slowly, than near the midpoint of its motion, where it is moving rapidly.) Consider a transition from the $v''=0$ vibrational level of the lower electronic state to a vibrational level of the upper electronic state. From the figure and the preceding discussion, we see that ψ''_{vib} for $v''=0$ and ψ'_{vib} for $v'=3$ are both large at the same value of R; hence these functions have maximum overlap, and the $v=0\rightarrow3$ vibrational band of the electronic absorption transition should be the most intense band. Similarly, emission from the $v'=3$ level of the upper state will give an intense band for $v''=0$. We also note a large overlap between the $v'=3$ and the $v''=5$ functions, and this emission band should be intense. From a semiclassical viewpoint, we can say that during the course of the electronic transition, the internuclear distance changes by a negligible amount, the fast-moving electrons changing their wave function in a time too small to allow the nuclei to move much. In the $v''=0$ level, the nuclei are most likely to be near R_e''; for the state ψ'_{el}, the vibrational level with the largest probability that the internuclear separation is R_e'' is $v'=3$. These considerations determining the relative intensities of the vibrational bands of an electronic transition are called the *Franck–*

Condon principle, and the square of the vibrational overlap integral (7.23) is the *Franck–Condon factor* for the transition.

Consider the line frequencies of an electronic transition. The quantity $T_e' - T_e''$ is some constant for the pair of electronic states involved, and simply determines which region of the electromagnetic spectrum the transition falls. The positions of the vibrational bands are determined by

$$G'(v') - G''(v'') = \tilde{\nu}_e'(v' + \tfrac{1}{2}) - \tilde{\nu}_e''(v'' + \tfrac{1}{2}) - \tilde{\nu}_e' x_e'(v' + \tfrac{1}{2})^2 + \tilde{\nu}_e'' x_e''(v'' + \tfrac{1}{2})^2 + \tilde{Y}_{00}' - \tilde{Y}_{00}'' \quad (7.25)$$

Note that the vibrational constants are different for the different electronic states. There is no restriction on the change in the vibrational quantum number, and we generally get many vibrational bands for a given electronic transition.

The rotational structure of the vibrational bands is determined by

R branch: $$F'(J+1) - F''(J) = (\tilde{B}_{v'}' + \tilde{B}_{v''}'')(J+1) + (\tilde{B}_{v'}' - \tilde{B}_{v''}'')(J+1)^2$$

Q branch (if present): $$F'(J) - F''(J) = (\tilde{B}_{v'}' - \tilde{B}_{v''}'')J + (\tilde{B}_{v'}' - \tilde{B}_{v''}'')J^2$$

P branch: $$F'(J-1) - F''(J) = -(\tilde{B}_{v'}' + \tilde{B}_{v''}'')J + (\tilde{B}_{v'}' - \tilde{B}_{v''}'')J^2$$

where (4.67) and (4.75) have been used, and where the centrifugal-distortion corrections have been omitted. (The convention is that an R-branch line is one for which the rotational quantum number of the upper electronic state is one greater than that of the lower electronic state; in emission spectroscopy, the final state of an R-branch line has a lower value of J than the initial state; in absorption spectroscopy, the reverse is true.) The coefficient of the term quadratic in J in these expressions is

$$\tilde{B}_{v'}' - \tilde{B}_{v''}'' = \tilde{B}_e' - \tilde{B}_e'' - \tilde{\alpha}_e'(v' + \tfrac{1}{2}) + \tilde{\alpha}_e''(v'' + \tfrac{1}{2})$$

In IR spectra, the term quadratic in J is small [Eq. (4.118)] because B_e is the same for the upper and lower levels of the transition. In electronic transitions, the dimensions of the molecule change substantially, and B_e' and B_e'' are generally quite different. For example, note the difference in R_e and B_e for the two states of CO listed in Table 4.1. The effect of the quadratic term is thus quite significant, and we get band-head formation for reasonably low values of J in electronic spectra; band heads are a highly characteristic feature of such spectra. In the majority of cases, the upper electronic state has a larger R_e and thus a smaller B_e than the lower state; this gives a band head in the R branch. Sometimes B_e' is greater than B_e'', giving a band head in the P branch.

Aside from vibration and rotation constants, an important piece of information available from electronic spectra is the dissociation energies of the states involved. In electronic absorption spectroscopy, most of the diatomic molecules will originate from the $v''=0$ level of the ground electronic state. The vibrational structure of the transition to a given excited electronic state will consist of a series of bands (called a *progression*) representing $v''\to v'$ changes of $0\to 0$, $0\to 1$, $0\to 2, \ldots$, $0\to v'_{\text{max}}$, where v'_{max} is the quantum number of the highest discrete vibrational level of the excited state; this series of bands with rotational structure will be followed by a continuum of absorption, corresponding to transitions from $v''=0$ to the continuum levels of the excited state. Figure 7.2 illustrates the situation. The quantity $a+b$ represents the frequency at the onset of the continuum in the absorption spectrum. The quantity c is the difference in separated-atom energies between the ground and excited electronic states. Usually an analysis of the spectrum will allow one to decide to what separated-atom states each electronic state dissociates; hence c will be available from known values of atomic-energy levels (as determined from atomic spectra). Therefore, the dissociation energy of the ground electronic state can be calculated as

$$D_0''=(a+b)-c \tag{7.26}$$

In many cases, the decreased spacing of the vibrational levels as v' increases causes the vibrational bands to overlap and merge together for high v', thereby obscuring the onset of the continuum. In this event, the ground-state dissociation energy is calculated by the following extrapolation method due to Birge and Sponer. Suppose we plot the differences between the energies of successive vibrational levels of an electronic state

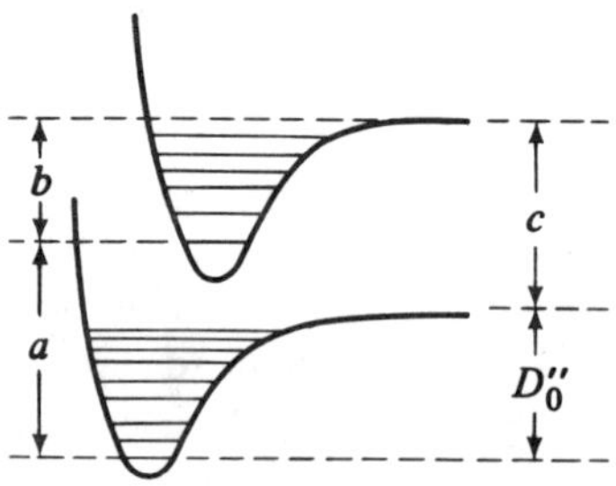

Fig. 7.2 Determination of dissociation energies from electronic band spectra.

as a function of the vibrational quantum number; that is, we set $\Delta G = G(v+1) - G(v)$, and plot ΔG versus v. Due to anharmonicity, the function ΔG decreases as v increases; when we reach the dissociation limit, we have a continuum of states, so that the spacing between adjacent levels has dropped to zero. (Of course, the energy difference between the last discrete vibrational level and the first continuum level is nonzero.) If we only know the first several $G(v)$ values, we can extrapolate the curve to the point where ΔG equals zero, to determine the dissociation limit. This is shown in Fig. 7.3. The dissociation energy is the sum of the vibrational energy-level differences (e.g., see Fig. 4.3):

$$D_0 = \sum_{v=0}^{v_{\max}} \Delta G + G(v_c) - G(v_{\max}) \tag{7.27}$$

where $v_{\max}$ is the quantum number of the highest discrete vibrational level, and v_c is the point of intersection of the ΔG curve with the v axis, indicating the beginning of the continuum levels. The area under the curve in Fig. 7.3 is a good approximation to the sum in (7.27), and thus we can estimate accurately the dissociation energy of the electronic state. [There is a possible pitfall in this procedure. According to the approximation (4.67), the vibrational energy contains terms linear and quadratic in v, and therefore ΔG will be a linear function of v. However, for high v, the higher-order anharmonicity correction $\nu_e y_e$, (4.79), becomes significant, so that the ΔG curve is not really linear. Thus to get an accurate extrapolation to the dissociation limit, one should have a substantial number of vibrational levels, not just the first few.] From the positions of the $0 \rightarrow 0$, $0 \rightarrow 1$, $0 \rightarrow 2, \ldots$ vibrational bands of the electronic spectrum, we can get the values of $G'(v'+1) - G'(v')$ for $v' = 0, 1, \ldots$, and then we can do a Birge–Sponer extrapolation of these intervals to determine D_0', the dissociation energy of the upper electronic state. The quantity b in Fig. 7.2 is D_0', and a is the $v = 0 \rightarrow 0$ frequency. We thus have

$$D_0'' = h\nu_{0\text{-}0} + D_0' - c$$

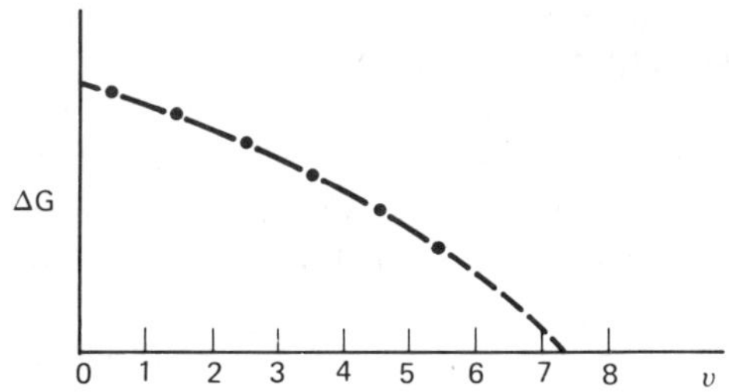

Fig. 7.3 Birge–Sponer extrapolation to determine D_0.

where ν_{0-0} is the frequency of the 0–0 band origin.[1]

Besides a transition to a continuum level of an excited electronic state, dissociation can occur by another mechanism in electronic absorption spectroscopy. If the potential-energy curve of an excited electronic state A that has a minimum in $U_A(R)$ happens to be intersected by the $U(R)$ curve of an unstable excited state B with no minimum in U, then a vibrational level of A whose energy lies near the point of intersection of U_A and U_B has a substantial probability to make a radiationless transition to state B, which then dissociates. This phenomenon is called *predissociation*. Predissociation shortens the lifetimes of those vibrational levels of A that are involved, and therefore by the uncertainty principle gives broad vibrational bands with rotational fine structure washed out.

7.3 ELECTRONIC SPECTRA OF POLYATOMIC MOLECULES

For polyatomic molecules, (7.2) and (7.6) give the wave numbers and transition moment for an electronic transition. Besides the S selection rule (7.7), there are electronic selection rules stating between which electronic symmetry species transitions are allowed; these are derived using group theory (Section 9.11). Equation (7.13) applies to polyatomics, except that $\mathbf{P}_{\text{el}}$ is now a function of the $3N-6$ (or $3N-5$) normal coordinates Q_i.

The rotational selection rules for electronic transitions are complicated by the fact that the molecular geometry can differ for the two electronic states, so that a molecule may be a symmetric top in one state and an asymmetric top in another. Polyatomic molecules have larger moments of inertia than diatomics, and their rotational levels are therefore closer together. Rotational structure in electronic transitions of medium-sized and large molecules is usually not resolved. This is because the spectrometer resolution may not be high enough, or because pressure and Doppler broadening merge the rotational fine-structure lines. Electronic spectra of polyatomics are often observed in solution, and here there is no rotational fine structure. For details on rotational fine structure, see *Herzberg, Electronic Spectra of Polyatomic Molecules*, pp. 183–271.

To determine the vibrational structure of electronic transitions of polyatomics, we can make the same approximation [Equation (7.21)] as for diatomics of replacing P_{el} by some sort of average P_{el}, which is independent of the Q_i's. The intensities of vibrational bands in an electronic transition then depend on the vibrational overlap integral, which is like

[1]For more on the determination of diatomic-molecule dissociation energies, see A. G. Gaydon, *Dissociation Energies and Spectra of Diatomic Molecules*, Third Edition, Chapman and Hall, London, 1968.

(7.23), except that the vibrational wave functions depend on the normal coordinates Q_i. Thus, as for diatomics, the Franck–Condon principle determines the vibrational band intensities. However, a polyatomic has many vibrational modes, and the vibrational band structure of electronic transitions is much more complicated than for diatomics. In solution, the vibrational structure may be washed out, giving just a single broad absorption for a given electronic transition.

The integral $\langle\psi'_{\text{vib}}|\psi''_{\text{vib}}\rangle$ may vanish because of symmetry considerations. For example, the CO_2 normal mode ν_3 in Fig. 6.2 has eigenvalue -1 for the inversion operation. Hence (Section 6.4), the ν_3 factor in the vibrational wave function is an even or odd function of the normal coordinate Q_3, depending on whether v_3 is even or odd. For a change of $1,3,5,\ldots$ in the vibrational quantum number v_3, the functions ψ'_{vib} and ψ''_{vib} have different parities and their product is an odd function, so that $\langle\psi'_{\text{vib}}|\psi''_{\text{vib}}\rangle$ vanishes. Thus we have the selection rule $\Delta v_3 = 0,2,4,\ldots$ for electronic transitions in CO_2.

As a measure of the intensity of an electronic absorption transition, the transition's *oscillator strength* f is defined as the ratio of the probability for the transition to the probability for a transition of a three-dimensional isotropic harmonic oscillator in which the vibrational quantum number v_x or v_y or v_z changes from 0 to 1; the harmonic oscillator is defined to have mass and charge equal to the electron's mass and charge, and vibrational frequency equal to the frequency of the electronic transition in question. The transition probability per unit radiation density is given by the Einstein coefficient $B_{n\to m}$ of (3.49). For a one-dimensional harmonic oscillator of charge e, Eq. (3.61) gives $\langle 1|\hat{d}_x|0\rangle = e(2\alpha)^{-1/2}$, where α is given by (1.134). Substitution into (3.49) and multiplication by 3 gives for a $0\to 1$ three-dimensional harmonic-oscillator transition

$$B_{n\to m} = (2\pi/3\hbar^2)(3he^2/8\pi^2\nu_{mn}m) \tag{7.28}$$

where ν_{mn} is the frequency of the electronic absorption transition being considered. The total probability for an electronic absorption transition from a given vibration–rotation state of the initial electronic state involves summation of the square of the transition moment (7.6) over all vibration–rotation states of the excited electronic state. We use the approximation of replacing the bracketed integral in (7.6) by an average value for the transition, and thereby remove it from the integral over nuclear coordinates. The quantity $|\mathbf{d}_{mn}|^2$ in (3.49) then involves the following sum over the excited-state nuclear wave functions:

$$\sum_{\psi'_N} \langle\psi''_N|\psi'_N\rangle\langle\psi'_N|\psi''_N\rangle \tag{7.29}$$

Using the sum rule (1.52) with $\hat{A}=\hat{B}=1$, we see that (7.29) equals $\langle\psi_N''|\psi_N''\rangle$, which equals 1, because the initial nuclear wave function is normalized. Hence the nuclear wave functions are eliminated, and the oscillator strength involves only the bracketed electronic integral in (7.6). Substitution of $\langle\psi_{el}'|\hat{\mathbf{d}}_{el}|\psi_{el}''\rangle$ into (3.49) and division by (7.28) then gives as the oscillator strength of an electronic absorption transition

$$f=(8\pi^2\nu_{mn}m/3he^2)|\langle\psi_{el}'|\hat{\mathbf{d}}_{el}|\psi_{el}''\rangle|^2 \tag{7.30}$$

An important type of electronic transition is a *Rydberg transition*. Consider H_2O as an example. If we use an extended basis set of AOs to calculate Hartree–Fock MOs for H_2O, we find that the main contributions to the occupied ground-state MOs come from AOs whose principal quantum numbers are $n=1$ and $n=2$. However, we can expect that for all but the lowest few excited electronic states of H_2O, the excited electron will be in an MO whose main contributions involve oxygen AOs with $n\geqslant 3$. The size of an AO increases roughly as n^2; for example, for a hydrogenlike atom, one finds $\langle r\rangle=(a_0/2Z)[3n^2-l(l+1)]$. (See *Bethe and Salpeter*, pp. 13–14, 17, for the derivation.) Thus an electron in such an excited orbital will be at a large average distance from the other electrons and the nuclei, and the rest of the H_2O molecule will look pretty much like an Ne^+ core to the excited electron. We can roughly view the excited electron as being in the appropriate approximately hydrogenlike $n\geqslant 3$ AO of the Ne united atom. Such an excited MO is called a *Rydberg* MO, and electronic states of H_2O arising from excitation to a Rydberg orbital are called *Rydberg states*. The electronic excitation energy of a Rydberg state with one electron excited to a Rydberg MO with united-atom quantum numbers n and l can be approximated by the following empirical formula [compare with the equation following (3.78)]:

$$T_0'-T_0''=I-\frac{R_\infty}{(n-\delta)^2} \tag{7.31}$$

In (7.31), T_0'' is the ground-state electronic energy including zero-point vibrational energy; R_∞ is the Rydberg constant for infinite nuclear mass (Appendix Table A.1); the parameter δ (the *Rydberg defect*) depends on the value of l (and on the symmetry species and electronic spin of the excited electronic state); the parameter I depends on which occupied MO the electron has been excited from—as n goes to infinity, we get the ionized molecule, so I equals the ionization energy for the occupied MO from which the excited electron is removed.

For H_2O the highest occupied MO is $1b_1$ (Section 1.20), and many excited electronic states with a $1b_1$ electron excited to *ns*, *np*, and *nd* Rydberg orbitals with $n \geqslant 3$ have been observed. For example, the wave numbers of the $v=0 \rightarrow 0$ bands of absorption transitions to $^1B_1$ states in which $4s, 5s, \ldots, 9s$ Rydberg orbitals are occupied are fitted to within several cm^{-1} by (7.31) with $I = 101{,}780$ cm^{-1}, $\delta = 1.05$; transitions to $^1B_1$ states with $3p, 4p, 5p$ Rydberg MOs occupied are fitted with $I = 101{,}780$ cm^{-1}, $\delta = 0.7$. The first ionization energy of H_2O is therefore 12.62 eV (Appendix Table A.3).

Excited MOs whose main contributions come from AOs with principal quantum numbers equal to the atomic valence-shell quantum numbers are called *sub-Rydberg* (or *valence*) MOs. For example, the excited (antibonding) π MOs (symbolized by π^*) of conjugated molecules for which $C2p\pi$ AOs make the main contribution are valence MOs. Another example is the $\sigma_u^* 1s$ MO of H_2. Of course, there is not a sharp dividing line between Rydberg and valence MOs. One way to decide how much Rydberg character an MO has is to calculate $\langle r \rangle$ for it.

As noted in Section 1.20, an electron in an excited MO generally cannot be considered to be localized in a particular bond. For example, consider CH_4. Let b_1 denote a localized C—H_1 bond orbital, and i the inner-shell C1*s* orbital. We can start with a localized description of the ground state with MOs i, b_1, b_2, b_3, b_4 each doubly occupied. If we now suppose that we have excited an electron from the localized MO b_1 to the localized C—H_1 antibonding MO b_1^*, we might write as the wave function for the excited state

$$|i\bar{i}b_1 b_1^* b_2 \overline{b_2} b_3 \overline{b_3} b_4 \overline{b_4}| \tag{7.32}$$

However, all four bonds b_1, b_2, b_3, b_4 have the same average energy, and we can write three other excited wave functions in which the excitation is in one of the other C—H bonds. These four approximate wave functions are degenerate; the interelectronic interactions neglected in writing MO wave functions act as a perturbation, and we must use degenerate perturbation theory. This leads to correct zeroth-order functions in which the four wave functions are equally mixed, so that the excitation is not localized in one bond, but distributed through all four bonds. If we consider a less symmetrical molecule than methane, it is true that the bonds will no longer be exactly equivalent; functions with excitations in different bonds will not have exactly the same energy, but their energies will be nearly equal (since single-bond energies are all roughly similar); for near degeneracy, we must still set up a secular equation between the near-degenerate functions (recall Fermi resonance), and we are led to correct zeroth-order functions in which the excitation is distributed over much of the molecule. Thus when an electron is excited, we must generally view it as being transferred from one delocalized (canonical) MO to another delocalized MO. The same reasoning applies to ionization; upon ionization, the electron is generally removed from a canonical MO.[2]

There are, however, cases where we can consider the excitation to be *approximately* localized in a particular bond. Consider an organic compound with a single C═C (or C═O) double bond. Even though the molecule may not be strictly planar, in most cases the carbon skeleton will be roughly planar and we can classify the MOs as approximately σ or π. We will have one occupied canonical π MO, corresponding to the pi bond of the $C_1{=}C_2$ group, and this MO will have the approximate form $C_1 2p\pi + C_2 2p\pi$. The lowest vacant canonical pi MO (π^*) will have the approximate form $C_1 2p\pi - C_2 2p\pi$. Thus the lowest-frequency electronic absorption, $\pi \rightarrow \pi^*$, will be *approximately* localized in the double bond, and we can expect its frequency to be about the same in most monoolefins. In fact, one finds that monoolefins typically have a λ_{max} (the longest-wavelength electronic absorption) near 185 nm. In aldehydes and ketones, the $n \rightarrow \pi^*$ transition, in which an electron is excited from the nonbonding (n) lone-pair MO on oxygen to the π^* MO of the double bond, has a lower frequency than the $\pi \rightarrow \pi^*$ transition. Approximately isolated groups like C═C, C═O, and N═N that give rise to characteristic electronic absorption frequencies are called *chromophores*. Of course, when we have conjugated double bonds, the pi MOs are delocalized over much of the molecule, and the $\pi \rightarrow \pi^*$ λ_{max} depends strongly on the chain length, as discussed in Section 1.21. Although chemists often assume that the highest occupied canonical MO in aldehydes, ketones, ethers, and alcohols has the character of a lone-pair nonbonding MO localized on oxygen, MINDO calculations indicate significant contributions to this MO from AOs of other atoms, these contributions increasing as the size of the molecule increases. Thus canonical "n" MOs have a significant amount of delocalization, a fact that is verified by significant differences in first ionization potentials for different organic oxygen compounds; see M. J. S. Dewar and S. D. Worley, *J. Chem. Phys.*, **50**, 654 (1969).

For an excited $\pi\pi^*$ configuration, we have the possibilities $S=0$ and $S=1$ for the total spin quantum number; hence we get two electronic terms, a singlet and a triplet. (If one or both of the pi energy levels involved are degenerate, as in C_6H_6, we get more than two electronic terms from the $\pi\pi^*$ configuration.) Lots of different notations are in use in

[2]A possible exception to this statement is the removal of an inner-shell electron from O_2. There is evidence that the electron is ejected from a $1s$ AO localized on one oxygen atom, rather than from the $\sigma_g 1s$ or $\sigma_u^* 1s$ MO; see P. S. Bagus and H. F. Schaeffer, *J. Chem. Phys.*, **56**, 224 (1972).

electronic spectroscopy. One common notation is N for the π^2 ground state, and T and V for the triplet and singlet $\pi\pi^*$ terms, respectively.

Molecular electronic spectroscopy can provide information on vibrational parameters (frequencies and force constants), rotational parameters (moments of inertia and therefore molecular geometries), electronic excitation energies, ionization potentials, and dissociation energies for ground and excited electronic states. Moreover, a knowledge of excited electronic states is important in understanding the course of photochemically induced reactions.

We now consider the theoretical calculation of excited-state wave functions. This is more difficult than ground-state calculations because we are dealing with open-shell configurations. The Hartree–Fock equations for a state of an open-shell configuration have a more complicated form than for closed shells, and there exist close to a dozen different approaches to excited-state Hartree–Fock calculations. As noted earlier, the Hartree–Fock wave function for a closed-shell state is a single determinant, but for open-shell states, we may have to take a linear combination of a few Slater determinants to get a Hartree–Fock function that is an eigenfunction of $\hat{S}^2$ and $\hat{S}_z$ and has the correct spatial symmetry.

Another difference between open- and closed-shell states is that configuration interaction is generally much more important for open-shell states. Near-Hartree–Fock functions for ground-state closed-shell systems give a good account of many molecular properties; the true wave function is reasonably well approximated by the Hartree–Fock function. However, for open-shell excited states, the Hartree–Fock wave function is not such a good approximation to the true wave function, and configuration interaction must generally be included if we want a reasonably accurate wave function. The reason for this difference lies in *Brillouin's theorem*. Brillouin showed in 1934 that the Hamiltonian matrix element H_{ij} between a closed-shell Hartree–Fock ψ_i and any singly excited determinantal function ψ_j is zero. (A *singly excited* ψ_j is one in which a single electron is placed in a different spin-orbital as compared to ψ_i.) Hence in the configuration-interaction wave function (1.300) for a closed-shell ground-state, we mix in only excited configurations with two (or more) electrons in higher orbitals; such functions correspond to large energy differences from the ground state, and so make relatively small contributions to the true wave function. In contrast, for an open-shell excited state, the matrix elements H_{ij} do not vanish for single excitations, and such functions make significant contributions to the true wave function.

[For open-shell systems, an extension of the Hartree–Fock method, called the unrestricted Hartree–Fock method, is sometimes used. For a closed-shell state, the lowest Hartree–Fock energy is generally obtained when the spatial functions of paired electrons are the same. Thus for the Be $1s^22s^2$ ground state, the two $1s$ electrons occupy the same spatial orbital, as do the $2s$ electrons. The two $1s$ Be electrons each interact with a $1s$ electron of spin opposite to it, and with two $2s$ electrons, one of the same spin and one with opposite spin; so these $1s$ electrons are fully equivalent. Consider, however, the Li $1s^22s$ configuration, with the $2s$ electron having spin α. The $1s\alpha$ electron interacts with a $2s$ electron of the same spin, but the $1s\beta$ electron interacts with a $2s$ electron of opposite spin. The $1s\alpha$–$2s\alpha$ and $1s\beta$–$2s\alpha$ interactions differ, since there is zero probability for two electrons of the *same* spin to be at the same point in space (Section 1.16). Thus the two $1s$ electrons in Li are not exactly equivalent, and we might expect them to occupy somewhat different spatial orbitals. In fact, one finds that use of slightly different spatial orbitals for the Li $1s$ electrons gives a slightly lower variational energy than when the same spatial orbital is used for both electrons. When the same spatial orbital is used for paired electrons, we have the *restricted* (or conventional) Hartree–Fock (RHF) method; when paired electrons are allowed to have different spatial orbitals, we have the *unrestricted* (or spin-unrestricted) Hartree–Fock (UHF) method. (The term "Hartree–Fock" without any adjective refers to the RHF method.) The RHF Li wave function has the form $|1s\,\overline{1s}\,2s|$, whereas the UHF function has the form $|1s_\alpha\,\overline{1s_\beta}\,2s|$, where the spatial functions $1s_\alpha$ and $1s_\beta$ differ somewhat. (An objection to the UHF function is that although it is an eigenfunction of $\hat{S}_z$, it is not an eigenfunction of $\hat{S}^2$. However, it is nearly an eigenfunction of $\hat{S}^2$, and for many purposes one can ignore this defect. Moreover, one can apply a certain operator to the UHF function to extract from it an eigenfunction of $\hat{S}^2$.) UHF calculations are, of course, harder than RHF calculations, and the RHF approach is much more commonly used with open-shell systems. The UHF method is most useful when one is interested in spin densities, as in electron spin resonance. For results of UHF calculations, see *Schaefer*, pp. 22–27, 95–101.]

Because of the necessity for extensive configuration interaction, far fewer *ab initio* excited-state calculations have been done than ground-state calculations. However, excited states of a number of small and medium-sized molecules have been calculated by *ab initio* methods. An example is the SCF CI calculation of *s-trans*-butadiene done by Hosteny, Dunning, Gilman, Pipano, and Shavitt [see *Schaefer*, pp. 352–359; T. H. Dunning, R. P. Hosteny, and I. Shavitt, *J. Am. Chem. Soc.*, **95**, 5067 (1973)]. These workers used a large contracted Gaussian basis set and did all calculations at the experimental geometry. After finding the SCF ground-state function, they used the calculated virtual (unoccupied) pi orbitals to form ground and excited-state CI wave functions; in these CI functions, the 26 sigma electrons were kept in their ground-state orbitals, but one, two, or three of the four pi electrons were put in the excited virtual pi orbitals. Because of the large basis set used, 14 virtual pi orbitals were found, so excitation to virtual pi orbitals gave large numbers of configurations. For example, the wave functions of ^1A_g states included 2752 configurations. The calculated excitation energies are in pretty good agreement with experiment. For example, the lowest excited triplet and singlet states were calculated to lie at 3.4 and 7.0 eV, respectively, compared to experimental values of 3.2 and 5.9 eV, respectively. Several excited-state calculations on H_2O have been done; see W. A. Goddard and W. J. Hunt, *Chem. Phys. Letters*, **24**, 464 (1974).

Very few *ab initio* calculations of oscillator strengths have been done due to the difficulties of the calculations. Oscillator strengths for atomic transitions calculated with RHF wave functions are found to differ from experimental f values by factors of 2 or 3; when CI is included, good agreement with experiment is obtained. *Ab initio* RHF oscillator strengths have been calculated for a number of diatomic molecules, and found to differ from experimental values by factors of 3 or 4. For details, see *Schaefer*, pp. 102–109, 168–176.

Much semiempirical work on excited states of pi-electron systems has been done. As noted in Section 1.21, the HMO method can be used to predict the longest absorption wavelengths of benzenoid hydrocarbons, but cannot predict their complete electronic spectra. The PPP method (with CI included, of course) has been notably successful in predicting electronic transition energies of conjugated compounds. For example, 17 PPP-predicted singlet–singlet excitation energies (which range from 4 to 8 eV) for the molecules benzene, pyridine, aniline, *p*-benzoquinone, and phenol showed an average absolute error of only 0.15 eV; the calculated oscillator strengths were typically in error by a factor of 2. [See M. L. Bailey, *Theor. Chim. Acta*, **13**, 56 (1969); **16**, 309 (1970).]

The original CNDO/2 method was found to be rather poor for calculating electronic spectra, but Del Bene and Jaffe modified the CNDO/2 parametrization so as to yield good singlet–singlet excitation energies; with inclusion of CI, their modification of CNDO/2 gave generally good results for pi-electron systems, but not better than obtained with the PPP method. (See *Murrell and Harget*, pp. 46–47.)

The *Jahn–Teller theorem* is important in considering the electronic states of polyatomic molecules. Jahn and Teller proved in 1937 that a nonlinear polyatomic molecule cannot have an equilibrium (minimum-energy) nuclear configuration that corresponds to an orbitally degenerate electronic term. Orbital degeneracy arises from molecular symmetry (Section 1.19), and the Jahn–Teller theorem can lead to a lower symmetry than one might otherwise have expected.

For example, consider the anion $C_6H_6^-$, prepared by reduction of benzene. For a regular-hexagonal geometry, the pi energy-level pattern is as in Fig. 1.3. (As noted in Section 1.21, although the values of the pi energies in Fig. 1.3 are not correct, the pattern of pi energies is correct.) The added electron goes in one of the MOs φ_4 or φ_5, and is in a doubly degenerate energy level with symmetry species e_{2u}. Hence the symmetric-geometry nuclear configuration of $C_6H_6^-$ has the ground electronic term symbol $^2E_{2u}$ (since we can ignore the electrons in closed shells). Since we have an orbitally degenerate term, the Jahn–Teller theorem tells us that the regular-hexagonal geometry cannot correspond to the minimum-energy point on the U potential-energy surface of $C_6H_6^-$; rather, U_e will occur at a point corresponding to a geometry that is not a regular hexagon; the distortion from a regular hexagon removes the degeneracy of the MOs φ_4 and φ_5 and gives an orbitally nondegenerate electronic term.

Another example is C_6H_6. If an electron is excited from the highest occupied to the lowest vacant pi MO in Fig. 1.3, then we have two partly filled degenerate pi energy levels, and it turns out that two of the electronic terms arising from this excited configuration are orbitally degenerate—see (9.177). Thus these two electronic terms will not have a regular-hexagonal equilibrium geometry.

The Jahn–Teller theorem says nothing about the magnitude of the departure of the equilibrium geometry from a symmetric geometry, and in many cases, the actual departures are less than the magnitudes of zero-point vibrations. For more on the Jahn–Teller theorem, see *Herzberg, Electronic Spectra of Polyatomic Molecules*, pp. 40–51; F. S. Ham, *Int. J. Quantum Chem.*, **5S**, 191 (1971).

7.4 PHOTOELECTRON SPECTROSCOPY

In ordinary spectroscopy one measures the frequency (or wavelength) and therefore the energy of *photons* absorbed or emitted (or scattered, as in the Raman effect) by molecules. In *photoelectron spectroscopy* (PES) one measures the energy of *electrons* emitted by molecules when they are photoionized by the absorption of high-energy (UV or x-ray) photons. If M stands for a molecule, the photoionization process can be symbolized by

$$M + h\nu \rightarrow M^+ + e^- \tag{7.33}$$

Conservation of energy shows the kinetic energy $E(e^-)$ of the emitted electron is

$$E(e^-) = h\nu + E(\mathrm{M}) - E(\mathrm{M}^+) \tag{7.34}$$

If $h\nu$ is known, then measurement of $E(e^-)$ gives the energy difference between the molecule M and the ion M^+.

In UV PES, approximately monochromatic radiation is passed through the sample in the vapor phase at low pressure. (If the sample has a very low vapor pressure, one uses a heated sample chamber.) The most commonly used UV photon source is a helium gas discharge tube; most of the radiation is due to the intense emission line at 584 Å (21.22 eV) corresponding to the transition from the $1s2p\ {}^1P_1$ He I (i.e., neutral He) level to the ground state. [By modifying the conditions of the discharge, one can, if desired, produce He II (i.e., He^+) radiation at 304 Å (40.8 eV).] The kinetic energies of the photoelectrons resulting from ionization are measured by passing them through electric or magnetic fields; the relative numbers of electrons emitted as a function of $E(e^-)$ are measured with an electron-multiplier tube. UV PES work began in the early 1960s with work of Turner and co-workers in England, and Terenin and co-workers in Russia. Commercial UV photoelectron spectrometers are now available.

The ion M^+ in (7.33) will be produced in its ground electronic state or in one of its excited electronic states, depending on which MO of M the electron was removed from. We can use Fig. 7.1, except that we must now think of the upper curve as an electronic state of M^+ (not M). (Since it always requires energy to ionize a stable molecule, the electronic energy of every state of M^+ lies above the ground electronic energy of M.) The Franck–Condon arguments of Section 7.2 show that the probabilities of M^+ being produced in various vibrational states are proportional to the squares of the vibrational overlap integrals (7.23). Rotational fine structure is not resolved in PES, so we need not consider rotational energy. We therefore write

$$E(\mathrm{M}^+) - E(\mathrm{M}) = h\nu - E(e^-) = [E_{\mathrm{el}}(\mathrm{M}^+) - E_{\mathrm{el}}(\mathrm{M})] + [E_{\mathrm{vib}}(\mathrm{M}^+) - E_{\mathrm{vib}}(\mathrm{M})]$$

Ordinarily M will be in its ground electronic and vibrational state. The lines in the photoelectron spectrum will thus be due to M^+ being produced in different electronic and vibrational states. The photoelectron spectrum consists of a number of bands; each band corresponds to removal of an electron from a different MO of M and production of a different electronic state of M^+. For diatomics and for polyatomics that are not too large, vibrational structure is resolved. The strongest vibrational line in a band is given by the "vertical" transition from M to M^+; for example, the $v = 0 \rightarrow 3$

line would be strongest in Fig. 7.1. The energy difference $E(\mathrm{M}^+) - E(\mathrm{M})$ for the $v = 0 \rightarrow 0$ transition of a given band is called the *adiabatic ionization potential* for that band, and is the true ionization potential. The energy difference for the strongest line of a band is the *vertical ionization potential*. If vibrational structure is not resolved, the peak of a PES band gives the vertical ionization potential, and the onset of the band gives the adiabatic ionization potential.

The He I photoelectron spectrum of N_2 is shown in Fig. 7.4. Photoelectron energy increases to the right, and therefore $E(\mathrm{M}^+) - E(\mathrm{M})$ increases to the left. Of course, it is the latter quantity that is of interest. The ground-state MO configuration is

$$KK(\sigma_g 2s)^2(\sigma_u^* 2s)^2(\pi_u 2p)^4(\sigma_g 2p)^2$$

The three bands in the spectrum indicate that the 21.22 eV photons have enough energy to remove an electron from each of the three highest MOs.

The first band corresponds to removal of a $\sigma_g 2p$ electron to produce N_2^+ in its ground electronic state $X\,^2\Sigma_g^+$; this band begins at 15.6 eV, which is the adiabatic ionization potential. The spacing between the $v(N_2^+) = 0$ and 1 lines is 0.267 eV, corresponding to $\tilde{\nu}_0 = 2150\ \mathrm{cm}^{-1}$ for this state of N_2^+. This is less than the vibrational frequency for ground-state N_2 (2345 cm^{-1}), as expected for removal of a bonding electron. (Vibrational frequency decreases with decreasing force constant.) The rather small decrease in $\tilde{\nu}_0$ indicates the $\sigma_g 2p$ MO to be only weakly bonding; this is confirmed by the $v = 0$ peak being the strongest, which indicates there is not much change in R_e when a $\sigma_g 2p$ electron is removed.

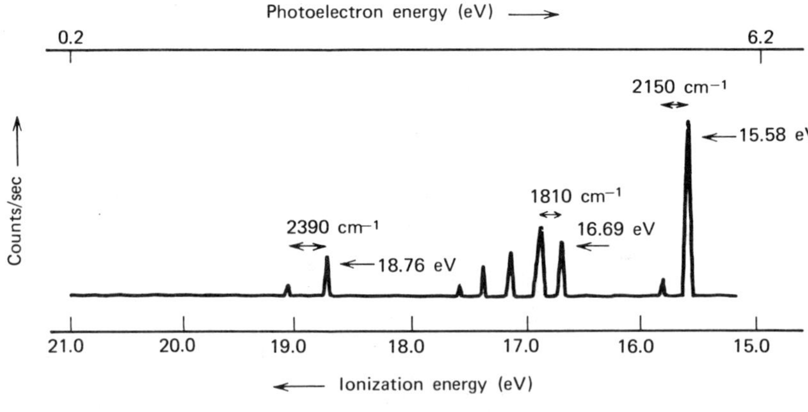

Fig. 7.4 Sketch of He I photoelectron spectrum of N_2.

The second band corresponds to removal of a $\pi_u 2p$ electron to give the N_2^+ $A^2\Pi_u$ state. Here $\tilde{\nu}_0$ is 1810 cm^{-1}, and the substantial drop from $\tilde{\nu}_0(N_2)$ shows the $\pi_u 2p$ MO to be strongly bonding. Here the $v=1$ line is the strongest, which indicates a significant change in bond length. The third band results from removal of a $\sigma_u^* 2s$ electron to give the $B\,^2\Sigma_u^+$ state of N_2^+; the slight increase in $\tilde{\nu}_0$ indicates removal of an electron from a weakly antibonding MO.

The most studied photoelectron spectrum is that of benzene. There is no question that the highest occupied MO level is the e_{1g} pair of pi MOs φ_2 and φ_3 in Fig. 1.3. A near-Hartree–Fock calculation of benzene [W. C. Ermler and C. W. Kern, *J. Chem. Phys.*, **58**, 3458 (1973)] found the doubly degenerate $3e_{2g}$ sigma MO energy level to lie very slightly (0.1 eV) above the lowest pi MO φ_1 of Fig. 1.3, the $1a_{2u}$ MO. (Since the Hartree–Fock limit was not reached, this conclusion is not final.) The benzene PE spectrum is difficult to analyze because of overlapping bands, but careful study indicates that the second band in the spectrum is probably due to ionization from the $3e_{2g}$ sigma MO. The ionization energies of the three highest MOs are $1e_{1g}(\pi)$, 9.3 eV; $3e_{2g}(\sigma)$, 11.8 eV; $1a_{2u}(\pi)$, 12.5 eV. Interleaving of sigma and pi MOs is also found in several other aromatic compounds. Ionization energies have been determined for all the C_6H_6 MOs; see T. A. Carlson and C. P. Anderson, *Chem. Phys. Letters*, **10**, 561 (1971); C. Fridh, L. Åsbrink, and E. Lindholm, *ibid.*, **15**, 282 (1972); U. Gelius *et al., Physica Scripta*, **3**, 237 (1971).

The preceding discussion shows the value of PES for checking the conclusions of MO theory. For quantitative comparison, we want to calculate adiabatic ionization potentials theoretically. The rigorously correct way to do this is to calculate wave functions and electronic energies for the appropriate states of M^+ and subtract the calculated electronic energy of the M ground state. To calculate adiabatic ionization potentials, one would have to do the excited-state calculations for several geometries to find the equilibrium geometries, and then add in the zero-point energies. Because of the difficulties involved in excited-state calculations, they are commonly done at just the one point corresponding to the ground-state equilibrium geometry; the calculation then gives vertical, rather than adiabatic, ionization potentials. (Usually, the two do not differ by very much.) The number of rigorous *ab initio* ionization-potential calculations is small. Meyer [W. Meyer, *Int. J. Quantum Chem.*, **5S**, 341 (1971)] did large-scale *ab initio* CI calculations on the H_2O ground state and on the five states of H_2O^+ that correspond to removal of one electron from one of the five occupied H_2O MOs in (1.299). The calculated CI vertical ionization potentials in eV for the MOs are (experimental PES values in parentheses) $1b_1$, 12.3 (12.6); $3a_1$, 14.5 (14.7); $1b_2$, 18.7 (18.6); $2a_1$, 32.2

(32.2); $1a_1$, 539.6 (539.7). Agreement with experiment is excellent. Meyer also calculated ionization potentials using near-Hartree–Fock wave functions for the states involved; without CI, agreement with experiment is only fairly good, with an average absolute error of 1.3 eV. For N_2, the three lowest vertical ionization potentials have been calculated from N_2^+ and N_2 Hartree–Fock wave functions. [P. E. Cade, K. D. Sales, A. C. Wahl, *J. Chem. Phys.*, **44**, 1973 (1966).] The calculated vertical ionization potentials in eV (obtained by subtraction of the N_2 energy from the N_2^+ energies, both calculated at R_e for N_2) are (experimental values, obtained by addition of the N_2 zero-point energy 0.15 eV to the observed PES values, in parentheses): $\sigma_g 2p$, 16.0 (15.7); $\pi_u 2p$, 15.7 (17.1); $\sigma_u^* 2s$, 19.9 (18.9). Agreement with experiment is reasonably good, but note the incorrect ordering of the MOs. Full agreement with experiment requires CI.

Because calculation of excited-state wave functions is hard, an approximate method of finding ionization potentials based on *Koopmans' theorem* is usually used. Koopmans proved in 1933 that the Hartree–Fock orbital energy ϵ_i of an orbital in a *closed*-shell electronic state is a good approximation to the ionization potential for removal of an electron from that orbital. (Tjalling C. Koopmans is a renowned mathematical economist; in his youth he wrote a couple of physics papers.) More precisely, Koopmans showed that if we take the closed-shell Hartree–Fock determinant ψ for the ground state of M, remove one spin-orbital i of orbital energy ϵ_i from this determinant to get a new determinant ψ^+, and calculate the energy of M^+ using the determinant ψ^+, we get an energy that lies $|\epsilon_i|$ above the energy of ψ. This procedure uses the Hartree–Fock orbitals of M to calculate the energy of M^+. But we know that the orbitals of M will change upon ionization, and it is only an approximation to use the M orbitals to describe M^+. To find the actual orbitals of M^+, we must do a Hartree–Fock calculation on M^+. However, the change in orbitals should not be large upon ionization, so we can use the orbital energies ϵ_i of M to estimate vertical ionization energies of M, and this saves us the trouble of doing calculations on M^+. For the N_2 ground state, Cade, Sales, and Wahl found the following Koopmans' theorem vertical ionization energies in eV (experimental values in parentheses): $\sigma_g 2p$, 17.4 (15.7); $\pi_u 2p$, 17.1 (17.1); $\sigma_u^* 2s$, 20.9 (18.9). The agreement with experiment is not as good as when actual RHF energies of the N_2^+ states are used. For H_2O, Meyer's near Hartree–Fock ψ gives the Koopmans' theorem ionization potentials as 13.9, 15.9, 19.5, 36.8, and 559.5 eV.

As to semiempirical calculation of ionization potentials, Koopmans' theorem CNDO/2 and INDO ionization potentials are not so good, being usually about 4 eV too high. However, MINDO/2 Koopmans' theorem ionization potentials are generally quite accurate, typical errors being $\frac{1}{4}$ or $\frac{1}{2}$ eV, which is more accurate than Hartree–Fock ionization potentials.

Use of He I and He II photons in PES allows determination of binding energies of molecular valence orbitals. X-ray photons are needed to knock out inner-shell molecular electrons. X-ray PES (XPES) was developed in the early 1960s by Siegbahn in Sweden. The sample in XPES is a gas or a solid. Mg or Al K_α x rays of energy 1254 eV or 1487 eV are most commonly used. The $1s$ energies differ considerably from one atom to another (due to the change in nuclear charge); hence the XPES spectrum allows both qualitative and quantitative elemental analysis of a sample. For this reason, XPES is often called *electron spectroscopy for chemical analysis* (ESCA). ESCA spectrometers are commercially available. The ESCA spectrum gives the molecular valence ionization energies, as well as the inner-shell energies, but the resolution is not as good as in UV PES, due to the 1 eV width of the x-ray photon source; hence vibrational structure is not resolved in ESCA.

The $1s$ energy of a given atom is found to depend somewhat on the chemical environment of the atom in the molecule. For example, the molecules CH_4, CH_3F, CH_2F_2, CHF_3, and CF_4 have ESCA carbon $1s$ energies of 290.8, 293.6, 296.3, 299.1, and 301.8 eV, respectively. If we take CH_4 as a standard, then these five compounds have C$1s$ ionization potential *chemical shifts* of 0, 2.8, 5.5, 8.3, and 11.0 eV. The XPES spectrum of $CF_3COOCH_2CH_3$ shows four C$1s$ peaks of nearly equal intensity at ionization energies of 285, 286, 289, and 292 eV, corresponding to the four kinds of carbon atoms. The ESCA chemical shifts and peak heights thus provide structural information on the compound, much as in NMR spectroscopy (Chapter 8). ESCA chemical shifts in nearly half the elements have been studied. *Ab initio* calculations do show the carbon inner-shell canonical MOs in a molecule with nonequivalent carbons to usually be localized on each carbon. For example, for $CH_3C{\equiv}CH$, the $1a_1$ MO is calculated to be almost completely localized on C_3 with orbital energy -308.3 eV, the $2a_1$ MO is almost completely localized on C_2 with energy -307.2 eV, and the $3a_1$ MO is localized on C_1 with energy -306.7 eV; see M. D. Newton and W. N. Lipscomb, *J. Am. Chem. Soc.*, **89**, 4261 (1967).

Core $1s$ ionization potentials are calculated by the same methods discussed above for valence ionization potentials. (Note the value for the $1a_1$ H_2O MO, which is the O$1s$ orbital.) For example, SCF calculations with a medium-sized basis set on the above-mentioned fluorocarbons gave predicted Koopmans' theorem $1s$ chemical shifts of 2.9, 6.1, 9.4, and 12.7 eV; the trend is correctly predicted; see C. R. Brundle, M. B. Robin, and H. Basch, *J. Chem. Phys.*, **53**, 2196 (1970). Since semiempirical methods like

CNDO, INDO, and MINDO treat only the valence electrons, they cannot be used to predict core binding energies. Jolly has developed an empirical method of estimating carbon $1s$ ESCA chemical shifts, which takes the shift as the sum of empirical parameters characteristic of the atoms bonded to the carbon in question; see W. L. Jolly, *J. Am. Chem. Soc.*, **92**, 3260 (1970).

Some books and review articles on PES are the following: D. W. Turner *et al., Molecular Photoelectron Spectroscopy*, Wiley-Interscience, London, 1970; K. Siegbahn *et al., ESCA Applied to Free Molecules*, North-Holland, Amsterdam, 1969; S. D. Worley, *Chem. Rev.*, **71**, 295 (1971); A. D. Baker, *Accounts Chem. Res.*, **3**, 17 (1970); J. M. Hollander and W. L. Jolly, *ibid.*, **3**, 193 (1970).

The July 21, 1970 *Wall Street Journal* carried a front-page story headlined "Weird Water: Debate Over Mysterious Fluid Splits Scientific World." The story was about "polywater," a supposed viscous polymeric form of water prepared in minute amounts by condensation of water vapor in capillary tubes. The controversy came to an end with the publication of ESCA and IR analyses of "polywater." [R. E. Davis, D. L. Rousseau, and R. D. Board, *Science*, **171**, 167 (1971); D. L. Rousseau, *ibid.*, **171**, 171.] The ESCA spectrum (obtained from 10^{-6} g of sample) showed "polywater" to be rich in sodium, oxygen, and carbon, with lesser, but significant amounts of potassium, nitrate, chloride, and sulfate. The IR spectrum was nearly the same as that of sodium lactate, a primary constituent of human sweat, and nearly the same as the IR spectrum of sweat itself. Thus "polywater" turned out to be a mixture of salts (mainly sodium lactate), with a small amount of ordinary water. Rousseau speculated that the likely source of the biological contamination that produced "polywater" was "the absorption by the capillaries of aerosol particles released by humans. The presence of an aerosol cloud surrounding individuals (not too dissimilar from the cloud surrounding Pigpen in 'Peanuts') is a well-known phenomenon encountered in several areas of scientific endeavor." Davis concluded that "polywater is polycrap." (R. E. Davis, *Chem. Eng. News*, Sept. 28, 1970, pp. 73, 79.)

PROBLEMS

7.1 (a) Derive the selection rule (7.9). (b) Prove that the vector (7.12) points along the molecular axis.

7.2 The onset of the continuum of the Schumann–Runge absorption bands of O_2 is at 1750.5 Å. The lower electronic state of the transition is the ground state and dissociates to ground-state oxygen atoms; the excited state involved dissociates to a $2p^4\ {}^3P_2$ oxygen atom and a $2p^4\ {}^1D_2$ oxygen atom. Find D_0 and D_e for the O_2 ground state. (The atomic energy levels of O are given in C. E. Moore, *Atomic Energy Levels*, Natl. Bur. Stds. Circular 467, vol. I, 1949.)

7.3 What is the outstanding difference between the high-resolution electronic absorption spectrum of $O^{16}O^{18}$ and that of $O^{16}O^{16}$?

7.4 The 0–0 wave numbers of members of one of the Rydberg series in the electronic absorption spectrum of H_2S are 65,963, 75,121, 78,945, 80,789, 81,866, 82,534 cm^{-1}. Find the first ionization potential of H_2S.

7.5 Show that in atomic units the oscillator strength for an electronic transition between states m and n is $f_{mn} = \frac{2}{3}(E_m - E_n)|\mathbf{d}_{mn}|^2$.

7.6 (a) Show that for a particle in a one-dimensional box, the oscillator strengths are

$$f = (64/3\pi^2)m^2n^2/(m^2 - n^2)^3, \qquad m - n \text{ odd}$$

and 0 for $m - n$ even. Save time by using the result of Problem 3.3. (b) The experimentally observed oscillator strengths for the λ_{max} electronic absorption transitions of the compounds (1.304) with $k = 0, 1, 2, 3$ are 0.34, 0.53, 0.62, 1.55. Use the FE MO method to predict the λ_{max} f values for these compounds.

7.7 (a) Does the Jahn–Teller theorem predict that C_5H_5 (cyclopentadienyl) is distorted from a regular pentagon in its ground electronic state? (b) Is the equilibrium geometry of CH_4^+ a regular tetrahedron? (The CH_4 MOs are listed in Problem 7.9.)

7.8 To what transition is the 304 Å He II radiation due?

7.9 An SCF MO calculation gives the following orbital energies for the occupied CH_4 subshells $1a_1$, $2a_1$, and $1t_2$: -11.20, -0.93, -0.54 hartrees, respectively. Answer the following questions about the He I PE spectrum of CH_4. (a) How many bands will be observed? (b) Give their approximate ionization potentials.

7.10 The same as Problem 7.9 for the He II PES of CH_4.

7.11 An SCF MO calculation gives the following energies for the occupied H_2O MOs (Section 1.20) $1a_1$, $2a_1$, $1b_2$, $3a_1$, $1b_1$: -20.56, -1.28, -0.62, -0.47, -0.40 hartrees, respectively. For the He I PES of H_2O, answer the following questions. (a) Give the band locations. (b) For each band, state whether or not the adiabatic and vertical ionization potentials will coincide. (c) For each band, state whether the vibrational frequencies of the upper state should be less than, about the same as, or greater than those in the lower state.

7.12 Consider the ESCA Mg K_α spectrum of each of the following compounds. State how many inner-shell peaks will occur. For inner-shell peaks due to the same element, state the relative intensities. Give the ionization energy approximate region (in eV) where inner-shell peaks of each element occur. (a) C_6H_6; (b) C_2H_5OH; (c) CH_3OCH_3; (d) CH_3F; (e) $(CH_3)_3CH$; (f) pyrrole; (g) CO_2. [$1s$ orbital energies can be estimated from Fig. 11.2 in *Levine* or from R. Latter, *Phys. Rev.*, **99**, 510 (1955).]

CHAPTER 8

MAGNETIC RESONANCE SPECTROSCOPY

8.1 MAGNETIC RESONANCE

Electrons and many nuclei have magnetic dipole moments associated with a nonzero spin angular momentum. Application of a static external magnetic field results in a series of energy levels corresponding to the quantum-mechanically allowed orientations of the spin and associated magnetic moment of the electron or nucleus. Transitions between these levels can be observed by allowing the spin magnetic moment to interact with the oscillating magnetic field of electromagnetic radiation of frequency corresponding to the level separation. These transitions are often treated from a classical viewpoint. Recall (Section 1.17) that classically the magnetic moment precesses about the direction of the applied static magnetic field; when the frequency of the oscillating magnetic field is equal to the frequency of this precession, the system is in *resonance* and can absorb energy from the oscillating magnetic field. Thus we have the terms *nuclear magnetic resonance* (NMR) and *electron spin resonance* (ESR) to describe such transitions for nuclei and electrons, respectively. In this chapter, NMR and ESR will be treated using mainly the quantum-mechanical viewpoint.

8.2 NUCLEAR MAGNETIC MOMENTS

A nucleus has an intrinsic ("spin") angular momentum **I** of magnitude $\sqrt{I(I+1)}\,\hbar$, where the nuclear-spin quantum number I has the possible values $0, \frac{1}{2}, 1, \ldots$ (Section 4.8). Just as the possession of a spin angular momentum **S** gives the electron a magnetic dipole moment proportional to

S (Section 1.15), a nucleus with nonzero spin has a magnetic dipole moment $\boldsymbol{\mu}_N$ that is proportional to **I**; we write

$$\boldsymbol{\mu}_N = g_N \frac{e}{2m_p c}\mathbf{I} \tag{8.1}$$

Equation (8.1) [which may be compared to (1.246)] defines g_N, the *nuclear g factor*. Note that the protonic charge and mass, e and m_p, are used in (8.1), rather than the charge and mass of the nucleus in question. Whereas the electronic g factor can be predicted accurately from theory, theories of nuclear structure are not sufficiently developed to allow the prediction of g_N for nuclei. Experimental values of I, g_N, and percent abundance for some nuclei are given in the Appendix, Table A.6.

The observed g_N values are by no means integers. Even the free neutron and proton g values are not simple numbers—experimental evidence shows neutrons and protons to have a structure. One might have expected the neutron g value to be zero since a magnetic field arises from moving charges and the neutron is uncharged. The negative g value for the neutron can be explained by supposing it to consist of equal amounts of positive and negative charge, with the negative charge having a larger average distance from the neutron's center. Although no firm model of neutron structure has yet been established, data of high-energy electron–deuteron and electron–proton scattering experiments have been interpreted[1] as showing the neutron's negative charge (which is about $-0.5e$) to lie between an inner core containing most of the positive charge and an outer shell of positive charge. Some nuclei have negative g values due to the negative g value of the neutron. All nuclei with even numbers of both protons and neutrons have zero spin. (This results from the pairing of nucleonic angular momenta and holds only for the nuclear ground state, which is the nuclear state chemists deal with.)

The magnitude of the magnetic moment of a nucleus is

$$|\boldsymbol{\mu}_N| = |g_N|\frac{e}{2m_p c}[I(I+1)]^{1/2}\hbar = |g_N|\,\beta_p\sqrt{I(I+1)} \tag{8.2}$$

where the *nuclear magneton* β_p (commonly denoted by β_N) is defined as

$$\beta_p \equiv \frac{e\hbar}{2m_p c} = 5.051\times 10^{-24}\ \text{erg/gauss} \tag{8.3}$$

The nuclear magneton is 1/1836th the Bohr magneton, due to the replacement of the electron's mass with the proton's mass. Nuclear magnetic moments are much smaller than the electron's magnetic moment.

[1] D. N. Olson, H. F. Schopper, and R. R. Wilson, *Phys. Rev. Letters*, **6**, 286 (1961).

As with any angular momentum, the z component of **I** is equal to $M_I\hbar$, where M_I ranges from $+I$ to $-I$ in integral steps. The possible values of the z component of $\boldsymbol{\mu}_N$ are thus

$$\beta_p g_N I,\ \beta_p g_N(I-1),\ldots,\ -\beta_p g_N I \tag{8.4}$$

We shall use $|M_I\rangle$ to symbolize the nuclear-spin eigenfunction of $\hat{I}_z$ with eigenvalue $M_I\hbar$:

$$\hat{I}_z|M_I\rangle = M_I\hbar|M_I\rangle \tag{8.5}$$

The quantity $g_N I$ is commonly called "the nuclear magnetic moment" and is listed as such in tables; actually, $g_N I$ is not the magnitude of $\boldsymbol{\mu}_N$, but rather is the maximum z component of $\boldsymbol{\mu}_N$ in units of nuclear magnetons.

8.3 NUCLEAR MAGNETIC RESONANCE

NMR transitions are between energy levels that correspond to different orientations of the nuclear magnetic dipole moment in an applied magnetic field **B**. The classical energy of interaction between an isolated nuclear magnetic moment $\boldsymbol{\mu}_N$ and **B** is[2]

$$-\boldsymbol{\mu}_N\cdot\mathbf{B} = -g_N\beta_p\hbar^{-1}\mathbf{I}\cdot\mathbf{B} = -g_N\beta_p\hbar^{-1}BI_z \tag{8.6}$$

where we chose the z axis to coincide with the direction of the uniform applied field, and where I_z is the z component of **I**.

Introduction of quantum-mechanical operators gives as the Hamiltonian for the interaction between a nuclear magnetic moment and a magnetic field:

$$\hat{H} = -g_N\beta_p\hbar^{-1}B\hat{I}_z = -\gamma_N B\hat{I}_z \tag{8.7}$$

$$\gamma_N \equiv g_N\beta_p/\hbar \tag{8.8}$$

γ_N is called the *magnetogyric ratio*, since it is the ratio of the classical angular frequency of precession to the magnitude of the applied field. The eigenfunctions and eigenvalues of $\hat{H}$ follow from (8.5):

$$\hat{H}|M_I\rangle = -g_N\beta_p BM_I|M_I\rangle, \qquad M_I = I, I-1,\ldots,-I \tag{8.9}$$

The energy levels are evenly spaced, and diverge as the applied magnetic field increases. For g_N positive, the negative M_I values lie highest.

[2] *Halliday and Resnick*, Eq. (33–12). Most NMR references use **H** instead of **B**; see the discussion following (1.269).

Since $\hat{H}$ is Hermitian, the nuclear-spin eigenfunctions are orthonormal:

$$\langle M_I'|M_I''\rangle = \delta_{M_I'M_I''} \tag{8.10}$$

where the notation on the left implies a summation over the spin variable M_I from $-I$ to $+I$ [compare (1.238)].

To determine the NMR spectrum, the selection rules must be found. The probability that a system originally in state ψ_n that is exposed to the time-dependent perturbation $\hat{H}'(t)$ from $t=0$ to t_1 will make a transition to ψ_m is given by (3.17), (3.18), and (3.22) as

$$\hbar^{-2}\left|\int_0^{t_1} e^{i\omega_{mn}t}\langle\psi_m|\hat{H}'(t)|\psi_n\rangle\, dt\right|^2 \tag{8.11}$$

where $\omega_{mn}=(E_m-E_n)/\hbar$. If a nuclear magnetic moment sitting in a static magnetic field $\mathbf{B}$ is exposed to electromagnetic radiation of frequency ν, then $\boldsymbol{\mu}_N$ will interact with the magnetic field $\mathbf{B}_1$ of the radiation; the time-dependent perturbation is

$$\hat{H}'(t) = -\mathbf{B}_1\cdot\hat{\boldsymbol{\mu}}_N = -\hbar^{-1}\beta_p g_N\left(B_{1x}\hat{I}_x + B_{1y}\hat{I}_y + B_{1z}\hat{I}_z\right) \tag{8.12}$$

Each component of $\mathbf{B}_1$ is given by an expression like (3.25); the spatial variation of $\mathbf{B}_1$ over the molecule can be neglected and we write

$$B_{1x} = B_{1x}^0\cos(2\pi\nu t) \tag{8.13}$$

with similar expressions for B_{1y} and B_{1z}. Substitution of (8.12) and (8.13) into (8.11) shows the transition probability to involve the integrals

$$\begin{aligned}
&\langle\psi_m|\hat{I}_x|\psi_n\rangle \propto \langle\psi_m|\,\hat{\mu}_{Nx}|\psi_n\rangle\\
&\langle\psi_m|\hat{I}_y|\psi_n\rangle \propto \langle\psi_m|\,\hat{\mu}_{Ny}|\psi_n\rangle\\
&\langle\psi_m|\hat{I}_z|\psi_n\rangle \propto \langle\psi_m|\,\hat{\mu}_{Nz}|\psi_n\rangle
\end{aligned} \tag{8.14}$$

NMR transitions are thus of magnetic-dipole type [Equation (3.57)].

Suppose the static field $\mathbf{B}$ is in the z direction. The wave functions in (8.14) are then the eigenfunctions of $\hat{I}_z$ [Eq. (8.9)]. If ψ_m and ψ_n have the spin quantum numbers M_I' and M_I'', respectively, then

$$\langle\psi_m|\hat{I}_z|\psi_n\rangle = \langle M_I'|\hat{I}_z|M_I''\rangle = M_I''\hbar\langle M_I'|M_I''\rangle = M_I''\hbar\delta_{M_I'M_I''} \tag{8.15}$$

Thus the z component of the radiation's magnetic field [which multiplies $\hat{I}_z$ in (8.12)] cannot induce NMR transitions. To evaluate the remaining two integrals of (8.14), we use the ladder operators $\hat{I}_+$ and $\hat{I}_-$ [Eq. (1.169)]:

$$\hat{I}_+ = \hat{I}_x + i\hat{I}_y, \qquad \hat{I}_- = \hat{I}_x - i\hat{I}_y \tag{8.16}$$

According to (1.170) and (1.171), these operators convert an eigenfunction of $\hat{I}_z$ with eigenvalue $M_I\hbar$ to an eigenfunction of $\hat{I}_z$ with eigenvalue $(M_I \pm 1)\hbar$:

$$\hat{I}_+|M_I\rangle = c|M_I + 1\rangle, \qquad \hat{I}_-|M_I\rangle = d|M_I - 1\rangle \tag{8.17}$$

where c and d are certain constants. The first integral in (8.14) becomes

$$2\langle M_I'|\hat{I}_x|M_I''\rangle = \langle M_I'|\hat{I}_+ + \hat{I}_-|M_I''\rangle \tag{8.18}$$

$$= c\langle M_I'|M_I'' + 1\rangle + d\langle M_I'|M_I'' - 1\rangle \tag{8.19}$$

$$= c\delta_{M_I', M_I''+1} + d\delta_{M_I', M_I''-1} \tag{8.20}$$

The x component of the radiation's magnetic field thus causes NMR transitions obeying the selection rule

$$\Delta M_I = \pm 1 \tag{8.21}$$

From $\hat{I}_y = (\hat{I}_+ - \hat{I}_-)/2i$, it follows that (8.21) also hold for NMR transitions induced by B_{1y}.

From (8.9), the energy levels of an isolated nuclear magnetic moment in a static field **B** are

$$E = -g_N\beta_p BM_I, \qquad M_I = I, I-1, \ldots, -I \tag{8.22}$$

The selection rule (8.21) then gives for the frequency of an NMR transition

$$\nu = g_N\beta_p B/h = \gamma_N B/2\pi \tag{8.23}$$

[Equation (8.23) is for g_N positive; for g_N negative, a minus sign must be added.] Although there are $2I+1$ nuclear-spin energy levels, they are equally spaced, and the selection rule allows only transitions between adjacent levels; hence we get a single NMR absorption frequency.

The NMR transition frequency depends on the magnitude B of the applied magnetic field and has no significance unless B is specified. A magnetic field readily attainable is 10,000 gauss; at this value of B, Eq. (8.23) gives the resonance frequency of a proton as 42.6 MHz. NMR transitions thus fall in the radio-frequency (rf) portion of the electromagnetic spectrum. NMR resonance frequencies in MHz for some nuclei at

10,000 gauss are

H^1	H^2	B^{11}	C^{13}	N^{14}	F^{19}	P^{31}
42.577	6.536	13.663	10.708	3.077	40.076	17.253

NMR transitions were first observed by Rabi and co-workers in 1938, using a molecular-beam apparatus.[3] NMR transitions in bulk matter were first detected in 1945 by two independent groups, one headed by Bloch, the other by Purcell; these two men shared the 1952 Nobel physics prize.[4]

We now consider NMR transitions in bulk matter. The sample can be a solid, liquid, or gas. Comparatively little gas-phase work has been done, due to the weakness of the absorption signal. NMR in solids will be considered separately in Section 8.7. The most common application of NMR is to liquid samples.

Nuclei with $I=0$ have no magnetic moment and give no NMR absorptions; such common nuclei as C^{12}, O^{16}, and S^{32} cannot be studied by NMR. Nuclei with $I \geqslant 1$ have electric quadrupole moments (Section 8.10), which broaden the NMR absorption and eliminate the fine details of the spectrum, which are of chemical significance. Most chemical NMR work is with nuclei with $I=\frac{1}{2}$; of these, the most extensively studied is H^1, the proton.

One way to observe an NMR absorption is to apply a constant field B and vary the frequency ν of the electromagnetic radiation to which the sample is exposed until (8.23) is satisfied; another way is to expose the sample to a fixed-frequency electromagnetic signal and slowly vary the applied magnetic field until (8.23) is satisfied. The electromagnetic signal is generated (in the same manner a radio or TV station generates its carrier frequency) by an rf transmitter using conventional electronic tubes. Since it is difficult to build a variable-frequency rf oscillator of high stability, the method of fixed rf frequency and variable applied magnetic field is generally used. The rf-oscillator frequency is held stable by a quartz crystal and is known to high accuracy. Many commercial proton-NMR spectrometers use 60 MHz as the rf frequency; this corresponds to $B=14{,}100$ gauss.

A simplified block-diagram of an NMR spectrometer is shown in Fig. 8.1. The magnetic field B_0 is applied using an electromagnet or permanent

[3] For details of Rabi's work, see P. Kusch, *Physics Today*, February 1966, p. 23.

[4] In his acceptance speech [*Science*, **118**, 431 (1953)], Purcell said, "I remember, in the winter of our first experiments, just seven years ago, looking on snow with new eyes. There the snow lay around my doorstep—great heaps of protons quietly precessing in the earth's magnetic field. To see the world for a moment as something rich and strange is the private reward of many a discovery."

magnet; subsidiary coils placed about the poles of the magnet vary B_0 over a narrow range. The current in these coils also drives the recorder chart paper on which the NMR spectrum is traced out. The coils attached to the fixed-frequency rf transmitter are arranged so that the oscillating field $\mathbf{B}_1$ is perpendicular to $\mathbf{B}_0$. The absorption signal is detected by a coil placed at right angles to the rf coil and to $\mathbf{B}_0$; when the value of B_0 is such that (8.23) is satisfied, there is a net absorption of energy and the nuclear spins flip; this reorientation of nuclear magnetic moments induces a current in the detector coil (just as moving a magnet near a coil of wire induces a current in the wire).

The liquid sample is placed in a narrow glass tube. One of the problems in NMR is broadening of the lines due to small inhomogeneities in $\mathbf{B}_0$. Spinning the sample tube rapidly causes the nuclei to "see" an averaged field and narrows the NMR lines (see Section 8.9).

Although NMR spectra are recorded at fixed rf frequency and variable applied field, the separations between closely spaced NMR lines are often expressed as the equivalent separations in hertz, by multiplying the magnetic-field differences by $g_N\beta_p/h$. For protons

$$g_p\beta_p/h = 4257.7 \text{ Hz/G}$$

To improve the sensitivity of the spectrometer in detecting weak signals, one can use time averaging. Instead of being fed into the recorder, the

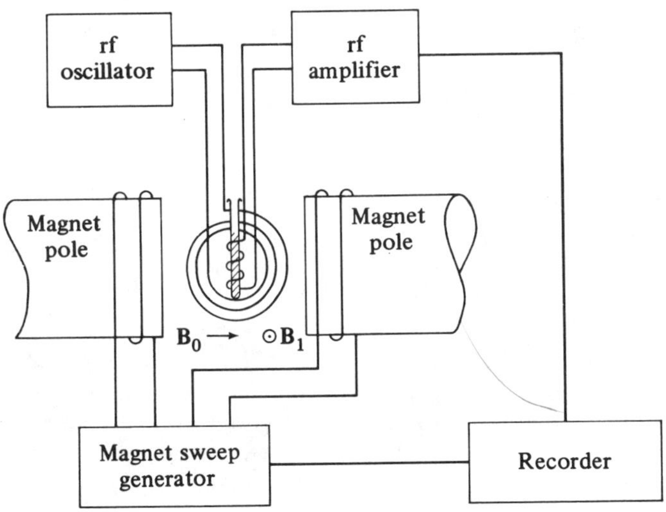

Fig. 8.1 Block diagram of an NMR spectrometer.

output voltage is sampled at many closely spaced points, converted electronically to digital form, and stored in a computer. The spectrum is scanned repeatedly, with the results of each sweep being added to those of the previous sweeps. Since noise voltage is randomly positive and negative, whereas signal voltage is always of fixed sign, the noise tends to cancel, whereas the signal is enhanced. This *computer of averaged transients* (CAT) technique is also applicable in other branches of spectroscopy.

It might be thought that after a sufficient number of sweeps, the random noise would cancel out completely, but this is not so. Suppose we toss a coin 10 times and store in the computer a $+1$ for each head, and a -1 for each tail. A typical outcome might be 6 heads and 4 tails, to give a stored total of $+2$. If we make 1000 tosses, a typical outcome might be 510 heads and 490 tails, to give a stored total of $+20$. Although the 1000-toss experiment shows a percentage of heads much nearer to 50 percent (51 percent versus 60 percent in the example), the difference between the number of heads and tails is larger in the 1000-toss experiment. For N tosses, probability theory shows the standard deviation of the number of heads from $0.5N$ to be $\frac{1}{2}N^{1/2}$. Thus, averaging over N sweeps gives an increase in noise proportional to $N^{1/2}$, but an increase in signal proportional to N.

The disadvantage of repetitive sweeps is that they take a long time. A technique to overcome this is *Fourier transform* (FT) *spectroscopy*. Here, instead of sweeping slowly through the spectrum, the magnetic field is held fixed, and a very intense, short-duration (10^{-5} sec) pulse of rf radiation is applied to the sample. We will not go into the details of the effect of the pulse on the NMR signal, but the following simplified discussion gives the main idea. Suppose we have a sample that absorbs electromagnetic radiation at certain frequencies $\nu_1, \nu_2, \ldots$. We expose it to a pulse of radiation. Since the pulse has a short duration, it does not have a single well-defined frequency. The functions (1.35) are a complete (orthogonal) set and we shall expand the pulse shape $f(t)$ (where t is time) in terms of them, according to (1.37):

$$f(t) = \int_{-\infty}^{\infty} b(\omega) e^{i\omega t} \, d\omega$$

where we defined ω as $k/\hbar$, and $b(\omega)$ as $\hbar c(\hbar\omega)$. The functions $f(t)$ and $b(\omega)$ are called *Fourier transforms* of each other. Since $e^{i\omega t} = \cos\omega t + i \sin\omega t$, the pulse f can be Fourier-analyzed into a superposition of all frequencies $\omega/2\pi$ with amplitudes $b(\omega)$. In effect, the pulse exposes the sample to a continuous range of frequencies, rather than a single frequency. The sample will absorb radiation from the pulse at frequencies $\nu_1, \nu_2, \ldots$, thereby decreasing $b(\omega_1), b(\omega_2), \ldots$. Let $f_s(t)$ be the pulse shape after it passes through the sample; $f_s(t)$ is given by an integral like the above, but with amplitudes $b(\omega)$ modified to $b_s(\omega)$ by the sample's absorption. We measure $f_s(t)$ and we want to find $b_s(\omega)$ to find the changes in amplitudes at the various frequencies; these changes give the sample's absorption spectrum. To find b from f, one uses an extension of the procedure used to derive (1.32); one finds (*Sokolnikoff and Redheffer*, pp. 70–71)

$$b_s(\omega)=\frac{1}{2\pi}\int_{-\infty}^{\infty} f_s(t)e^{-i\omega t}\,dt$$

The Fourier transformation of f_s to obtain b_s is performed by a small on-line digital computer. By "on-line" we mean the computer is connected to the spectrometer output, and functions as part of the spectrometer system. (At optical frequencies, a prism acts as an analog computer to Fourier-analyze an irregular waveform into its component frequencies.) Thus in FT NMR, the pulse allows us to get the whole spectrum at once, rather than sweeping slowly through the spectrum as in conventional NMR. The Fourier transforms of the NMR responses resulting from successive pulses are time averaged by the computer to give an improved signal. A thousand scans of a spectrum may take 60 days by conventional NMR, but can be done in 15 minutes by FT NMR. For details, see T. C. Farrar and E. D. Becker, *Pulse and Fourier Transform NMR*, Academic Press, 1971. The FT method is also used in IR spectroscopy; see E. D. Becker and T. C. Farrar, *Science*, **178**, 361 (1972). FT IR spectroscopy can give a spectrum from 10^{-8} g of sample. FT NMR and IR spectrometers are commercially available.

8.4 NMR CHEMICAL SHIFTS

In 1951 Arnold, Dharmatti, and Packard observed the proton NMR spectrum of liquid ethanol to consist of three lines; the separation between these lines was found to increase linearly with increasing applied field B_0. The three lines are attributed to the three chemically different protons in CH_3CH_2OH. NMR studies where conditions are such as to resolve lines due to identical nuclei in chemically different environments are given the adjective *high-resolution*. Let us consider why the chemical environment of a nucleus affects its NMR transition frequency.

The quantity B in (8.9) is the magnitude of the magnetic field at the nucleus in question. For an isolated nucleus, B is just the magnitude of the applied field. For a nucleus in a molecule, however, the magnetic field at the nucleus is due not only to the applied external field, but to the magnetic fields of the electrons and the other nuclei. In this section we consider the effects of the electrons, postponing consideration of the effects of other nuclei to Section 8.5.

All electrons have spin angular momentum and all but s electrons have orbital angular momentum; magnetic dipole moments are associated with these angular momenta. However, in all but a few molecules, the electronic ground states are nondegenerate with all electron spins paired, and there is no net electronic spin or orbital angular momentum. Hence in the absence of an external magnetic field, the electrons usually contribute essentially nothing to the magnetic field felt by a nucleus.

However, when an external magnetic field is applied, the molecular electronic wave function is perturbed, giving rise to an electronic contribution $\mathbf{B}'$ to the magnetic field at a nucleus. In general, $\mathbf{B}'$ has different values at different nuclei. The field $\mathbf{B}'_i$ at nucleus i produced by the perturbed electrons is proportional to the strength of the applied external field $\mathbf{B}_0$, and is given by $\mathbf{B}'_i = -\boldsymbol{\sigma}_i \mathbf{B}_0$, where $\boldsymbol{\sigma}_i$ is a symmetric second-rank tensor (matrix) with nine components (six of which are independent). [Recall that the response of a molecule to an external *electric* field also involves a tensor, the polarizability $\boldsymbol{\alpha}$ in (4.139).] The above equation for $\mathbf{B}'_i$ defines the *shielding tensor* $\boldsymbol{\sigma}_i$ at nucleus i. As usual (Section 5.2), there is a particular set of molecule-fixed axes x, y, z at nucleus i for which $\boldsymbol{\sigma}_i$ is diagonal, the diagonal values being the principal values $\sigma_{xx,i}$, $\sigma_{yy,i}$, $\sigma_{zz,i}$. The molecules in the liquid or gas sample are rotating rapidly with respect to the applied space-fixed field $\mathbf{B}_0$. The effect of this rotation is to average over $\boldsymbol{\sigma}$, so that $\langle \mathbf{B}'_i \rangle = -\langle \boldsymbol{\sigma}_i \rangle \mathbf{B}_0$. The average of $\boldsymbol{\sigma}_i$ turns out to be the average of its principal values, namely $\frac{1}{3}(\sigma_{xx,i} + \sigma_{yy,i} + \sigma_{zz,i}) = \frac{1}{3}\operatorname{Tr}\boldsymbol{\sigma}_i \equiv \sigma_i$. Thus $\mathbf{B}'_i = -\sigma_i \mathbf{B}_0$, and the actual field at nucleus i is $\mathbf{B}_i = \mathbf{B}_0 + \mathbf{B}'_i$, or

$$B_i = B_0(1 - \sigma_i) \tag{8.24}$$

where the scalar proportionality constant σ_i is called the *shielding constant* (or *screening constant*) for nucleus i.

To evaluate σ_i theoretically, one must use perturbation theory. Ramsey did so (see *Murrell and Harget*, pp. 121–123), and found that σ is the sum of two terms, a positive term σ^d (called the *diamagnetic* contribution, since it decreases the applied field), and a negative term σ^p (the *paramagnetic* contribution). The term σ^p involves the usual perturbation-theory sum over excited states, and therefore is difficult to calculate; however, one can use a variation–perturbation approach (Section 1.10) in σ^p calculations. For molecular protons, σ^d exceeds σ^p, and σ_{H} is positive.

For a single nucleus i with shielding constant σ_i, the Hamiltonian (8.7) becomes

$$\hat{H}_i = -g_i \beta_p B_0 \hbar^{-1}(1 - \sigma_i)\hat{I}_{z,i} \tag{8.25}$$

where B_0 is the magnitude of the applied static field. The eigenfunctions of (8.25) are the eigenfunctions of $\hat{I}_{z,i}$ and its eigenvalues are

$$-g_i\beta_p B_0(1-\sigma_i)M_{I,i} \tag{8.26}$$

For a molecule with several nuclei with nonzero spin, each nuclear magnetic moment interacts with $\mathbf{B}_0$ and the Hamiltonian is

$$\hat{H} = -\beta_p B_0 \hbar^{-1} \sum_i (1-\sigma_i) g_i \hat{I}_{z,i} = \sum_i \hat{H}_i \tag{8.27}$$

The Hamiltonian (8.27) is the sum of Hamiltonians $\hat{H}_i$ for the individual nuclei; therefore, the nuclear-spin wave function is the product of eigenfunctions of the $\hat{H}_i$'s and the energy is the sum of eigenenergies of the $\hat{H}_i$'s:

$$\psi = \prod_i |M_{I,i}(i)\rangle, \qquad M_{I,i} = -I_i, \ldots, +I_i \tag{8.28}$$

$$E = -\beta_p B_0 \sum_i g_i(1-\sigma_i)M_{I,i} \tag{8.29}$$

Consider the selection rules. The interaction between the rf field $\mathbf{B}_1$ and the nuclear magnetic moments is [Equation (8.12)]

$$\hat{H}'(t) = -\mathbf{B}_1 \cdot \sum_i \hat{\boldsymbol{\mu}}_i \tag{8.30}$$

Using (8.28), (8.30), (8.12), and (8.20), one finds the selection rules to be

$$\left.\begin{array}{ll} \Delta M_{I,i} = \pm 1 & \\ \Delta M_{I,j} = 0, & j \neq i \end{array}\right\} i = 1, 2, \ldots \tag{8.31}$$

The selection rules allow only one nuclear magnetic moment to change its orientation in a transition. The frequencies of the NMR lines are then

$$\nu_i = h^{-1}\beta_p B_0 g_i(1-\sigma_i), \qquad i = 1, 2, \ldots \tag{8.32}$$

Identical nuclei in chemically equivalent environments in the molecule have the same values of g_i and σ_i and have the same NMR transition frequency; identical nuclei in chemically different environments have the same value of g_i but different values of σ_i and (since σ_i is generally small) have NMR transition frequencies lying close together; nonidentical nuclei have different values of g_i and their NMR transition frequencies are generally well separated.

As an example, consider CH_3CH_2OH. The O^{16} and C^{12} nuclei have zero spin and can be ignored. There are three kinds of protons, and each kind has its own value of σ. (It might be argued that the three CH_3 protons are not quite in chemically equivalent environments; however, the low barrier to internal rotation about the single bonds assures their equivalence and the equivalence of the two CH_2 protons—see also Section 8.9.) The OH hydrogen is bonded to a highly electronegative atom and we might expect the electron density to be lower about this proton than about the other protons of the molecule; this means a lower diamagnetic shielding and a lower value of σ for the OH proton. Experimentally it is found that for *pure* liquid ethanol

$$\sigma_{OH} < \sigma_{CH_2} < \sigma_{CH_3} \tag{8.33}$$

in apparent accord with the electronegativity argument. *However*, in the gas phase and in dilute CCl_4 solutions, σ for the OH proton in ethanol is *greater* than for the CH_2 and CH_3 protons. Thus there must be other factors besides electronegativity determining the shifts in the isolated ethanol molecule. The variation in position of the OH resonance of ethanol is due to hydrogen bonding in the pure liquid.

From (8.32) it is seen that at a fixed value of B_0, the resonance frequency is *highest* for the OH proton (which has the lowest σ) in *pure* ethanol. However, NMR spectra are observed with the rf frequency fixed at the value $\nu_{\text{spectrometer}}$, while B_0 is varied; under these conditions the value of B_0 required for resonance of nucleus i is

$$B_{0,i} = \frac{h\nu_{\text{spectrometer}}}{\beta_p g_i(1-\sigma_i)} \tag{8.34}$$

Hence $B_{0,i}$ is *lowest* for the OH proton in pure ethanol. (See Fig. 8.2.) A sweep from low to high B_0 at fixed rf frequency corresponds to a sweep from high to low rf frequency at fixed B_0. Figure 8.2 shows the NMR spectrum of pure liquid ethanol. (With better resolution than in this figure, the spectrum shows a fine structure due to spin–spin coupling; see Section 8.5.)

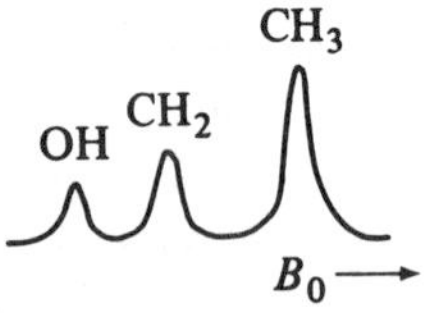

Fig. 8.2 The NMR spectrum of pure liquid CH_3CH_2OH observed at a fixed rf frequency and varying applied field B_0.

The change in NMR transition frequency at fixed B_0 (or equivalently, the change in B_0 at fixed $\nu_{\text{spectrometer}}$) due to the shielding of a nucleus by the electrons of the molecule is called the *chemical shift*. The chemical shift is observed in liquid and gas samples, but not in solids (Section 8.7). The shielding constant is small compared to 1; for protons in diamagnetic molecules, σ_i ranges from 1×10^{-5} to 4×10^{-5}, while for heavier nuclei it may be 10^{-4} or 10^{-3}.

Actually, σ_i cannot be determined with high accuracy. To do so requires accurate measurements of both $\nu_{\text{spectrometer}}$ and B_0; however, it is not feasible to make a highly accurate measurement of B_0. Because of this difficulty, chemical shifts are measured relative to the shift in some standard compound. (A good estimate of σ_i for a set of equivalent protons in a given compound can be obtained as follows: Calculations using Ramsey's formula predict σ to be 27×10^{-6} for the protons in H_2. Measurement of a compound's proton shift relative to the shift in H_2 then allows the calculation of σ for the compound.)

It follows from (8.34) that the separation between the resonances of two nuclei with different shielding constants is proportional to $\nu_{\text{spectrometer}}$. (In an NMR experiment at constant B_0 and variable rf frequency, the separation is proportional to B_0.) To express chemical shifts in a form independent of the value of $\nu_{\text{spectrometer}}$, the quantity δ_i is defined as

$$\delta_i \equiv (\sigma_{\text{ref}} - \sigma_i)\times 10^6 \text{ ppm} \tag{8.35}$$

where σ_{ref} and σ_i are the shielding constants for the standard reference nucleus and for nucleus i. The factor 10^6 is included to give numbers of convenient magnitude; all chemical-shift δ values in this chapter will be given in parts per million (ppm). The definition (8.35) makes δ_i positive for nuclei less shielded than the reference nucleus, that is, for nuclei that resonate at lower values of B_0 than the reference nucleus. By convention, NMR spectra are presented with B_0 increasing from left to right on the page.

From (8.34) and (8.35), we have

$$\delta_i = \frac{h\nu_{\text{spectrometer}}}{\beta_p g_i}\left(\frac{B_{0,\text{ref}} - B_{0,i}}{B_{0,\text{ref}}B_{0,i}}\right)\times 10^6 \text{ ppm} \tag{8.36}$$

(We have $g_{\text{ref}} = g_i$, since the reference nucleus is always the same kind as nucleus i.) Since σ_i is generally small, the first fraction in (8.36) is nearly equal to $B_{0,i}$ [Eq. (8.34)], and to a good approximation we can write

$$\delta_i = \frac{B_{0,\text{ref}} - B_{0,i}}{B_{0,\text{ref}}}\times 10^6 \text{ ppm} \tag{8.37}$$

For proton NMR work in organic compounds, the reference standard is $Si(CH_3)_4$, tetramethylsilane (TMS); all the protons are equivalent and TMS gives a single resonance line. A small amount of TMS is added to the liquid whose NMR spectrum is to be run; the positions of the sample's NMR lines are then observed relative to the TMS line. Equation (8.34) takes into account only the shielding of a nucleus due to electrons in the same molecule as that nucleus; there is also shielding from the other molecules in the liquid sample and from the sample-tube molecules. Mixing the TMS with the sample exposes the sample and reference molecules to the same magnetic field, so that there is no need to apply corrections for the bulk diamagnetic susceptibility of the tube or sample. The shielding constant σ_{ref} for the TMS protons is about 31×10^{-6} and is larger than that of protons in nearly all organic compounds; δ ranges from 0 to 16 for protons in organic compounds. The bare-proton δ is 31. Certain inorganic compounds have larger σ's than TMS; for example, the proton in HI has $\sigma=44\times10^{-6}$.

An alternative scale, the τ scale, is also used to express proton chemical shifts. In this scale the TMS protons are assigned a shift of exactly 10 parts per million and the τ value of a line relative to TMS is then defined as

$$\tau_i \equiv 10.000 - \delta_i \tag{8.38}$$

For organic compounds, τ ranges from 10 to -6. The chart paper of commercial NMR spectrometers is generally calibrated in both δ and τ values. There is no need for the two competing scales, and it would be desirable for one of them to be abandoned. The International Union of Pure and Applied Chemistry has recommended the δ scale.

To add to the confusion, some workers take δ_i as the negative of the definition (8.35). Some workers use factors of 10^5 or 10^7 instead of 10^6. Some workers use δ_i to mean the shift in hertz; their δ_i is equal to (8.35) multiplied by $\nu_{\text{spectrometer}}$. Prior to the introduction (by Tiers in 1958) of TMS as the reference for proton shifts, a variety of reference compounds (such as benzene, cyclohexane, and water) were used. When examining reported shifts, the definitions used should be noted.

Observed chemical shifts are affected by intermolecular interactions. Ideally δ_i should be determined in the gas phase at low pressure, but this is generally impractical to do. The next best approach is to use a dilute solution of the compound in an inert solvent. For organic compounds, the most widely used solvents are $CDCl_3$ and CCl_4. For alcohols, the OH proton shift varies very considerably with concentration of the solution, as a result of intermolecular hydrogen bonding.

The chemical shifts of protons in a given type of chemical group vary over a narrow range from compound to compound. NMR spectroscopy has become a tool as useful to the organic chemist as IR spectroscopy. The number of proton peaks, their relative intensities, and their chemical shifts aid greatly in a structural determination. The ratios of the areas under the absorption peaks due to different sets of equivalent nuclei give the ratios of the numbers of equivalent protons in each set. For example, the areas under the three CH_3CH_2OH peaks have the ratios 3:2:1. As an approximation to the area ratios, the peak-height ratios may be used. However, NMR line widths vary for different nuclei in the same molecule, so that peak-height ratios are not reliable indicators of relative intensities. Commercial spectrometers have electronic integrators and can display the integral of the spectrum to give accurate relative intensities.

Figure 8.3 lists typical ranges of δ and τ for organic groups.[5]

The proton NMR spectra of thousands of compounds have been tabulated.

Because of the difficulty in evaluating the paramagnetic contribution, *ab initio* calculations of NMR shielding constants are few in number and limited to rather small molecules. Jaszunski and Sadlej used a variational perturbation approach with an *ab initio* SCF wave function of H_2O and found $\sigma_H = 28.3$ ppm. [M. Jaszunski and A. J. Sadlej, *Theor. Chim. Acta*, **27**, 135 (1972).] The known proton screening constant in H_2 is 26.6 ppm,[6] and gaseous H_2O shows a proton chemical shift of 3.6 ppm relative to gaseous H_2; hence the experimental H_2O proton shielding constant is 30.2 ppm.

Pople has developed an approximate approach which takes σ as the sum of contributions due to electron currents induced by the applied field in the various atoms and bonds of the molecule. The theory gives reasonably good results in many cases. (See *Emsley, Feeney, and Sutcliffe*, Chapter 4.)

8.5 NMR SPIN–SPIN SPLITTINGS

When the NMR spectrum of ethanol is examined under higher resolution than in Fig. 8.2, each of the three lines is found to be split into a number of closely spaced lines (Fig. 8.10). Unlike chemical shifts, these splittings are independent of the applied field B_0. They are due to neigh-

[5] See also M. W. Dietrich and R. E. Keller, *Anal. Chem.*, **36**, 258 (1964); F. A. Bovey, *Chem. Eng. News*, Aug. 30, 1965, p. 98; K. Nukada *et al.*, *Anal. Chem.*, **35**, 1892 (1963); *Mathieson*, pp. 179–186.

[6] The diamagnetic contribution for H_2 is easily calculated theoretically, and it turns out that the paramagnetic contribution is related to something called the spin–rotation interaction constant, which can be measured experimentally.

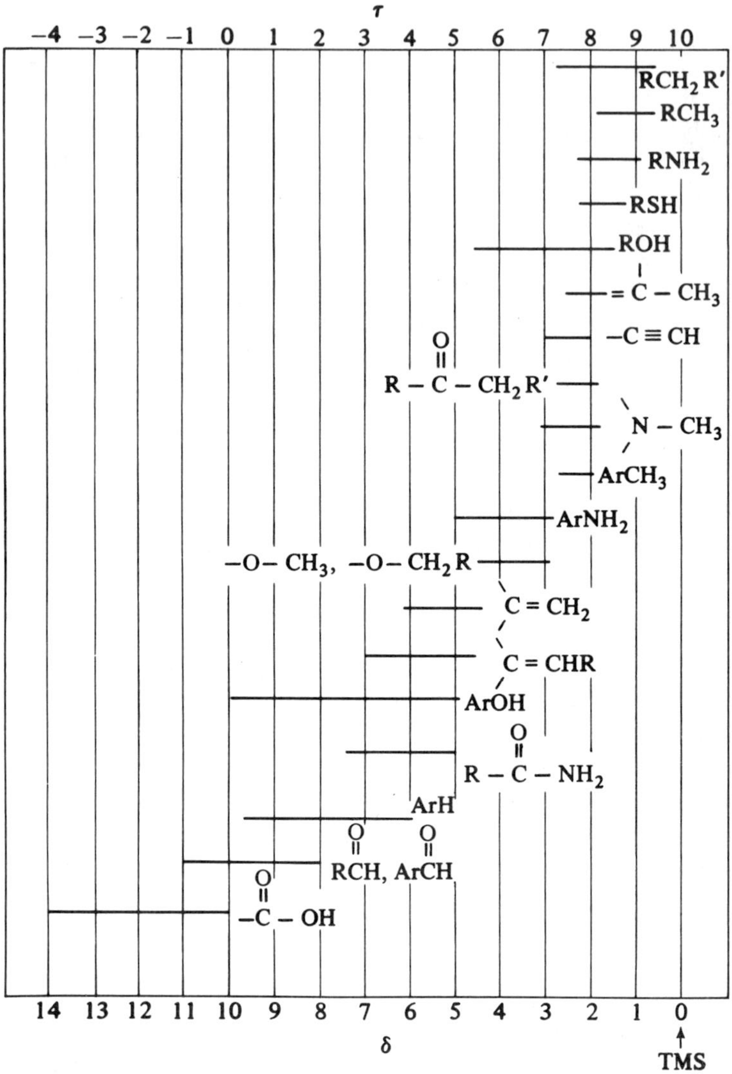

Fig. 8.3 Proton chemical shifts for organic compounds in dilute solutions of CCl_4 or $CDCl_3$. R and Ar stand for alkyl and aryl groups.

boring magnetic nuclei in the same molecule. We now consider these splittings. As in Section 8.4, the discussion is limited to liquids and gases.

Consider a molecule with two nuclei with spin $\frac{1}{2}$ and no other nuclei with nonzero spin. The magnetic field due to the magnetic dipole moment of nucleus 2 will interact with the magnetic dipole moment $\boldsymbol{\mu}_1$ of nucleus 1, giving the additional term in the Hamiltonian

$$-\mathbf{B}_2 \cdot \hat{\boldsymbol{\mu}}_1 = -g_N \beta_p \hbar^{-1} \mathbf{B}_2 \cdot \hat{\mathbf{I}}_1 \tag{8.39}$$

where $\mathbf{B}_2$ is the magnetic field at nucleus 1 produced by $\boldsymbol{\mu}_2$. Since $\mathbf{B}_2$ is proportional to $\boldsymbol{\mu}_2$ and therefore to $\mathbf{I}_2$, we can write (8.39) as

$$(h/\hbar^2) J_{12} \hat{\mathbf{I}}_1 \cdot \hat{\mathbf{I}}_2 \tag{8.40}$$

where J_{12} is a proportionality constant called the *nuclear spin–spin coupling constant*.[7] Enough Planck's constants have been sprinkled about (8.40) to give J_{12} units of hertz. There is no connection between J_{12} and the quantum number J for electrons in atoms. (The description just given of the origin of NMR spin–spin couplings in liquids is grossly oversimplified. For a correct discussion, see Section 8.7.)

With the inclusion of spin–spin interaction, the NMR Hamiltonian (8.27) for a molecule with two magnetic nuclei in chemically different environments becomes

$$\hat{H} = -\frac{g_N \beta_p B_0}{\hbar}(1-\sigma_1)\hat{I}_{z1} - \frac{g_N \beta_p B_0}{\hbar}(1-\sigma_2)\hat{I}_{z2} + \frac{hJ_{12}}{\hbar^2}\hat{\mathbf{I}}_1 \cdot \hat{\mathbf{I}}_2 \tag{8.41}$$

[For simplicity, the two nuclei have been assumed to be of the same species, but the following discussion is applicable to two different kinds of nuclei if the two g_N's in (8.41) are replaced by g_1 and g_2.]

Because of the spin–spin coupling term, (8.41) is not separable into the sum of Hamiltonians for the individual nuclei, and the corresponding Schrödinger equation is not separable. To deal with the problem, we shall use the method of expanding the unknown wave functions in terms of a known complete set of functions. The eigenvalues and eigenfunctions (eigenvectors) are obtained as the solutions of (2.68) [or (2.77)] and (2.67).

To use the method, we need a complete set of well-behaved nuclear-spin basis functions φ_n. If the spin–spin coupling term is omitted from (8.41), we get the Hamiltonian

$$-g_N \beta_p B_0 \hbar^{-1}(1-\sigma_1)\hat{I}_{z1} - g_N \beta_p B_0 \hbar^{-1}(1-\sigma_2)\hat{I}_{z2} \tag{8.42}$$

The eigenfunctions of (8.42) are, according to (8.28),

$$|M_{I1}(1)\rangle |M_{I2}(2)\rangle, \qquad M_{I1} = \pm\tfrac{1}{2}, \quad M_{I2} = \pm\tfrac{1}{2} \tag{8.43}$$

We shall use α and β to stand for nuclear-spin wave functions with

[7] For tabulations of NMR chemical shifts and coupling constants, see F. A. Bovey, *NMR Data Tables for Organic Compounds*, vol. 1, Interscience, New York, 1967; W. Brügel, *Nuclear Magnetic Resonance Spectra and Chemical Structure*, vol. 1, Academic Press, New York, 1967.

$M_I = +\frac{1}{2}$ and $-\frac{1}{2}$, respectively. (The same symbols were used previously to designate electronic spin functions with $m_s = \pm\frac{1}{2}$; the context will indicate whether α and β mean electronic or nuclear spin functions.) Since the operator (8.42) is Hermitian, its eigenfunctions (8.43) form a complete set of well-behaved functions for the problem of two nuclei with spin $\frac{1}{2}$. Moreover, since there are only a finite number of functions in this complete set, the secular determinant in (2.77) is of finite order and easy to deal with. The complete set is then

$$\varphi_1 = \alpha(1)\alpha(2), \qquad \varphi_2 = \alpha(1)\beta(2)$$
$$\varphi_3 = \beta(1)\alpha(2), \qquad \varphi_4 = \beta(1)\beta(2) \tag{8.44}$$

where the numbers in parentheses refer to nuclei 1 and 2. It follows from (8.10) that the φ's are orthonormal

$$\langle\varphi_i|\varphi_j\rangle = \delta_{ij} \tag{8.45}$$

To set up the secular determinant, we need the effect of $\hat{H}$ on the functions (8.44). We write (8.41) as

$$\hat{H} = -c_1\hbar^{-1}\hat{I}_{z1} - c_2\hbar^{-1}\hat{I}_{z2} + hJ_{12}\hbar^{-2}\hat{\mathbf{I}}_1 \cdot \hat{\mathbf{I}}_2 \tag{8.46}$$

$$c_1 \equiv g_N\beta_p B_0(1-\sigma_1), \qquad c_2 \equiv g_N\beta_p B_0(1-\sigma_2) \tag{8.47}$$

The effects of the first two terms in (8.46) on the φ's follow from (8.5), which states that

$$\hat{I}_z\alpha = \tfrac{1}{2}\hbar\alpha, \qquad \hat{I}_z\beta = -\tfrac{1}{2}\hbar\beta \tag{8.48}$$

For example,

$$-c_1\hbar^{-1}\hat{I}_{z1}[\alpha(1)\alpha(2)] = -c_1\hbar^{-1}\alpha(2)\hat{I}_{z1}\alpha(1) = -\tfrac{1}{2}c_1\alpha(1)\alpha(2) \tag{8.49}$$

To find the effect of the third term of (8.46), we write

$$\hat{\mathbf{I}}_1 \cdot \hat{\mathbf{I}}_2 = \hat{I}_{x1}\hat{I}_{x2} + \hat{I}_{y1}\hat{I}_{y2} + \hat{I}_{z1}\hat{I}_{z2} \tag{8.50}$$

Corresponding to (1.244) and (1.245) for an electron's spin operators and functions, we have the following equations for a spin-$\frac{1}{2}$ nucleus's spin operators and functions:

$$\hat{I}_x\alpha = \tfrac{1}{2}\hbar\beta, \qquad \hat{I}_x\beta = \tfrac{1}{2}\hbar\alpha \tag{8.51}$$
$$\hat{I}_y\alpha = \tfrac{1}{2}i\hbar\beta, \qquad \hat{I}_y\beta = -\tfrac{1}{2}i\hbar\alpha \tag{8.52}$$

We then have

$$hJ_{12}\hbar^{-2}\hat{\mathbf{I}}_1 \cdot \hat{\mathbf{I}}_2\varphi_1 = \tfrac{1}{4}hJ_{12}[\beta(1)\beta(2) - \beta(1)\beta(2) + \alpha(1)\alpha(2)]$$

$$= \tfrac{1}{4}hJ_{12}\varphi_1 \tag{8.53}$$

Similarly

$$hJ_{12}\hbar^{-2}\hat{\mathbf{I}}_1 \cdot \hat{\mathbf{I}}_2\varphi_2 = \tfrac{1}{4}hJ_{12}(2\varphi_3 - \varphi_2) \tag{8.54}$$

$$hJ_{12}\hbar^{-2}\hat{\mathbf{I}}_1 \cdot \hat{\mathbf{I}}_2\varphi_3 = \tfrac{1}{4}hJ_{12}(2\varphi_2 - \varphi_3) \tag{8.55}$$

$$hJ_{12}\hbar^{-2}\hat{\mathbf{I}}_1 \cdot \hat{\mathbf{I}}_2\varphi_4 = \tfrac{1}{4}hJ_{12}\varphi_4 \tag{8.56}$$

Therefore,

$$\hat{H}\varphi_1 = (-\tfrac{1}{2}c_1 - \tfrac{1}{2}c_2 + \tfrac{1}{4}hJ_{12})\varphi_1 \tag{8.57}$$

$$\hat{H}\varphi_2 = (-\tfrac{1}{2}c_1 + \tfrac{1}{2}c_2 - \tfrac{1}{4}hJ_{12})\varphi_2 + \tfrac{1}{2}hJ_{12}\varphi_3 \tag{8.58}$$

$$\hat{H}\varphi_3 = (\tfrac{1}{2}c_1 - \tfrac{1}{2}c_2 - \tfrac{1}{4}hJ_{12})\varphi_3 + \tfrac{1}{2}hJ_{12}\varphi_2 \tag{8.59}$$

$$\hat{H}\varphi_4 = (\tfrac{1}{2}c_1 + \tfrac{1}{2}c_2 + \tfrac{1}{4}hJ_{12})\varphi_4 \tag{8.60}$$

(Note that φ_1 and φ_4 are eigenfunctions of $\hat{H}$, so that we already have two of the solutions of the Schrödinger equation.)

The matrix elements of $\hat{H}$ are now easily evaluated; for example,

$$\langle\varphi_2|\hat{H}|\varphi_2\rangle = (-\tfrac{1}{2}c_1 + \tfrac{1}{2}c_2 - \tfrac{1}{4}hJ_{12})\langle\varphi_2|\varphi_2\rangle + \tfrac{1}{2}hJ_{12}\langle\varphi_2|\varphi_3\rangle$$

$$= -\tfrac{1}{2}c_1 + \tfrac{1}{2}c_2 - \tfrac{1}{4}hJ_{12}$$

Evaluation of the remaining matrix elements gives for the secular equation (2.77):

$$\begin{vmatrix} [-\tfrac{1}{2}(c_1+c_2) + \tfrac{1}{4}hJ_{12} - E_i] & 0 & 0 & 0 \\ 0 & [\tfrac{1}{2}(c_2-c_1) - \tfrac{1}{4}hJ_{12} - E_i] & \tfrac{1}{2}hJ_{12} & 0 \\ 0 & \tfrac{1}{2}hJ_{12} & [\tfrac{1}{2}(c_1-c_2) - \tfrac{1}{4}hJ_{12} - E_i] & 0 \\ 0 & 0 & 0 & [\tfrac{1}{2}(c_1+c_2) + \tfrac{1}{4}hJ_{12} - E_i] \end{vmatrix} = 0 \tag{8.61}$$

The secular determinant is in block-diagonal form and factors into three determinants. Two of the roots of the secular equation are

$$E_1 = -\tfrac{1}{2}(c_1 + c_2) + \tfrac{1}{4}hJ_{12}, \qquad E_4 = \tfrac{1}{2}(c_1 + c_2) + \tfrac{1}{4}hJ_{12} \tag{8.62}$$

[These energies also follow directly from (8.57) and (8.60).] The remaining two energies are the roots of a quadratic equation:

$$E_i^2 + \tfrac{1}{2}hJ_{12}E_i - \tfrac{1}{4}(c_1 - c_2)^2 - \tfrac{3}{16}h^2J_{12}^2 = 0 \tag{8.63}$$

$$E_{2,3} = -\tfrac{1}{4}hJ_{12} \mp \tfrac{1}{2}\left[(c_1 - c_2)^2 + h^2J_{12}^2\right]^{1/2} \tag{8.64}$$

The four normalized wave functions ψ_1, ψ_2, ψ_3, ψ_4 that correspond to these four energies are found by solving the appropriate simultaneous equations; the details are left as an exercise (Problem 8.8); the wave functions are

$$\begin{aligned} &\psi_1 = \alpha(1)\alpha(2), \qquad \psi_2 = a\alpha(1)\beta(2) - b\beta(1)\alpha(2) \\ &\psi_3 = b\alpha(1)\beta(2) + a\beta(1)\alpha(2), \qquad \psi_4 = \beta(1)\beta(2) \end{aligned} \tag{8.65}$$

where

$$\begin{aligned} a &\equiv \frac{\left\{\left[h^2J_{12}^2 + (c_1 - c_2)^2\right]^{1/2} + (c_1 - c_2)\right\}^{1/2}}{2^{1/2}\left[h^2J_{12}^2 + (c_1 - c_2)^2\right]^{1/4}} \\ b &\equiv \frac{\left\{\left[h^2J_{12}^2 + (c_1 - c_2)^2\right]^{1/2} - (c_1 - c_2)\right\}^{1/2}}{2^{1/2}\left[h^2J_{12}^2 + (c_1 - c_2)^2\right]^{1/4}} \end{aligned} \tag{8.66}$$

From (8.62), (8.64), and (8.47), the energies of these wave functions are

$$\begin{aligned} E_1 &= -g_N\beta_p B_0(1 - \tfrac{1}{2}\sigma_1 - \tfrac{1}{2}\sigma_2) + \tfrac{1}{4}hJ_{12} \\ E_2 &= -\tfrac{1}{4}hJ_{12} - \tfrac{1}{2}h\left[J_{12}^2 + g_N^2\beta_p^2B_0^2h^{-2}(\sigma_2 - \sigma_1)^2\right]^{1/2} \\ E_3 &= -\tfrac{1}{4}hJ_{12} + \tfrac{1}{2}h\left[J_{12}^2 + g_N^2\beta_p^2B_0^2h^{-2}(\sigma_2 - \sigma_1)^2\right]^{1/2} \\ E_4 &= g_N\beta_p B_0(1 - \tfrac{1}{2}\sigma_1 - \tfrac{1}{2}\sigma_2) + \tfrac{1}{4}hJ_{12} \end{aligned} \tag{8.67}$$

Since (2.77) involves no approximations, the results (8.65) and (8.67) are exact, as may be verified directly by showing the functions (8.65) to be eigenfunctions of $\hat{H}$ with the eigenvalues (8.67).

In the limit $J_{12} \to 0$, the energy levels (8.67) go over to those of two independent nuclear spins [Equation (8.29)], and the functions (8.65) go over to the functions (8.44), as shown in Fig. 8.4.

To determine the NMR spectrum of the coupled spin-$\frac{1}{2}$ nuclei, the selection rules are needed. They are determined according to (8.14) by integrals of the form $\langle \psi_i | \hat{I}_{x1} + \hat{I}_{x2} | \psi_j \rangle$, where the applied rf field is in the x direction. The ψ's are linear combinations of the φ functions (8.44), and we have seen [(8.31)] that the selection rule for the φ's is such that only one spin changes at a time. The integral $\langle \varphi_k | \hat{I}_{x1} + \hat{I}_{x2} | \varphi_l \rangle$ is nonzero for k and l having the pairs of values 1,2; 1,3; 2,4; 3,4; and vanishes for the k and l values 1,4; 2,3; 2,2; 3,3. From (8.65), the NMR selection rules for two coupled protons are

$$\psi_1 \leftrightarrow \psi_2 \qquad \psi_1 \leftrightarrow \psi_3 \qquad \psi_2 \leftrightarrow \psi_4 \qquad \psi_3 \leftrightarrow \psi_4 \qquad (8.68)$$
$$\psi_1 \not\leftrightarrow \psi_4 \qquad \psi_2 \not\leftrightarrow \psi_3$$

The relative intensities of the four allowed transitions are proportional to $|\langle \psi_i | \hat{I}_{x1} + \hat{I}_{x2} | \psi_j \rangle|^2$. (It might be thought necessary to consider also the effect of the relative populations of the levels on the transition intensities, but the separation between nuclear-spin energy levels is much less than kT, so that the very slight differences in population can be ignored here.)

Using (8.65), (8.51), (8.52), and (8.45) to find the relative intensities and

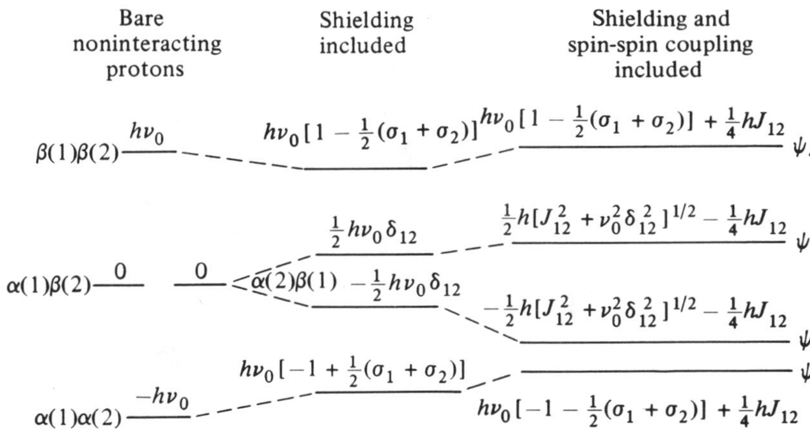

Fig. 8.4 NMR energy levels for the AB case. ν_0 and δ_{12} are defined in (8.70).

(8.67) to find the transition frequencies at fixed B_0, we get

Transition	Frequency	Relative Intensity
$\psi_1 \to \psi_2$	$\nu_0(1-\frac{1}{2}\sigma_1-\frac{1}{2}\sigma_2)-\frac{1}{2}J_{12}-\frac{1}{2}[J_{12}^2+\nu_0^2\delta_{12}^2]^{1/2}$	$(a-b)^2$
$\psi_3 \to \psi_4$	$\nu_0(1-\frac{1}{2}\sigma_1-\frac{1}{2}\sigma_2)+\frac{1}{2}J_{12}-\frac{1}{2}[J_{12}^2+\nu_0^2\delta_{12}^2]^{1/2}$	$(a+b)^2$
$\psi_1 \to \psi_3$	$\nu_0(1-\frac{1}{2}\sigma_1-\frac{1}{2}\sigma_2)-\frac{1}{2}J_{12}+\frac{1}{2}[J_{12}^2+\nu_0^2\delta_{12}^2]^{1/2}$	$(a+b)^2$
$\psi_2 \to \psi_4$	$\nu_0(1-\frac{1}{2}\sigma_1-\frac{1}{2}\sigma_2)+\frac{1}{2}J_{12}+\frac{1}{2}[J_{12}^2+\nu_0^2\delta_{12}^2]^{1/2}$	$(a-b)^2$

(8.69)

where the relative shift δ_{12} and the bare-nucleus frequency ν_0 are

$$\delta_{12} \equiv \sigma_2-\sigma_1, \qquad \nu_0 = g_N\beta_p B_0 h^{-1} \tag{8.70}$$

Replacement of J_{12} by $-J_{12}$ gives four lines of the same frequencies as (8.69); the observed spectrum is independent of the sign of J_{12}. The spectrum of two coupled spin-$\frac{1}{2}$ nuclei consists of four lines symmetrically placed about the frequency $\nu_0(1-\frac{1}{2}\sigma_1-\frac{1}{2}\sigma_2)$, as shown in Fig. 8.5. The wave functions (8.65) can be expressed in terms of $\nu_0\delta_{12}$ using

$$c_1-c_2 = h\nu_0\delta_{12} \tag{8.71}$$

Figures 8.6 and 8.7 show the variation in patterns as a function of the ratio J_{12} to $\nu_0\delta_{12}$. For $J_{12}=0$, we have $a=1$ and $b=0$, and we get only two lines of equal intensity, as found in Section 8.4. For $J_{12} \ll \nu_0\delta_{12}$, we have $b \ll a$, and the outer two lines are slightly weaker than the inner two lines; the spectrum consists of two well-separated doublets. For $J_{12} \gg \nu_0\delta_{12}$, we have $(a+b)^2 \gg (a-b)^2$; the two inner lines of the pattern have nearly merged and the outer lines are quite weak. For $\nu_0\delta_{12}=0$, we have $a=b$, so that the two outer lines disappear; also, the two inner lines have merged. Thus for two chemically equivalent coupled protons ($\sigma_1=\sigma_2=\sigma$), the

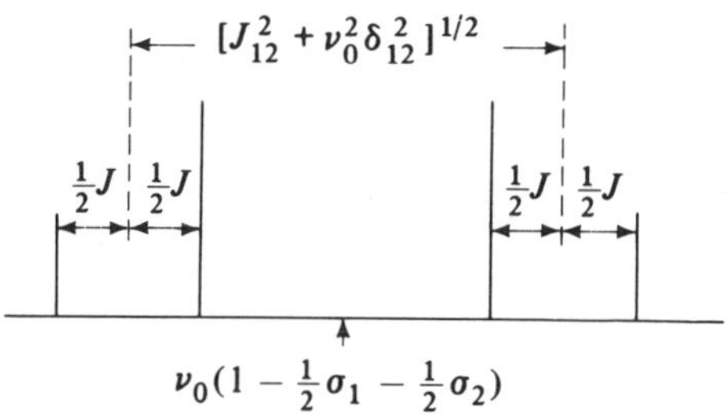

Fig. 8.5 Pattern of the NMR spectrum of two coupled nuclei of spin $\frac{1}{2}$.

NMR spectrum consists of a single line at frequency $\nu_0(1-\sigma)$. As we shall see, in many cases (but *not* in all cases), we get no spin–spin line splittings from the interaction of chemically equivalent nuclei.

The nature of the NMR spectrum thus depends on the relative magnitudes of $\nu_0\delta_{12}$ and J_{12}. Typically, J_{12} ranges from 0 to ± 20 Hz for proton–proton couplings. (Proton–F^{19} coupling constants range up to 100 Hz.) The magnitude of J_{12} decreases as the number of chemical bonds separating nuclei 1 and 2 increases. Protons bonded to the same carbon atom in a saturated open-chain compound are separated by two bonds and generally have J_{12} in the range -10 to -15 Hz; the geminal coupling constant in CH_4 is -12 Hz. (The determination of the absolute signs of coupling constants is discussed in Section 8.7.) Protons bonded to adjacent carbon atoms of a saturated open-chain compound are separated by three bonds and generally have J_{12} in the range $+5$ to $+8$ Hz. For protons separated by four or more single bonds, J_{12} is usually negligible. Protons

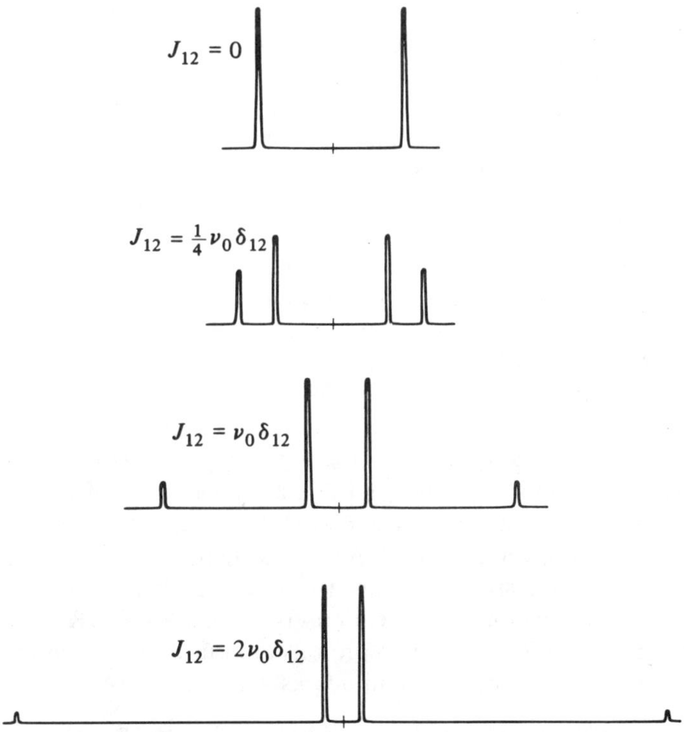

Fig. 8.6 The NMR spectrum of two nuclei of spin $\frac{1}{2}$, as a function of $J_{12}/\nu_0\delta_{12}$. In this figure, $v_0\delta_{12}$ is held fixed while J_{12} varies.

separated by multiple bonds may show appreciable coupling over many bonds; an example is $CH_3-C\equiv C-C\equiv C-CH_3$, where the coupling constant for the protons separated by seven bonds is 1.3 Hz. Coupling constants aid in determining the structures of geometrical isomers; for example, the coupling constants for protons in olefins usually lie in the following ranges[8]:

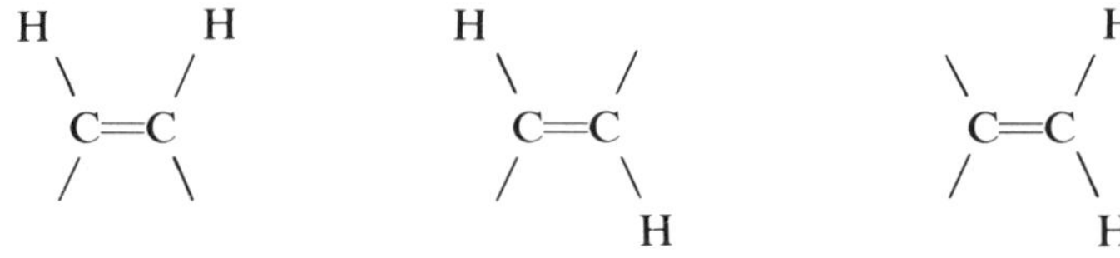

(8.72)

cis:	trans:	gem:
+5 to +12 Hz	+12 to +19 Hz	−3.5 to +2.5 Hz

A standard terminology is used to designate the various possibilities for $J_{12}/\nu_0\delta_{12}$. Letters from the same part of the alphabet label nuclei for which J_{12} is comparable in magnitude to $\nu_0\delta_{12}$, whereas letters from different parts of the alphabet label nuclei for which $J_{12} \ll \nu_0\delta_{12}$. For a molecule with two magnetic nuclei such that $J_{12} \ll \nu_0\delta_{12}$, the spectrum is designated as AX and consists of two doublets with all lines of essentially equal intensity; the separation between the centers of the doublets is essentially $\nu_0\delta_{12}$. When J_{12} becomes of the same order of magnitude as $\nu_0\delta_{12}$ or greater than $\nu_0\delta_{12}$, the lines are no longer of equal intensity and the separation between the doublets is not a good approximation to $\nu_0\delta_{12}$; this case is designated AB. With $\nu_0\delta_{12}=0$, the two nuclei are equivalent and give the single-line A_2 spectrum. Of course, there is no sharp dividing line between the AB and AX cases; as shown in Figs. 8.6 and 8.7, with J_{12} equal to only $\frac{1}{4}\nu_0\delta_{12}$ there are very substantial deviations from the equal-intensity AX case, and the spectrum is best designated as AB. The designation AX may be applied when $\nu_0\delta_{12}$ is at least 8 or 10 times J_{12}. A typical value (Fig. 8.3) of δ_{12} for nonequivalent protons might be 2×10^{-6}; a typical value of J_{12} for protons is 10 Hz; for these values and a 60-MHz spectrometer, we have $\nu_0\delta_{12} = 12J_{12}$, and the spectrum is of AX type. Quite frequently in organic compounds, the spin–spin splittings are much smaller than the chemical shifts. One of many notable exceptions is for CHR=CHR′ compounds; here the chemical shifts of the two protons are quite similar, so that $\nu_0\delta_{12}$ is small and the same order of magnitude as J_{12}.

[8] *Emsley, Feeney, and Sutcliffe*, Section 10.4. For a survey of spin–spin coupling data, see the article by A. A. Bothner-By in *Advances in Magnetic Resonance*, vol. 1, J. S. Waugh, ed., Academic Press, New York, 1965.

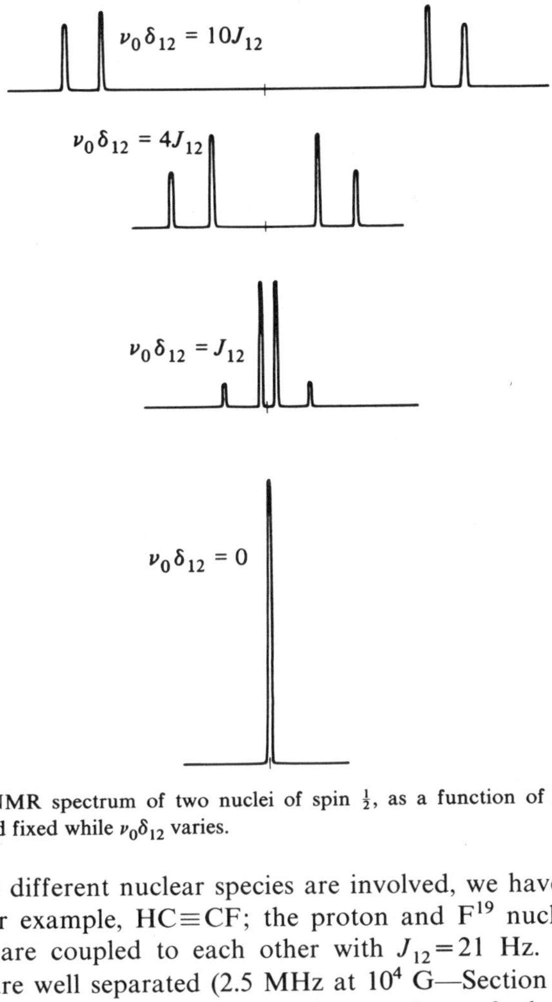

Fig. 8.7 The NMR spectrum of two nuclei of spin $\frac{1}{2}$, as a function of $J_{12}/\nu_0\delta_{12}$. In this figure, J_{12} is held fixed while $\nu_0\delta_{12}$ varies.

When two different nuclear species are involved, we have the AX case. Consider, for example, HC≡CF; the proton and F^{19} nucleus each have spin $\frac{1}{2}$ and are coupled to each other with $J_{12}=21$ Hz. The F and H resonances are well separated (2.5 MHz at 10^4 G—Section 8.3). To apply the preceding analysis to this compound, we replace $\nu_0\delta_{12}$ by $\nu_{0,\mathrm{H}}(1-\sigma_\mathrm{H})-\nu_{0,\mathrm{F}}(1-\sigma_\mathrm{F})$, where $\nu_{0,\mathrm{H}}$ and $\nu_{0,\mathrm{F}}$ are the bare-proton and bare-F^{19} nuclear resonance frequencies. The proton portion of the NMR spectrum shows a doublet with the two lines of equal intensity; the same is true of the F^{19} portion of the spectrum.

For a molecule with three spin-$\frac{1}{2}$ nuclei, the secular determinant is 8×8; we omit detailed analysis.[9] The possible cases are A_3, AB_2, AX_2, ABC,

[9] For the treatment of this and other cases, see *Emsley, Feeney, and Sutcliffe*, Chapter 8; *Pople, Schneider, and Bernstein*, Chapter 6; J. D. Roberts, *An Introduction to the Analysis of Spin–Spin Splitting in High-Resolution Nuclear Magnetic Resonance Spectra*, Benjamin, New York, 1962; *Mathieson*.

ABX, and AMX. The AMX case is three nonequivalent nuclei with all spin–spin couplings much less than the corresponding chemical-shift differences. The appearance of a three-spin spectrum depends, in general, on the relative (but not the absolute) signs of the coupling constants.

For a system of four nuclear spins, the analysis becomes extremely complicated. One important case is A_2X_2, which is two pairs of chemically equivalent nuclei with $\nu_0\delta_{AX} \gg J_{AX}$. Actually, there are two possible A_2X_2 cases, according to whether we have one or two A–X coupling constants. For 1,1-difluoroallene (Fig. 8.8a), there is a single value of J_{AX}, whereas for 1,1-difluoroethylene there are two different A–X coupling constants: J_{AX} and J'_{AX}; this latter case is best designated AA′XX′.

Chemically equivalent nuclei have the same chemical shift (generally as a result of molecular symmetry or internal rotation about a single bond). The members of a set of chemically equivalent nuclei that all have the same coupling constant to any other single nucleus in the molecule are *magnetically equivalent*. The protons in $CH_2{=}C{=}CF_2$ (and in CH_2F_2) are magnetically equivalent (as well as chemically equivalent). The protons in $CH_2{=}CF_2$ are chemically equivalent, but not magnetically equivalent. A good discussion of the subtleties of spin-system classification is A. Ault, *J. Chem. Educ*., **47**, 812 (1970).

In general, NMR spectra are sufficiently complex that each case requires an individual analysis. Under certain conditions, however, the spectrum is simple enough to analyze without extensive calculations. One such case is a molecule containing n chemically equivalent nuclear spins and no other nuclei with nonzero spin; here, the spin–spin couplings give no splitting of the NMR line, even if the coupling constants are not all equal. Thus the NMR spectrum of C_6H_6 consists of a single line, even though there are three different coupling constants: J_{ortho}, J_{meta}, and J_{para}. Other examples are $Si(CH_3)_4$, $CH_2{=}CH_2$, and CH_4.

The proof that A_n compounds give a single NMR line is as follows. The NMR Hamiltonian for such compounds is

$$\hat{H} = -h\nu_0(1-\sigma_A)\hbar^{-1}\sum_{i=1}^{n}\hat{I}_{zA_i} + h\hbar^{-2}\sum_{j=1}^{n}\sum_{i>j}J_{ij}\hat{\mathbf{I}}_{A_i}\cdot\hat{\mathbf{I}}_{A_j}$$

$$\hat{H} = \hat{H}^0 + \hat{H}' \tag{8.73}$$

where the definitions of $\hat{H}^0$ and $\hat{H}'$ are clear. The total z component $\hat{T}_z$ of nuclear-spin angular momentum is

$$\hat{T}_z = \sum_{i=1}^{n}\hat{I}_{zA_i} \tag{8.74}$$

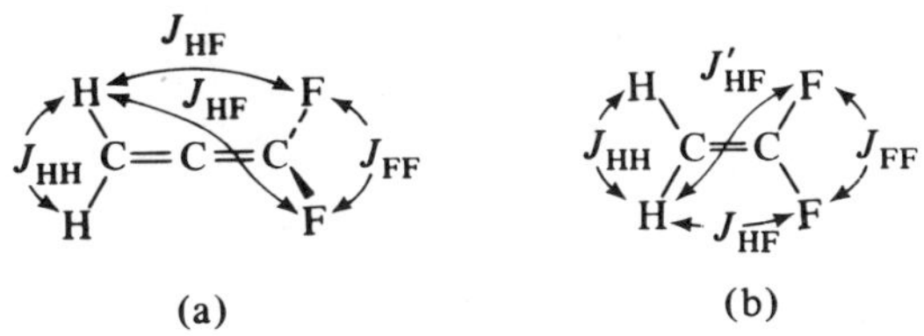

Fig. 8.8 Examples of the A_2X_2 and $AA'XX'$ cases.

$\hat{T}_z$ commutes with $\hat{H}$ (Problem 8.10), so that the nuclear-spin eigenfuctions ψ can be chosen as eigenfunctions of $\hat{T}_z$ and therefore of $\hat{H}^0$, which is proportional to $\hat{T}_z$. Since the ψ's are eigenfunctions of both $\hat{H}$ and $\hat{H}^0$, they must also be eigenfunctions of $\hat{H}'$, which equals $\hat{H} - \hat{H}^0$. Let

$$\hat{H}^0\psi = E^0\psi, \qquad \hat{H}'\psi = E'\psi \tag{8.75}$$

$$\hat{H}\psi = (E^0 + E')\psi = E\psi \tag{8.76}$$

For an rf field along the x axis, the transition probability between states k and l involves the square of the integral

$$\langle k|\sum_i \hat{I}_{xA_i}|l\rangle = \langle k|\hat{T}_x|l\rangle \tag{8.77}$$

Since $\hat{T}_x$ commutes with $\hat{H}'$ (Problem 8.10), the matrix element (8.77) vanishes unless $E'_k = E'_l$ [Eq. (1.51)]. Hence the frequency for an allowed transition is $(E_k^0 - E_l^0)/h$ and does not involve the coupling constants J_{ij}, which occur in E'. The selection rule (8.31) that only one nuclear spin change by ± 1 then gives the spectrum as a single line of frequency $\nu_0(1 - \sigma_A)$.

For a molecule containing several groups of chemically equivalent nuclei, the spectrum is simple provided two conditions are met: (1) Between any two given sets of chemically equivalent nuclei, there is only one coupling constant; that is, all chemically equivalent nuclei are also magnetically equivalent. (2) All the coupling constants are small compared to the differences in chemical shifts. When condition (1) is met, it can be shown that the couplings between nuclei belonging to the same set of equivalent nuclei can be ignored in deriving the spectrum (we will prove this at the end of this section); for example, in 1,1-difluoroallene the coupling constants J_{HH} and J_{FF} do not enter into the expressions for the transition frequencies (although they do occur in the energy-level expressions). When condition (2) is met, we can apply first-order perturbation theory to derive the spectrum, rather than using the exact treatment with the secular equation (2.77). For example, consider the A_mX_n spectrum with but one A–X coupling constant. The Hamiltonian is

$$\hat{H} = -h\nu_0(1-\sigma_A)\hbar^{-1}\sum_{i=1}^{m}\hat{I}_{zA_i} - h\nu_0(1-\sigma_X)\hbar^{-1}\sum_{j=1}^{n}\hat{I}_{zX_j}$$

$$+h\hbar^{-2}\sum_{i=1}^{m}\sum_{j>i}J_{A_iA_j}\hat{\mathbf{I}}_{A_i}\cdot\hat{\mathbf{I}}_{A_j} + h\hbar^{-2}\sum_{i=1}^{n}\sum_{j>i}J_{X_iX_j}\hat{\mathbf{I}}_{X_i}\cdot\hat{\mathbf{I}}_{X_j}$$

$$+hJ_{AX}\hbar^{-2}\sum_{i=1}^{m}\sum_{j=1}^{n}\hat{\mathbf{I}}_{A_i}\cdot\hat{\mathbf{I}}_{X_j} \tag{8.78}$$

where $\nu_0 = g_N\beta_p B_0 h^{-1}$. Although the J_{AA} and J_{XX} terms do affect the energy levels, since condition (1) is met, these terms have no effect on transition frequencies and will be ignored. We therefore take $\hat{H}^0$ as the first two terms in (8.78) and $\hat{H}'$ as the last term. According to condition (2), the perturbation $\hat{H}'$ is small. The unperturbed wave functions and energy levels are [(8.28) and (8.29)]

$$\psi^{(0)} = |M_{I,A_1}\rangle\cdots|M_{I,A_m}\rangle|M_{I,X_1}\rangle\cdots|M_{I,X_n}\rangle \tag{8.79}$$

$$E^{(0)} = -h\nu_0(1-\sigma_A)\sum_{i=1}^{m}M_{I,A_i} - h\nu_0(1-\sigma_B)\sum_{j=1}^{n}M_{I,X_j}$$

The perturbation-theory first-order energy correction is (see also Problem 8.35)

$$E^{(1)} = \frac{hJ_{AX}}{\hbar^2}\sum_{i=1}^{m}\sum_{j=1}^{n}\langle\psi^{(0)}|\hat{\mathbf{I}}_{A_i}\cdot\hat{\mathbf{I}}_{X_j}|\psi^{(0)}\rangle$$

Using

$$\hat{\mathbf{I}}_{A_i}\cdot\hat{\mathbf{I}}_{X_j} = \hat{I}_{zA_i}\hat{I}_{zX_j} + \tfrac{1}{2}(\hat{I}_+)_{A_i}(\hat{I}_-)_{X_j} + \tfrac{1}{2}(\hat{I}_-)_{A_i}(\hat{I}_+)_{X_j} \tag{8.80}$$

and the orthonormality of the spin eigenfunctions, we find

$$E^{(1)} = hJ_{AX}\sum_{i=1}^{m}\sum_{j=1}^{n}M_{I,A_i}M_{I,X_j} \tag{8.81}$$

The selection rule (8.31) allows only one nucleus to change its M_I value in a transition, and that change is ± 1. Hence for an NMR transition of one of the A nuclei, the line frequencies (at fixed B_0) are

$$\nu_0(1-\sigma_A) + J_{AX}\sum_{j=1}^{n}M_{I,X_j} \tag{8.82}$$

Suppose the X nuclei have spin $\frac{1}{2}$. The possible values of each quantum number M_{I,X_j} are $\pm\frac{1}{2}$. The sum

$$\sum_{j=1}^{n} M_{I,\mathrm{X}_j} \tag{8.83}$$

takes on $n+1$ different values, which range in integral steps from $+n/2$ to $-n/2$. Thus with $n=3$, the sum (8.83) equals $\frac{3}{2}$ for the X nuclear-spin function $\alpha\alpha\alpha$, equals $\frac{1}{2}$ for $\alpha\alpha\beta$, $\alpha\beta\alpha$, and $\beta\alpha\alpha$, equals $-\frac{1}{2}$ for $\alpha\beta\beta$, $\beta\alpha\beta$, $\beta\beta\alpha$, and equals $-\frac{3}{2}$ for $\beta\beta\beta$. With $n=3$, transitions involving the A nuclei give four equally spaced lines centered about $\nu_0(1-\sigma_\mathrm{A})$, the separation between adjacent lines being J_{AX}. The relative intensities of the lines of this quartet are given by the number of X spin functions giving each of the different values of the sum (8.83); these numbers are, as we have seen, 1, 3, 3, 1. Similar considerations apply for the portion of the NMR spectrum due to transitions of X nuclei; we get a multiplet centered about $\nu_0(1-\sigma_\mathrm{X})$.

For the general case $\mathrm{A}_m\mathrm{X}_n$, the spin function of the X nuclei has n factors, each of which can be either α or β (assuming we have nuclei of spin $\frac{1}{2}$). Since n is fixed, the value of the sum (8.83) is determined by the number of X nuclei having spin α; let this number be k. To determine the relative intensities of the lines of the A nuclei, the number of different ways of assigning spin α to k of the X nuclei must be found. In other words, we want the number of different ways of picking out k objects from a set of n numbered objects—this is by definition ${}_nC_k$, the number of combinations of n things taken k at a time, and is given by[10]

$${}_nC_k = \frac{n!}{k!(n-k)!} \tag{8.84}$$

The numbers (8.84) occur as the coefficients in the binomial expansion:

$$(a+b)^n = \sum_{k=0}^{n} {}_nC_k a^{n-k} b^k \tag{8.85}$$

where $0! \equiv 1$. An easy way to write down the binomial coefficients for a given n is to use Pascal's triangle:

$$\begin{array}{ccccccccccc|c}
 & & & & & & & & & & & n \\
 & & & & & 1 & & & & & & 0 \\
 & & & & 1 & & 1 & & & & & 1 \\
 & & & 1 & & 2 & & 1 & & & & 2 \\
 & & 1 & & 3 & & 3 & & 1 & & & 3 \\
 & 1 & & 4 & & 6 & & 4 & & 1 & & 4 \\
1 & & 5 & & 10 & & 10 & & 5 & & 1 & 5 \\
\cdot & \cdot & \cdot & \cdot & \cdot & \cdot & \cdot & \cdot & \cdot & \cdot & \cdot & \cdots
\end{array} \tag{8.86}$$

[10] *Sokolnikoff and Redheffer*, p. 598.

Each entry is the sum of the two numbers to the right and left of it in the preceding line.

Summarizing, the NMR spectrum of an A_mX_n compound when conditions (1) and (2) are satisfied and when A and X each have spin $\frac{1}{2}$ consists of a set of $n+1$ lines centered about $\nu_0(1-\sigma_A)$ with the spacing J_{AX} between adjacent lines and with relative intensities ${}_nC_k$, $k=0,1,\ldots,n$, and a set of $m+1$ lines centered about $\nu_0(1-\sigma_X)$ with the spacing J_{AX} between adjacent lines and with relative intensities ${}_mC_k$, $k=0,1,\ldots,m$. When conditions (1) and (2) are met, the NMR spectrum is said to be *first order*.

Consider acetaldehyde, CH_3CHO. Figure 8.3 shows σ for the methyl and CHO protons to differ substantially, so that $\nu_0\delta_{AX} \gg J_{AX}$. The low barrier to internal rotation causes condition (1) to be satisfied. Hence the first-order analysis of the preceding paragraphs is applicable. We have an A_3X case and the spectrum consists of a doublet (from the methyl protons) whose lines are of equal intensity and a quartet (from the CHO proton) whose lines have the intensity ratios 1:3:3:1; the doublet and quartet are well separated and show the same splitting (Fig. 8.9).

Frequently a set of equivalent nuclei is coupled to two other sets of nuclei. An example is CH_3CH_2OH; the CH_2 protons are separated by three bonds from both the CH_3 and the OH protons and are therefore spin coupled to both these sets of nuclei. The CH_3 protons and the OH proton are separated by four single bonds from each other and the coupling between them is negligible. The chemical shifts of the three kinds of protons differ substantially from one another. The system is of type A_3M_2X with $J_{AM} \neq 0$, $J_{MX} \neq 0$, and $J_{AX}=0$. The first-order analysis shows that the CH_3 protons give a triplet with spacing J_{AM}, and the OH proton gives a triplet with spacing J_{MX}. The calculation of the CH_2 part of the spectrum can be broken into two steps: The CH_3 protons split the CH_2 line into a quartet with spacing J_{AM}, and then each line of this hypothetical quartet is split by the OH proton into two lines with spacing J_{MX}—see Fig. 8.10. Of course, the two steps in the calculation could have been done in reverse order, but the net result is the same.

The simplified analysis is valid *only* when conditions (1) and (2) are met. Otherwise, the line intensities deviate considerably from the binomial pattern, the line frequencies differ considerably from those of the first-order analysis, and the first-order analysis may even predict the wrong number of NMR lines. (Even for ethanol, which was discussed above as an A_3M_2X system, a good NMR spectrometer will show that there are small but significant deviations from a first-order spectrum.)

For an A_n molecule, the NMR spectrum does not allow determination of the coupling constant(s) between the A nuclei. The values of such coupling constants can be found by using isotopically substituted mole-

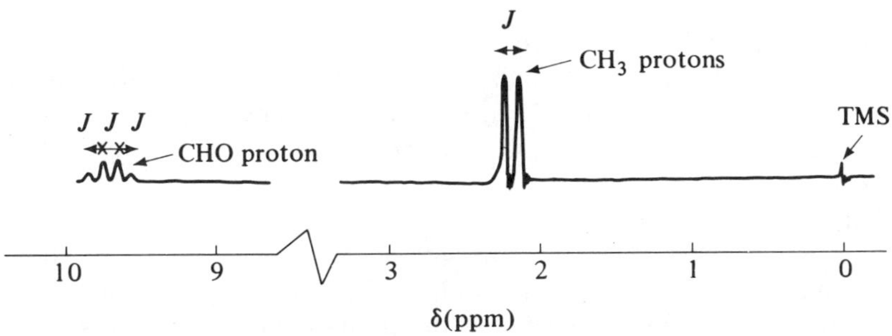

Fig. 8.9 NMR spectrum of CH_3CHO. The total area of the doublet peaks is three times the total area of the quartet peaks. The coupling constant $J=2.9$ Hz, which is substantially less than the usual value 5 to 8 Hz for vicinal coupling constants in compounds with essentially free rotation. The splittings are drawn as equal to about one-tenth part-per-million, which corresponds to a spectrometer frequency of 29 MHz.

cules. One method uses the C^{13} nucleus, whose spin is $\frac{1}{2}$, and whose natural abundance is 1.1 percent. Consider, for example, 1,1,2,2-tetrabromoethane. The predominant species $C^{12}HBr_2$—$C^{12}HBr_2$ gives the single peak expected for an A_2 spectrum. The species $C^{12}HBr_2$—$C^{13}HBr_2$ (present in about 2.2 percent natural abundance) is an A_2X system with X being the C^{13} nucleus; there are two different A–X coupling constants in this species (we might use the designation AA′X), so that condition (1) is not satisfied, and the spectrum depends on the value of $J_{AA'}$, the proton–proton coupling constant. Observation of the proton NMR spectrum of the mono-C^{13} species showed the proton–proton coupling constant to be 2.9 Hz in this molecule.

When the A_n protons are bonded to the same carbon atom, use of C^{13} satellites does not yield the proton–proton coupling constant(s). Here, one

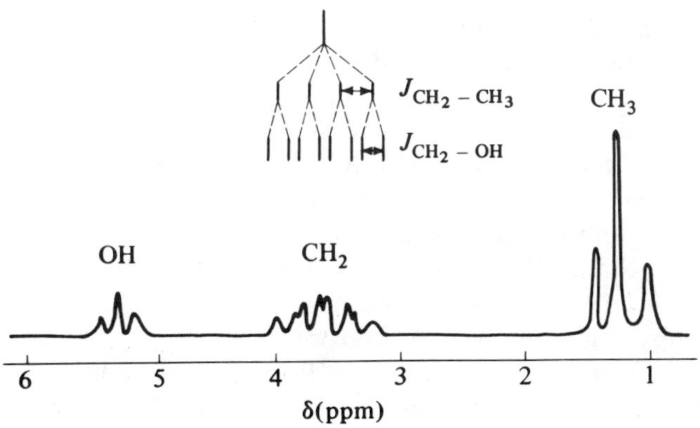

Fig. 8.10 High-resolution proton NMR spectrum of pure liquid CH_3CH_2OH.

can prepare partially deuterated samples for NMR analysis. For example, CH_3D is an A_3X case (although the D atom is in essentially the same chemical environment as the H atoms, the difference in g_N values makes this an A_3X case), and J_{H-D} can be found from the splittings in the proton NMR spectrum. Then J_{H-H} is calculated as $J_{H-D}g_H/g_D$, with the result quoted earlier that $|J_{H-H}| = 12$ Hz.

The NMR spectrum of a system with several nuclei can be so complex as to make analysis nearly impossible. When the coupling constants are small compared to the chemical-shift differences, the spectrum is simple, as just discussed. Chemical shifts are proportional to the applied field B_0 [(8.32)], whereas spin–spin splittings are independent of B_0 [Eq. (8.40)]. Hence the use of a spectrometer with a high value of B_0 (and a correspondingly high value of $\nu_{\text{spectrometer}}$) can frequently simplify a complicated spectrum. In addition to 60-MHz proton NMR spectrometers, 100-, 220-, and 300-MHz proton instruments are commercially available. The principal difficulty is to get a very strong, very homogeneous magnetic field; an electromagnet whose wires are cooled to liquid-helium temperatures to make the wire a superconductor is used in the 220- and 300-MHz instruments.

A second method for simplifying NMR spectra is *double resonance*. In a double-resonance experiment, the system is simultaneously exposed to electromagnetic radiation of two different frequencies. For example, an AX system can be exposed to an rf magnetic field B_1 at the frequency of the A resonance, while an rf field B_2 is simultaneously applied at the frequency of the X resonance; for large amplitudes of B_2, one finds that the A doublet collapses to a single line—the A and X spins are said to be *decoupled*. When B_2 is strong, but not strong enough to cause decoupling, the integrated intensity of the A resonance is affected. This is the *nuclear Overhauser effect* (NOE); its magnitude depends on the distance between the A and X nuclei, and therefore the NOE can be used to deduce molecular configurations and conformations. (The field B_2 changes the energy-level population distribution of the X spins, and the magnetic dipole–magnetic dipole interaction between X and A then leads to a change in A populations and therefore a change in the A absorption intensity.) For more on double resonance and the NOE, see *Becker*, Chapter 8; J. H. Noggle and R. E. Schirmer, *The Nuclear Overhauser Effect*, Academic Press, 1971. (Double resonance is useful in other branches of spectroscopy, e.g., microwave spectroscopy.)

A third method for simplifying proton NMR spectra is deuterium substitution. We see from Table A.6 that g_D/g_H is about $\frac{1}{6}$; hence an H—D spin–spin coupling is about one-sixth the corresponding H—H coupling constant, so that partial deuteration of a compound allows condition (2) to be satisfied. Moreover, the deuterium resonances do not appear in the proton portion of the spectrum.

The solution of the NMR secular equation is simplified by the usual method (Section 1.10) of using basis functions that are eigenfunctions of operators that commute with the Hamiltonian. The operator $\hat{T}_z$ for the total z component of nuclear spin is

$$\hat{T}_z = \sum_{i=1}^{N} \hat{I}_{zi} \tag{8.87}$$

where we have N nuclear spins. One finds (Problem 8.10) that $\hat{T}_z$ commutes with the NMR Hamiltonian, so that matrix elements between basis functions with different eigenvalues of $\hat{T}_z$ vanish. Thus, the functions of (8.44) are eigenfunctions of $\hat{I}_{z1} + \hat{I}_{z2}$ with eigenvalues $2\hbar$, 0, 0, and $-2\hbar$; the only nonzero off-diagonal matrix elements are thus between φ_2 and φ_3, as found previously [Eq. (8.61)]. For molecules with symmetry, group theory can be used to simplify the NMR secular equation.[11]

To aid in the analysis of NMR spectra, computer programs have been developed which solve the nuclear-spin secular equation.[12] The input to these programs is the observed spectral data and initial guesses for the chemical shifts and coupling constants; by an iterative process, accurate parameters are calculated. Programs are available from the Quantum Chemistry Program Exchange (Section 1.21).

We conclude this section by outlining the proof that couplings within each set of equivalent nuclei can be ignored in deriving the spectrum of a compound with but one coupling constant between each pair of different sets of equivalent nuclei. We shall use the Hamiltonian (8.78), which is for an A_mX_n molecule, but the argument does not depend on condition (2) being satisfied and applies to the A_mB_n case as well. We abbreviate the five terms on the right of (8.78) as follows:

$$\hat{H} = \hat{H}^0(\mathrm{A}) + \hat{H}^0(\mathrm{X}) + \hat{H}'(\mathrm{A}) + \hat{H}'(\mathrm{X}) + \hat{H}'(\mathrm{AX}) \tag{8.88}$$

We can show (Problem 8.11) that $\hat{H}'(\mathrm{A})$ commutes with each of the five terms of $\hat{H}$ and therefore commutes with $\hat{H}$. Hence the eigenfunctions of $\hat{H}$ are also eigenfunctions of $\hat{H}'(\mathrm{A})$ with eigenvalues $E'(\mathrm{A})$. For an applied rf field in the x direction, the transition probability between states k and l is proportional to the square of the following integral involving the

[11] E. B. Wilson, Jr., *J. Chem. Phys.*, **27**, 60 (1957).

[12] See the article by J. D. Swalen in *Progress in Nuclear Magnetic Resonance Spectroscopy*, vol. 1, J. W. Emsley, J. Feeney, and L. H. Sutcliffe, eds., Pergamon Press, London, 1966, pp. 205–250.

operator for the x component of the total nuclear-spin magnetic moment:

$$\left\langle k \left| \sum_{i=1}^{m} \hat{\mu}_{x,\mathrm{A}_i} + \sum_{j=1}^{n} \hat{\mu}_{x,\mathrm{X}_j} \right| l \right\rangle = \beta_p \hbar^{-1} \langle k | g_\mathrm{A} \hat{T}_x(\mathrm{A}) + g_\mathrm{X} \hat{T}_x(\mathrm{X}) | l \rangle \quad (8.89)$$

where

$$\hat{T}_x(\mathrm{A}) = \sum_{i=1}^{m} \hat{I}_{x\mathrm{A}_i} \qquad \hat{T}_x(\mathrm{X}) = \sum_{j=1}^{n} \hat{I}_{x\mathrm{X}_j} \quad (8.90)$$

$\hat{H}'(\mathrm{A})$ commutes with $\hat{T}_x(\mathrm{A})$ and with $\hat{T}_x(\mathrm{X})$ (Problem 8.10), so that the transition-moment integral vanishes unless the states k and l have the same eigenvalue $E'(\mathrm{A})$ [Eq. (1.51)]. Thus the coupling constants $J_{\mathrm{A}_i\mathrm{A}_j}$, which occur in $E'(\mathrm{A})$, do not appear in the expression for the transition frequencies. The same argument shows the transition frequencies to be independent of $J_{\mathrm{X}_i\mathrm{X}_j}$. Q.E.D.

8.6 NMR SPECTRA OF NUCLEI OTHER THAN PROTONS

Although the proton is the most extensively studied nucleus, NMR spectra of other nuclei have been investigated.

C^{13} NMR spectra provide information on the skeleton of organic compounds. The low isotopic abundance (1.1 percent) and the rather small magnetic dipole moment of C^{13} make the spectra difficult to study with a conventional spectrometer. However, FT NMR (Section 8.3) makes natural-abundance C^{13} studies relatively easy to do. The availability of FT NMR spectrometers has led to a rapid growth of C^{13} work.

Since the probability of having two C^{13} nuclei in the same molecule is negligible, the natural-abundance C^{13} NMR spectrum of a compound is the superposition of the spectra of species with one carbon substituted by C^{13}; C^{13}–C^{13} spin–spin splitting is not observed. The observed range of C^{13} shifts is 600 ppm. Commonly used reference compounds include carbon disulfide and benzene. The convention used to define δ for C^{13} compounds is opposite to that in (8.35) for protons; that is, $\delta_i = (\sigma_i - \sigma_{\mathrm{ref}}) \times 10^6$ ppm for C^{13} nuclei. However, Stothers has proposed that TMS be used as the C^{13} reference compound, and that δ_C be redefined in accord with (8.35). Spin–spin splittings are due to proton–C^{13} or F^{19}–C^{13} coupling. The coupling constants between protons and C^{13} nuclei not directly bonded to each other are usually negligible. To simplify the spectrum, one usually eliminates proton–C^{13} spin–spin splitting by using double resonance to decouple the proton spins. Thus the C^{13} spectrum usually consists of singlet lines due to the various nonequivalent carbons.

Applications of C^{13} NMR have included studies of molecular structure and stereochemistry, reaction mechanisms, and biochemical pathways; amino acids, proteins, and nucleotides have been studied. For more on C^{13} NMR, see J. B. Stothers, *Carbon-13 NMR Spectroscopy*, Academic Press, 1972; G. C. Levy and G. L. Nelson, *Carbon-13 Nuclear Magnetic Resonance for Organic Chemists*, Wiley-Interscience, 1972.

F^{19} NMR spectra have been extensively studied. The observed range of chemical shifts is 1000 ppm. F^{19}–F^{19} coupling constants range up to 400 Hz, while F^{19}–H^1 coupling constants range up to 100 Hz; these couplings are generally transmitted through more bonds than proton–proton couplings. An application of F^{19} spectroscopy is to SF_4, whose spectrum shows two well-separated triplets of equal intensity, indicating an A_2X_2 spectrum. This rules out such symmetries as $\mathcal{T}_d$, $\mathcal{D}_{2d}$, and $\mathcal{D}_{4h}$, each of which would give a single-line A_4 spectrum. Subsequent microwave work proved the geometry of SF_4 to be $\mathcal{C}_{2v}$.

A number of B^{11} ($I=\frac{3}{2}$) NMR studies have been made. Because of the B^{11} electric quadrupole moment (Section 8.10), the lines are broad and small coupling constants are not observed. An application of B^{11} NMR is to boron hydrides. (*Emsley, Feeney, and Sutcliffe*, Section 12.1.3.)

Other nuclei studied by NMR spectroscopy include Si^{29}, P^{31}, N^{14}, and N^{15}.

8.7 ORIGIN OF NMR SPIN–SPIN COUPLING

Spin–spin coupling is observed in the NMR spectra of both solids and liquids, but its origin is quite different for the two phases. Solids will be considered first, using a qualitative, semiclassical approach, rather than the correct quantum-mechanical treatment.

Suppose we have a single crystal of HF in an applied magnetic field B_0, which points in the z direction (Fig. 8.11). Consider first one particular HF molecule in the crystal. The F^{19} and H^1 nuclei each have spin $\frac{1}{2}$ and their magnetic moments precess about the direction of B_0 (Section 1.17). The time averages of μ_x and μ_y are zero for each magnetic moment, and we need only consider the interaction between the z components of the two magnetic moments. The z component μ_{zF} of the F^{19} moment produces a certain magnetic field at the proton. Consider a magnetic dipole moment $\boldsymbol{\mu}$ located at the origin and pointing along the z axis, and let $\mathbf{r}$ be the vector from the origin to point (x,y,z). The magnetic field $\mathbf{B(r)}$ produced by $\boldsymbol{\mu}$ is[13]

$$\mathbf{B}(\mathbf{r}) = -\frac{\boldsymbol{\mu}}{r^3} + \frac{3(\mathbf{r}\cdot\boldsymbol{\mu})\mathbf{r}}{r^5} \tag{8.91}$$

[13] R. P. Feynman *et al.*, *The Feynman Lectures on Physics*, vol. II, Addison-Wesley, 1964, Eqs. (14.35) and (14.36).

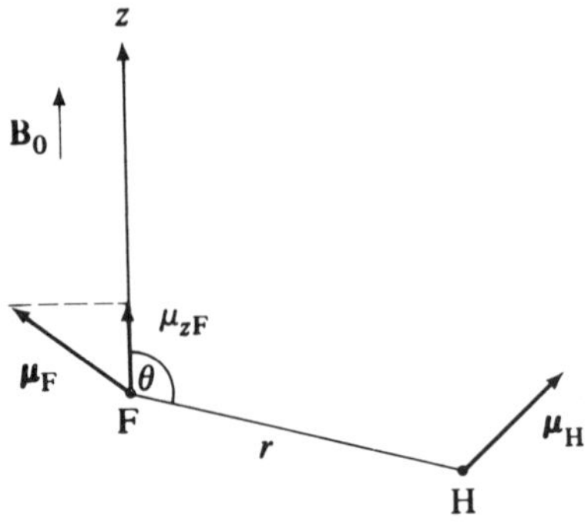

Fig. 8.11 Interaction between proton and F^{19} magnetic moments in an HF crystal in a magnetic field $\mathbf{B}_0$.

The z component of (8.91) shows that at the proton, the z component of the magnetic field produced by μ_{zF} is $\mu_{zF}(3\cos^2\theta - 1)/r^3$, where r is the proton–F^{19} internuclear distance and θ is the angle the line joining H^1 and F^{19} makes with the z axis.

From (8.4) the possible values of μ_{zF} are $\pm\frac{1}{2}g_F\beta_p$, so that the magnetic field felt by the proton is

$$B_0(1-\sigma_H) \pm \frac{g_F\beta_p}{2r^3}(3\cos^2\theta - 1) \tag{8.92}$$

where σ_H is the proton's shielding constant. (For the purposes of our qualitative treatment, we ignore the fact that $\boldsymbol{\sigma}_H$ is a tensor.) Typically, $B_0 = 2\times10^4$ gauss and $\sigma_H = 2\times10^{-5}$, so that $B_0\sigma_H = 0.4$ gauss; this may be compared to $g_F\beta_p/2r^3 = 13$ gauss at $r = 1$ Å. The spin–spin couplings in solids are much greater than the chemical shifts, which is just the reverse of the situation in liquids.

An expression similar to (8.92) can be written for the magnetic field at the F^{19} nucleus.

The magnetic fields at H^1 and F^{19} interact with μ_{zH} and μ_{zF} to give four energy levels corresponding to the following four possibilities for $M_{I,H}$ and $M_{I,F}$:

$$\tfrac{1}{2},\tfrac{1}{2} \qquad \tfrac{1}{2},-\tfrac{1}{2} \qquad -\tfrac{1}{2},\tfrac{1}{2} \qquad -\tfrac{1}{2},-\tfrac{1}{2}$$

The selection rule (8.31) allows only one spin to change in a transition. Hence the proton portion of the NMR spectrum consists of a doublet due to the transitions

$$\tfrac{1}{2},\tfrac{1}{2}\rightarrow-\tfrac{1}{2},\tfrac{1}{2} \quad \text{and} \quad \tfrac{1}{2},-\tfrac{1}{2}\rightarrow-\tfrac{1}{2},-\tfrac{1}{2} \tag{8.93}$$

where the proton spin is given first in each pair. The center of the doublet is at $\nu_{0,\mathrm{H}}(1-\sigma_{\mathrm{H}})$, where $\nu_{0,\mathrm{H}}=g_p\beta_p B_0/h$. The spacing of the doublet is

$$g_p g_{\mathrm{F}}\beta_p^2(3\cos^2\theta-1)/hr^3 \tag{8.94}$$

The F^{19} portion of the spectrum shows a similar doublet. [Equation (8.94) is correct for two dissimilar nuclei, but a quantum-mechanical treatment shows that for two identical spin-$\frac{1}{2}$ nuclei in a solid, (8.94) should be multiplied by $\frac{3}{2}$.]

Up to here, we have considered only the interaction between a proton and the F^{19} nucleus bonded to it. There are also weaker interactions between a given proton and more-distant nuclei in the crystal. These interactions superimpose doublets with smaller splittings on the main proton doublet. The net effect is a doublet whose lines are greatly broadened. For this reason, NMR studies of solids are called *broad-line* NMR. Because of this broadening and the smallness of chemical-shift differences as compared to spin–spin splittings, different resonance lines due to chemically nonequivalent protons are not observed in NMR spectroscopy of solids (unless special techniques are used).

The doublet spacing (8.94) depends on the magnitude and orientation of the internuclear distance. A study of the variation of the spacing with change of orientation of the single crystal allows the calculation of r and θ. If one is dealing with a powdered sample instead of a single crystal, the NMR resonance is further broadened by the superposition of signals corresponding to all possible values of θ in (8.94); here, only bond distances and not bond orientations can be determined. Since proton positions are difficult to determine by x-ray diffraction of crystals, broad-line NMR can aid in the determination of crystal structures.

Now consider NMR spin–spin splitting in liquids and gases. In these states, the molecules are undergoing rapid rotation, and the magnetic field experienced by one nucleus due to the magnetic dipole moment of another nucleus must be replaced by an average over all possible values of θ in (8.92) (see Section 8.9). We have

$$\langle\cos^2\theta\rangle=\frac{\int_0^\pi\cos^2\theta\sin\theta\,d\theta\int_0^{2\pi}d\varphi}{\int_0^\pi\sin\theta\,d\theta\int_0^{2\pi}d\varphi}=\frac{1}{3} \tag{8.95}$$

Hence $\langle 3\cos^2\theta-1\rangle=0$, and the direct dipole–dipole coupling between two magnetic nuclei vanishes for liquids and gases. Thus the NMR lines of liquids and gases are narrow, allowing chemical shifts to be observed; we have high-resolution NMR.

Although the direct dipole–dipole interaction vanishes for liquids and gases, their NMR spectra *do* exhibit spin–spin splittings. The origin of these splittings was explained by Ramsey and Purcell in 1952; they showed these splittings to result from an indirect coupling of the nuclear spins, transmitted through the bonding electrons. Consider, for example, HF. The two bonding electrons must have opposite spins since they are in the same MO. Suppose the H nucleus has $M_I = +\frac{1}{2}$. The interaction between the electronic and nuclear spin magnetic moments makes it energetically more favorable for the electron near the H nucleus to have $m_s = -\frac{1}{2}$. The other electron of the bond will thus tend to have $m_s = +\frac{1}{2}$, and its spin magnetic moment will interact with the F spin magnetic moment, giving the observed spin–spin coupling. This indirect electron-coupled spin–spin interaction of the nuclei is small, as compared to the direct dipole–dipole interaction in solids, because the amount of electronic spin polarization is slight.

Further evidence for this mechanism of spin–spin coupling in liquids is the fact that the coupling dies off as the number of chemical bonds separating the two nuclei increases. Moreover, the fact that couplings between *trans* protons in olefins are greater than between *cis* protons is incompatible with a direct interaction between the two nuclear magnetic dipoles.

Using perturbation theory, one can obtain a theoretical expression (involving the usual sum over states) for the indirect spin–spin coupling constant J_{12}. Alternatively, one can directly calculate wave functions with the perturbation operator included in the Hamiltonian. *Ab initio* calculations of J_{12} are limited to quite small molecules. Many INDO calculations of coupling constants have been done. INDO correctly reproduces many observed trends in coupling constants, but quantitative agreement with experiment is only fair. Some J_{HH} values (in Hz) calculated by Pople, McIver, and Ostlund using INDO are (experimental values in parentheses) CH_4, -6.1 (-12.4); C_2H_4 (*cis*), 9.3 (11.7); C_2H_4 (*trans*), 25.2 (19.1); C_6H_6 (*ortho*), 8.2 (7.5); H_2CO, 31.9 (40.2). For details, see *Pople and Beveridge*, pp. 149–159; *Murrell and Harget*, pp. 128–134; *Emsley, Feeney, and Sutcliffe*, Chapter 5.

As mentioned earlier, NMR spectra ordinarily yield only relative and not absolute signs for coupling constants. Karplus pointed out in 1959 that theoretical calculations predict that C^{13}–H coupling constants for C^{13} and H^1 nuclei bonded to each other are almost certainly positive; this has been experimentally confirmed by an NMR study of $C^{13}H_3F$ oriented in a liquid-crystal solvent. [T. R. Krugh and R. A. Bernheim, *J. Chem. Phys.*, **52**, 4942 (1970).] One can thus determine absolute signs of proton–proton coupling constants by experimentally determining the relative signs of J_{H-H} and $J_{H-C^{13}}$ in the same molecule. This has been done for many molecules, with the results quoted earlier.

8.8 SPIN RELAXATION

Suppose we have molecules with spin-$\frac{1}{2}$ nuclei. In the absence of an applied magnetic field, spin-α and spin-β nuclei have equal energies, and therefore, according to the Boltzmann distribution law (3.51), the number N_α of nuclei with spin α equals the number N_β with spin β. Now we apply a uniform external magnetic field B_0 in the z direction. The instant after the field is turned on, we have equal populations; however, in the presence of B_0, the α spins have a lower energy [Equation (8.9)] than the β spins, so we have a nonequilibrium population distribution. The question then follows: How is the equilibrium population distribution with $N_\alpha > N_\beta$ achieved? One possibility might be for spontaneous emission of radiation from the β spin states. The Einstein spontaneous emission coefficient A_{mn} in (3.55) is proportional to ν_{mn}^3. At optical frequencies ($\sim 10^{14}$ Hz), spontaneous emission leads to lifetimes of $\sim 10^{-8}$ sec (except for metastable states). NMR energy differences are small, and ν_{mn} here is $\sim 10^8$ Hz for a typical B_0. Hence if spontaneous emission were the only process occurring, it would take thousands of years for the β spin population to be depleted to near its equilibrium value. There must be some other process that allows the excess excited β spins to "relax" to the ground state. In actuality, what happens is that the interactions between the nuclear spins and their surroundings allow for a nonradiative energy transfer to occur between the spins and their surroundings. Relaxation processes were originally studied in solids, where the nuclear spins transferred energy to the crystal lattice. Hence the term *spin–lattice relaxation* is applied to any relaxation process involving nonradiative energy transfer between spins and surroundings (the "lattice"); we are mainly interested in spin–lattice relaxation in liquids.

The magnetic dipole moment of each nucleus experiences fluctuating magnetic fields from one instant to the next, due to the magnetic moments of other nuclei (in the same molecule and in other molecules). This interaction allows energy transfer from nuclear spins to translational and rotational energy of the molecules, and vice versa. Let $W_{\alpha\beta}$ and $W_{\beta\alpha}$ be the spin–lattice relaxation transition probabilities per second for upward and downward radiationless transitions, respectively. Let $N_{\alpha,\text{eq}}$ and $N_{\beta,\text{eq}}$ be the equilibrium populations. At equilibrium, upward and downward transitions occur at equal rates, and therefore we have $N_{\alpha,\text{eq}} W_{\alpha\beta} = N_{\beta,\text{eq}} W_{\beta\alpha}$. (We ignore spontaneous emission, which has a negligible probability.) Since the equilibrium populations of the α and β levels are unequal, the two W's must be unequal. Because NMR energy-level differences are small, $N_{\beta,\text{eq}}$ is only slightly less than $N_{\alpha,\text{eq}}$, and the two W's are nearly equal. From the definition of the W's, we have

$$dN_{\beta}/dt = -N_{\beta}W_{\beta\alpha} + N_{\alpha}W_{\alpha\beta} \tag{8.96}$$

Let $N = N_{\alpha} + N_{\beta}$ be the total number of spins, and let n be the population difference: $n = N_{\alpha} - N_{\beta}$. Adding and subtracting these last two equations, we get

$$N_{\alpha} = \tfrac{1}{2}(N+n), \qquad N_{\beta} = \tfrac{1}{2}(N-n) \tag{8.97}$$

From (8.97), we have $dN_{\beta}/dt = -\frac{1}{2}dn/dt$; substitution of (8.97) into (8.96) gives after multiplication by -2

$$dn/dt = -n(W_{\alpha\beta} + W_{\beta\alpha}) + N(W_{\beta\alpha} - W_{\alpha\beta}) \tag{8.98}$$

When population equilibrium is reached, we have $dn/dt = 0$; hence (8.98) gives for n_{eq}, the population difference at equilibrium,

$$n_{eq} = N(W_{\beta\alpha} - W_{\alpha\beta})/(W_{\alpha\beta} + W_{\beta\alpha})$$

Equation (8.98) becomes

$$dn/dt = -(n - n_{eq})(W_{\alpha\beta} + W_{\beta\alpha}) \tag{8.99}$$

Integrating (8.99) and using the boundary condition that the initial population difference at $t = 0$ was n_0 to evaluate the integration constant, we get

$$n - n_{eq} = (n_0 - n_{eq})e^{-(W_{\alpha\beta} + W_{\beta\alpha})t} \tag{8.100}$$

After a time $T_1 \equiv 1/(W_{\alpha\beta} + W_{\beta\alpha}) \approx 1/2W_{\alpha\beta} \approx 1/2W_{\beta\alpha}$, the population-difference deviation from its final equilibrium value, $n - n_{eq}$, has fallen to $1/e$ of its initial deviation. The time T_1 is called the *spin–lattice relaxation time*; T_1 is a measure of how fast the α and β populations approach equilibrium. In liquids, T_1 ranges from 10^{-2} to 10^2 sec. The quantity $1/W_{\alpha\beta}$ is the average lifetime τ_{α} of state α (Section 3.2), and $1/W_{\beta\alpha} = \tau_{\beta}$. Thus T_1 is related to the lifetimes of states α and β; hence by the energy–time uncertainty principle we expect the half width $\Delta\nu$ of NMR lines to be inversely proportional to T_1. In particular, $\Delta E T_1 = h\Delta\nu T_1 \approx \hbar$, so we expect $\Delta\nu \approx 1/2\pi T_1$. However, other factors may contribute to line width besides spin–lattice relaxation. In particular, the fluctuating magnetic fields experienced by the nuclear spins from the surrounding nuclear spins can (in addition to causing radiationless transitions) vary the energies of the α and β spins [Equation (8.9)] and therefore vary their energy difference, giving further line broadening. A theoretical treatment shows that NMR lines have a Lorentz shape [Equation (3.86)]. Letting $\Delta\nu$ be the observed NMR resonance line half width at half maximum (for a perfectly homogeneous applied field B_0), we define the *spin–spin relaxation time* T_2 by $T_2 = 1/(2\pi\Delta\nu)$. In general, the observed width $\Delta\nu$ contains a contribution from spin–lattice radiationless transitions between the α and β states (T_1), plus a contribution due to shifting of the α and β energy levels. However, in nonviscous liquids, and in gases, the rapid rotation of the molecules averages the fluctuating magnetic fields to zero, so only T_1 contributes to the width; in this case we have $T_2 = T_1$. In viscous liquids (e.g., polymers) and in solids, the fluctuating fields contribute to the width, and T_2 is less than T_1; the lines are broader than in nonviscous liquids.

In the above treatment we assumed the absence of an applied rf field B_1. If B_1 is present (as it is when we observe the spectrum), it causes stimulated absorption and emission, and this modifies the above equations. We can, however, keep B_1 quite small, so that the stimulated absorption and emission probabilities are small with respect to the spin–lattice transition probabilities; in this case, the preceding equations apply. If we make B_1 large enough so that the stimulated absorption and emission rates are of comparable magnitude with the spin–lattice relaxation rate, this affects the observed resonance line as follows. From Section 3.2, B_1 causes a net stimulated absorption proportional to the population difference n. If spin–lattice relaxation is not fast enough (as compared to the stimulated transition rate) to maintain a Boltzmann distribution, B_1 will deplete the population of the lower level significantly below its equilibrium value; this diminishes n and therefore diminishes the intensity of the NMR resonance line; moreover, the high rate of stimulated absorption and emission shortens the lifetimes of the states and broadens the lines beyond the $1/2\pi T_2$ value. This weakening and broadening of the signal is called *saturation*. (See Problem 8.40.)

There are several ways to measure T_1 and T_2. If T_1 is greater than, say, 10 sec, one can place the sample in the B_0 field of the spectrometer and immediately make a series of rapid scans of a resonance line, keeping B_1 small. The absorption signal is proportional to n, so the approach of n to n_{eq} can be followed, and T_1 obtained from (8.100). If T_2 is less than, say, $\frac{1}{2}$ sec, one can obtain T_2 directly from the measured line width. However, if T_2 is greater than $\frac{1}{2}$ sec, then the true line width is quite narrow, and in fact is narrower than the width produced by inhomogeneities in the applied field B_0. For T_1 and T_2 not in the above ranges, one uses various pulse techniques to measure them (*Becker*, Section 9.7).

NMR proton relaxation times for water in living tissues are found to be shorter than in ordinary water, and the lines correspondingly broader. Some workers have interpreted this fact as indicating that water in living cells may be in a semicrystalline state, rather than a liquid state; the semicrystallinity is hypothesized to result from binding of water onto proteins. However, Allerhand and Trull believe this conclusion is premature, and state that the shortening of relaxation times indicates that "some tissue water molecules are less mobile than in pure water, as one would expect." See A. Allerhand and E. A. Trull, *Ann. Rev. Phys. Chem.*, **21**, 317 (1970); this article reviews many biological applications of NMR.

8.9 NMR AND RATE PROCESSES

In NMR experiments one frequently deals with a nucleus moving among different environments and thus subject to rapidly varying magnetic fields. Examples already mentioned are nuclei in a sample tube spinning in a slightly inhomogeneous magnetic field, nuclei in liquids and gases (which are subject to time-varying fields due to a varying magnetic dipole–dipole interaction), and nuclei undergoing internal rotation. The treatment of this situation is involved and will not be given here. Instead, we shall summarize the conclusions of the treatment.[14]

For simplicity, consider a single nucleus moving back and forth between two different environments. Let these environments have magnetic fields B_1 and B_2 and let ν_1 and ν_2 be the corresponding nuclear resonance frequencies. Let ν_{exch} be the frequency at which the nucleus is being exchanged between the two environments. There are two limiting cases. For very slow exchange, $\nu_{exch} \ll |\nu_2 - \nu_1|$, the NMR spectrum has two resonance lines, one at ν_1 and one at ν_2. For very fast exchange, $\nu_{exch} \gg |\nu_2 - \nu_1|$, the NMR spectrum has a single line at frequency $\frac{1}{2}(\nu_1 + \nu_2)$. Figure 8.12 shows the transition between these two extremes.

NMR can thus be used to study rate processes whose frequencies are of the same order of magnitude as the separation between the corresponding NMR line frequencies; this corresponds to species with lifetimes in the range 1 to 10^{-3} sec. An illustration is the F^{19} NMR spectrum of SF_4. At $-100°C$, the spectrum consists of two triplets, as expected for a molecule with two different kinds of F atoms (Section 8.6). At room temperature, however, only a single peak is obtained, indicating that rapid exchange of the fluorine atoms is occurring.

NMR is applicable to studying internal rotation in certain molecules. Consider N,N-dimethylnitrosamine, $(CH_3)_2NNO$. The molecule is expected to be planar (except for the hydrogen atoms) due to a contribution from a resonance structure with a nitrogen–nitrogen double bond. We also expect the barrier to internal rotation about the nitrogen–nitrogen bond to be considerably higher than for most single bonds. At room temperature,

[14] See *Carrington and McLachlan*, Chapter 12.

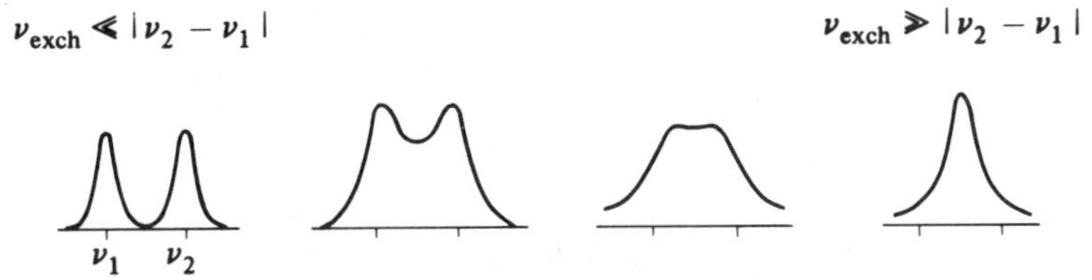

Fig. 8.12 NMR line shapes for a nucleus undergoing exchange between two different environments that are equally populated. The exchange frequency ν_{exch} increases from left to right.

the proton NMR spectrum of this molecule shows two well-separated lines, due to the *cis* and *trans* methyl protons. (Spin–spin coupling between *cis* and *trans* protons is negligible, and the quadrupole moment of the N^{14} nucleus eliminates N^{14}–proton coupling.) The *cis* and *trans* protons can be exchanged by internal rotation about the nitrogen–nitrogen bond:

The rate of this exchange increases with increasing temperature, and as the temperature of the compound is raised, the two separate NMR peaks gradually broaden and ultimately merge at 180°C to form a single peak. Study of this temperature-dependent change yields a rotational barrier of 23 kcal/mole.

For most substituted ethanes, which have a low barrier to internal rotation (~3 kcal/mole), the rate of internal rotation is sufficiently fast at room temperature (and well below room temperature) to give only a single peak for the nuclei exchanged by the internal rotation.

Exchange of nuclei between different environments affects spin–spin couplings. An example is ethanol. When a small amount of aqueous HCl is added to pure CH_3CH_2OH, the OH triplet becomes a single line and the CH_2 multiplet becomes a quartet. This removal of the coupling between OH and CH_2 protons results from the rapid, acid-catalyzed exchange of OH protons between different ethanol molecules.

8.10 NUCLEAR QUADRUPOLE MOMENTS

Nuclei with $I \geqslant 1$ have an electric quadrupole moment eQ (Section 5.8). As mentioned in Section 5.8, the presence of a nuclear quadrupole moment couples the nuclear spin angular momentum and the molecular rotational angular momentum. This coupling has important consequences for the NMR spectrum. Consider an intramolecular spin–spin interaction between a proton and a nucleus, A, that has a quadrupole moment. If the quadrupole coupling between $\mathbf{I}_A$ and the molecular rotation is strong, the nuclear spin of A tends to follow the molecular rotation, and $M_{I,A}$ fluctuates rapidly among all its possible values (from $+I_A$ to $-I_A$). The effect is the same as if the proton were being rapidly exchanged between molecules having various values of $M_{I,A}$. As noted in Section 8.9, the spin–spin coupling vanishes for rapid exchange; hence the presence of A in the molecule has no effect on the appearance of the proton NMR spectrum, in this case. If the quadrupole coupling for A is very weak, then the "exchange" rate is slow and the spin–spin interaction between A and molecular protons gives a splitting of the lines in the proton NMR spectrum. For quadrupole couplings of intermediate strength, the proton NMR spectrum will show line-broadening (Fig. 8.12).

For N^{14}-containing compounds, the proton–N^{14} spin–spin interaction usually leads to broadening, rather than splitting, of proton resonance lines. Thus the NH_2 proton resonance of amides is a single broad line, rather than the triplet ($I=1$ for N^{14}) that would occur if the N^{14} quadrupole coupling were negligible. For Cl-, Br-, and I-containing compounds, the quadrupole coupling is generally large enough (note the values of Q for these nuclei) to lead to no proton–halogen splitting or broadening of the proton NMR lines; the presence of these nuclei has no effect on the proton NMR spectrum.

Different values of the nuclear-spin quantum number M_I correspond to different orientations of the nuclear electric quadrupole moment in the inhomogeneous electric field of the electrons and other nuclei of the molecule, and therefore correspond to different energies. In *nuclear quadrupole resonance* (NQR) spectroscopy, one observes transitions between these levels when rf radiation of appropriate frequency is applied. Unlike NMR, one does not apply an external field; the internal molecular electric field serves to produce the energy levels. NQR can be observed only in solids. From NQR, one obtains quadrupole coupling constants (which can also be gotten from splittings in pure-rotational spectra.) For details of NQR, see J. A. S. Smith, *J. Chem. Educ.*, **48**, 39, A77, A147, A243 (1971).

8.11 ELECTRON SPIN RESONANCE

The second major field of magnetic resonance spectroscopy is electron spin resonance (ESR), first observed by Zavoisky in 1945. Synonyms for

ESR include electron magnetic resonance (EMR) and electron paramagnetic resonance (EPR). ESR is more limited than NMR, since it can only be used to study species with one or more unpaired electron spins. Common examples of such species are the following:

(a) *Paramagnetic transition-metal ions in solutions and in crystals.* Ions such as Mn^{++}, Ti^{+++}, Fe^{+++}, and Cu^{++} have unpaired *d* electrons. Study of the ESR spectra of transition-metal complexes such as $[Ti(H_2O)_6]^{+++}$ and $[MnF_6]^{4-}$ has provided information on the nature of the ligand–metal bonding.

(b) *Radiation-produced radicals trapped in crystals.* Exposure of a crystal to γ rays or high-energy electrons can produce trapped free radicals. Thus γ irradiation of a crystal of malonic acid, $CH_2(COOH)_2$, produces the trapped free radical $\cdot CH(COOH)_2$, which can be studied by ESR spectroscopy. Irradiation of CH_4 at 4°K produces trapped $\cdot CH_3$ radicals. The simplest free radical, the hydrogen atom, has been produced and studied in several solids, for example, by photolysis of H_2O in Ar at 4°K.

(c) *Organic molecules in triplet states.* Since singlet–triplet transitions are forbidden, the lowest-lying triplet state of a molecule with a singlet ground state is metastable, with a relatively long lifetime. When naphthalene molecules present in low concentration in a durene crystal are exposed to UV radiation, one gets triplet-state naphthalene molecules whose ESR spectrum can be observed. Some organic molecules having triplet ground states have been studied. For example, UV photolysis of $(C_6H_5)_2CN_2$ produces the triplet-ground-state species $C_6H_5\dot{\underset{\cdot}{C}}C_6H_5$, which can be stabilized by trapping in a crystal of benzophenone.

(d) *Organic radicals in solution.* Many stable organic free radicals can be prepared in solution. The first example was Gomberg's preparation in 1900 of the triphenylmethyl radical, $(C_6H_5)_3C\cdot$. The ESR spectra of ion radicals derived from aromatic hydrocarbons such as benzene, naphthalene, and anthracene have been extensively studied. Negative ions such as $C_6H_6^-$, $C_{10}H_8^-$, and $C_{14}H_{10}^-$ are prepared by reducing the corresponding hydrocarbon with an alkali metal using tetrahydrofuran as the solvent. Positive ions such as $C_{14}H_{10}^+$ are prepared by oxidizing the hydrocarbon with H_2SO_4 or $SbCl_5$. Although alkyl radicals are not long-lived, the liquid-phase ESR spectra of such radicals as $\cdot CH_3$ and $\cdot C_2H_5$ have been studied by continuously bombarding the corresponding liquid hydrocarbons with electrons.

The theory of liquid-phase ESR spectra is much simpler than that of solid-phase spectra. The theory of the ESR spectra of transition-metal complexes is complicated by the necessity to consider the contribution to the electron's magnetic moment made by its orbital angular momentum, as well as by its spin angular momentum. We shall confine our discussion to the simplest case, that of solutions of organic free radicals containing one unpaired electron, category (d). For discussions of the ESR spectra of categories (a), (b), and (c), the reader is referred elsewhere.[15]

Before considering an unpaired electron in a molecule, we shall look at the interaction of an isolated electron having spin magnetic moment $\boldsymbol{\mu}_S$ with an applied external magnetic field **B**; the Hamiltonian is [(1.268) and (1.246)]

$$\hat{H} = -\mathbf{B} \cdot \hat{\boldsymbol{\mu}}_S = g_e \beta_e \hbar^{-1} \mathbf{B} \cdot \hat{\mathbf{S}} = g_e \beta_e \hbar^{-1} B \hat{S}_z \tag{8.101}$$

where the z axis is taken parallel to **B**. The *Bohr magneton* β_e is defined as $\beta_e = e\hbar/2mc$, where m is the electron's mass. The eigenfunctions of $\hat{H}$ are the electronic spin functions α and β, and its eigenvalues are

$$\pm \tfrac{1}{2} g_e \beta_e B \tag{8.102}$$

An absorption transition between these two levels can occur if the electron is exposed to electromagnetic radiation of frequency

$$\nu = g_e \beta_e B h^{-1} \tag{8.103}$$

Although $g_e (= 2.0023)$ is the same order of magnitude as nuclear g values, the smallness of the electronic mass makes β_e much larger than β_p; the ESR resonance frequency for $B = 10{,}000$ gauss is 28,000 MHz, which is a microwave frequency, as compared to a radio frequency for a nuclear magnetic moment in the same field. As in NMR spectroscopy, the frequency of the electromagnetic radiation in ESR spectroscopy is held fixed, while the applied magnetic field is varied until the resonance condition is satisfied. Common frequencies for ESR spectrometers are 9.5 and 35 GHz, which correspond to applied magnetic fields of 3400 and 12,500 gauss.

The setup for ESR spectroscopy is a cross between NMR and microwave techniques (Section 5.8). The source is a frequency-stabilized klystron, whose frequency is measured as in microwave spectroscopy. The microwave radiation is transmitted down a waveguide to a resonant cavity (a hollow metal enclosure), which contains the sample. The cavity is between the poles of an electromagnet, whose field is varied until resonance is achieved. Absorption of microwave power at resonance is observed using the same kind of crystal detector as in microwave spectroscopy. Sensitivity is enhanced, as in microwave spectroscopy, by the use of modulation: The magnetic field applied to the sample is modulated at, say, 100 kHz, thus producing a 100-kHz signal at the crystal when an absorption is reached. The spectrum is recorded on chart paper.

[15] *Carrington and McLachlan*, Chapters 7–10.

The system used for amplification and detection of an ESR signal is such that the first derivative of the absorption line is recorded. The shape of an ESR line in solution is usually Lorentzian [Equation (3.86)]. The Lorentzian shape resembles a Gaussian (except that it falls off more slowly). Differentiation of the Gaussian shape (3.89) gives $-2cd(\nu-\nu_0)\exp[-d(\nu-\nu_0)^2]$, which has the form of the $v=1$ harmonic-oscillator function [(1.133) and (1.137)] with $x=\nu-\nu_0$. Thus the first-derivative of an absorption resembles Fig. 1.1b with the origin at ν_0. (See also Problem 8.22.)

The sample in ESR spectroscopy is often placed in a quartz tube (glass has paramagnetic impurities), which is inserted into the resonant cavity through holes. The sample is usually a solid or liquid; some gas-phase ESR work has been done.

To determine the magnetic field at which resonance occurs and the spacings between lines of a spectrum, the ESR spectrum of a standard sample is superimposed on the spectrum being observed. A commonly used standard is the stable solid radical 1,1-diphenyl-2-picrylhydrazyl (DPPH)—the last radical in Fig. 8.15.

The large value of β_e as compared to β_p makes ESR a far more sensitive technique than NMR; under favorable conditions, ESR can detect as little as 10^{-12} moles of a radical (10^{12} spins).

For an unpaired electron in a molecule, (8.103) must be modified to include the internal magnetic fields acting on the electron. For a sufficiently dilute solution of an organic radical containing one unpaired electron, the only significant source of internal magnetic fields is the magnetic moments of nonzero-spin nuclei within the radical; for less dilute solutions, intermolecular interactions between spin magnetic moments would have to be considered.

Consider a single electron with magnetic moment $\boldsymbol{\mu}_e$ interacting with a nucleus having spin magnetic moment $\boldsymbol{\mu}_N$. Let $\boldsymbol{\mu}_N$ be at the coordinate origin and let $\mathbf{r}$ be the vector from the origin to (x,y,z). The magnetic field $\mathbf{B}(\mathbf{r})$ produced at (x,y,z) by $\boldsymbol{\mu}_N$ is given by (8.91) with $\boldsymbol{\mu}$ replaced by $\boldsymbol{\mu}_N$. The interaction energy E between $\mathbf{B}(\mathbf{r})$ and an electron located at $\mathbf{r}$ is $-\boldsymbol{\mu}_e\cdot\mathbf{B}(\mathbf{r})$:

$$E=\frac{\boldsymbol{\mu}_e\cdot\boldsymbol{\mu}_N}{r^3}-\frac{3(\boldsymbol{\mu}_e\cdot\mathbf{r})(\boldsymbol{\mu}_N\cdot\mathbf{r})}{r^5},\qquad r\neq 0 \tag{8.104}$$

To derive the Hamiltonian operator corresponding to (8.104), we replace $\boldsymbol{\mu}_e$ and $\boldsymbol{\mu}_N$ by the corresponding operators. The resulting Hamiltonian contains the electron's spatial coordinates $\mathbf{r}$, as well as electronic and nuclear spin coordinates. In this chapter, we are dealing only with spin wave functions and spin Hamiltonians, which do not contain electronic spatial coordinates. Therefore, the Hamiltonian corresponding to (8.104) must be averaged over the electronic spatial wave function $\psi(x,y,z)$ to obtain the spin Hamiltonian as

$$\hat{\mu}_e \cdot \hat{\mu}_N \int' \frac{|\psi|^2}{r^3} dx\,dy\,dz - 3\int' \frac{|\psi|^2(\hat{\mu}_e \cdot \mathbf{r})(\hat{\mu}_N \cdot \mathbf{r})}{r^5} dx\,dy\,dz \quad (8.105)$$

There is, however, a difficulty, in that (8.104) applies only for $r \neq 0$; at $r \approx 0$, the approximation of viewing the nucleus as a point magnetic dipole breaks down. We must therefore exclude small values of r from (8.105) by doing the r integration between the limits ϵ and ∞ and then taking the limit of the resulting expression as $\epsilon \to 0$; this procedure is indicated by the primes on the integrals. The expression (8.105) thus is not the complete interaction between the electron's dipole moment and the nucleus's magnetic field, since it excludes that part of the interaction arising from electron probability density at the nucleus. (In atoms, s electrons have a nonzero ψ at $r=0$; in planar molecules, σ electrons have nonzero probability density at a nucleus.) To derive this part of the interaction, we adopt the model of the nucleus as a tiny rotating sphere of radius ϵ, with the nuclear charge uniformly distributed on the surface. Outside the nucleus, this charge distribution gives rise to the magnetic field of a magnetic dipole moment $\boldsymbol{\mu}_N$. Classical electromagnetic theory shows the magnetic field within the nuclear sphere to be uniform with the value[16]

$$\mathbf{B} = (2/\epsilon^3)\boldsymbol{\mu}_N$$

The desired interaction energy between the electron probability density within the nuclear sphere and the nuclear magnetic field is then

$$E = -\int_{\text{nucleus}} \mathbf{B} \cdot \boldsymbol{\mu}_e |\psi|^2 \, dx\,dy\,dz \quad (8.106)$$

where the integral is over the volume of the nucleus. The electron probability density varies only slightly over the nuclear volume, so that (8.106) becomes

$$E = -|\psi(0)|^2 \int_{\text{nucleus}} \mathbf{B} \cdot \boldsymbol{\mu}_e \, dx\,dy\,dz \quad (8.107)$$

[16] R. A. Ferrell, *Am. J. Phys.*, **28**, 484 (1960).

where $\psi(0)$ is the value of the electronic spatial wave function at $r=0$. We have

$$E = -|\psi(0)|^2 \frac{2}{\epsilon^3} \boldsymbol{\mu}_N \cdot \boldsymbol{\mu}_e \int_{\text{nucleus}} dx\, dy\, dz$$

$$E = -(8\pi/3)|\psi(0)|^2 \boldsymbol{\mu}_e \cdot \boldsymbol{\mu}_N \tag{8.108}$$

Although (8.108) has been derived on the basis of a specific nuclear model, it can be shown[16] to be independent of any nuclear model. The term in the spin Hamiltonian corresponding to (8.108) is

$$\hat{H}_c = -(8\pi/3)|\psi(0)|^2 \hat{\boldsymbol{\mu}}_e \cdot \hat{\boldsymbol{\mu}}_N \tag{8.109}$$

This term is called the *Fermi contact term*. That part of the electron–nucleus magnetic-dipole interaction represented by (8.104) depends on the angular coordinates of the electron and is therefore anisotropic; in contrast, the Fermi contact energy (8.108) is isotropic. The contact term plays an important role in the electron-coupled nuclear spin–spin interactions seen in the NMR spectra of liquids.

The Fermi contact term is often stated to be a purely quantum-mechanical interaction, having no classical analog. This is not so. The preceding derivation shows the contact interaction to be readily understandable within the framework of classical electromagnetism.

In this chapter, we are dealing only with spin wave functions and Hamiltonians. In the complete molecular Hamiltonian, which involves electronic spatial coordinates and momenta in addition to spin coordinates, the contact term in the Hamiltonian would have to be written in a form such that the integral over the electronic spatial wave function ψ,

$$\int \psi^* \hat{H}_c' \psi \, dx\, dy\, dz$$

would yield the expression (8.109). The form of $\hat{H}_c'$ that will accomplish this purpose is

$$\hat{H}_c' = -(8\pi/3)\hat{\boldsymbol{\mu}}_e \cdot \hat{\boldsymbol{\mu}}_N \delta(\mathbf{r}) \tag{8.110}$$

where $\delta(\mathbf{r}) = \delta(x)\delta(y)\delta(z)$ is the three-dimensional Dirac delta function (Section 1.2). [Equation (8.110) is for one electron. For a many-electron molecule, we have a sum of terms like (8.110), one term for each electron.]

The interaction between the electron and nuclear spin magnetic moments is the sum of the terms (8.105) and (8.109). Our discussion is limited to liquids. For liquids, the rapid rotation of the molecules causes the interaction energy due to the term (8.105) to average to zero—this is shown by arguments similar to those used at the beginning of Section 8.7 to prove that the dipole–dipole interaction between two nuclei of a molecule vanishes for liquids. Hence we consider only (8.109). We are treating organic free radicals with one unpaired electron; here, the electronic magnetic dipole moment results from the electron's spin **S** [Eq. (1.246)]:

$$\boldsymbol{\mu}_e = -g_e\beta_e\hbar^{-1}\mathbf{S} \tag{8.111}$$

From (8.111) and (8.1), the Fermi contact term (8.109) is then

$$\hat{H}_c = g_e g_N \beta_e \beta_p \frac{8\pi}{3\hbar^2}|\psi(0)|^2\hat{\mathbf{S}}\cdot\hat{\mathbf{I}} \tag{8.112}$$

where $\hat{\mathbf{S}}$ and $\hat{\mathbf{I}}$ are the electronic and nuclear spin operators. The contact term is independent of the external magnetic field applied to the molecule, and has the form

$$\hat{H}_c = \frac{hA}{\hbar^2}\hat{\mathbf{S}}\cdot\hat{\mathbf{I}} \tag{8.113}$$

$$A \equiv \frac{8\pi}{3h} g_e g_N \beta_e \beta_p |\psi(0)|^2 \tag{8.114}$$

where A, the *hyperfine coupling constant*, has units of hertz. (Recall that the interaction between electronic and nuclear magnetic moments gives rise to the *hyperfine structure* of atomic spectra—Section 1.17.) We derived (8.114) from (8.109), which is for a one-electron system. For the many-electron systems one usually deals with, one obtains a modified expression for (8.114)—see *Murrell and Harget*, pp. 108–109.

For the system of a single unpaired electron and a single nucleus, the spin Hamiltonian in an applied field $\mathbf{B}_0$ is [(8.6), (8.101), and (8.113)]

$$\hat{H} = g_e\beta_e\hbar^{-1}\mathbf{B}_0\cdot\hat{\mathbf{S}} - g_N\beta_p\hbar^{-1}\mathbf{B}_0\cdot\hat{\mathbf{I}} + hA\hbar^{-2}\hat{\mathbf{S}}\cdot\hat{\mathbf{I}} \tag{8.115}$$

Taking $\mathbf{B}_0$ in the z direction and using (8.80), we have

$$\hat{H} = g_e\beta_e\hbar^{-1}B_0\hat{S}_z - g_N\beta_p\hbar^{-1}B_0\hat{I}_z + hA\hbar^{-2}\hat{S}_z\hat{I}_z + \tfrac{1}{2}hA\hbar^{-2}(\hat{S}_+\hat{I}_- + \hat{S}_-\hat{I}_+) \tag{8.116}$$

For simplicity, let the nucleus be a proton. To carry out an exact quantum-mechanical calculation, we would take the four electron–proton basis spin functions

$$\varphi_1 = \alpha(e)\alpha(p), \qquad \varphi_2 = \alpha(e)\beta(p)$$
$$\varphi_3 = \beta(e)\alpha(p), \qquad \varphi_4 = \beta(e)\beta(p) \tag{8.117}$$

and set up the secular equation (2.77), as in Section 8.5. However, for organic free radicals, A usually lies in the range 0 to 500 MHz, as compared to $g_e\beta_e B_0/h = 14{,}000$ MHz at $B_0 = 5000$ gauss. Hence the terms in (8.116) containing A are small and can be treated using first-order perturbation theory. The zeroth-order wave functions are (8.117), which are eigenfunctions of $\hat{S}_z$ and $\hat{I}_z$. The energies, correct through first order, are

$$\langle \varphi_i | \hat{H} | \varphi_i \rangle, \qquad i = 1, 2, 3, 4 \tag{8.118}$$

Using the orthonormality of the spin functions, we find

$$E_1 = \tfrac{1}{2} g_e \beta_e B_0 - \tfrac{1}{2} g_p \beta_p B_0 + \tfrac{1}{4} hA$$
$$E_2 = \tfrac{1}{2} g_e \beta_e B_0 + \tfrac{1}{2} g_p \beta_p B_0 - \tfrac{1}{4} hA$$
$$E_3 = -\tfrac{1}{2} g_e \beta_e B_0 - \tfrac{1}{2} g_p \beta_p B_0 - \tfrac{1}{4} hA \tag{8.119}$$
$$E_4 = -\tfrac{1}{2} g_e \beta_e B_0 + \tfrac{1}{2} g_p \beta_p B_0 + \tfrac{1}{4} hA$$

ESR transitions are due to the interaction of the electron's spin magnetic moment $\boldsymbol{\mu}_S$ with the oscillating field $\mathbf{B}_1$ of the incident microwave radiation. The experimental setup is such that $\mathbf{B}_1$ is perpendicular to $\mathbf{B}_0$, so that the selection rules are determined by the integral

$$-B_1 \langle \varphi_i | \hat{\mu}_{S,x} | \varphi_j \rangle = g_e \beta_e \hbar^{-1} B_1 \langle \varphi_i | \hat{S}_x | \varphi_j \rangle \tag{8.120}$$

Use of (1.244) shows that the allowed ESR transitions are between levels 1 and 3 and between levels 2 and 4. The electron's spin flips, while the nuclear spin remains unchanged. The transition frequencies are

$$\nu_{3\to 1} = \frac{g_e \beta_e B_0}{h} + \tfrac{1}{2} A, \qquad \nu_{4\to 2} = \frac{g_e \beta_e B_0}{h} - \tfrac{1}{2} A \tag{8.121}$$

We have a doublet with separation A; the components of the doublet are of equal intensity, since the energy-level spacings are much less than kT. The observed spectrum is independent of the sign of A.

Let us calculate A for the simplest free radical, the hydrogen atom. Equation (1.228) gives for the ground state

$$|\psi(0)|^2 = 1/\pi a^3 \tag{8.122}$$

where a (not to be confused with the hyperfine coupling constant) equals $\hbar^2/\mu e^2 = 0.529465$ Å, where μ is the reduced mass. The A_H for the ground-state hydrogen atoms is then [Equation (8.114)]

$$A_H = \frac{8\pi}{3h} g_e g_p \beta_e \beta_p \frac{1}{\pi a^3} = 1420.5 \text{ MHz} \tag{8.123}$$

[For $B_0 = 0$, the spin Hamiltonian (8.116) contains only the hyperfine interaction between electron and nuclear spin magnetic moments. The first-order perturbation-theory treatment used to obtain the levels (8.119) is, of course, invalid, and the levels must be found from the secular equation, using the basis functions (8.117). The result (Problem 8.24) is a nondegenerate level with energy $-3hA_H/4$, and a triply degenerate level with energy $hA_H/4$. A transition between these hydrogen-atom hyperfine levels gives a single line at the frequency $A_H = 1420.5$ MHz, which is close to the experimentally observed hyperfine splitting of 1420.4 MHz (Section 1.17).[17] The 0.1 MHz discrepancy is due to corrections to (8.123) arising from the finite size of the proton and other factors.]

The similarity of the preceding first-order ESR treatment to the first-order NMR treatment of two coupled protons is evident. For an unpaired electron interacting with n equivalent nuclei of spin $\frac{1}{2}$, the hyperfine coupling term in the spin Hamiltonian is

$$\frac{hA}{\hbar^2} \sum_{i=1}^{n} \hat{\mathbf{S}} \cdot \hat{\mathbf{I}}_i \tag{8.124}$$

Carrying over the ideas of Section 8.5 (the derivation is formally the same as the derivation of the A portion of the NMR spectrum of an AX_n molecule), we see that the ESR spectrum in this case consists of a multiplet of $n+1$ lines centered at $g_e \beta_e B_0 h^{-1}$; the spacing between adjacent lines is A, and the relative intensities are given by the binomial coefficients (8.86). For example, the ESR spectrum of $C_6H_6^-$ is shown in Fig. 8.13.

For an unpaired electron interacting with both a set of m equivalent protons and a set of n equivalent protons, the first-order analysis is carried out in two stages, like the treatment in Section 8.5 of the CH_2 portion of

[17] Radio-astronomy searches (Project Ozma) for communications from intelligent extraterrestrial beings have been carried out at this frequency; see *Interstellar Communication*, A. G. W. Cameron, ed., Benjamin, New York, 1963; *Communication With Extraterrestrial Intelligence*, C. Sagan, ed. MIT Press, 1973.

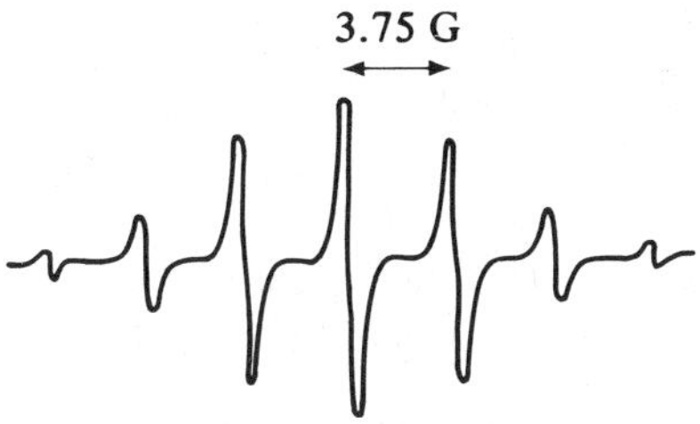

Fig. 8.13 The first-derivative ESR spectrum of the benzene anion radical $C_6H_6^-$. The relative intensities are 1:6:15:20:15:6:1.

the ethanol proton NMR spectrum. Figure 8.14 shows the predicted spectrum for an electron simultaneously coupled to two protons with $A = 62$ Hz and to three protons with $A = 75$ Hz; these are the approximate values for the ethyl radical. The four dotted lines have the binomial intensities 1:3:3:1. Each dotted line is split into three lines; multiplication of each of the 1:3:3:1 intensities by 1, 2, and 1 gives the final intensities. From left to right, these are 1:2:3:1:6:3:3:6:1:3:2:1.

The multiplet hyperfine pattern of ESR lines in organic radicals is most frequently due to electron–proton interactions, but other nuclei with nonzero spin may also cause hyperfine structure. Weak satellite lines arising from interactions between the unpaired electron and C^{13} nuclei are sometimes observed; C^{13} has $I = \frac{1}{2}$, and the analysis is straightforward. Nitrogen-containing radicals may show hyperfine splittings due to N^{14}, which has $I = 1$. The possible M_I values are 1, 0, and -1, so that an electron

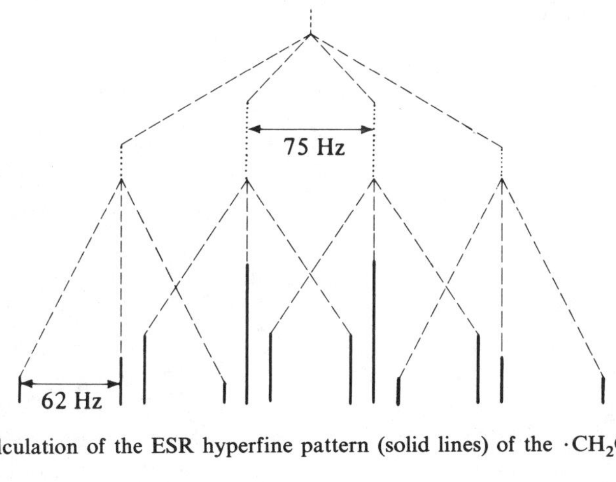

Fig. 8.14 Calculation of the ESR hyperfine pattern (solid lines) of the $\cdot CH_2CH_3$ radical in solution.

coupled to a single N^{14} nucleus shows a triplet hyperfine ESR pattern with the lines being of equal intensity. If the electron also interacts with a second N^{14} nucleus, each line of the triplet is split into three lines, giving, in general, nine lines. However, if the two N^{14} nuclei are equivalent (i.e., have the same coupling constant), five lines with the intensity ratios 1:2:3:2:1 are obtained, due to coincidence of some of the lines:

$$\begin{array}{ccccc} 1 & 1 & 1 & & \\ & 1 & 1 & 1 & \\ & & 1 & 1 & 1 \\ \hline 1 & 2 & 3 & 2 & 1 \end{array}$$

The frequency differences between ESR hyperfine components are much greater than NMR spin–spin splittings, so that the N^{14} quadrupole moment does not give line broadening in ESR spectra.

Often a radical is deuterated to aid in the assignment of coupling constants. Deuterium has $I=1$ and a much smaller g_N value than the proton. Thus replacement of an H^1 nucleus by H^2 changes both the number of hyperfine components and their spacings.

Since the simple first-order treatment is generally adequate for ESR spectra of organic radicals in solution, their ESR spectra are easy to analyze, provided the lines do not overlap. Frequently, however, the lines are not all resolved and the analysis is difficult and subject to error.

NMR spectroscopists express coupling constants (J values) in hertz; ESR spectroscopists usually express coupling constants of organic radicals in solution in units of gauss (G). Multiplication of A in (8.115) by $h/g_e\beta_e$ converts it from hertz to gauss; we have

$$g_e\beta_e/h = 2.80 \text{ MHz/G} \qquad (8.125)$$

(For organic radicals trapped in solids, workers usually express hyperfine coupling constants in hertz. Coupling constants of transition-metal ions are usually expressed in cm^{-1}.) It turns out that for organic radicals in solution, the electron's orbital angular momentum makes a very slight contribution to its magnetic moment; this contribution is allowed for by replacing the free-electron g_e value 2.0023 in (8.121) by an effective g value; observed g values for organic radicals in solution typically deviate from g_e by 0.1 percent. We won't worry about these slight deviations.

ESR coupling constants for some free radicals are given in Fig. 8.15.

As seen from (8.114), the coupling constant between the unpaired electron and a given nucleus of an organic radical in solution is zero unless

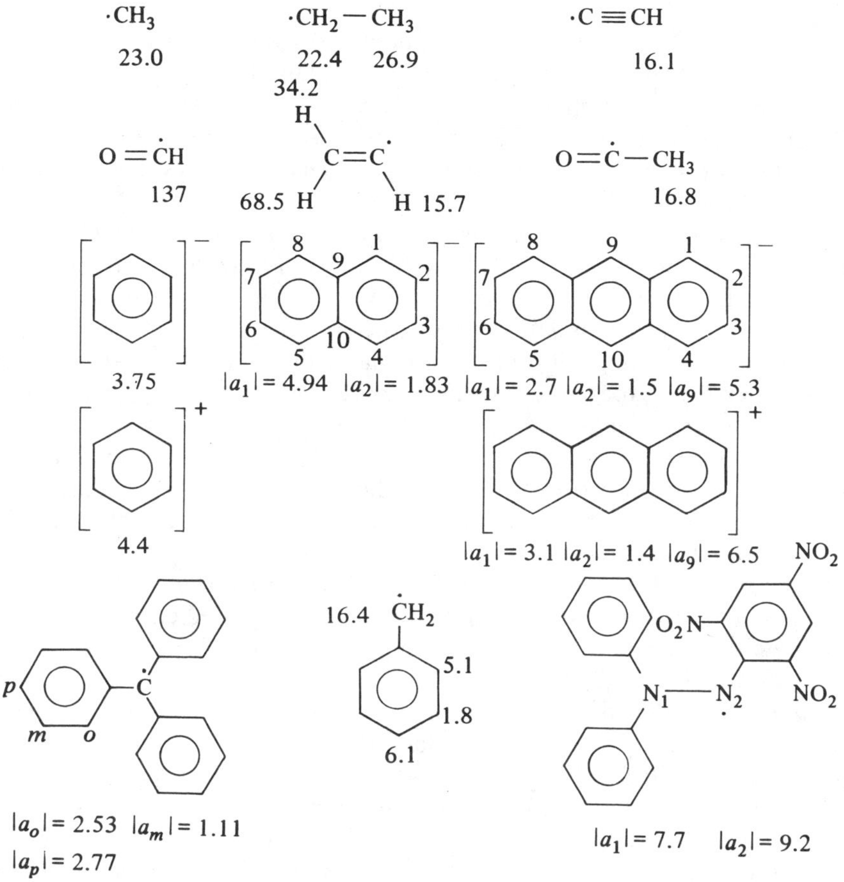

Fig. 8.15 Coupling constants of some organic radicals in solution. The constants are for electron-proton interactions, except for DPPH, whose proton hyperfine coupling constants are small (1 to 2 G). In all cases, the absolute values of the coupling constants in gauss are given; theory indicates certain of the constants to be negative. Data from: H. Fischer, *Magnetic Properties of Free Radicals*, Landolt-Börnstein Tables, New Series, Group II, vol. 1, Springer-Verlag, Berlin, 1965; I. C. Lewis and L. S. Singer, *J. Chem. Phys.*, **43**, 2712 (1965); H. J. Bielski and J. M. Gebicki, *Atlas of Electron Spin Resonance Spectra*, Academic Press, New York, 1967; M. K. Carter and G. Vincow, *J. Chem. Phys.*, **47**, 292 (1967); R. G. Lawler, J. R. Bolton, M. Karplus, and G. K. Fraenkel, *ibid.*, **47**, 2149 (1967).

there is some probability for this electron to be at the nucleus in question. In atoms, only s electrons have nonzero values of ψ at the nucleus. We saw that for the hydrogen $1s$ AO, the coupling constant A is 1420 MHz, which is equivalent to 507 G. For the vinyl radical, the observed coupling constants of 34, 68, and 16 G at the three hydrogens indicate a significant amount of delocalization of the MO of the unpaired electron.

For the triphenylmethyl radical $(C_6H_5)_3C\cdot$, the π MO of the unpaired electron is delocalized over the entire molecule, and there are couplings to six *ortho*, six *meta*, and three *para* hydrogens. The ESR spectrum has $7 \times 7 \times 4 = 196$ lines.

The nonzero values of the electron–proton coupling constants of such aromatic free radicals as $C_6H_6^-$, $C_{10}H_8^-$, and $(C_6H_5)_3C\cdot$ raise an important question. In these radicals, the odd electron is in a delocalized π MO, which is a linear combination of p_z carbon-atom AOs. However, the p_z AOs and the resulting π MO have a node in the molecular plane, and we would therefore expect the coupling constants between the ring protons and the unpaired electron to be zero. We can explain the observed nonzero coupling constants as follows. In Section 7.3 we noted the differing interaction between two electrons with the same spin, as compared to two electrons with opposite spin, and pointed out that in a molecule with unequal numbers of spin-α and spin-β electrons, use of slightly different sets of spatial MOs for spin-α and spin-β electrons (the UHF method) gives a better wave function and energy than when paired electrons occupy the same spatial MO. In radicals, the number of α electrons differs by one from the number of β electrons. In the radical's sigma MOs (which have nonzero probability density at the ring protons), two paired electrons will not occupy exactly the same spatial MO; therefore, the sum of the probability densities at a given proton for the sigma electrons with spin α will differ slightly from the corresponding sum for β-spin sigma electrons. Thus we get a net unpaired sigma electron density at the protons, thereby giving the required contact interaction. (Alternatively, one can work within the RHF plus CI approach, keeping the same spatial orbitals for paired electrons; here, the spin density at the protons in aromatic radicals arises from inclusion of configurations with one or more sigma electrons excited to higher MOs.)

It turns out that in a $C_6H_6^-$ radical ion for which the unpaired π electron has spin α, there is a slight excess of unpaired electrons with spin β at each of the ring protons. (When a particular nucleus of a radical has an excess of electrons with spin opposite to that of the unpaired electron, this nucleus is said to have a *negative spin density*.) It follows that the electron–proton hyperfine coupling constant is negative in $C_6H_6^-$ and similar aromatic hydrocarbons. The appearance of the liquid-phase ESR spectrum of a radical is independent of the sign of the proton coupling constants, but the theoretically predicted negative sign of the ring proton coupling constants in aromatic radicals has been verified in several cases by observation of the ESR spectra of organic radicals trapped in solids.

Another case of a negative electron–proton coupling constant is the methyl radical $\cdot CH_3$, where the unpaired electron is predominantly in a π MO perpendicular to the plane of the molecule.

Semiempirical VB and MO plus CI calculations by McConnell and others yielded the following theoretical equation for a_H, the coupling constant (in units of gauss) between the unpaired π electron and a ring proton of an aromatic radical[18]:

$$a_H = Q\rho_{\pi,C} \tag{8.126}$$

The quantity $\rho_{\pi,C}$ is the unpaired π-electron spin density at the carbon atom to which the hydrogen atom in question is bonded; $\rho_{\pi,C}$ is defined as ± 1 times the fractional number of unpaired π electrons on the carbon atom, with the sign being determined by whether the net unpaired spin at the carbon atom is in the same or opposite direction as the spin vector of the molecule. The term π-electron spin "density" is somewhat misleading in that ρ_π is not an electron probability density (which is measured in electrons/cm^3), but rather is a pure number. The semiempirical constant Q (no connection with nuclear quadrupole moments) is approximately -23 G.

We now consider some applications of the McConnell equation (8.126), using Hückel MO theory to estimate $\rho_{\pi,C}$. For the methyl radical, $\rho_{\pi,C}=1$, and a_H is predicted to be -23 G, in agreement with the observed proton coupling constant (Fig. 8.15).

For the naphthalene negative ion, Hückel MO theory puts the unpaired electron in the lowest unoccupied MO of the neutral naphthalene molecule; this MO is (*Levine*, Section 15.8)

$$\varphi_6 = 0.4253(f_1+f_4-f_5-f_8)+0.2629(-f_2-f_3+f_6+f_7) \tag{8.127}$$

where the f's are carbon $2p\pi$ AOs and the carbon atoms are numbered as in Fig. 8.15. The fractional number of unpaired electrons on carbon atom i is found by taking the square of the coefficient of f_i in φ_6. We find $\rho_{\pi,C}=0.1809$ for the α carbon atoms (1,4,5,8) and $\rho_{\pi,C}=0.0691$ for the β carbons (2,3,6,7). The predicted coupling constants are -4.2 and -1.6 G, in reasonable agreement with the observed values $(-)4.94$ and $(-)1.83$ G. (The ESR spectrum of $C_{10}H_8^-$ cannot tell us whether the larger coupling is to the α or β ring protons, but the ESR spectrum of α-deuteronaphthalene confirms the larger coupling constant for the α protons.) The predicted ratio of α to β coupling constants is 2.62, in very good agreement with the observed value 2.70, a striking confirmation of the predictions of simple MO theory.

[18] H. M. McConnell and D. B. Chesnut, *J. Chem. Phys.*, **28**, 107 (1958); see also A. Carrington, *Quart. Rev.*, **17**, 67 (1963).

For the naphthalene positive ion $C_{10}H_8^+$, the odd electron is in the MO φ_5, which has the same fractional populations as φ_6 (naphthalene is an alternant hydrocarbon). Hence the McConnell equation predicts the ESR spectra of $C_{10}H_8^+$ and $C_{10}H_8^-$ to be identical. Unfortunately, this prediction cannot be checked; oxidation of naphthalene by $SbCl_5$ yields not $C_{10}H_8^+$, but the dimeric radical ion $(C_{10}H_8)_2^+$. The McConnell equation and Hückel MO theory also predict the coupling constants for the anthracene positive and negative radical ions $C_{14}H_{10}^+$ and $C_{14}H_{10}^-$ to be the same; these radicals have been prepared, and as seen from Fig. 8.15, their coupling constants are quite similar (but not identical).

The McConnell equation is useful for aromatic radicals, but one would like to be able to predict coupling constants in all types of radicals. Pople, Beveridge, and Dobosh applied the INDO method using different orbitals for electrons with different spins to the calculation of a values. Using only a single adjustable parameter, they found moderately good agreement with experimental values in a wide variety of radicals. For details, see *Pople and Beveridge*, pp. 128–149; *Murrell and Harget*, pp. 103–119. *Ab initio* calculation of a is a rather tricky matter because (unlike the other molecular properties we have discussed) a depends on the value of the wave function at a single point, the nucleus in question. A very slight change in basis set, which might make an insignificant change in the variational energy, can make a substantial change in a.

As with NMR, ESR spectroscopy can be used to study rate processes. For example, if some naphthalene molecules are added to a solution of naphthalene negative ion, there is the possibility of the unpaired electron of a $C_{10}H_8^-$ ion jumping to a $C_{10}H_8$ molecule. If the rate of exchange is very slow compared to the hyperfine splitting, then the hyperfine structure of the $C_{10}H_8^-$ ion's ESR spectrum is maintained; if the rate of exchange is very fast, then the ESR spectrum shows a single line. Examination of the ESR spectrum has yielded the rate constants for this electron-transfer reaction in various solvents.

Many chemical reactions involve free radicals as intermediates; ESR spectroscopy is the most direct way of detecting their presence. However, the low concentrations and short lifetimes of these species make their detection difficult. By allowing the reaction mixture to flow rapidly through the resonant cavity, the ESR spectra of many free-radical reaction intermediates have been observed.

ESR spectroscopy has biological applications. ESR study of plant and animal tissues shows the presence of small but significant concentrations of free radicals. Exposure of a plant leaf to visible light raises the free-radical concentration in the leaf, indicating that free radicals are reaction intermediates in photosynthesis. ESR spectra of radicals produced by γ irradiation of DNA and of various enzymes have been studied. Another way to study macromolecules such as nucleic acids is by *spin labeling*, which is the bonding of a radical to the macromolecule of interest; the ESR spectrum then gives information about the macromolecule. See H. Swartz *et al.*, eds., *Biological Applications of Electron Spin Resonance Spectroscopy*, Wiley-Interscience, 1972.

PROBLEMS

8.1 (a) Show that in Gaussian units

$$1 \text{ G} = 1 \text{ g}^{1/2} \text{ sec}^{-1} \text{ cm}^{-1/2}$$

(b) Verify (8.3).

8.2 Calculate the magnitude of the magnetic dipole moment vectors of (a) the proton; (b) the neutron.

8.3 Graph the energy levels of the Br^{79} nuclear magnetic moment as a function of B for applied fields ranging from 0 to 10^4 G; label each line with its M_I value.

8.4 Calculate the NMR transition frequency at 10,000 G for (a) B^{10}; (b) I^{127}.

8.5 Calculate the magnetic fields used in 100-MHz and 220-MHz proton NMR spectrometers.

8.6 Calculate the ratio of equilibrium populations at 25°C of α and β spin states for protons in a 100-MHz NMR spectrometer.

8.7 Verify (8.31).

8.8 Verify (8.65) and (8.66).

8.9 Verify that the functions (8.65) are eigenfunctions of the Hamiltonian (8.46) with eigenvalues (8.67).

8.10 (a) Prove that the nuclear-spin operator $\hat{T}_z$ of (8.74) commutes with the Hamiltonian (8.73). (b) Prove that $\hat{T}_x$ commutes with $\hat{H}'$ in (8.73).

8.11 Prove that $\hat{H}'(\text{A})$ in (8.88) commutes with the Hamiltonian (8.88).

8.12 (a) The compound sapphyrine[19] is one of the few organic compounds having protons with a negative δ relative to TMS. For a 60-MHz spectrometer, calculate the separation in hertz between the resonances of the internal protons ($\delta = -6.0$ ppm) and the bridge protons ($\delta = 12.0$ ppm). (Ignore spin–spin splitting.) (b) For 60-MHz and for 100-MHz proton

[19] *Salem*, p. 206.

instruments, calculate in hertz the separation between two resonances differing by 1.00 in δ.

8.13 For a 60-MHz instrument, the proton NMR spectrum of HD shows lines separated by 43.5 Hz. (a) Give the number of lines and their relative intensities in the HD proton NMR spectrum. (b) Give the number of lines, their relative intensities, and their spacing for the HD deuteron NMR spectrum. (c) Calculate the proton–proton coupling constant in H_2.

8.14 Explain why the observed proton–proton coupling constant in $CHCl_2CHCl_2$ changes considerably with temperature.

8.15 Classify the following compounds as to type of NMR spectrum (e.g., AB_2, AX_3): (a) CH_4; (b) CH_3F; (c) CH_3Br; (d) $CH_2{=}O$; (e) $CH_2{=}CHBr$; (f) $CH_2{=}CHF$; (g) *cis*-CHCl=CHCl; (h) 1,2,3-trichlorobenzene; (i) *p*-bromochlorobenzene; (j) the *anti* (*trans*) conformation of CH_2BrCH_2Cl; (k) the *gauche* conformation of CH_2BrCH_2Cl; (l) CH_2BrCH_2Cl, assuming rapid internal rotation; (m) chlorobenzene.

8.16 The 60-MHz proton NMR spectrum of 2-bromo-5-chlorothiophene is shown in Fig. 8.16. Find δ_{12} and J_{12}. Draw the spectrum at 100 MHz.

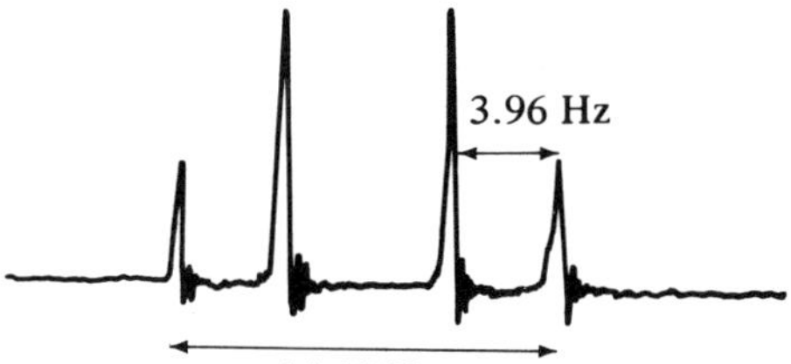

Fig. 8.16 Proton NMR spectrum of 2-bromo-5-chlorothiophene at 60 MHz.

8.17 Give the structural formula for each of the compounds whose 60-MHz proton NMR spectrum is shown in Fig. 8.17.

8.18 Which of the following would show the greatest proton NMR chemical shift relative to the bare proton? (a) H; (b) H_2; (c) HF.

8.19 Calculate the percent error in δ_{12} when a first-order analysis is applied to an AB spectrum when $\nu_0\delta_{12}/J_{12}$ has each of the following values[20]: (a) 2; (b) 4; (c) 7; (d) 10.

8.20 Calculate and draw the 60-MHz proton NMR spectrum of $HC(OCH_2CH_3)_3$ using the following proton shifts (in ppm) and coupling constants: $\delta_{CH_3}=1.37$, $\delta_{CH_2}=3.72$, $\delta_{CH}=5.30$, $J_{CH_2-CH_3}=7.0$ Hz. Give the relative intensities of all lines.

[20] A number of chemical shifts reported in the literature by organic chemists are inaccurate due to unjustified use of a first-order analysis.

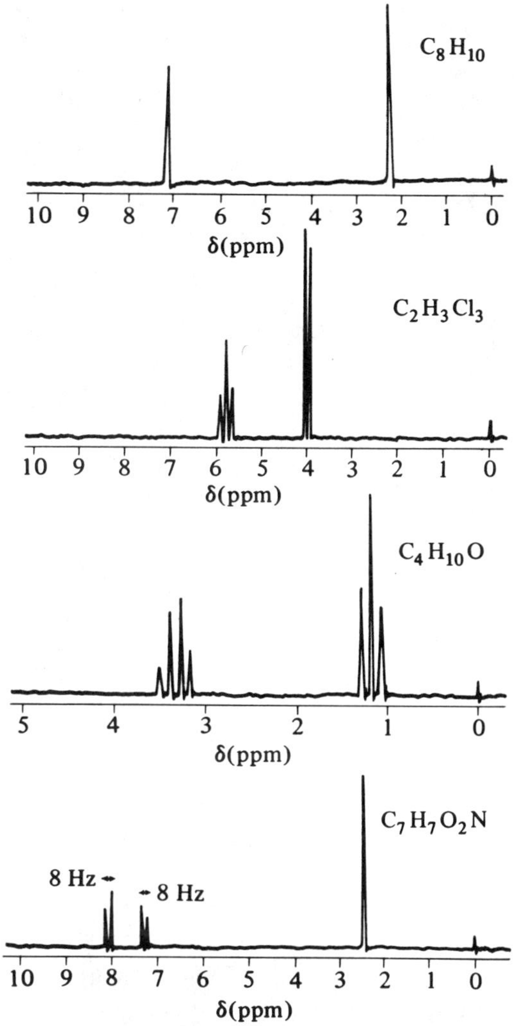

Fig. 8.17 NMR spectra for Problem 8.17. (Note the slight asymmetries in the multiplets, which result from deviations from the first-order case.)

8.21 A certain commercial ESR spectrometer operates at 24 GHz. What magnetic field is used?

8.22 (a) Sketch the first derivative of a Lorentz line shape. (b) Some ESR spectrometers record the second derivative (rather than the first derivative) of resonance lines. Sketch the second derivative of a Lorentzian line, indicating the position of ν_0.

8.23 To justify the first-order approximate calculation of the eigenvalues of the ESR Hamiltonian (8.116), we used the fact that $g_e\beta_e B_0/h \gg A$. Does the validity of the first-order treatment also require that $g_N\beta_N B_0/h \gg A$?

8.24 (a) Find the exact eigenvalues of the ESR Hamiltonian (8.116). (b) Show that when $(g_e\beta_e + g_N\beta_p)B_0/h \gg A$, the exact energies reduce to (8.119). (c) For a hydrogen atom in a field of 5000 G, calculate the percent error in each of the spin energy levels made by using (8.119). Do the same for a field of 500 G. (d) In the limit $B_0 \to 0$, what are the values of the spin energy levels?

8.25 Calculate the population ratio at 25°C of α and β spin states for an isolated electron in a field of 10,000 G. Compare with Problem 8.6.

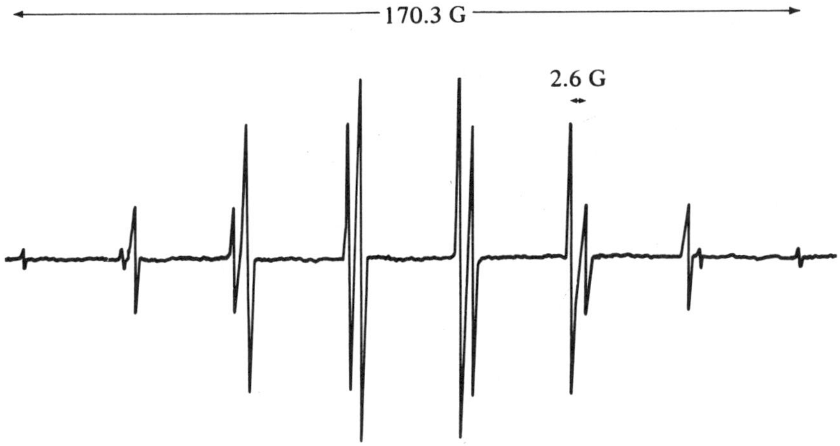

Fig. 8.18 The first-derivative ESR spectrum of the isopropyl radical.

8.26 Give the number of lines in the ESR spectrum of the anion radical of each of the following molecules. (Assume all lines are resolved.) (a) Naphthalene; (b) anthracene; (c) pentacene; (d) azulene; (e) *o*-xylene; (f) *m*-xylene; (g) *p*-xylene; (h) nitrobenzene; (i) *p*-fluoronitrobenzene.

8.27 The cyclobutyl radical C_4H_7 has the coupling constants 21.2, 36.7, and 1.1 G for the α, β, and γ protons, respectively. Calculate its ESR spectrum.

8.28 The ESR spectrum of the isopropyl radical $(CH_3)_2CH$ is shown in Fig. 8.18 for a frequency of 9.5 GHz. Calculate the coupling constants.

8.29 Calculate the hyperfine coupling constant for a deuterium atom.

8.30 For which of the following hydrocarbons does simple HMO theory predict substantial differences between the anion and cation ESR spectra? (a) phenanthrene; (b) azulene; (c) anthracene; (d) pyrene.

8.31 Is the ESR spectrum of $C_6H_6^-$ (Fig. 8.13) evidence for or against a substantial Jahn–Teller distortion?

8.32 Look up (in one of the references to Fig. 8.15) the proton hyperfine coupling constants for the anions of each of the following hydrocarbons: naphthalene, anthracene, tetracene (naphthacene), pentacene, phenanthrene, pyrene, coronene, and azulene. Look up (in one of the books that tabulate HMOs) the HMO coefficients for these hydrocarbons and calculate the unpaired spin densities for these anions. (a) How well do simple HMO theory and the McConnell equation predict the observed ratios of coupling constants in a given hydrocarbon radical? (b) Treat Q in (8.126) as an adjustable parameter and calculate for each radical the value of Q that gives good agreement with the observed coupling constants.

8.33 Consider an unpaired electron interacting with n nuclei with spins $I_1, \ldots, I_n$ and coupling constants $A_1, \ldots, A_n$. (a) Show that (assuming the first-order analysis holds) the frequencies of the ESR lines are

$$\nu = g_e \beta_e B_0 h^{-1} + \sum_{i=1}^{n} A_i M_{I_i}$$

(b) Show that if all the nuclei are equivalent, the number of ESR lines is $2nI+1$. (c) From the result of part (b), show that if there are m sets of equivalent nuclei, with the jth set having n_j nuclei with spin I_j, then the number of ESR lines is

$$\prod_{j=1}^{m} (2n_j I_j + 1)$$

8.34 For a hydrogen atom in an external field of 10,000 G, draw a figure that shows the effect on the original $1s$ energy level of including first the electron Zeeman term, then the nuclear Zeeman term, and finally the hyperfine coupling term in the Hamiltonian.

8.35 In the first-order approximation to NMR spectra when $\hat{H}^0$ is taken as the first two terms in (8.78), the nuclear-spin energy levels are, in general, degenerate. Show for the A_2X_2 case (where A and X are protons) that the functions (8.79) are the correct zeroth-order wave functions for the perturbation $\hat{H}'$, where $\hat{H}'$ is the last term in (8.78).

8.36 Verify (8.119).

8.37 Name five molecular properties that are second-rank tensors.

8.38 Consider the molecule

$$C_3H_7CH_a{=}CH_bCH_2CH_2CH_c{=}CH_dC_3H_7$$

If H_a and H_b are *trans* to each other, and H_c and H_d are *trans* to each other, we say the configuration is *trans–trans*. The natural-abundance C^{13} FT NMR spectrum with protons decoupled shows 12 lines. How many lines would occur for each of the following configurations? (a) *trans–trans*; (b) *cis–trans*; (c) *cis–cis*.

8.39 A sample of benzene was placed in the magnetic field of an NMR spectrometer and (using a small value of B_1) the spectrum was scanned repeatedly at intervals of 8.0 sec. The following relative intensities were observed on successive scans: 2, 11, 16, 19, 21, $22\frac{1}{2}$, $23\frac{1}{2}$. Ultimately, the intensity reached 25. Find the spin–lattice relaxation time for benzene. (The first scan was made 6 sec after the sample was placed in the spectrometer. Is it necessary to know this?)

8.40 Let n_{eq} and $n_{\text{eq,rad}}$ be the equilibrium spin population differences between the α and β levels in the absence and presence of radiation, respectively. Let $P_{\alpha\beta} = P_{\beta\alpha} \equiv P$ be the probabilities per second for stimulated absorption and emission of radiation. (a) Show that

$$n_{\text{eq,rad}} = \frac{n_{\text{eq}}}{1+2P/(W_{\alpha\beta}+W_{\beta\alpha})} = \frac{n_{\text{eq}}}{1+2PT_1}$$

Thus if $P_{\alpha\beta} \ll W_{\alpha\beta}$, then $n_{\text{eq,rad}} \approx n_{\text{eq}}$. If $P_{\alpha\beta} \sim W_{\alpha\beta}$, then $n_{\text{eq,rad}}$ is significantly less than n_{eq} and saturation is observed. (b) P is proportional to the squares of what two quantities? (c) What happens to $n_{\text{eq,rad}}$ as $B_1 \to \infty$?

CHAPTER 9

GROUP THEORY

9.1 GROUPS

We have used symmetry in discussing molecular electronic structure and molecular vibrations and rotations. The full application of symmetry to such problems uses the branch of mathematics called *group theory*, the rudiments of which we shall present in this chapter. The initial development of group theory began in the late eighteenth century. Major advances were made by the mathematicians Évariste Galois (who was killed in a duel in 1832 at the age of 21) and Niels Abel (who died in 1829 of tuberculosis at the age of 26). The full formal definition of a group was given in 1854 by the mathematician Arthur Cayley.

The notion of a *group* is an abstract one. Consider a set of elements $A, B, C, \ldots$, no two of which are identical. The elements might be numbers, but there is no necessity that they be numbers. Suppose we have defined a rule of combination, symbolized by $*$, so that the "product" of any two elements A and B taken in a given order is uniquely determined as, say, P: $A * B = P$. (There is no necessity that the "product" taken in the reverse order, $B * A$, be equal to $A * B$.) The rule of combination is not necessarily arithmetical multiplication, but can be any well-defined rule. The set of elements $A, B, C, \ldots$ is said to form a *group* under the designated rule of combination if and only if the following four conditions are met:

(1) *Closure*. If A and B are any two members of the group, then their product $A * B$ must also be a member of the group.

(2) *Associativity*. The rule of combination must be such that the associative law holds: If A, B, and C are any three elements of the group, then $(A * B) * C = A * (B * C)$.

(3) *Identity*. The group must contain a single element I such that for any element A of the group, we have $A * I = I * A = A$. We call I the *identity*

element; "multiplication" by the identity element leaves an element unchanged.

(4) *Inverse*. Each element A of the group must have an inverse A^{-1} that is also a member of the group. By the term *inverse*, we mean that $A*A^{-1}=A^{-1}*A=I$, where I is the identity element.

In enumerating the set of elements $A,B,C,\ldots$ that form the group, we assume that they are all different. The number of elements in the group is called its *order*. In the closure requirement, we do not exclude the possibility that A and B might be the same element, that is, we require $A*A$ to be a member of the group. Note that the group elements are not arranged in any particular order.

Let us look at some examples. Consider the set of integers from 1 through 10, and let the rule of combination be addition. Do we have a group? The answer is no: The closure requirement is not met since, for example, $8+7=15$, and 15 is not a member of the set of elements $1,2,\ldots,10$.

Now consider the set of all positive integers $1,2,3,\ldots$ with the rule of combination being ordinary multiplication. Closure is satisfied since the product of any two positive integers is a positive integer. Ordinary multiplication is associative, so that the associativity condition is met. [One tends to take associativity for granted, but it does not always hold. Thus exponentiation is not an associative operation; for example, $(2^3)^2=64$, but $2^{(3^2)}=512$.] The identity element is 1, and condition (3) is met. However, the only member of the set with an inverse is 1 (which is its own inverse), so that condition (4) is not met. We do not have a group.

Next, consider the set of all real numbers except zero, and let the rule of combination be ordinary multiplication. The product of two nonzero real numbers is a nonzero real number, so that closure is satisfied. Associativity is satisfied. The identity element is 1. Finally, every element has an inverse, the inverse being the ordinary arithmetical reciprocal of the number. The set of all nonzero real numbers forms a group under ordinary multiplication. This group is of infinite order. (If zero were included in the set, we would not have a group, since zero would have no inverse.)

A group where the rule of combination is not arithmetical multiplication is the set of all integers, positive, negative, and zero, with the rule of combination being ordinary addition. Closure is satisfied. Addition is associative. The identity element is zero: $A+0=0+A=A$. The inverse of each element is its negative: $A+(-A)=-A+A=0$.

A group whose elements are not numbers is the set of all symmetry operations of a molecule, with the rule of combination being successive performance of the two operations being combined. If $\hat{R}$ and $\hat{S}$ are two symmetry operations, each of them leaves the molecule in a position physically indistinguishable from the original one; hence successive performance of $\hat{R}$ and $\hat{S}$ must leave the molecule in a physically indistinguishable position, so that the product $\hat{S}\hat{R}$ is one of the symmetry operations of the molecule. Thus closure is satisfied. Associativity holds for any operators [Equation (1.62)]. The identity element is the symmetry operation $\hat{E}$ (do nothing). Clearly, each symmetry operation has an inverse, the inverse being the operation that undoes the effect of the original operation. Thus the inverse of a counterclockwise 120° rotation ($\hat{C}_3$) is a clockwise 120° rotation, which is equivalent to a counterclockwise 240° rotation: $\hat{C}_3^{-1} = \hat{C}_3^2$, since $\hat{C}_3\hat{C}_3^2 = \hat{C}_3^2\hat{C}_3 = \hat{E}$. The group postulates are satisfied. Note that it is the symmetry *operations*, not the symmetry elements, that are the members (elements) of the molecular point group. The groups $\mathcal{C}_{\infty v}$, $\mathcal{D}_{\infty h}$, and $\mathcal{K}_h$ are of infinite order, whereas the remaining groups in Section 1.18 are of finite order.

9.2 MULTIPLICATION TABLES

The power of group theory derives from its abstractness. The mathematical theorems about groups are proved without reference to the exact nature of the group elements or the rule of combination; only the four group postulates are used. All the significant properties of a group are contained in its multiplication table, which is simply a listing of the products of all pairs of group elements. If the group elements are $A, B, C, \ldots$, then the multiplication table has the form (from here on, we shall drop the $*$ and write $A*B$ as AB)

	A	B	C	$\cdots$
A	AA	AB	AC	$\cdots$
B	BA	BB	BC	$\cdots$
C	CA	CB	CC	$\cdots$
.	.	.	.	$\cdots$
.	.	.	.	$\cdots$

Note the convention that in forming each product, the element on the side of the table is put on the left. In the definition of a group, it was not postulated that $AB = BA$. Groups such that $AB = BA$ for all pairs of elements are called *commutative* or *Abelian*.

The multiplication table for the group $\mathcal{C}_{3v}$ is given in Table 9.1. The three symmetry planes σ_a, σ_b, σ_c, are defined in Fig. 9.1 as making angles of 30°, 150°, and 270°, respectively, with the positive x axis. (The subscripts have no reference to the principal axes of inertia.) Consider the entry in the fourth row and second column; this product is $\hat{\sigma}_a\hat{C}_3$. The symmetry operation on the right of the product is performed first. The effect of $\hat{\sigma}_a\hat{C}_3$ is shown in Fig. 9.1. We use the convention that a symmetry operation moves the molecule and leaves the coordinate axes unmoved. The symmetry planes σ_a, σ_b, and σ_c are defined relative to the coordinate axes, and therefore do not move when a symmetry operation is performed. From Fig. 9.1, we have $\hat{\sigma}_a\hat{C}_3=\hat{\sigma}_b$, as given in the table. Note that $\hat{C}_3\hat{\sigma}_a=\hat{\sigma}_c\neq\hat{\sigma}_b$, so that $\mathcal{C}_{3v}$ is non-Abelian. Each reflection is its own inverse; the inverse of $\hat{C}_3$ is $\hat{C}_3^2$, and vice versa. (Of course, if A is the inverse of B, then B is the inverse of A.)

Table 9.1 Multiplication Table for the Group $\mathcal{C}_{3v}$

	$\hat{E}$	$\hat{C}_3$	$\hat{C}_3^2$	$\hat{\sigma}_a$	$\hat{\sigma}_b$	$\hat{\sigma}_c$
$\hat{E}$	$\hat{E}$	$\hat{C}_3$	$\hat{C}_3^2$	$\hat{\sigma}_a$	$\hat{\sigma}_b$	$\hat{\sigma}_c$
$\hat{C}_3$	$\hat{C}_3$	$\hat{C}_3^2$	$\hat{E}$	$\hat{\sigma}_c$	$\hat{\sigma}_a$	$\hat{\sigma}_b$
$\hat{C}_3^2$	$\hat{C}_3^2$	$\hat{E}$	$\hat{C}_3$	$\hat{\sigma}_b$	$\hat{\sigma}_c$	$\hat{\sigma}_a$
$\hat{\sigma}_a$	$\hat{\sigma}_a$	$\hat{\sigma}_b$	$\hat{\sigma}_c$	$\hat{E}$	$\hat{C}_3$	$\hat{C}_3^2$
$\hat{\sigma}_b$	$\hat{\sigma}_b$	$\hat{\sigma}_c$	$\hat{\sigma}_a$	$\hat{C}_3^2$	$\hat{E}$	$\hat{C}_3$
$\hat{\sigma}_c$	$\hat{\sigma}_c$	$\hat{\sigma}_a$	$\hat{\sigma}_b$	$\hat{C}_3$	$\hat{C}_3^2$	$\hat{E}$

A useful theorem states that in a given row or a given column of the multiplication table of a group, each element appears only once; thus each row and each column of the table is some permutation of the group elements. The proof is as follows: Suppose the element F appeared twice in the row having B as the left member of the product. We would have

$$BD=F, \qquad BG=F$$

where D and G are two different members of the group. Multiplication of each of these equations by B^{-1} on the left gives (since the group is not necessarily Abelian, we must put the factor B^{-1} on the same side of each member of the original equation; if $BD=F$, it does not necessarily follow that $B^{-1}BD=FB^{-1}$):

$$B^{-1}BD=B^{-1}F, \qquad B^{-1}BG=B^{-1}F$$

$$ID=B^{-1}F, \qquad IG=B^{-1}F$$

$$D=B^{-1}F, \qquad G=B^{-1}F$$

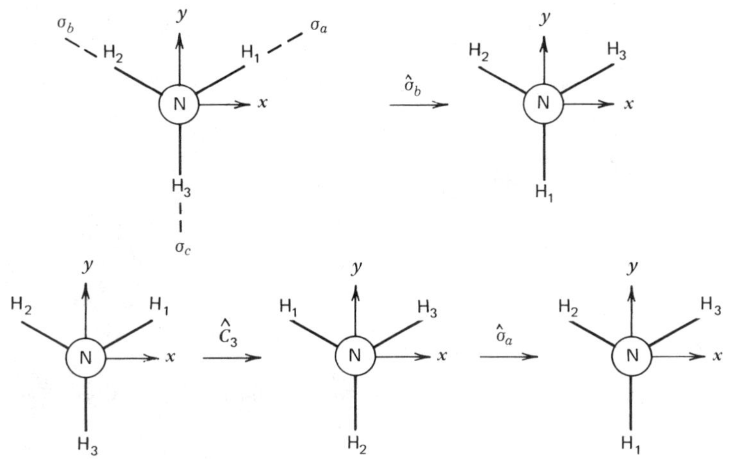

Fig. 9.1 The ammonia symmetry operations $\hat{\sigma}_b$ and $\hat{\sigma}_a\hat{C}_3$. The nitrogen atom lies above the xy plane.

Since the product $B^{-1}F$ is uniquely defined, we have $D=G$. But we postulated the group elements to be all different, so that D and G cannot be identical. Q.E.D.

Let us consider group multiplication tables in an abstract way, divorced from any particular group. Consider a group of order 1; what does its multiplication table look like? The single element must be the identity element I, and the multiplication table has the single entry $II=I$.

For a group of order 2, one element is the identity element I, and the other element we shall call A. The only possible form of the multiplication table of a group of order 2 is

	I	A
I	I	A
A	A	I

(9.1)

This table is derived as follows. We have $IA=AI=A$, and $II=I$. The only remaining product is AA, which by closure must be either A or I. If AA were equal to A, then A would occur twice in column 2 (and twice in row 2), thereby violating the theorem proved above.

Similar reasoning (Problem 9.11) gives as the only possible multiplication table for a group of order 3

$$\begin{array}{c|ccc} & I & A & B \\ \hline I & I & A & B \\ A & A & B & I \\ B & B & I & A \end{array} \tag{9.2}$$

For groups of order 4, there are two possible forms for the multiplication table (Problem 9.12).

From (9.1), (9.2), and Problem 9.12, we see that all groups of orders 1 through 4 are Abelian.

If two groups have the same order, and if the forms of their multiplication tables are the same, the groups are *isomorphic*. More specifically, if group 1 has elements $A_1, B_1, \ldots, G_1$, and group 2 has elements $A_2, B_2, \ldots, G_2$, and if the elements can be put into one-to-one correspondence

$$A_1 \leftrightarrow A_2 \qquad B_1 \leftrightarrow B_2 \quad \cdots \quad D_1 \leftrightarrow D_2 \quad \cdots \quad G_1 \leftrightarrow G_2$$

such that if $A_1 B_1 = D_1$, then $A_2 B_2 = D_2$, where A_1, B_1 and A_2, B_2 are any pairs of group elements, then groups 1 and 2 are isomorphic. Isomorphic groups have the same *structure*, but they may differ in the specific nature of their elements and the rule of combination.

Since there is only one possible structure for the multiplication table of a group of order 2, all groups of order 2 are isomorphic. Thus the point groups $\mathcal{C}_s$, $\mathcal{C}_i$, and $\mathcal{C}_2$ are isomorphic; also, the group of elements $+1$ and -1 with the rule of combination being ordinary multiplication is isomorphic to each of $\mathcal{C}_s$, $\mathcal{C}_i$, and $\mathcal{C}_2$.

If a subset of the elements of a group forms a group with respect to the same rule of combination as the original group, we say the subset of elements forms a *subgroup* of the original group. For example, the elements $\hat{E}$, $\hat{C}_3$, and $\hat{C}_3^2$ of the group $\mathcal{C}_{3v}$ form a subgroup of $\mathcal{C}_{3v}$; this subgroup is the point group $\mathcal{C}_3$. The proper rotations (including $\hat{E}$) of a point group form a subgroup called the *rotational subgroup*. Besides $\mathcal{C}_3$, the group $\mathcal{C}_{3v}$ has five other subgroups. There is the (trivial) subgroup of order 1, whose only element is $\hat{E}$; there are the three subgroups of order 2, each containing $\hat{E}$ and one of the reflections; there is the (trivial) subgroup of order 6, which is identical to the original group.

The element P of a group is said to be *conjugate* to the element Q of the same group if P and Q are related by $P = X^{-1}QX$, where X is one of the group elements (not necessarily different from P or Q). Note that if

$P = X^{-1}QX$, then multiplication on the left by X, followed by multiplication on the right by X^{-1}, gives $Q = XPX^{-1}$; now X^{-1} is some element of the group, which we shall call Y. We have $Q = Y^{-1}PY$. Hence if P is conjugate to Q, then Q is conjugate to P. Moreover (Problem 9.10), if P is conjugate to Q, and R is also conjugate to Q, then P and R are conjugate to each other. We can thus divide the elements of a group into subsets of mutually conjugate elements; each such subset is called a *class*.

As an example, we shall find the classes of $\mathcal{C}_{3v}$. To find the elements that are in the same class as $\hat{E}$, we form all possible products of the form $\hat{X}^{-1}\hat{E}\hat{X}$; since $\hat{X}^{-1}\hat{E}\hat{X} = \hat{E}\hat{X}^{-1}\hat{X} = \hat{E}^2 = \hat{E}$, the group element $\hat{E}$ is in a class by itself. Next, we take $\hat{\sigma}_a$; using the group multiplication table (Table 9.1), we find

$$\hat{E}^{-1}\hat{\sigma}_a\hat{E} = \hat{\sigma}_a \qquad \hat{C}_3^{-1}\hat{\sigma}_a\hat{C}_3 = \hat{\sigma}_c$$

$$\left(\hat{C}_3^2\right)^{-1}\hat{\sigma}_a\hat{C}_3^2 = \hat{\sigma}_b \qquad \hat{\sigma}_a^{-1}\hat{\sigma}_a\hat{\sigma}_a = \hat{\sigma}_a$$

$$\hat{\sigma}_b^{-1}\hat{\sigma}_a\hat{\sigma}_b = \hat{\sigma}_c \qquad \hat{\sigma}_c^{-1}\hat{\sigma}_a\hat{\sigma}_c = \hat{\sigma}_b$$

Thus $\hat{\sigma}_a$, $\hat{\sigma}_b$, and $\hat{\sigma}_c$ form a class of $\mathcal{C}_{3v}$. (As a check, we can form all products of the form $\hat{X}^{-1}\hat{\sigma}_b\hat{X}$ or $\hat{X}^{-1}\hat{\sigma}_c\hat{X}$.) Finally,

$$\hat{E}^{-1}\hat{C}_3\hat{E} = \hat{C}_3 \qquad \hat{C}_3^{-1}\hat{C}_3\hat{C}_3 = \hat{C}_3$$

$$\left(\hat{C}_3^2\right)^{-1}\hat{C}_3\hat{C}_3^2 = \hat{C}_3 \qquad \hat{\sigma}_a^{-1}\hat{C}_3\hat{\sigma}_a = \hat{C}_3^2$$

$$\hat{\sigma}_b^{-1}\hat{C}_3\hat{\sigma}_b = \hat{C}_3^2 \qquad \hat{\sigma}_c^{-1}\hat{C}_3\hat{\sigma}_c = \hat{C}_3^2$$

so that $\hat{C}_3$ and $\hat{C}_3^2$ form a class. $\mathcal{C}_{3v}$ has the following three classes:

$$(1)\ \hat{E}; \qquad (2)\ \hat{\sigma}_a, \hat{\sigma}_b, \hat{\sigma}_c; \qquad (3)\ \hat{C}_3, \hat{C}_3^2$$

Note that the members of each class are closely related symmetry operations.

For an Abelian group, each element is in a class by itself, since $X^{-1}AX = AX^{-1}X = A$. Since rotations about the same axis commute with each other, the group $\mathcal{C}_n$ is Abelian and has n classes, each class consisting of one symmetry operation.

Since inversion commutes with every other symmetry operation (this will be proved in Section 9.3), point groups of molecules with a center of symmetry have $\hat{i}$ in a class by itself.

9.3 COORDINATE TRANSFORMATIONS AND MATRICES

Our convention is that a symmetry operation $\hat{R}$ changes the locations of points in space, while the coordinate axes remain fixed. In contrast, in Section 1.2 we considered a change (proper or improper rotation) of coordinate axes, while the points in space remained fixed: Let $x'y'z'$ be a set of axes derived from the xyz axes by a proper or improper rotation. Consider a point fixed in space. We found that its coordinates in the $x'y'z'$ system are related to its coordinates in the xyz system by (1.120) or (2.29):

$$\mathbf{x}' = \mathbf{L}\mathbf{x} \tag{9.3}$$

where $\mathbf{L}$ is the orthogonal matrix of direction cosines of the rotated coordinate axes.

Let the symmetry operation $\hat{R}$ take a general point P with coordinates (x,y,z) to the location (x_R,y_R,z_R), leaving the xyz axes fixed. We want to relate (x_R,y_R,z_R) to (x,y,z). To do this, we carry out the symmetry operation $\hat{R}$ in two steps: (a) We first apply the (proper or improper) rotation $\hat{R}$ to all points in space *and* to the axes. (b) We then rotate the axes back to their original location, leaving fixed the points in space. Figure 9.2 illustrates these steps for a $\hat{C}_8(z)$ operation. Step (a) clearly leaves the coordinates of the general point P unchanged. Step (b) is a coordinate-axis rotation (proper or improper) and (9.3) applies. However, the rotation in step (b) is the *inverse* of $\hat{R}$. Hence (9.3) becomes

$$\mathbf{x}_R = \mathbf{L}(\hat{R}^{-1})\mathbf{x} \tag{9.4}$$

where $\mathbf{L}(\hat{R}^{-1})$ is the direction-cosine matrix of the coordinate system obtained by applying $\hat{R}^{-1}$ to the xyz axes, and where $\mathbf{x}_R$ is a column vector with elements x_R, y_R, z_R. We shall call the direction-cosine matrix in (9.4) the $\mathbf{R}$ matrix:

$$\mathbf{R} \equiv \mathbf{L}(\hat{R}^{-1}) \tag{9.5}$$

$$\mathbf{x}_R = \mathbf{R}\mathbf{x} \tag{9.6}$$

$$\begin{pmatrix} x_R \\ y_R \\ z_R \end{pmatrix} = \begin{pmatrix} r_{11} & r_{12} & r_{13} \\ r_{21} & r_{22} & r_{23} \\ r_{31} & r_{32} & r_{33} \end{pmatrix} \begin{pmatrix} x \\ y \\ z \end{pmatrix} \tag{9.7}$$

Let us find the matrix $\mathbf{R}$ for the various symmetry operations. The

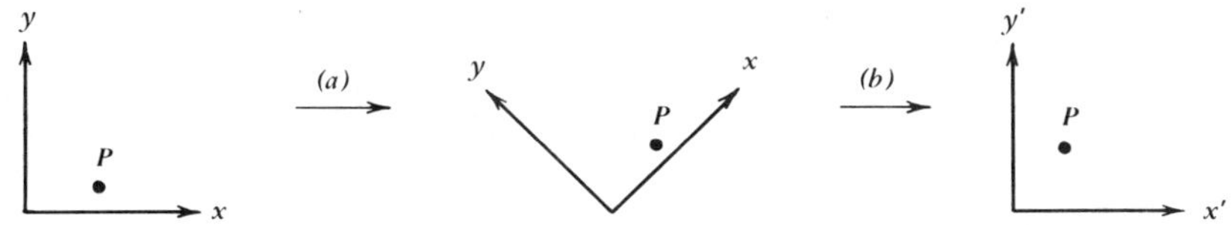

Fig. 9.2 Two-step performance of $\hat{C}_8(z)$.

identity operation $\hat{E}$ corresponds to the coordinate transformation

$$x_R = x, \qquad y_R = y, \qquad z_R = z \tag{9.8}$$

Comparing (9.8) with (9.7), we have as the matrix corresponding to $\hat{E}$:

$$\begin{pmatrix} 1 & 0 & 0 \\ 0 & 1 & 0 \\ 0 & 0 & 1 \end{pmatrix} \tag{9.9}$$

For the inversion operation $\hat{i}$, we have

$$x_R = -x, \qquad y_R = -y, \qquad z_R = -z \tag{9.10}$$

The matrix corresponding to $\hat{i}$ is thus

$$\begin{pmatrix} -1 & 0 & 0 \\ 0 & -1 & 0 \\ 0 & 0 & -1 \end{pmatrix} \tag{9.11}$$

This matrix is simply -1 times the unit matrix; therefore, (9.11) commutes with every third-order square matrix. We shall shortly show that the matrices corresponding to the various symmetry operations multiply in exactly the same way the corresponding symmetry operations do; hence the symmetry operation $\hat{i}$ commutes with every other symmetry operation.

Next, consider a proper counterclockwise rotation about the z axis by an angle α. $\hat{R}^{-1}$ is a clockwise rotation by angle α about the z axis. After $\hat{R}^{-1}$ is applied to the axes, the x' axis makes an angle α with the x axis, an angle $\alpha + 90°$ with the y axis, and an angle $90°$ with the z axis (Fig. 9.3a). Since $\cos(\alpha + 90°) = \cos\alpha \cos 90° - \sin\alpha \sin 90° = -\sin\alpha$, the direction cosines of the x' axis are $\cos\alpha$, $-\sin\alpha$, and $\cos 90° = 0$. This gives the top row of the **R** matrix. After $\hat{R}^{-1}$, the y' axis makes angles of $90° - \alpha$, α, and $90°$ with the x, y, and z axes; the direction cosines of y' are $\cos(90° - \alpha) = \sin\alpha$, $\cos\alpha$, and 0. This is the second row of **R**. The z' axis coincides with

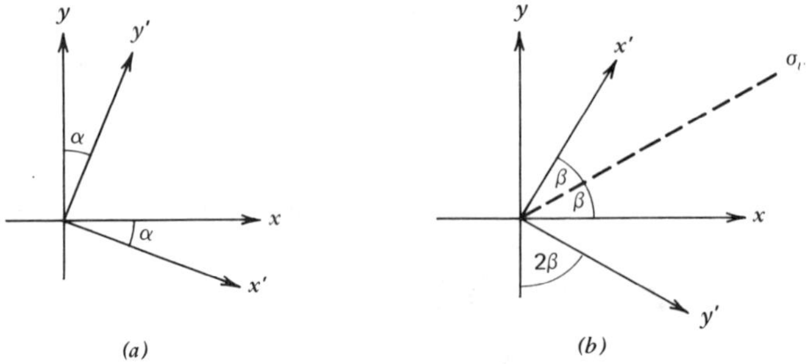

Fig. 9.3 Effects on coordinate axes of (*a*) rotation about the *z* axis; (*b*) reflection in a vertical plane.

the z axis, and its direction cosines are 0, 0, and 1. For a $\hat{C}_n(z)$ rotation, the angle α equals $2\pi/n$, and the $\mathbf{R}$ matrix is therefore

$$\begin{pmatrix} \cos(2\pi/n) & -\sin(2\pi/n) & 0 \\ \sin(2\pi/n) & \cos(2\pi/n) & 0 \\ 0 & 0 & 1 \end{pmatrix} \tag{9.12}$$

For $[\hat{C}_n(z)]^m$, the angle α is $2\pi m/n$, so that the matrix corresponding to this operation is

$$\begin{pmatrix} \cos(2\pi m/n) & -\sin(2\pi m/n) & 0 \\ \sin(2\pi m/n) & \cos(2\pi m/n) & 0 \\ 0 & 0 & 1 \end{pmatrix} \tag{9.13}$$

The operation $\hat{S}_n(z)$ has the same effect on the x and y coordinates as $\hat{C}_n(z)$, but replaces z by $-z$. Hence the matrix corresponding to $\hat{S}_n(z)$ is

$$\begin{pmatrix} \cos(2\pi/n) & -\sin(2\pi/n) & 0 \\ \sin(2\pi/n) & \cos(2\pi/n) & 0 \\ 0 & 0 & -1 \end{pmatrix} \tag{9.14}$$

The matrix corresponding to $[\hat{S}_n(z)]^m$ is

$$\begin{pmatrix} \cos(2\pi m/n) & -\sin(2\pi m/n) & 0 \\ \sin(2\pi m/n) & \cos(2\pi m/n) & 0 \\ 0 & 0 & (-1)^m \end{pmatrix} \tag{9.15}$$

The matrix corresponding to reflection in the *xy* plane is

$$\begin{pmatrix} 1 & 0 & 0 \\ 0 & 1 & 0 \\ 0 & 0 & -1 \end{pmatrix} \tag{9.16}$$

Now consider a $\hat{\sigma}_v$ reflection in a plane that contains the z axis and makes an angle β with the x axis. This operation is its own inverse, and Fig. 9.3*b* shows the effect on the axes. The direction cosines of the x' axis are $\cos 2\beta$, $\cos(90° - 2\beta) = \sin 2\beta$, and 0. Those of the y' axis are $\cos(90° - 2\beta) = \sin 2\beta$, $\cos(180° - 2\beta) = -\cos 2\beta$, and 0. The **R** matrix is therefore

$$\begin{pmatrix} \cos 2\beta & \sin 2\beta & 0 \\ \sin 2\beta & -\cos 2\beta & 0 \\ 0 & 0 & 1 \end{pmatrix} \tag{9.17}$$

We now show that *the matrices corresponding to the symmetry operations multiply the same way the symmetry operations do.* Consider the symmetry operations $\hat{R}$ and $\hat{S}$, and let $\hat{S}\hat{R} = \hat{T}$. The operation $\hat{R}$ applied to all points in space brings the point originally at (x,y,z) to (x_R, y_R, z_R), according to (9.7):

$$\mathbf{x}_R = \mathbf{R}\mathbf{x}$$

We then apply $\hat{S}$, which brings the point at (x_R, y_R, z_R) to (x_{SR}, y_{SR}, z_{SR}), according to

$$\mathbf{x}_{SR} = \mathbf{S}\mathbf{x}_R = \mathbf{S}\mathbf{R}\mathbf{x} \tag{9.18}$$

The symmetry operation $\hat{T}$ brings the point originally at (x,y,z) to (x_T, y_T, z_T):

$$\mathbf{x}_T = \mathbf{T}\mathbf{x} \tag{9.19}$$

However, since $\hat{T}$ equals $\hat{S}\hat{R}$, it must have the same effect as $\hat{S}\hat{R}$ on the point with coordinates **x**. Therefore, $\mathbf{x}_{SR} = \mathbf{x}_T$, and (9.18) and (9.19) give

$$\mathbf{SRx} = \mathbf{Tx} \qquad \text{or} \qquad (\mathbf{SR} - \mathbf{T})\mathbf{x} = \mathbf{0} \tag{9.20}$$

The only way (9.20) can be satisfied for every possible set of initial coordinates (x,y,z) is for **SR** to equal **T**. More formally, (2.11) and (2.12) give the *i*th element of the column matrix $(\mathbf{SR} - \mathbf{T})\mathbf{x}$ as

$$\sum_k \left[\left(\sum_j s_{ij} r_{jk} - t_{ik} \right) x_k \right] = 0$$

where $x_1=x$, $x_2=y$, and $x_3=z$. Since x, y, and z are linearly independent, the coefficient of x_k (in parentheses) must vanish. This proves each element of **T** to equal the corresponding element of **SR**. Hence **SR** = **T** and the matrices multiply the same way the symmetry operations do.

We conclude this section by considering the operator $\hat{O}_R$ that corresponds to the symmetry operation $\hat{R}$. From (1.271), the definition of $\hat{O}_R$ is

$$\hat{O}_R f(x_R, y_R, z_R) = f(x, y, z) \tag{9.21}$$

where the effect of $\hat{R}$ on a general point in space (x,y,z) is given by (9.6) as $\mathbf{Rx}=\mathbf{x}_R$. We shall write (9.21) as

$$\hat{O}_R f(\mathbf{x}_R) = f(\mathbf{x}) \tag{9.22}$$

where $f(\mathbf{x})$ indicates that the function f depends on the components (x,y,z) of $\mathbf{x}$. Application of $\mathbf{R}^{-1}$ to (9.6) gives $\mathbf{x}=\mathbf{R}^{-1}\mathbf{x}_R$. Hence (9.22) becomes

$$\hat{O}_R f(\mathbf{x}_R) = f(\mathbf{R}^{-1}\mathbf{x}_R) \tag{9.23}$$

The point (x_R, y_R, z_R) is a general point in space, and there is no harm done if we drop the subscripts from (9.23) to give

$$\hat{O}_R f(\mathbf{x}) = f(\mathbf{R}^{-1}\mathbf{x}) \tag{9.24}$$

As an illustration of (9.24), let $\hat{R}=\hat{C}_4(z)$. We have $\hat{C}_4(z)^{-1}=[\hat{C}_4(z)]^3$. Using (9.7) and (9.13) with $m/n=\frac{3}{4}$, we find that $\hat{\mathbf{R}}^{-1}$ brings a point at (x,y,z) to $(y,-x,z)$. Hence (9.24) gives

$$\hat{O}_{C_4(z)} f(x,y,z) = f(y,-x,z)$$

For example, let f be a $2p_x$ atomic orbital:

$$f = 2p_x = Nxe^{-k(x^2+y^2+z^2)^{1/2}}$$

where N and k are constants. We have

$$\hat{O}_{C_4(z)} 2p_x = Nye^{-k(y^2+x^2+z^2)^{1/2}} = 2p_y$$

as noted previously (Section 1.18).

The operators $\hat{O}_R$ are linear, and they multiply the same way the corresponding symmetry operations multiply; that is, if $\hat{R}\hat{S}=\hat{T}$, then $\hat{O}_R\hat{O}_S=\hat{O}_T$. See Problem 9.13.

9.4 REPRESENTATIONS

We showed in the last section how to find the matrices which acting on the column vector with elements x, y, z will carry out the coordinate transformations corresponding to the symmetry operations of a point group. We further showed that these matrices multiply in the same way as the corresponding symmetry operations, that is, if $\hat{S}\hat{R}=\hat{T}$, then $\mathbf{SR}=\mathbf{T}$, where $\mathbf{R}$, $\mathbf{S}$, and $\mathbf{T}$ are the matrices corresponding to the symmetry operations $\hat{R}$, $\hat{S}$, and $\hat{T}$. If we make a multiplication table of the matrices corresponding to the symmetry operations of a point group, this table must have the same structure as the symmetry-operations multiplication table; hence the matrices form a group that is isomorphic to the point group of symmetry operations.

Take the group $\mathcal{C}_{3v}$, for example. From Fig. 9.1 and Eqs. (9.9), (9.12), (9.13), and (9.17), the matrices corresponding to the six group symmetry operations are

$$\hat{E}:\begin{pmatrix}1&0&0\\0&1&0\\0&0&1\end{pmatrix}=\mathbf{I}\qquad \hat{C}_3:\begin{pmatrix}-\frac{1}{2}&-\frac{1}{2}\sqrt{3}&0\\ \frac{1}{2}\sqrt{3}&-\frac{1}{2}&0\\0&0&1\end{pmatrix}=\mathbf{A}$$

$$\hat{C}_3^2:\begin{pmatrix}-\frac{1}{2}&\frac{1}{2}\sqrt{3}&0\\ -\frac{1}{2}\sqrt{3}&-\frac{1}{2}&0\\0&0&1\end{pmatrix}=\mathbf{B}\qquad \hat{\sigma}_a:\begin{pmatrix}\frac{1}{2}&\frac{1}{2}\sqrt{3}&0\\ \frac{1}{2}\sqrt{3}&-\frac{1}{2}&0\\0&0&1\end{pmatrix}=\mathbf{C}$$

$$\hat{\sigma}_b:\begin{pmatrix}\frac{1}{2}&-\frac{1}{2}\sqrt{3}&0\\ -\frac{1}{2}\sqrt{3}&-\frac{1}{2}&0\\0&0&1\end{pmatrix}=\mathbf{D}\qquad \hat{\sigma}_c:\begin{pmatrix}-1&0&0\\0&1&0\\0&0&1\end{pmatrix}=\mathbf{F}\qquad(9.25)$$

We can verify directly that these matrices multiply in the same manner as the corresponding symmetry operations. Carrying out the matrix multiplications, we get Table 9.2, which has the same structure as Table 9.1.

The matrices of (9.25) are not the only matrices that multiply in the same manner as the $\mathcal{C}_{3v}$ group operations, as we shall shortly see. Any set of nonnull square matrices that multiply in the same way as the elements of a given group is said to form a *representation* of that group. The order of the matrices is called the *dimension* of the representation. To form a representation, the correspondence between matrices and group elements need not be one-to-one; rather, a given matrix can correspond to more

Table 9.2 Multiplication Table for Matrices Forming a Representation of $\mathcal{C}_{3v}$

	I	A	B	C	D	F
I	I	A	B	C	D	F
A	A	B	I	F	C	D
B	B	I	A	D	F	C
C	C	D	F	I	A	B
D	D	F	C	B	I	A
F	F	C	D	A	B	I

than one group element. If the correspondence is one-to-one, the matrices form a group that is isomorphic (Section 9.2) to the group for which they form a representation.

Let us find more representations of $\mathcal{C}_{3v}$. Starting with the six matrices (9.25), we can get an infinite number of three-dimensional representations by the following device. We form the matrices

$$\mathbf{P}^{-1}\mathbf{IP},\ \mathbf{P}^{-1}\mathbf{AP},\ \mathbf{P}^{-1}\mathbf{BP},\ \mathbf{P}^{-1}\mathbf{CP},\ \mathbf{P}^{-1}\mathbf{DP},\ \mathbf{P}^{-1}\mathbf{FP} \tag{9.26}$$

where $\mathbf{P}$ is any nonsingular square matrix of order 3. The proof that the matrices (9.26) form a representation of $\mathcal{C}_{3v}$ is simple. We know, for example, that $\mathbf{AC}=\mathbf{F}$ (Table 9.2). For the corresponding product of the matrices (9.26), we have

$$(\mathbf{P}^{-1}\mathbf{AP})(\mathbf{P}^{-1}\mathbf{CP})=\mathbf{P}^{-1}\mathbf{A}(\mathbf{PP}^{-1})\mathbf{CP}=\mathbf{P}^{-1}\mathbf{ACP}=\mathbf{P}^{-1}\mathbf{FP}$$

The same argument holds for every other matrix product; having shown that the matrices of (9.26) multiply the same way the corresponding matrices (9.25) multiply, we have proved that (9.26) is a representation of $\mathcal{C}_{3v}$. Two representations whose matrices are related by the same similarity transformation (Section 2.1) are said to be *equivalent*. The representations (9.25) and (9.26) are equivalent. Obviously, equivalent representations have the same dimension.

Aside from the infinite number of representations equivalent to (9.25), there are other representations of $\mathcal{C}_{3v}$. The matrices (9.25) are each in block-diagonal form. If we partition them into submatrices of the form (2.30), then the submatrices $\mathbf{S}_{12}$ and $\mathbf{S}_{21}$ will be column and row matrices whose elements are all zero. The product of any two of these partitioned matrices will have the form

$$\mathbf{ST}=\begin{pmatrix} \mathbf{S}_{11}\mathbf{T}_{11} & \mathbf{0} \\ \mathbf{0} & \mathbf{S}_{22}\mathbf{T}_{22} \end{pmatrix}$$

Take the specific product $\mathbf{AC}=\mathbf{F}$. We have

$$\mathbf{AC}=\begin{pmatrix} \mathbf{A}_{11}\mathbf{C}_{11} & \mathbf{0} \\ \mathbf{0} & \mathbf{A}_{22}\mathbf{C}_{22} \end{pmatrix}=\mathbf{F}=\begin{pmatrix} \mathbf{F}_{11} & \mathbf{0} \\ \mathbf{0} & \mathbf{F}_{22} \end{pmatrix} \tag{9.27}$$

$$\mathbf{A}_{11}=\begin{pmatrix} -\frac{1}{2} & -\frac{1}{2}\sqrt{3} \\ \frac{1}{2}\sqrt{3} & -\frac{1}{2} \end{pmatrix} \quad \mathbf{C}_{11}=\begin{pmatrix} \frac{1}{2} & \frac{1}{2}\sqrt{3} \\ \frac{1}{2}\sqrt{3} & -\frac{1}{2} \end{pmatrix} \quad \mathbf{F}_{11}=\begin{pmatrix} -1 & 0 \\ 0 & 1 \end{pmatrix}$$

$$\mathbf{A}_{22}=(1) \qquad \mathbf{C}_{22}=(1) \qquad \mathbf{F}_{22}=(1)$$

Since equality of matrices means equality of corresponding matrix elements, equal matrices must have their corresponding submatrices equal; (9.27) implies that

$$\mathbf{A}_{11}\mathbf{C}_{11}=\mathbf{F}_{11}, \qquad \mathbf{A}_{22}\mathbf{C}_{22}=\mathbf{F}_{22}$$

The same argument holds for any product of two of the matrices (9.25). Therefore, corresponding block-diagonal submatrices of the matrices (9.25) multiply in the same way as the original matrices. We have thus found two further representations of $\mathcal{C}_{3v}$; these are formed by the matrices

$$\mathbf{I}_{11}=\begin{pmatrix} 1 & 0 \\ 0 & 1 \end{pmatrix} \quad \mathbf{A}_{11}=\begin{pmatrix} -\frac{1}{2} & -\frac{1}{2}\sqrt{3} \\ \frac{1}{2}\sqrt{3} & -\frac{1}{2} \end{pmatrix} \quad \mathbf{B}_{11}=\begin{pmatrix} -\frac{1}{2} & \frac{1}{2}\sqrt{3} \\ -\frac{1}{2}\sqrt{3} & -\frac{1}{2} \end{pmatrix}$$

$$\mathbf{C}_{11}=\begin{pmatrix} \frac{1}{2} & \frac{1}{2}\sqrt{3} \\ \frac{1}{2}\sqrt{3} & -\frac{1}{2} \end{pmatrix} \quad \mathbf{D}_{11}=\begin{pmatrix} \frac{1}{2} & -\frac{1}{2}\sqrt{3} \\ -\frac{1}{2}\sqrt{3} & -\frac{1}{2} \end{pmatrix} \quad \mathbf{F}_{11}=\begin{pmatrix} -1 & 0 \\ 0 & 1 \end{pmatrix} \tag{9.28}$$

and by the matrices (scalars)

$$\mathbf{I}_{22}=(1), \quad \mathbf{A}_{22}=(1), \quad \mathbf{B}_{22}=(1), \quad \mathbf{C}_{22}=(1), \quad \mathbf{D}_{22}=(1), \quad \mathbf{F}_{22}=(1) \tag{9.29}$$

Any representation of dimension greater than one, all of whose matrices can be converted by the same similarity transformation (i.e., by left multiplication by $\mathbf{P}^{-1}$ and right multiplication by $\mathbf{P}$, where $\mathbf{P}$ is some nonsingular square matrix) into the same block-diagonal form, is said to be a *reducible* representation; otherwise the representation is *irreducible*. Thus the representation (9.25) is reducible, as are all representations equivalent to (9.25). The representation (9.29) is irreducible. The representation (9.28) can be shown to be irreducible (see the next section).

Clearly, every group possesses the irreducible representation in which every group element corresponds to a one-dimensional unit matrix; this representation is called the *totally symmetric* one.

In a reducible representation, a given irreducible representation may occur more than once. For example, in the reducible four-dimensional representation in which the matrix

$$\begin{pmatrix} 1 & 0 & 0 & 0 \\ 0 & 1 & 0 & 0 \\ 0 & 0 & 1 & 0 \\ 0 & 0 & 0 & 1 \end{pmatrix}$$

corresponds to each group element, the irreducible totally symmetric representation occurs four times. By forming block-diagonal matrices that contain the matrices of irreducible representations as submatrices, we can form a representation that contains each irreducible representation any desired number of times.

A reducible representation is said to be the *direct sum* of the irreducible representations of which it is made up. In the standard notation (Section 9.5) for point-group irreducible representations, the $\mathcal{C}_{3v}$ representation (9.29) is called A_1 and the representation (9.28) is called E. If we denote the reducible representation (9.25) by Γ, then

$$\Gamma = A_1 \oplus E$$

where the $\oplus$ sign indicates the direct sum. Letting Δ denote the reducible representation in which each group element is represented by the fourth-order unit matrix, we write $\Delta = A_1 \oplus A_1 \oplus A_1 \oplus A_1$, or more compactly: $\Delta = 4A_1$.

9.5 CHARACTER TABLES

The connection between group theory and quantum mechanics will be established in Section 9.7, where the importance of the representations of a group will become evident. Instead of dealing with the matrices of a representation, for many purposes the information provided by their traces is sufficient. If in a certain representation the matrix $\mathbf{D}(\hat{R})$ corresponds to the symmetry operation $\hat{R}$, then the trace of $\mathbf{D}(\hat{R})$ is called the *character* of $\hat{R}$ for that representation. For example, in the totally symmetric representation (9.29) the character of each of the six symmetry operations of $\mathcal{C}_{3v}$ is 1. For the representation (9.25), the character of $\hat{E}$ is 3, the characters of $\hat{C}_3$ and $\hat{C}_3^{-1}$ are each 0, and the characters of the three reflections are each 1. Since the identity operation $\hat{E}$ is always represented by a unit matrix, the character of $\hat{E}$ equals the dimension of the representation.

Two symmetry operations $\hat{R}$ and $\hat{S}$ that are in the same class (Section 9.2) are related by $\hat{R}=\hat{T}^{-1}\hat{S}\hat{T}$, where $\hat{T}$ is some symmetry operation of the point group. Let $\mathbf{R}$, $\mathbf{S}$, and $\mathbf{T}$ be the matrices that correspond to the operations $\hat{R}$, $\hat{S}$, and $\hat{T}$ in some particular representation. We then have $\mathbf{R}=\mathbf{T}^{-1}\mathbf{S}\mathbf{T}$. Since similar matrices have equal traces (Section 2.1), we conclude that in any given representation, the characters of symmetry operations in the same class are equal.

We shall see that it is the nonequivalent irreducible representations that are of quantum-mechanical significance. For these representations, the following theorems hold[1]:

(1) The number of nonequivalent irreducible representations of a group is equal to the number of classes in the group.

(2) The sum of the squares of the dimensions of all the nonequivalent irreducible representations of a group is equal to the order of the group.

(3) The sum of the squares of the absolute values of the characters in any irreducible representation is equal to the order of the group.

(4) The characters of two nonequivalent irreducible representations i and j satisfy

$$\sum_{\hat{R}} \chi_i^*(\hat{R})\chi_j(\hat{R})=0 \tag{9.30}$$

where $\chi_i(\hat{R})$ and $\chi_j(\hat{R})$ are the characters of the symmetry operation $\hat{R}$ in the representations i and j, and where the sum runs over the h symmetry operations of the group. (The characters are not necessarily real numbers.)

Theorems (3) and (4) can be combined in the single equation

$$\sum_{\hat{R}} \chi_i^*(\hat{R})\chi_j(\hat{R})=h\delta_{ij} \tag{9.31}$$

where h is the order of the group.

Equation (9.31) implies that two nonequivalent irreducible representations cannot have the same set of characters.

[1] For the proofs, see *Hamermesh*, pp. 101–111.

The converse of Theorem (3) is valid:

(5) If the characters for some particular representation i satisfy

$$\sum_{\hat{R}} |\chi_i(\hat{R})|^2 = h$$

then i is irreducible.

We illustrate these theorems by applying them to $\mathcal{C}_{3v}$. For the representation (9.28), we find

$$\sum_{\hat{R}} |\chi_i(\hat{R})|^2 = 2^2 + (-1)^2 + (-1)^2 + 0^2 + 0^2 + 0^2 = 6 \tag{9.32}$$

Since the order of $\mathcal{C}_{3v}$ is 6, Theorem (5) tells us that (9.28) is irreducible. We saw in Section 9.2 that $\mathcal{C}_{3v}$ has three classes; hence by Theorem (1), $\mathcal{C}_{3v}$ has three nonequivalent irreducible representations. Calling the dimensions of these three representations l_1, l_2, and l_3, we have from Theorem (2):

$$l_1^2 + l_2^2 + l_3^2 = 6 \tag{9.33}$$

The only set of positive integers that satisfies (9.33) is 1, 1, 2. Thus, besides the irreducible nonequivalent representations (9.28) and (9.29), there is one further such representation, and its dimension is 1. This representation can be found using (9.31) in conjunction with the characters of (9.28) and (9.29). However, the fastest way to find it is to evaluate the determinants of the matrices (9.28). From (2.17), it follows that the determinants of the matrices of any representation multiply in the same way as the matrices multiply. Hence these determinants form a one-dimensional representation. Evaluating the determinants of the matrices (9.28), we get as the third nonequivalent irreducible representation of $\mathcal{C}_{3v}$:

$$(1), \quad (1), \quad (1), \quad (-1), \quad (-1), \quad (-1) \tag{9.34}$$

where the matrices (9.34) correspond to the operations $\hat{E}$, $\hat{C}_3$, $\hat{C}_3^2$, $\hat{\sigma}_a$, $\hat{\sigma}_b$, and $\hat{\sigma}_c$, respectively.

The nonequivalent irreducible representations (9.28), (9.29), and (9.34) can be used to illustrate (9.31). For example, letting i be (9.28) and j be (9.34), we find

$$\sum_{\hat{R}} \chi_i^*(\hat{R})\chi_j(\hat{R}) = 2\cdot 1 - 1\cdot 1 - 1\cdot 1 + 0\cdot(-1) + 0\cdot(-1) + 0\cdot(-1) = 0$$

The characters of the group symmetry operations for each nonequivalent irreducible representation are listed in the *character table* of the group.

Taking the traces of the matrices in (9.28), (9.29), and (9.34), we find the following character table for $\mathcal{C}_{3v}$:

$\mathcal{C}_{3v}$	$\hat{E}$	$\hat{C}_3$	$\hat{C}_3^2$	$\hat{\sigma}_a$	$\hat{\sigma}_b$	$\hat{\sigma}_c$
(9.29)	1	1	1	1	1	1
(9.34)	1	1	1	−1	−1	−1
(9.28)	2	−1	−1	0	0	0

(9.35)

The character table can be abbreviated by use of the theorem that symmetry operations in the same class have the same character in any given representation. Thus the $\mathcal{C}_{3v}$ character table is written as

$\mathcal{C}_{3v}$	$\hat{E}$	$2\hat{C}_3$	$3\hat{\sigma}_v$
A_1	1	1	1
A_2	1	1	−1
E	2	−1	0

(9.36)

where $3\hat{\sigma}_v$ stands for the three symmetry reflections, and $2\hat{C}_3$ stands for the operations $\hat{C}_3$ and $\hat{C}_3^2$ $(=\hat{C}_3^{-1})$.

We have introduced certain labels in (9.36) for the three nonequivalent irreducible representations of $\mathcal{C}_{3v}$. The conventions (formulated by Mulliken) for such labels are the following:

(a) One-dimensional irreducible representations are labeled either A or B, usually according to whether the character of a $2\pi/n$ (proper or improper) rotation about the symmetry axis of highest order (n) is $+1$ or -1, respectively. (In certain cases, the character of this rotation is a complex number; this situation will be discussed in Section 9.7.)

(b) Two-dimensional irreducible representations are labeled E (not to be confused with the identity operation $\hat{E}$).

(c) Three-dimensional irreducible representations are labeled T. (The alternative symbol F is often used.)

(d) Four-dimensional and five-dimensional irreducible representations (which occur for $\mathcal{I}_h$) are labeled G and H, respectively.

(e) If the molecule has a center of symmetry, a subscript g or u is added to the label according to whether the character of $\hat{i}$ is positive or negative, respectively. The inversion operation is always represented by $+1$ or -1 times a unit matrix[2]; hence the character of $\hat{i}$ is either $+l$ or $-l$, where l is the dimension of the representation.

[2] For the proof, see *Schonland*, pp. 97–98, 263–264.

(f) If the molecule has a σ_h plane but no center of symmetry (groups $\mathcal{C}_{nh}$ and $\mathcal{D}_{nh}$ with n odd), the labels are primed or double primed according to whether the character of $\hat{\sigma}_h$ is positive or negative. The situation is similar to molecules with a center of symmetry in that $\hat{\sigma}_h$ will be represented by either $+1$ or -1 times a unit matrix.[2]

(g) If necessary, numerical subscripts are added to distinguish non-equivalent irreducible representations not distinguished by rules (a)–(e). The totally symmetric representation is always numbered and listed first.

Character tables for many point groups are listed in Section 9.12 at the end of this chapter. As an example, consider the character table for $\mathcal{D}_{6h}$, the point group of benzene. This group is of order 24. The symmetry operations are found to be divided into the 12 classes: (1) $\hat{E}$; (2) $\hat{C}_6$, $\hat{C}_6^5$; (3) $\hat{C}_6^2$, $\hat{C}_6^4$; (4) $\hat{C}_6^3$; (5) $\hat{C}_{2,a}$, $\hat{C}_{2,b}$, $\hat{C}_{2,c}$; (6) $\hat{C}'_{2,a}$, $\hat{C}'_{2,b}$, $\hat{C}'_{2,c}$; (7) $\hat{\sigma}_h$; (8) $\hat{\sigma}_{v,a}$, $\hat{\sigma}_{v,b}$, $\hat{\sigma}_{v,c}$; (9) $\hat{\sigma}_{d,a}$, $\hat{\sigma}_{d,b}$, $\hat{\sigma}_{d,c}$; (10) $\hat{S}_6$, $\hat{S}_6^5$; (11) $\hat{S}_3$, $\hat{S}_3^5$; (12) $\hat{i}$.

The operations of classes (2), (3), (4), (10), and (11) are performed with respect to the C_6 axis, which is also a C_3, C_2, S_6, S_3, and S_2 axis. Note that the second member of class (11) is $\hat{S}_3^5$, and not $\hat{S}_3^2$; the operation $\hat{S}_3^2$ is the same as $\hat{C}_3^2$. The operations $\hat{S}_6^2$, $\hat{S}_6^3$, $\hat{S}_6^4$, and $\hat{S}_6^6$ are not listed since they are the same as $\hat{C}_6^2$, $\hat{i}$, $\hat{C}_6^4$, and $\hat{E}$, respectively. The operations of classes (5) and (6) are rotations about the six C_2 axes perpendicular to the C_6 axis; in benzene, the convention is that unprimed C_2 axes pass through opposite pairs of carbon atoms, whereas the primed C_2 axes pass between the carbon atoms. The two kinds of vertical symmetry planes are distinguished by labeling them either v or d; the convention for benzene is shown in Fig. 9.4. The operations of class (3) are listed in the character table as $2\hat{C}_3$, since they are both 120° rotations, one counterclockwise, the other clockwise; similar abbreviations are used for the other proper and improper rotations. Since there are 12 classes, there are 12 irreducible representations; their dimensions satisfy Theorem (2):

$$l_1^2 + l_2^2 + \cdots + l_{12}^2 = 24$$

By trial and error, the only solution to this equation is found to be eight one-dimensional and four two-dimensional representations, as listed in the character table. There is no standard order for listing the classes. The irreducible representations should, however, always be listed in the order given in Section 9.12; this order determines the numbering of the vibrational modes (see Section 9.9). The significance of the symbols x, y, z, R_x, R_y, R_z will be explained in Section 9.9.

Consider an Abelian group of order h. Since the group is Abelian, it has h classes (Section 9.2) and therefore h irreducible representations. Theorem

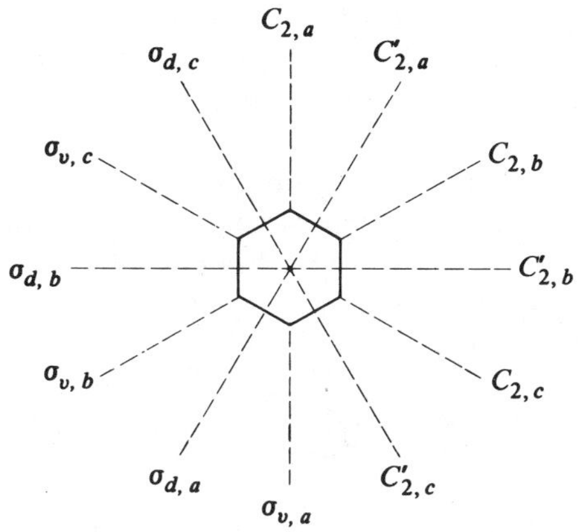

Fig. 9.4 Labeling of symmetry elements in benzene.

(2) then gives

$$l_1^2 + l_2^2 + \cdots + l_h^2 = h$$

The only solution to this equation is

$$l_1 = l_2 = \cdots = l_h = 1$$

The irreducible representations of an Abelian group are all one-dimensional.

Note from the character tables that groups having no threefold or higher proper or improper axes have only one-dimensional irreducible representations. This is because all such groups are Abelian.

An Abelian group has each element in a class by itself. The converse of this theorem is also true: A group with each element in a class by itself is Abelian. The proof is simple: By hypothesis $C^{-1}AC = A$ for all elements A and C; left multiplication by C gives $AC = CA$. Q.E.D. The number of symmetry operations in each class of a point group is indicated by an integer in front of the symbol for the symmetry operation, and it is therefore easy to see whether a group is Abelian by looking at the top line of the character table.

Theorem (2) shows that there are only a finite number of nonequivalent irreducible representations of a group of finite order. Any reducible representation must either be the direct sum of two or more irreducible representations or be convertible to such a direct sum by the performance of a similarity transformation on its matrices. In the former case, it is easy to see by inspection what irreducible representations make up the reducible representation; in the latter case, this is not obvious, since the matrices will not be in block-diagonal form. We now derive a theorem that enables us to readily find how many times each irreducible representation occurs in any given reducible representation. If the reducible-representation matrices are in block-diagonal form, then obviously their traces are the sums of the traces of the matrices of the irreducible representations contained in the reducible representation; if the irreducible representation j occurs $a_{j\Gamma}$ times ($a_{j\Gamma}=0$ or 1 or $2\cdots$) in the block-diagonal reducible representation Γ, and if $\chi_j(\hat{R})$ and $\chi_\Gamma(\hat{R})$ are the characters of the symmetry operation $\hat{R}$ in these two representations, we have

$$\chi_\Gamma(\hat{R})=\sum_j a_{j\Gamma}\chi_j(\hat{R}) \tag{9.37}$$

where the sum runs over the various nonequivalent irreducible representations. Even if the matrices of Γ are not in block-diagonal form, (9.37) still holds, since the traces of the matrices of Γ are unaltered by the similarity transformation that brings these matrices into block-diagonal form. We now multiply this equation by $\chi_i^*(\hat{R})$, where i is one of the nonequivalent irreducible representations, and sum over the h symmetry operations $\hat{R}$:

$$\sum_{\hat{R}}\chi_i^*(\hat{R})\chi_\Gamma(\hat{R})=\sum_j\sum_{\hat{R}}a_{j\Gamma}\chi_i^*(\hat{R})\chi_j(\hat{R})$$

Application of (9.31) gives

$$\sum_{\hat{R}}\chi_i^*(\hat{R})\chi_\Gamma(\hat{R})=\sum_j a_{j\Gamma}h\delta_{ij}=ha_{i\Gamma}$$

$$a_{i\Gamma}=\frac{1}{h}\sum_{\hat{R}}\chi_i^*(\hat{R})\chi_\Gamma(\hat{R}) \tag{9.38}$$

where h is the order of the group. This is the desired theorem.

Since symmetry operations in the same class have the same character in a given representation and are grouped together in the character table, it is convenient to rewrite (9.38) in a form where the sum goes over the various classes, rather than over the individual symmetry operations. If h_t is the number of symmetry operations in class t, then (9.38) becomes

$$a_{i\Gamma}=\frac{1}{h}\sum_t h_t\chi_i^*(\hat{R}_t)\chi_\Gamma(\hat{R}_t) \tag{9.39}$$

where $\hat{R}_t$ is any member of class t.

Note that to apply (9.39), we do not need the matrices of the nonequivalent irreducible representations; only their traces (which are given in the character table) are needed.

As a simple illustration of (9.39), let Γ be the $\mathcal{C}_{3v}$ representation (9.25). The h_t values are the "coefficients" of the symmetry operations listed at the top of the $\mathcal{C}_{3v}$ character table (9.36); these numbers are 1, 2, and 3. The $\chi_i(\hat{R}_t)$ values are taken from the character table for each irreducible representation; for example, for A_1, they are 1, 1, and 1. The $\chi_\Gamma(\hat{R}_t)$ values are found from the traces of the matrices (9.25); these values are 3, 0, and 1 for matrices of each class. Hence (9.39) gives

$$a_{A_1\Gamma}=\tfrac{1}{6}(1\cdot 1\cdot 3+2\cdot 1\cdot 0+3\cdot 1\cdot 1)=1$$

$$a_{A_2\Gamma}=\tfrac{1}{6}[1\cdot 1\cdot 3+2\cdot 1\cdot 0+3\cdot(-1)\cdot 1]=0$$

$$a_{E\Gamma}=\tfrac{1}{6}[1\cdot 2\cdot 3+2\cdot(-1)\cdot 0+3\cdot 0\cdot 1]=1$$

Of course, these results are obvious from inspection of (9.25), (9.28), and (9.29).

9.6 BASIS FUNCTIONS

Consider some set of linearly independent functions $F_1, F_2, \ldots, F_j, \ldots, F_n$. Let the symmetry operations of some particular point group be $\hat{R}, \hat{S}, \ldots, \hat{W}, \ldots$. Corresponding to these operations are the operators $\hat{O}_R, \hat{O}_S, \ldots, \hat{O}_W, \ldots$, which multiply the same way the corresponding symmetry operations multiply; in particular, let $\hat{R}\hat{S}=\hat{W}$, so that

$$\hat{O}_R\hat{O}_S=\hat{O}_W \tag{9.40}$$

Suppose the effect of each of the operators $\hat{O}_R, \hat{O}_S, \ldots$ on any F_j is to transform F_j into some linear combination of the functions $F_1, F_2, \ldots, F_n$:

$$\hat{O}_R F_j=\sum_{i=1}^{n} r_{ij}F_i, \qquad j=1,2,\ldots,n \tag{9.41}$$

$$\hat{O}_S F_j=\sum_{i=1}^{n} s_{ij}F_i, \qquad j=1,2,\ldots,n \tag{9.42}$$

$$\cdot \quad \cdot \quad \cdot \quad \cdot \quad \cdot \quad \cdot \quad \cdot$$

$$\hat{O}_W F_j=\sum_{i=1}^{n} w_{ij}F_i, \qquad j=1,2,\ldots,n \tag{9.43}$$

$$\cdot \quad \cdot \quad \cdot \quad \cdot \quad \cdot \quad \cdot \quad \cdot$$

where the coefficients $r_{ij}, s_{ij}, \ldots, w_{ij}, \ldots$ are certain constants. Let

$$\mathbf{R}, \mathbf{S}, \ldots, \mathbf{W}, \ldots \tag{9.44}$$

be square matrices of order n whose elements are the coefficients in (9.41)–(9.43): $(\mathbf{R})_{ij} = r_{ij}$, $(\mathbf{S})_{ij} = s_{ij}, \ldots$. We assert that the matrices $\mathbf{R}, \mathbf{S}, \ldots, \mathbf{W}, \ldots$ form a representation of the point group. The proof follows:

$$\hat{O}_W F_j = \hat{O}_R(\hat{O}_S F_j) = \hat{O}_R\left(\sum_{i=1}^{n} s_{ij} F_i\right) = \sum_{i=1}^{n} s_{ij} \hat{O}_R F_i$$

$$\hat{O}_W F_j = \sum_{k=1}^{n} \sum_{i=1}^{n} r_{ki} s_{ij} F_k \tag{9.45}$$

where (9.41) with the subscripts i and j changed to k and i, respectively, has been used. Interchanging the dummy subscripts i and k in (9.45) and subtracting it from (9.43), we get

$$\sum_{i=1}^{n}\left(w_{ij} - \sum_{k=1}^{n} r_{ik} s_{kj}\right) F_i = 0$$

Since the F_i are linearly independent, it follows that all the coefficients in the preceding equation vanish:

$$w_{ij} = \sum_{k=1}^{n} r_{ik} s_{kj}$$

Hence we have the matrix equation $\mathbf{RS} = \mathbf{W}$, so that the matrices (9.44) multiply the same way the symmetry operations do and form a representation of the point group. The functions $F_1, F_2, \ldots, F_n$ are said to form a *basis* for the representation (9.44), which consists of the matrices that describe how these functions transform upon application of the symmetry operators. Any member of the set $F_1, \ldots, F_n$ is said to *belong* to the representation (9.44). We denote the representation (9.44) by Γ_F; it may be reducible or irreducible.

We have shown that *any set of n linearly independent functions that are transformed into linear combinations of one another by the symmetry operators of a group forms a basis for an n-dimensional representation of the group*. The representation matrices consist of the coefficients describing the effect of the symmetry operations on the basis functions.

For example, each member of the set of functions x, y, z is transformed into a linear combination of x, y, z by the symmetry operators of a point group, according to (9.24) and (9.6). Hence the set x, y, z forms a basis for a three-dimensional representation of the point group; the matrix representing the symmetry operation $\hat{R}$ is the coefficient matrix $\mathbf{R}$ of (9.7).

Let the functions $G_1,\ldots,G_n$ be linear combinations of the functions $F_1,\ldots,F_n$, which form a basis for Γ_F:

$$G_k = \sum_{j=1}^{n} a_{jk} F_j, \qquad k=1,\ldots,n \tag{9.46}$$

where the a_{jk}'s are certain coefficients. The F_j are transformed into linear combinations of one another by the symmetry operators and the G_k are linear combinations of the F_j; hence the G_k are transformed into linear combinations of one another by the symmetry operators and they therefore form a basis for some representation Γ_G. We want to find the relation between the matrices of Γ_F and Γ_G.

The equations of (9.46) can be written as the matrix equation

$$(G_1 \quad \cdots \quad G_n) = (F_1 \quad \cdots \quad F_n)\mathbf{A} \qquad \text{or} \qquad \mathbf{G} = \mathbf{FA} \tag{9.47}$$

where $\mathbf{A}$ is the square matrix with elements a_{jk}, and $\mathbf{G}$ and $\mathbf{F}$ are defined as row vectors with elements $G_1,\ldots,G_n$ and $F_1,\ldots,F_n$, respectively. The effect of the symmetry operator $\hat{O}_R$ on the function G_k is

$$\hat{O}_R G_k = \sum_{j=1}^{n} a_{jk} \hat{O}_R F_j = \sum_{i=1}^{n} \sum_{j=1}^{n} F_i r_{ij} a_{jk}, \qquad k=1,\ldots,n \tag{9.48}$$

where (9.41), (9.46), and the linearity of $\hat{O}_R$ have been used. Equation (9.48) can be written as the matrix equation

$$\left(\hat{O}_R G_1 \quad \cdots \quad \hat{O}_R G_k \quad \cdots \quad \hat{O}_R G_n\right) = (F_1 \quad \cdots \quad F_i \quad \cdots \quad F_n)\mathbf{RA}$$

Multiplication of (9.47) by $\mathbf{A}^{-1}$ on the right gives

$$(F_1 \quad \cdots \quad F_n) = (G_1 \quad \cdots \quad G_n)\mathbf{A}^{-1}$$

Hence

$$\left(\hat{O}_R G_1 \quad \cdots \quad \hat{O}_R G_n\right) = (G_1 \quad \cdots \quad G_n)\mathbf{A}^{-1}\mathbf{RA} \tag{9.49}$$

Equation (9.41) reads

$$\left(\hat{O}_R F_1 \quad \cdots \quad \hat{O}_R F_n\right) = (F_1 \quad \cdots \quad F_n)\mathbf{R} \tag{9.50}$$

Comparison of (9.49) and (9.50) shows that the matrices of the representation Γ_G are $\mathbf{A}^{-1}\mathbf{RA}, \mathbf{A}^{-1}\mathbf{SA}, \ldots$; these matrices are related to the matrices

$\mathbf{R}, \mathbf{S}, \ldots$ of Γ_F by a similarity transformation; hence Γ_F and Γ_G are equivalent representations. A change of basis functions according to $\mathbf{G} = \mathbf{FA}$ causes a similarity transformation of the representation matrices with the matrix $\mathbf{A}$. By reversing the above procedure, we see that the converse is true: If Γ_F and Γ_G are equivalent representations, a set of basis functions for Γ_G can be obtained by taking the linear combinations (9.47) of basis functions of Γ_F.

The nomenclature in this section is similar to that used in Section 2.3. This is because in both cases we are dealing with the mathematics of vector spaces.

9.7 GROUP THEORY AND QUANTUM MECHANICS

Having developed the mathematics of group theory, we now apply it to molecular quantum mechanics. As usual, we use the Born–Oppenheimer approximation.

Consider either the electronic wave functions or the vibrational wave functions of a molecule. These functions satisfy the Schrödinger equation

$$\hat{H}\psi_{j,\nu} = E_\nu \psi_{j,\nu} \tag{9.51}$$

where $\hat{H}$, $\psi_{j,\nu}$, and E_ν are either the electronic or the vibrational Hamiltonian, wave functions and energies. The subscript ν labels the energy levels and the subscript j distinguishes the wave functions belonging to each level E_ν; the νth level is n-fold degenerate: $j = 1, 2, \ldots, n$.

The electronic Hamiltonian (assuming we are dealing with the electronic Schrödinger equation for the equilibrium nuclear configuration) and the vibrational Hamiltonian are each specified relative to the equilibrium nuclear configuration, which defines the molecular point group. The point group consists of certain symmetry operations $\hat{R}, \hat{S}, \hat{T}, \ldots$; corresponding to these operations are the operators $\hat{O}_R, \hat{O}_S, \hat{O}_T, \ldots$, which operate on functions of the spatial coordinates. These latter operators form a group that is isomorphic to the molecular point group.

Now consider the effect of applying $\hat{O}_R$ to the Schrödinger equation:

$$\hat{O}_R\left(\hat{H}\psi_{j,\nu}\right) = \hat{O}_R\left(E_\nu \psi_{j,\nu}\right), \qquad j = 1, 2, \ldots, n \tag{9.52}$$

Since a symmetry operation sends the molecular framework into a configuration physically indistinguishable from the original one, it does not affect either the electronic or the vibrational Hamiltonian, which are defined relative to the molecular framework.[3] Hence

$$\hat{O}_R\left(\hat{H}\psi_{j,\nu}\right)=\hat{H}\hat{O}_R\psi_{j,\nu} \tag{9.53}$$

Since the wave functions form a complete set, we conclude that $\hat{O}_R$ commutes with the electronic Hamiltonian and with the vibrational Hamiltonian:

$$\left[\hat{O}_R,\hat{H}\right]=0 \tag{9.54}$$

Thus the electronic and vibrational wave functions can be chosen to be eigenfunctions of $\hat{O}_R$, a fact we have used many times in discussions of these functions.

Use of (9.53) and the linearity of $\hat{O}_R$ in (9.52) gives

$$\hat{H}\left(\hat{O}_R\psi_{j,\nu}\right)=E_\nu\left(\hat{O}_R\psi_{j,\nu}\right) \tag{9.55}$$

Since $\hat{O}_R\psi_{j,\nu}$ is an eigenfunction of $\hat{H}$ with eigenvalue E_ν, it must be some linear combination of the n eigenfunctions belonging to the level E_ν:

$$\hat{O}_R\psi_{j,\nu}=\sum_{i=1}^{n}\psi_{i,\nu}r_{ij,\nu},\qquad j=1,2,\ldots,n \tag{9.56}$$

where the $r_{ij,\nu}$'s are a set of n^2 constants that describe the effect of $\hat{O}_R$ on each of the functions $\psi_{j,\nu}$, $j=1,2,\ldots,n$. [These $r_{ij,\nu}$'s are, of course, not necessarily the same as the r_{ij}'s of (9.7).] We define the matrix $\mathbf{R}_\nu$ as the nth-order square matrix with elements $r_{ij,\nu}$. Similarly, let $\mathbf{S}_\nu,\mathbf{T}_\nu,\ldots$ be the matrices giving the effect of $\hat{O}_S,\hat{O}_T,\ldots$ on the n wave functions of the degenerate level. Since the n linearly independent wave functions

$$\psi_{1,\nu},\quad \psi_{2,\nu},\quad \ldots,\quad \psi_{n,\nu}$$

of the n-fold degenerate vibrational or electronic level are transformed into linear combinations of one another by the symmetry operations [Eq. (9.56)], we conclude (Section 9.6) that these functions form a basis for some particular representation Γ_ν of the molecular point group.

There are two possibilities: Γ_ν is either reducible or irreducible.

Suppose Γ_ν is reducible. Then the matrices $\mathbf{R}_\nu,\mathbf{S}_\nu,\mathbf{T}_\nu,\ldots$ either are in the same block-diagonal form or can be put in the same block-diagonal form by a similarity transformation. To see the significance of this, we write

[3] For a careful discussion, see *Schonland*, pp. 117–123; also *Hamermesh*, pp. 86–88.

(9.56) as

$$\left(\hat{O}_R\psi_{1,\nu} \quad \cdots \quad \hat{O}_R\psi_{n,\nu}\right)=\left(\psi_{1,\nu} \quad \cdots \quad \psi_{n,\nu}\right)\begin{pmatrix} r_{11,\nu} & r_{12,\nu} & \cdots & r_{1n,\nu} \\ r_{21,\nu} & r_{22,\nu} & \cdots & r_{2n,\nu} \\ \cdot & \cdot & \cdots & \cdot \\ r_{n1,\nu} & r_{n2,\nu} & \cdots & r_{nn,\nu} \end{pmatrix} \tag{9.57}$$

with similar equations for the effects of $\hat{O}_S, \hat{O}_T, \ldots$. If $\mathbf{R}_\nu$ is in block-diagonal form, then the set of n degenerate wave functions can be divided into subsets such that the members of each subset are transformed into linear combinations of one another by $\hat{O}_R$; for example, if $n=5$ and if $\mathbf{R}_\nu$ has the form

$$\begin{pmatrix} r_{11,\nu} & r_{12,\nu} & 0 & 0 & 0 \\ r_{21,\nu} & r_{22,\nu} & 0 & 0 & 0 \\ 0 & 0 & r_{33,\nu} & r_{34,\nu} & r_{35,\nu} \\ 0 & 0 & r_{43,\nu} & r_{44,\nu} & r_{45,\nu} \\ 0 & 0 & r_{53,\nu} & r_{54,\nu} & r_{55,\nu} \end{pmatrix} \tag{9.58}$$

then $\hat{O}_R$ acting on either $\psi_{1,\nu}$ or $\psi_{2,\nu}$ gives a linear combination of $\psi_{1,\nu}$ and $\psi_{2,\nu}$, and $\hat{O}_R$ acting on $\psi_{3,\nu}$, $\psi_{4,\nu}$, or $\psi_{5,\nu}$ gives a linear combination of these three functions. If each matrix $\mathbf{R}_\nu, \mathbf{S}_\nu, \mathbf{T}_\nu, \ldots$ is in this same block-diagonal form, then each of the operators $\hat{O}_R, \hat{O}_S, \hat{O}_T, \ldots$ sends a degenerate wave function into a linear combination of only members of the same subset. Now take the case that $\mathbf{R}_\nu, \mathbf{S}_\nu, \ldots$ are not in block-diagonal form, but can be put in this form by a similarity transformation using the matrix $\mathbf{A}$ and its inverse. In this case, the representation Γ_ν is equivalent to some representation Γ_ν' whose matrices are in block-diagonal form; as shown in Section 9.6, the n linear combinations

$$\left(\psi_{1,\nu}' \quad \cdots \quad \psi_{n,\nu}'\right)=\left(\psi_{1,\nu} \quad \cdots \quad \psi_{n,\nu}\right)\mathbf{A}$$

form a basis for the block-diagonal representation Γ_ν'. One set of n linearly independent wave functions is just as good as another in discussing a degenerate level, so we can deal with the set $\psi_{1,\nu}', \ldots, \psi_{n,\nu}'$ instead of the set $\psi_{1,\nu}, \ldots, \psi_{n,\nu}$.

We conclude that when the representation generated by the degenerate wave functions is reducible, the degenerate wave functions can always be chosen so that they divide up into subsets whose members transform among themselves under the action of the symmetry operators $\hat{O}_R, \hat{O}_S, \ldots$. Under these conditions, there would be no reason from symmetry alone for the wave functions of different subsets to have the same energy; thus it is unlikely (but not absolutely impossible) for the representation to be reducible. If the representation generated by the n degenerate wave functions is reducible, we say the degeneracy is *accidental*. An example of accidental degeneracy would be the degeneracy of vibrational levels if one fundamental frequency happened to be exactly an integral multiple of another fundamental frequency.

We shall therefore proceed on the assumption that the representation generated by the degenerate wave functions of each level is irreducible. Thus *the wave functions of each electronic or vibrational level transform according to an irreducible representation of the molecular point group.* For each degenerate level, we can use any linear combination of the n wave functions of the level, but the representations generated by different linear combinations will be equivalent to each other (Section 9.6); hence we need only consider the *nonequivalent* irreducible representations. Each electronic and vibrational energy level must correspond to one of the nonequivalent irreducible representations of the molecular point group.

From (9.57) we see that the order of the matrices of the representation generated by the n degenerate wave functions is equal to n; hence *the degree of degeneracy of an energy level is equal to the dimension of the irreducible representation to which its wave functions belong.* For example, looking at the character table of $\mathcal{C}_{2v}$ we see that all irreducible representations are one-dimensional; hence every vibrational level of a $\mathcal{C}_{2v}$ molecule is nondegenerate, as is every electronic energy level. Thus without solving the Schrödinger equation, we can obtain the possible degeneracies simply by inspection of the character table of the molecular point group—A and B levels are nondegenerate; E levels are doubly degenerate; T levels are triply degenerate. Furthermore, we can obtain qualitative information on the appearance of the electronic and vibrational wave functions from the character table, since the behavior of $\psi_{i,\nu}$ with respect to the symmetry operators $\hat{O}_R, \hat{O}_S, \ldots$ is given by the matrices of the irreducible representation that corresponds to the energy level ν. For one-dimensional representations, the characters are also the matrices of the irreducible representations and the effect of the symmetry operations is given directly by the characters. Thus an electronic or vibrational energy level corresponding to the A_1 representation of $\mathcal{C}_{2v}$ has its wave function unchanged by each of the symmetry operations of $\mathcal{C}_{2v}$, whereas an A_2 wave function is unchanged by 180° rotation about the C_2 axis, but changes sign on reflection in either symmetry plane. Recall that we assigned electronic and vibrational wave functions to various *symmetry species* (Sections 1.19 and 6.3) according to their behavior with respect to the symmetry operations; we now see that these symmetry species are the irreducible representations of the molecular point group.

Certain point groups have one-dimensional representations with complex characters. For such groups we have an exception to the statement that the representation to which the wave functions of a degenerate level belong is irreducible, as we now show.

Consider $\mathcal{C}_n$. This is an Abelian group of order n and therefore has n one-dimensional irreducible representations. These are easily found. Let ω be the scalar that represents the operation $\hat{C}_n$; since $\hat{C}_n^n(=\hat{E})$ must be represented by 1 in one-dimensional representations, we have $\omega^n = 1$. Hence [Equation (1.76)]

$$\omega = e^{i2\pi k/n}, \qquad k = 0, 1, \ldots, n-1 \tag{9.59}$$

Since $\hat{C}_n^m = (\hat{C}_n)^m$, the operation $\hat{C}_n^m$ is represented by ω^m, and we have found the n one-dimensional irreducible representations.

Looking at the character tables for $\mathcal{C}_n$, we see that the one-dimensional representations with complex characters are grouped in pairs, such that one set of characters is the complex conjugate of the other set, and each pair of one-dimensional representations is labeled E, indicating doubly degenerate energy levels. The reason for this practice is as follows. The complex conjugate of the Schrödinger equation is $(\hat{H}\psi)^* = E\psi^*$. The Hamiltonian $\hat{H}$ contains no complex quantities [provided no magnetic field is applied—the operator $\hat{L}_z$ in (1.269) contains the factor $\hbar/i$], so that

$$\hat{H}\psi^* = E\psi^* \tag{9.60}$$

The complex conjugate of a wave function is thus a wave function belonging to the same energy level. (E.g., the $2p_{+1}$ and $2p_{-1}$ hydrogen-atom functions differ in having the phi factors $e^{i\varphi}$ and $e^{-i\varphi}$; their degeneracy is removed by application of a magnetic field—the Zeeman effect.) If ψ is real, the complex conjugate gives the same function, and ψ belongs to one of the representations with real characters. If ψ is complex and transforms according to

$$\hat{O}_{C_n}\psi = b\psi \tag{9.61}$$

where b is the character of $\hat{C}_n$ in one of the one-dimensional irreducible representations with complex characters, then taking the complex conjugate, we have

$$\hat{O}_{C_n}\psi^* = b^*\psi^* \tag{9.62}$$

so that ψ^* belongs to a one-dimensional irreducible representation whose characters are the complex conjugates of those of the representation to which ψ belongs. Since ψ and ψ^* belong to the same energy level (in the absence of a magnetic field), from a physical point of view their two one-dimensional representations are to be considered as one two-dimensional representation. (The groups $\mathcal{C}_{nh}$ and $\mathcal{S}_{2n}$ also have one-dimensional representations with complex characters.)

9.8 GROUP THEORY AND ELECTRONIC STRUCTURE

We now apply group theory to the study of molecular electronic wave functions. As found in the last section, the dimensions of the irreducible representations of the molecular point group give us the possible degeneracies of E_{el}, and the matrices of the irreducible representations tell us how ψ_{el} behaves upon application of the symmetry operations of the group. It should be noted that the Schrödinger equation (9.51) does not include electronic spin; thus the degeneracies found from the character tables do not include the contribution of electron spin to the degeneracy of E_{el}.

The simplest "molecules" are atoms, which belong to point group $\mathcal{K}_h$ (often called the *full rotation–reflection group*). The character table (which we omit) contains irreducible representations of dimensions $1,3,5,\ldots$; these representations correspond to energy levels with electronic orbital angular-momentum quantum number $l=0,1,2,\ldots$; we have the $(2l+1)$-fold degeneracy associated with different values of the quantum number m. (For the hydrogen atom, states of different l but the same n are degenerate; this can be viewed as an accidental degeneracy.) Instead of the Mulliken notation, the angular-momentum notation $S,P,D,F,\ldots$ is used to label the irreducible representations of $\mathcal{K}_h$.

Linear molecules belong to either $\mathcal{C}_{\infty v}$ or $\mathcal{D}_{\infty h}$; the character tables are given in Section 9.12. The operation $\hat{C}^{\alpha}(z)$ is a rotation by an angle α about the C_∞ axis. In addition to the Mulliken notation, the spectroscopic symbols $\Sigma,\Pi,\Delta,\ldots$ are also given. As we saw in Section 1.19, apart from spin degeneracy, Σ electronic levels are nondegenerate (A), whereas $\Pi,\Delta,\ldots$ levels are doubly degenerate (E). The symmetry of Σ representations with respect to reflection in a σ_v plane containing the molecular axis is indicated by the usual $+$ or $-$ superscript. For $\mathcal{D}_{\infty h}$ molecules, the symmetry with respect to $\hat{i}$ is indicated by a g or u subscript.

For nonlinear molecules, each electronic wave function is classified according to the irreducible representation (symmetry species) to which it belongs; the symmetry properties of ψ_{el} follow accordingly. For example, for the equilibrium nuclear configuration of benzene (symmetry $\mathcal{D}_{6h}$), the character table tells us that there are only nondegenerate and doubly degenerate electronic energy levels (apart from spin degeneracy). The electronic wave functions are either symmetric or antisymmetric with respect to inversion, according to whether they belong to a g or u representation. The symmetry behavior of ψ_{el} with respect to other symmetry operators is also determined by the irreducible representations.

Most commonly, ψ_{el} is approximated as an MO function. The best MOs are the eigenfunctions of the Hartree–Fock operator $\hat{F}$; from (1.291) and (1.292), we see that if the interelectronic interactions are neglected (i.e., if the Coulomb and exchange operators are omitted from $\hat{F}$), then $\hat{F}$ is invariant with respect to symmetry operations of the molecular point group, just as the true many-electron Hamiltonian is. Hence we might expect that just as each many-electron wave function ψ_{el} can be assigned to an irreducible representation of the molecular point group, each of the one-electron MO functions can be so assigned. This statement is true, but we omit its rigorous proof.[4] Thus the degeneracies and symmetry properties of the MO functions are also determined by the irreducible representations of the molecular point group.

When approximate MOs are used, each approximate MO should, of course, belong to the same irreducible representation as the MO it is approximating. In the LCAO-MO method, the MOs are approximated as linear combinations of some set of AOs:

$$f_1, f_2, \ldots, f_m \tag{9.63}$$

Since each MO belongs to some irreducible representation of the molecular point group, we must find linear combinations of the AOs that transform according to the irreducible representations; group theory enables us to do this. A symmetry operation sends each nucleus either into itself or into an atom of the same type; a symmetry operation will thus transform each AO into some linear combination of the AOs (9.63). Therefore (Section 9.6), the AOs form a basis for some representation Γ_{AO} of the point group of the molecule. This representation (as any reducible representation) will be the direct sum of certain of the irreducible representations $\Gamma_1, \Gamma_2, \ldots, \Gamma_k$ (not necessarily all different) of the molecular point group:

$$\Gamma_{AO} = \Gamma_1 \oplus \Gamma_2 \oplus \cdots \oplus \Gamma_k \tag{9.64}$$

(The representation Γ_{AO} is said to *contain* the irreducible representations $\Gamma_1, \Gamma_2, \ldots, \Gamma_k$.)

[4] See C. C. J. Roothaan, *Rev. Mod. Phys.*, **23**, 69 (1951): **32**, 179 (1960).

Let $\mathbf{R}_{AO}$ be the matrix representing the symmetry operation $\hat{R}$ in the representation Γ_{AO}. Equation (9.64) means that there is a similarity transformation that transforms $\mathbf{R}_{AO}$ to $\overline{\mathbf{R}}_{AO}$, where $\overline{\mathbf{R}}_{AO}$ is in block-diagonal form, the blocks being the matrices $\mathbf{R}_1, \mathbf{R}_2, \ldots, \mathbf{R}_k$ of the irreducible representations $\Gamma_1, \Gamma_2, \ldots, \Gamma_k$. Let $\overline{\Gamma}_{AO}$ denote the block-diagonal representation equivalent to Γ_{AO} and let $\mathbf{A}$ be the matrix of the similarity transformation that converts the matrices of Γ_{AO} to $\overline{\Gamma}_{AO}$: $\overline{\mathbf{R}}_{AO} = \mathbf{A}^{-1}\mathbf{R}_{AO}\mathbf{A}$. We form the following linear combinations of the AOs:

$$g_j = \sum_{i=1}^{m} f_i a_{ij}, \qquad j = 1, 2, \ldots, m \tag{9.65}$$

where the a_{ij}'s are the elements of $\mathbf{A}$. As shown in Section 9.6, the functions $g_1, \ldots, g_m$ form a basis for the block-diagonal representation $\overline{\Gamma}_{AO}$; hence

$$(\hat{O}_R g_1 \quad \cdots \quad \hat{O}_R g_m) = (g_1 \quad \cdots \quad g_m) \begin{pmatrix} \mathbf{R}_1 & \mathbf{0} & \cdots & \mathbf{0} \\ \mathbf{0} & \mathbf{R}_2 & \cdots & \mathbf{0} \\ \vdots & \vdots & \ddots & \vdots \\ \mathbf{0} & \mathbf{0} & \cdots & \mathbf{R}_k \end{pmatrix} \tag{9.66}$$

with similar equations for the effects of the other symmetry operators. Let the dimensions of $\Gamma_1, \Gamma_2, \ldots, \Gamma_k$ be $l_1, l_2, \ldots, l_k$. Equation (9.66) shows that the l_1 functions $g_1, g_2, \ldots, g_{l_1}$ transform among themselves according to the irreducible representation Γ_1, the l_2 functions $g_{l_1+1}, g_{l_1+2}, \ldots, g_{l_1+l_2}$ transform among themselves according to the irreducible representation Γ_2, and so on. We have thus found the desired linear combinations. These AO linear combinations that belong to irreducible representations of the molecular point group are called *symmetry-adapted orbitals* or *symmetry orbitals*. Each MO must be a linear combination of the symmetry-adapted orbitals that belong to the same irreducible representation as the MO; the coefficients of the various symmetry orbitals making up the MO are found by solving a secular equation (which amounts to minimizing the variational integral). By using the symmetry-adapted orbitals, we reduce the $m \times m$ secular determinant to several secular determinants of lower order, thereby simplifying the calculation (see also Section 9.11). If all the irreducible representations in (9.64) are different (this can only happen with a small basis set of AOs), then symmetry alone will give the MOs. Thus if a certain one-dimensional irreducible representation occurs once in (9.64), the corresponding symmetry-adapted orbital must be the MO that transforms according to that irreducible representation; if a certain two-dimensional irreducible representation occurs once in (9.64), the two degenerate MOs that transform according to that representation must be linear combinations of the two corresponding symmetry-adapted functions and vice versa—but any linear combinations of degenerate MOs are valid MOs, so that we can take the MOs to be the symmetry-adapted orbitals. If any irreducible representation occurs more than once in (9.64), we must solve a secular equation to find the MOs that transform according to that representation. (All this will become clearer after we consider some examples.)

The symmetry-adapted functions $g_1, g_2, \ldots$ can be found according to (9.65) from the matrix of the similarity transformation that brings the matrices of Γ_{AO} to block-diagonal form. Rather than attempting to find this matrix, we give the following procedure for finding symmetry-adapted functions g_i from the functions f_i (the AOs): We take one of the AOs f, apply the symmetry operator $\hat{O}_R$, multiply by the complex conjugate of the character of $\hat{R}$ in the irreducible representation j, and finally sum over the operators of the point group; the result is a function g that transforms according to the jth irreducible representation:

$$g = \frac{l_j}{h} \sum_{\hat{R}} \chi_j^*(\hat{R}) \hat{O}_R f \tag{9.67}$$

The factor l_j/h, where h is the order of the group and l_j is the dimension of the jth irreducible representation, has been included in (9.67) for convenience. Application of this procedure to the functions f_i gives us (unnormalized) symmetry-adapted functions g_i. This procedure is applicable to generating sets of functions that form bases for irreducible representations from any set of functions that form a basis for a reducible representation. The proof of the procedure (9.67) for one-dimensional representations is outlined in Problem 9.22; we omit its general proof.[5] Symmetry-adapted functions produced by (9.67) that belong to the same irreducible representation are not, in general, orthogonal.

Our first example will be H_2. Of course, this application is trivial, but it will illustrate the method. For a minimal-basis-set MO calculation (Section 1.20) of the H_2 ground electronic state, we consider only the valence AOs $1s_a$ and $1s_b$ of atoms a and b. Each symmetry operation of $\mathcal{D}_{\infty h}$ either leaves the AO $1s_a$ unaffected or converts it to $1s_b$ (Fig. 9.5); we have

$\hat{R}$:	$\hat{E}$	$2\hat{C}^{\alpha}$	$\infty\hat{\sigma}_v$	$\hat{i}$	$\hat{\sigma}_h$	$2\hat{S}^{\alpha}$	$\infty\hat{C}_2$
$\hat{O}_R 1s_a$:	$1s_a$	$1s_a$	$1s_a$	$1s_b$	$1s_b$	$1s_b$	$1s_b$

(9.68)

[5] See *Schonland*, Section 6.6, or *Hamermesh*, Section 3-18.

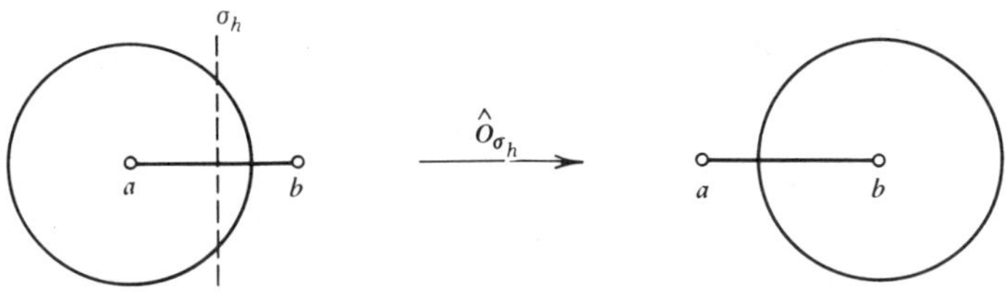

Fig. 9.5 Effect of a $\mathscr{D}_{\infty h}$ symmetry operator on the $1s_a$ AO.

Interchange of a and b in (9.68) gives the effect of the symmetry operations on $1s_b$. The first three operators listed leave $1s_a$ and $1s_b$ unaffected:

$$\hat{O}_R(1s_a \quad 1s_b)=(1s_a \quad 1s_b)\begin{pmatrix} 1 & 0 \\ 0 & 1 \end{pmatrix} \tag{9.69}$$

Hence these three operators are represented in Γ_{AO} by the unit matrix of order 2. The remaining operators interchange $1s_a$ and $1s_b$:

$$\hat{O}_R(1s_a \quad 1s_b)=(1s_a \quad 1s_b)\begin{pmatrix} 0 & 1 \\ 1 & 0 \end{pmatrix} \tag{9.70}$$

Each of these operators is represented by the square matrix in (9.70). The two-dimensional representation Γ_{AO} thus has the characters

$\hat{E}$	$2\hat{C}^{\alpha}$	$\infty\hat{\sigma}_v$	$\hat{i}$	$\hat{\sigma}_h$	$2\hat{S}^{\alpha}$	$\infty\hat{C}_2$
2	2	2	0	0	0	0

(9.71)

Inspection of the $\mathscr{D}_{\infty h}$ character table shows that no two-dimensional irreducible representation has the characters (9.71); hence Γ_{AO} must be reducible to two one-dimensional representations. The only pair of one-dimensional representations of $\mathscr{D}_{\infty h}$, the sum of whose characters is (9.71), is Σ_g^+ and Σ_u^+. [Equation (9.39) applies to groups of finite order; to apply it to $\mathscr{D}_{\infty h}$, we would have to take the limit of (9.39) for $\mathscr{D}_{nh}$ as n goes to infinity. It is simpler to decompose Γ_{AO} by inspection.] We have

$$\Gamma_{\text{AO}}=\Sigma_g^+\oplus\Sigma_u^+ \tag{9.72}$$

Because the group is of infinite order, (9.67) is cumbersome to use, and it is simplest to find the symmetry-adapted functions by inspection. Each symmetry operation either leaves $1s_a$ and $1s_b$ alone or transforms them into each other, and the following functions are easily seen to transform according to Σ_g^+ and Σ_u^+:

$$\sigma_g^+: \quad 1s_a + 1s_b \tag{9.73}$$

$$\sigma_u^+: \quad 1s_a - 1s_b \tag{9.74}$$

(The convention is to use lowercase letters for the symmetry species of one-electron functions.) Since each irreducible representation occurs only once in (9.72), these symmetry-adapted orbitals are also the (unnormalized) MOs. As a check, using (9.73), (9.74), and (9.65), we find for the matrix **A** of the similarity transformation that reduces the matrices of Γ_{AO} to block-diagonal form

$$\mathbf{A} = \begin{pmatrix} 1 & 1 \\ 1 & -1 \end{pmatrix}$$

Using (2.18), we find $\mathbf{A}^{-1} = \frac{1}{2}\mathbf{A}$. Applying the similarity transformation to the matrices of Γ_{AO}, we find

$$\mathbf{A}^{-1}\begin{pmatrix} 1 & 0 \\ 0 & 1 \end{pmatrix}\mathbf{A} = \begin{pmatrix} 1 & 0 \\ 0 & 1 \end{pmatrix} \qquad \mathbf{A}^{-1}\begin{pmatrix} 0 & 1 \\ 1 & 0 \end{pmatrix}\mathbf{A} = \begin{pmatrix} 1 & 0 \\ 0 & -1 \end{pmatrix}$$

which verifies (9.72).

For the minimal basis set $1s_a$ and $1s_b$, group theory has given us the precise form (except for the orbital exponents, which must be found by a variational calculation) of the two lowest H_2 MOs: $1\sigma_g$ and $1\sigma_u$. [Since a σ MO is unchanged upon reflection in a plane containing the molecular axis, any σ one-electron function must be a σ^+ function; hence there is no need to use the superscript for σ MOs. We can, however, have Σ^+ and Σ^- many-electron electronic wave functions, and here the superscript must be included.]

To get better approximations to these two MOs, and to get higher MOs, we must include more AOs in the basis set. Let us take as the basis AOs

$$1s_a, 2s_a, 2p_{xa}, 2p_{ya}, 2p_{za}, 1s_b, 2s_b, 2p_{xb}, 2p_{yb}, 2p_{zb}$$

The z axis is the internuclear axis, and the z axes on atoms a and b point toward each other. Without using the full group-theory formalism, we can write down by inspection the following symmetry-adapted orbitals, which transform according to the indicated irreducible representations:

$$\sigma_g^+:\quad g_1 = 1s_a + 1s_b \quad g_2 = 2s_a + 2s_b \quad g_3 = 2p_{za} + 2p_{zb}$$

$$\sigma_u^+:\quad g_4 = 1s_a - 1s_b \quad g_5 = 2s_a - 2s_b \quad g_6 = 2p_{za} - 2p_{zb}$$

$$\pi_g:\quad g_7 = 2p_{xa} + 2p_{xb} \quad g_8 = 2p_{ya} + 2p_{yb}$$

$$\pi_u:\quad g_9 = 2p_{xa} - 2p_{xb} \quad g_{10} = 2p_{ya} - 2p_{yb}$$

The three lowest MOs of σ_g^+ symmetry, the $1\sigma_g$, $2\sigma_g$, and $3\sigma_g$ MOs, are each linear combinations of the symmetry-adapted functions g_1, g_2, and g_3:

$$\begin{aligned} 1\sigma_g &= c_1 g_1 + c_2 g_2 + c_3 g_3 \\ 2\sigma_g &= c_1' g_1 + c_2' g_2 + c_3' g_3 \\ 3\sigma_g &= c_1'' g_1 + c_2'' g_2 + c_3'' g_3 \end{aligned} \tag{9.75}$$

The coefficients in (9.75) must be found by solving a 3×3 secular equation. The $1\sigma_u$, $2\sigma_u$, and $3\sigma_u$ MOs are each linear combinations of g_4, g_5, and g_6, the coefficients being derived from another 3×3 secular equation. The $1\pi_g$ subshell consists of two degenerate MOs, and we can take g_7 and g_8 (or any linear combination of them) as the two $1\pi_g$ MOs. For the $1\pi_u$ MOs, we can use g_9 and g_{10}. Of course, if $3p$ and higher AOs were included in the basis set, it would be necessary to solve a secular equation to get the $1\pi_g$ and $1\pi_u$ MOs. The preceding discussion is applicable to determining the MOs of any homonuclear diatomic molecule.

Next we consider H_2O (Fig. 1.2). The minimal-basis-set AOs are

$$1s_{H_1}, 1s_{H_2}, 1s_O, 2s_O, (2p_x)_O, (2p_y)_O, (2p_z)_O \tag{9.76}$$

We will not bother to write out the four 7×7 matrices of Γ_{AO} since we only need their traces. Clearly, for each AO that is unchanged by the symmetry operator $\hat{O}_R$, we get a contribution of $+1$ to the trace of the matrix representing $\hat{R}$ in Γ_{AO}; for example, see (9.69). For each AO that is sent into minus itself by $\hat{O}_R$, we get a contribution of -1 to $\chi_{AO}(\hat{R})$. For each AO that is sent into some other AO by $\hat{O}_R$, we get a contribution of 0 to $\chi_{AO}(\hat{R})$; for example, see (9.70). [In certain cases, a symmetry operation may send an AO into a linear combination of itself and one or more other AOs; for example, a $\hat{C}_3(z)$ rotation in NH_3 sends a p_x orbital of nitrogen into a linear combination of p_x and p_y orbitals:

$$\hat{O}_{C_3} p_x = a p_x + b p_y \tag{9.77}$$

Here, the contribution to $\chi_{AO}(\hat{C}_3)$ is a. Since the functions $2p_x$, $2p_y$, and $2p_z$ transform like x, y, and z, the total contribution to $\chi_{AO}(\hat{R})$ of the three $2p$ orbitals on an atom unshifted by $\hat{R}$ is given by the trace of the matrix in (9.7). For example, for a minimal-basis-set calculation of NH_3, we have $\chi_{AO}(\hat{C}_3) = 2 + 0 = 2$, where "2" comes from the $1s$ and $2s$ nitrogen AOs and "0" is the trace of (9.12) for $n = 3$.] For H_2O, reflection in the molecular yz plane converts $(2p_x)_O$ to its negative and does not affect the other six basis AOs; hence

$$\chi_{AO}[\hat{\sigma}(yz)] = 6 - 1 = 5$$

Rotation about the $C_2(z)$ axis converts the oxygen $2p_x$ and $2p_y$ orbitals to their negatives, sends the hydrogen $1s$ orbitals into each other, and has no effect on the other three AOs; hence

$$\chi_{AO}[\hat{C}_2(z)] = 2(-1) + 2(0) + 3(1) = 1$$

Reflection in the xz plane converts the oxygen $2p_y$ orbital to its negative, interchanges the hydrogen $1s$ orbitals, and has no effect on the other four AOs; hence

$$\chi_{AO}[\hat{\sigma}(xz)] = -1 + 4 = 3$$

Of course, $\chi_{AO}(\hat{E}) = 7$. The characters for Γ_{AO} are

$\hat{E}$	$\hat{C}_2(z)$	$\hat{\sigma}(xz)$	$\hat{\sigma}(yz)$
7	1	3	5

(9.78)

Equation (9.38) and the $\mathcal{C}_{2v}$ character table give

$$\Gamma_{AO} = 4A_1 \oplus B_1 \oplus 2B_2 \tag{9.79}$$

From (9.67), the symmetry orbitals are generated by application of the following four operators to the AOs:

$$\tfrac{1}{4}\left[\hat{O}_E+\hat{O}_{C_2}+\hat{O}_{\sigma(xz)}+\hat{O}_{\sigma(yz)}\right]\equiv\hat{P}_{A_1} \tag{9.80}$$

$$\tfrac{1}{4}\left[\hat{O}_E+\hat{O}_{C_2}-\hat{O}_{\sigma(xz)}-\hat{O}_{\sigma(yz)}\right]\equiv\hat{P}_{A_2} \tag{9.81}$$

$$\tfrac{1}{4}\left[\hat{O}_E-\hat{O}_{C_2}+\hat{O}_{\sigma(xz)}-\hat{O}_{\sigma(yz)}\right]\equiv\hat{P}_{B_1} \tag{9.82}$$

$$\tfrac{1}{4}\left[\hat{O}_E-\hat{O}_{C_2}-\hat{O}_{\sigma(xz)}+\hat{O}_{\sigma(yz)}\right]\equiv\hat{P}_{B_2} \tag{9.83}$$

(These operators are examples of *projection operators*; they project out the symmetry-adapted functions from the basis AOs.) Application of $\hat{P}_{A_1}$ to $1s_{\text{O}}$ gives

$$\hat{P}_{A_1}1s_{\text{O}}=1s_{\text{O}}$$

indicating that $1s_{\text{O}}$ transforms according to A_1, which is obvious from the fact that this function is unchanged by each symmetry operator. If any of the operators $\hat{P}_{A_2}$, $\hat{P}_{B_1}$, $\hat{P}_{B_2}$ is applied to $1s_{\text{O}}$, the result is zero, as it should be. Similarly,

$$\hat{P}_{A_1}2s_{\text{O}}=2s_{\text{O}}, \qquad \hat{P}_{A_1}(2p_z)_{\text{O}}=(2p_z)_{\text{O}}$$

We have three of the a_1 symmetry orbitals. We further find

$$\hat{P}_{A_1}(2p_x)_{\text{O}}=\hat{P}_{A_1}(2p_y)_{\text{O}}=0$$

$$\hat{P}_{A_1}1s_{\text{H}_1}=\hat{P}_{A_1}1s_{\text{H}_2}=\tfrac{1}{2}(1s_{\text{H}_1}+1s_{\text{H}_2})$$

and the fourth a_1 symmetry-adapted orbital is $\tfrac{1}{2}(1s_{\text{H}_1}+1s_{\text{H}_2})$. The remaining symmetry orbitals are found from the relations

$$\hat{P}_{B_2}1s_{\text{H}_1}=\tfrac{1}{2}(1s_{\text{H}_1}-1s_{\text{H}_2}), \qquad \hat{P}_{B_1}(2p_x)_{\text{O}}=(2p_x)_{\text{O}}, \qquad \hat{P}_{B_2}(2p_y)_{\text{O}}=(2p_y)_{\text{O}}$$

The seven symmetry orbitals found from the seven basis functions are

$$a_1\!: \quad g_1=1s_{\text{O}} \quad g_2=2s_{\text{O}} \quad g_3=(2p_z)_{\text{O}} \quad g_4=\tfrac{1}{2}(1s_{\text{H}_1}+1s_{\text{H}_2}) \tag{9.84}$$

$$b_1\!: \quad g_5=(2p_x)_{\text{O}} \tag{9.85}$$

$$b_2\!: \quad g_6=(2p_y)_{\text{O}} \quad g_7=\tfrac{1}{2}(1s_{\text{H}_1}-1s_{\text{H}_2}) \tag{9.86}$$

In the minimal-basis-set approximation, the lowest four a_1 MOs are linear combinations of g_1 through g_4 (the coefficients being found from a 4×4 secular determinant), the $1b_1$ MO is g_5, and the $1b_2$ and $2b_2$ MOs are linear combinations of g_6 and g_7. In the ground electronic state, five of these seven MOs are occupied. (Results of SCF calculations on H_2O are discussed in *Levine*, Section 15.4.)

Next consider the π orbitals of benzene. These MOs will be constructed using the six $2p_z$ carbon AOs as the minimal basis set. These six AOs form the basis for a representation of $\mathcal{D}_{6h}$, which we shall call Γ_π. The characters are readily found by the methods discussed above; for example, each $2p_z$ AO goes into its negative upon reflection in the σ_h plane, so that $\chi_\pi(\hat{\sigma}_h)=-6$. We find for the characters of Γ_π (see Fig. 9.4 for the conventions used)

$\hat{E}$	$2\hat{C}_6$	$2\hat{C}_3$	$\hat{C}_2''$	$3\hat{C}_2$	$3\hat{C}_2'$	$\hat{\sigma}_h$	$3\hat{\sigma}_v$	$3\hat{\sigma}_d$	$2\hat{S}_6$	$2\hat{S}_3$	$\hat{i}$
6	0	0	0	-2	0	-6	2	0	0	0	0

Equation (9.39) and the $\mathcal{D}_{6h}$ character table give

$$\Gamma_\pi = A_{2u}\oplus B_{2g}\oplus E_{1g}\oplus E_{2u} \tag{9.87}$$

Therefore, two of the four lowest π energy levels are doubly degenerate and the other two are nondegenerate; moreover, since no irreducible representation occurs more than once in (9.87), we can obtain the correct linear combinations for the six lowest π MOs without solving a secular equation.

From (9.67), the projection operator for the representation E_{2u} is

$$\hat{P}_{E_{2u}} = \tfrac{2}{24}\Big[2\hat{O}_E - \hat{O}_{C_6} - \hat{O}_{C_6^5} - \hat{O}_{C_3} - \hat{O}_{C_3^2} + 2\hat{O}_{C_2''} - 2\hat{O}_{\sigma_h} + \hat{O}_{S_6} + \hat{O}_{S_6^5} + \hat{O}_{S_3} + \hat{O}_{S_3^5} - 2\hat{O}_i\Big]$$

The projection operators for A_{2u}, B_{2g}, and E_{1g} are easily written down. Let $f_1,\ldots,f_6$ designate the carbon $2p_z$ AOs, where the carbon atoms are numbered 1 through 6 going counterclockwise. We find for the (unnormalized) nondegenerate a_{2u} and b_{2g} π MOs

$$\hat{P}_{A_{2u}} f_1 = \tfrac{1}{6}(f_1+f_2+f_3+f_4+f_5+f_6) \tag{9.88}$$

$$\hat{P}_{B_{2g}} f_1 = \tfrac{1}{6}(f_1-f_2+f_3-f_4+f_5-f_6) \tag{9.89}$$

For one of the e_{1g} π MOs, we have

$$\hat{P}_{E_{1g}} f_1 = \tfrac{1}{6}(2f_1+f_2-f_3-2f_4-f_5+f_6) \tag{9.90}$$

To get the second e_{1g} π MO, we can operate on f_2; cyclic permutation of the subscripts in (9.90) gives

$$\hat{P}_{E_{1g}} f_2 = \tfrac{1}{6}(f_1 + 2f_2 + f_3 - f_4 - 2f_5 - f_6) \tag{9.91}$$

[If we were to operate on f_3 with the E_{1g} projection operator, we would get an MO that is some linear combination of (9.90) and (9.91).] The degenerate MOs (9.90) and (9.91) are not orthogonal, but we can easily get two orthogonal MOs from them; let us call these functions F and G. Since they differ only in the labeling of the carbons, we have

$$\int F^2\,d\tau = \int G^2\,d\tau$$

Now consider the linear combinations $F+G$ and $F-G$; we have (F and G are real)

$$\int (F+G)^*(F-G)\,d\tau = \int F^2\,d\tau - \int G^2\,d\tau - \int FG\,d\tau + \int GF\,d\tau = 0$$

The sum and difference of (9.90) and (9.91) thus give orthogonal e_{1g} MOs:

$$\tfrac{1}{2}(f_1 + f_2 - f_4 - f_5) \tag{9.92}$$

$$\tfrac{1}{6}(f_1 - f_2 - 2f_3 - f_4 + f_5 + 2f_6) \tag{9.93}$$

Similarly

$$\hat{P}_{E_{2u}} f_1 = \tfrac{1}{6}(2f_1 - f_2 - f_3 + 2f_4 - f_5 - f_6) \tag{9.94}$$

$$\hat{P}_{E_{2u}} f_2 = \tfrac{1}{6}(-f_1 + 2f_2 - f_3 - f_4 + 2f_5 - f_6) \tag{9.95}$$

Taking the sum and difference, we have as the orthogonal e_{2u} MOs:

$$\tfrac{1}{2}(f_1 - f_2 + f_4 - f_5) \tag{9.96}$$

$$\tfrac{1}{6}(f_1 + f_2 - 2f_3 + f_4 + f_5 - 2f_6) \tag{9.97}$$

Aside from the node in the molecular plane, the number of nodal planes in these pi MOs is zero for a_{2u}, three for b_{2g}, one for e_{1g}, and two for e_{2u}. Hence the order of the π MO energies in benzene is $a_{2u} < e_{1g} < e_{2u} < b_{2g}$ (Fig. 1.3).

9.9 GROUP THEORY AND MOLECULAR VIBRATIONS

The applications of group theory to molecular vibrations parallel its applications to molecular electronic structure. The vibrational wave functions of each vibrational level of a linear molecule transform according to one of the irreducible representations of either $\mathcal{D}_{\infty h}$ or $\mathcal{C}_{\infty v}$, depending on whether the molecule has a center of symmetry. The symmetry of the vibrational wave functions is given by the irreducible representation to which they belong, and the degeneracy of the vibrational level is given by the dimension of the irreducible representation. Vibrational wave functions of nonlinear molecules are similarly classified according to the irreducible representations of the molecular point group.

Just as group theory enables one to find symmetry-adapted orbitals, which simplify the solution of the MO secular equation, group theory enables one to find symmetry-adapted displacement coordinates, which simplify the solution of the vibrational secular equation. We first show that the matrices describing the transformation properties of any set of degenerate normal coordinates form an irreducible representation of the molecular point group. The proof is based on the potential-energy expression for vibration, (6.23) and (6.33):

$$U = U_e + 2\pi^2 \sum_{i=1}^{3N-6} \nu_i^2 Q_i^2$$

where $\nu_i = \lambda_i^{1/2}/2\pi$. We now rewrite this expression, grouping together degenerate normal coordinates. Let $Q_{1,\nu_k}, Q_{2,\nu_k}, \ldots, Q_{d_k,\nu_k}$ each correspond to the same vibrational frequency ν_k, and let the molecule have n different normal-mode frequencies. We have

$$U = U_e + 2\pi^2 \sum_{k=1}^{n} \sum_{j=1}^{d_k} \nu_k^2 Q_{j,\nu_k}^2 \tag{9.98}$$

As discussed in Section 6.3, the function U must remain unchanged upon application of any symmetry operator; this means that the sums of the squares of the transformed normal coordinates that have the same frequency must equal the sums of the squares of the corresponding untransformed coordinates:

$$\sum_{j=1}^{d_k} \left(\hat{O}_R Q_{j,\nu_k}\right)^2 = \sum_{j=1}^{d_k} Q_{j,\nu_k}^2 \tag{9.99}$$

The transformed normal coordinate $\hat{O}_R Q_{j,\nu_k}$ can be expressed as some linear combination of the $3N$ mass-weighted displacement coordinates q_i; using (6.21), we can express $\hat{O}_R Q_{j,\nu_k}$ as a linear combination of all the normal coordinates of the molecule. If in this linear combination, the coefficient of a normal coordinate whose vibrational frequency differed from that of Q_{j,ν_k} were nonzero, then the left side of (9.99) would not be the same function as the right side, which contains only normal coordinates of the same frequency as Q_{j,ν_k}. Hence, upon application of a symmetry operation, we must have each normal coordinate go into a linear combination of normal coordinates with the same vibrational frequency:

$$\hat{O}_R Q_{j,\nu_k} = \sum_{i=1}^{d_k} r_{ij,\nu_k} Q_{i,\nu_k}, \qquad j=1,2,\ldots,n \tag{9.100}$$

where the r_{ij,ν_k} are certain coefficients such that (9.99) is satisfied.

Let $\mathbf{R}_{\nu_k}$ be the matrix whose elements are r_{ij,ν_k}; let $\mathbf{S}_{\nu_k}$ be the matrix whose elements give the effect of the symmetry operator $\hat{O}_s$ on the Q_{j,ν_k}; and so on. Since the normal coordinates $Q_{1,\nu_k}, Q_{2,\nu_k},\ldots,Q_{d_k,\nu_k}$ are transformed into linear combinations of one another by the symmetry operations, they form a basis for a d_k-dimensional representation of the molecular point group (Section 9.6); the matrices of this representation are $\mathbf{R}_{\nu_k}, \mathbf{S}_{\nu_k},\ldots$. This representation is either reducible or irreducible. If it were reducible, then the d_k degenerate normal coordinates could be divided into two or more subsets such that members of one subset were transformed by symmetry operations into linear combinations of only the coordinates of the same subset; in this case, there would be no reason from symmetry alone for the vibrational modes of different subsets to have the same frequency. We thus expect this representation to be irreducible (barring accidental degeneracy).

For certain point groups, we have one-dimensional (irreducible) representations with complex characters. Suppose that the normal coordinate Q_1 transforms according to the one-dimensional (irreducible) representation Γ_1, some of whose characters are complex numbers. We then have for any symmetry operation $\hat{R}$

$$\hat{O}_R Q_1 = bQ_1 \tag{9.101}$$

where b is the character of $\hat{R}$ in Γ_1. Since some of the characters of Γ_1 are complex, the normal coordinate Q_1 must, according to (9.101), be complex —some of the coefficients l_{k1} in (6.19) must be complex. Consider now Q_1^*. The quantities u_{jk} and λ_m in (6.18) are real numbers; taking the complex conjugate of (6.18) shows that if the quantities l_{km} satisfy (6.18) for a particular value of λ_m, then so do their complex conjugates. Hence Q_1^* is a normal coordinate that corresponds to the same value of λ_m (and thus the same frequency) as Q_1. The complex conjugate of (9.101) reads

$$\hat{O}_R Q_1^* = b^* Q_1^*$$

so that the normal coordinate Q_1^* transforms according to the representation Γ_1^* whose characters are the complex conjugates of Γ_1. Since Q_1 and Q_1^* have the same vibrational frequency, from a physical viewpoint the two one-dimensional representations Γ_1 and Γ_1^* are considered as one two-dimensional "irreducible" representation. Of course, the physical nature of the vibration problem would lead us to use real coefficients l_{ik} in the normal coordinates; we can get real normal coordinates from Q_1 and Q_1^* by taking $(Q_1+Q_1^*)/2$ and $(Q_1-Q_1^*)/2i$. Just as with degenerate wave functions, any linear combination of degenerate normal coordinates is a normal coordinate corresponding to the same vibrational frequency; this conclusion follows from the linear homogeneous nature of the set of equations (6.18). The real linear combinations of Q_1 and Q_1^* transform according to the two-dimensional representation that is the direct sum of Γ_1 and Γ_1^*.

As noted earlier, point groups with no threefold or higher proper or improper axis have only one-dimensional representations; hence a necessary condition for a molecule to have degenerate vibrational modes is that it possess a C_n or an S_n axis with $n \geqslant 3$. Asymmetric tops have no degenerate vibrational modes.

The normal coordinates are certain linear combinations of the $3N$ displacement coordinates of the nuclei from their equilibrium positions; the correct linear combinations are found from the secular equation (6.17) and the associated equations (6.18) and (6.19). We now apply group theory to the problem.

We start with the $3N$ unit displacement vectors, where there are three mutually perpendicular vectors on each nucleus. Consider the effect of a symmetry operation on one of these $3N$ vectors. Each vector either will remain on the same nucleus, with or without its direction changed, or it will be moved to an identical nucleus, with or without its direction changed. (We consider the nuclei as fixed while examining the effect of a symmetry operation on a displacement vector; see Section 6.3.) Since any vector on a nucleus can be expressed as some linear combination of the three mutually perpendicular unit displacement vectors on that nucleus, a symmetry operation will send each of the $3N$ displacement vectors into some linear combination of these $3N$ vectors. Therefore, the unit displacement vectors form a basis for some $3N$-dimensional representation of the molecular point group; we shall call this representation Γ_{3N}.

Using (9.38), we can decompose Γ_{3N} into the direct sum of certain irreducible representations, and this decomposition is unique:

$$\Gamma_{3N}=\Gamma_1\oplus\Gamma_2\oplus\cdots\oplus\Gamma_n \tag{9.102}$$

Recall (Section 9.6) that the similarity transformation that throws the matrices of Γ_{3N} into block-diagonal form corresponds to taking certain linear combinations of the $3N$ unit displacement coordinates (which form the basis for Γ_{3N}); these linear combinations fall into subsets, each of whose members transforms according to one of the irreducible representations into which Γ_{3N} has been decomposed. These linear combinations of the $3N$ coordinates are the symmetry-adapted coordinates or, more simply, the *symmetry coordinates*. Since the reduction of Γ_{3N} is unique, each normal coordinate must be a linear combination of those symmetry coordinates that belong to the same irreducible representation as that normal coordinate. If no irreducible representation occurs more than once in (9.102), then the symmetry coordinates must be the normal coordinates; for irreducible representations that occur more than once in (9.102), the normal coordinates are found by solving a secular equation which involves only those symmetry coordinates that belong to the same irreducible representation as the normal coordinate. (Note how these statements parallel those made in Section 9.8 dealing with the application of group theory to molecular orbitals.)

As an example, consider the normal modes of H_2O. The $3N=9$ unit displacement vectors are shown in Fig. 9.6, which also shows the effect of the $\hat{C}_2$ symmetry operation on them. The effect of the $\hat{C}_2$ operation is as follows:

$$x_1 \to -x_2, \qquad y_1 \to -y_2, \qquad z_1 \to z_2$$

$$x_2 \to -x_1, \qquad y_2 \to -y_1, \qquad z_2 \to z_1$$

$$x_3 \to -x_3, \qquad y_3 \to -y_3, \qquad z_3 \to z_3$$

We have

$$\hat{C}_2 \begin{pmatrix} x_1 \\ y_1 \\ z_1 \\ x_2 \\ y_2 \\ z_2 \\ x_3 \\ y_3 \\ z_3 \end{pmatrix} = \begin{pmatrix} 0 & 0 & 0 & -1 & 0 & 0 & 0 & 0 & 0 \\ 0 & 0 & 0 & 0 & -1 & 0 & 0 & 0 & 0 \\ 0 & 0 & 0 & 0 & 0 & 1 & 0 & 0 & 0 \\ -1 & 0 & 0 & 0 & 0 & 0 & 0 & 0 & 0 \\ 0 & -1 & 0 & 0 & 0 & 0 & 0 & 0 & 0 \\ 0 & 0 & 1 & 0 & 0 & 0 & 0 & 0 & 0 \\ 0 & 0 & 0 & 0 & 0 & 0 & -1 & 0 & 0 \\ 0 & 0 & 0 & 0 & 0 & 0 & 0 & -1 & 0 \\ 0 & 0 & 0 & 0 & 0 & 0 & 0 & 0 & 1 \end{pmatrix} \begin{pmatrix} x_1 \\ y_1 \\ z_1 \\ x_2 \\ y_2 \\ z_2 \\ x_3 \\ y_3 \\ z_3 \end{pmatrix}$$

The above square matrix represents $\hat{C}_2$ in Γ_{3N}.

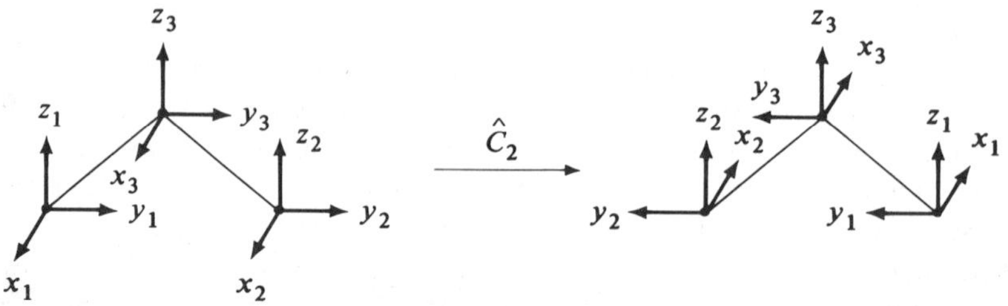

Fig. 9.6 Effect of the $\hat{C}_2$ symmetry operation on the $3N$ unit displacement vectors of H_2O. The yz plane is the molecular plane.

The other three matrices of Γ_{3N} can be found similarly, but there is no need to write them down in full, since we need only their traces to decompose Γ_{3N}. The following rules [similar to those following (9.76)] are easily seen to hold: For each unit displacement vector that is unmoved by the symmetry operation $\hat{R}$, we get a contribution of $+1$ to $\chi_{3N}(\hat{R})$; for each unit displacement vector that is moved to another nucleus by $\hat{R}$, we get a contribution of 0 to $\chi_{3N}(\hat{R})$; for each unit displacement vector that is sent into minus itself, we get a contribution of -1 to $\chi_{3N}(\hat{R})$. If a unit displacement vector is sent into a linear combination of the original unit vectors on the same nucleus, the contribution to $\chi_{3N}(\hat{R})$ is the coefficient of that unit vector in the linear combination; this case arises for rotation about threefold and higher axes.

Using these rules, we find as the characters for Γ_{3N}

$\hat{E}$	$\hat{C}_2(z)$	$\hat{\sigma}_v(xz)$	$\hat{\sigma}_v(yz)$
9	-1	1	3

(9.103)

(Since $\hat{E}$ is represented by a unit matrix, its character is always $3N$ in Γ_{3N}.) Equation (9.39) gives

$$\Gamma_{3N}=3A_1\oplus A_2\oplus 2B_1\oplus 3B_2 \tag{9.104}$$

As a check, it can be verified that the sums of the characters of the irreducible representations on the right of (9.104) equal the characters of Γ_{3N}.

The $3N$ degrees of freedom for nuclear motion are divided into 3 translational, 3 (or 2) rotational, and $3N-6$ (or $3N-5$) vibrational degrees of freedom. (The translations and rotations are often called *nongenuine vibrations*.) The 9 irreducible representations in (9.104) include the 3 translations and the 3 rotations. To find the symmetry species of the 3 vibrations, we must find the symmetry species of the translations and rotations.

The translational motion corresponds to a displacement of the molecule as a whole in an arbitrary direction; it can be depicted by a single vector showing the displacement of the center of mass. Let this vector have components x,y,z. We showed in Section 9.3 that under any symmetry operation, each of the functions x,y,z is transformed into a linear combination of x,y, and z. Hence (Section 9.6) the set of functions x,y,z forms a basis for some three-dimensional representation of the molecular point group; we shall call this representation Γ_{trans}. [The representation (9.25) is in fact Γ_{trans} for $\mathcal{C}_{3v}$.]

Let us find Γ_{trans} for H_2O. From (9.9), (9.12), and (9.17), the matrices corresponding to the symmetry operations of $\mathcal{C}_{2v}$ in the representation whose basis is the set of functions x,y,z are

$$\hat{E}:\begin{pmatrix} 1 & 0 & 0 \\ 0 & 1 & 0 \\ 0 & 0 & 1 \end{pmatrix} \qquad \hat{C}_2(z):\begin{pmatrix} -1 & 0 & 0 \\ 0 & -1 & 0 \\ 0 & 0 & 1 \end{pmatrix} \tag{9.105}$$

$$\hat{\sigma}_v(xz):\begin{pmatrix} 1 & 0 & 0 \\ 0 & -1 & 0 \\ 0 & 0 & 1 \end{pmatrix} \qquad \hat{\sigma}_v(yz):\begin{pmatrix} -1 & 0 & 0 \\ 0 & 1 & 0 \\ 0 & 0 & 1 \end{pmatrix} \tag{9.106}$$

The matrices are in block-diagonal form and we have

$$\Gamma_{\text{trans}} = B_1 \oplus B_2 \oplus A_1 \tag{9.107}$$

For $\mathcal{C}_{2v}$, each of the functions x, y, and z taken alone forms a basis for an irreducible representation; for example,

$$\begin{aligned} \hat{E}x &= (1)x, \qquad [\hat{C}_2(z)]x = (-1)x \\ [\hat{\sigma}_v(xz)]x &= (1)x, \qquad [\hat{\sigma}_v(yz)]x = (-1)x \end{aligned} \tag{9.108}$$

and x transforms according to the irreducible representation B_1. This fact is indicated in the character table by writing "x" on the same line as the B_1 characters. The functions y and z transform according to B_2 and A_1, respectively.

The three-dimensional representation Γ_{trans} may or may not be reducible. If it is irreducible (this can only happen for groups with at least one three-dimensional irreducible representation), then "(x,y,z)" is written on the same line as the characters of the irreducible representation Γ_{trans}; an example is $\mathcal{T}_d$, where Γ_{trans} is the irreducible representation T_2. For $\mathcal{C}_{3v}$, we saw that Γ_{trans} is reducible to the direct sum of the A_1 one-dimensional irreducible representation (9.29) and the E two-dimensional irreducible representation (9.28); the basis for (9.29) is the function z, and the basis for (9.28) is the set of functions x,y. In the $\mathcal{C}_{3v}$ character table, "z" is written on the A_1 line, and "(x,y)" is written on the E line. Note that "x,y" in a character table has a different meaning from "(x,y)"; the absence of parentheses indicates that x forms a basis for a one-dimensional irreducible representation and y forms a basis for the same irreducible representation as x, whereas the presence of parentheses indicates that x and y together form a basis for a two-dimensional irreducible representation.

We also must find the symmetry species of the rotations in H_2O. An arbitrary rigid rotation of the molecule can be resolved into rotations by various angles about the three axes x, y, and z. Figure 9.7 shows the displacement vectors for a rotation about the x axis in H_2O. We can represent this rotation by a single vector $\mathbf{R}_x$ pointing along the axis of rotation; the length of $\mathbf{R}_x$ is proportional to the angle of rotation and its direction is given by the direction the thumb points to when the fingers of the right hand curl in the direction of rotation. The effect of any symmetry operation on the nuclear displacement vectors of a rotation defines its effect on $\mathbf{R}_x$. Thus the $\hat{C}_2(z)$ operation in H_2O sends the displacement vectors for rotation about the x axis into their negatives, thereby reversing the direction of this rotation; hence $\mathbf{R}_x$, whose direction is defined relative to the direction of rotation, is reversed in direction (Fig. 9.7). The $\hat{\sigma}(yz)$ operation in H_2O has no effect on the displacement vectors for rotation about the x axis; hence $\mathbf{R}_x$ is unchanged by this symmetry operation. The effect of $\hat{\sigma}(yz)$ on $\mathbf{R}_x$ differs from its effect on a vector pointing along the x axis and representing translational motion; the latter vector is changed to its negative by a $\hat{\sigma}(yz)$ reflection. Vectors like $\mathbf{R}_x$ are called *axial vectors* or *pseudovectors*. Displacement vectors are examples of *polar* vectors.[6] The symmetry operations transform each of the three axial vectors representing molecular rotation about the x, y, and z axes into linear combinations of one another, thereby generating the representation Γ_{rot}, which corresponds to the molecular rotational modes.

The characters of Γ_{rot} are closely related to those of Γ_{trans}. Any symmetry operation is either a $\hat{C}_n$ or an $\hat{S}_n$ operation. Consider first the effect on, say, $\mathbf{R}_x$ of a $\hat{C}_n$ rotation about some axis, not necessarily the x axis. This $\hat{C}_n$ rotation will move the rotation displacement vectors in such a manner as to transform $\mathbf{R}_x$ into a vector $\mathbf{R}'_x$, where $\mathbf{R}'_x$ is the vector obtained by applying $\hat{C}_n$ directly to $\mathbf{R}_x$; an example is shown in Fig. 9.7. Thus for proper rotations, the matrices describing how $\mathbf{R}_x$, $\mathbf{R}_y$, and $\mathbf{R}_z$ transform are exactly those matrices that describe how ordinary (polar)

[6] See *Halliday and Resnick*, Supplementary Topic II. The cross product (e.g., angular momentum) of two polar vectors is a pseudovector.

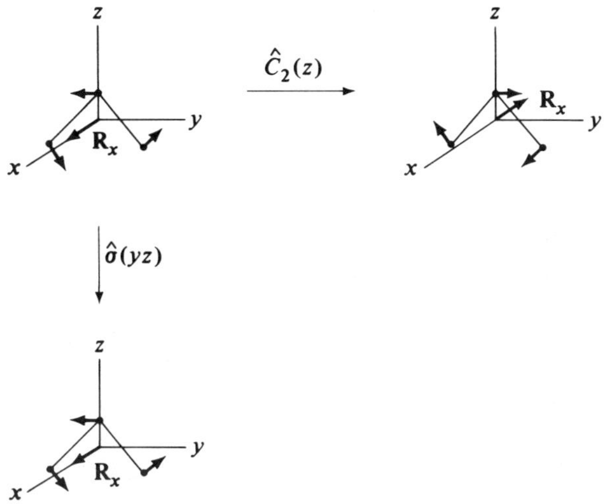

Fig. 9.7 Effect of $\hat{C}_2(z)$ and $\hat{\sigma}(yz)$ on a rotational mode of H_2O. The yz plane is the molecular plane.

vectors along the x, y, and z axes transform; the matrix in the representation Γ_{rot} corresponding to any $\hat{C}_n$ operation is the same as the matrix in Γ_{trans} that corresponds to that $\hat{C}_n$ operation, and the characters for proper rotations are the same for Γ_{rot} as for Γ_{trans}.

Now consider the effect on $\mathbf{R}_x$ of an $\hat{S}_n$ operation. We have $\hat{S}_n = \hat{\sigma}_h \hat{C}_n$. The $\hat{C}_n$ part of the $\hat{S}_n$ operation has the same effect on $\mathbf{R}_x$ as on a translation vector along the x axis. However, as shown in Fig. 9.8, the effect of a reflection on $\mathbf{R}_x$ is opposite to its effect on a translation vector along the x axis; an example is shown in Fig. 9.7 for H_2O. Thus the matrices describing the effect of an $\hat{S}_n$ operation on the rotation vectors are the negatives of the matrices describing the effect of $\hat{S}_n$ on the translation vectors. The character of any $\hat{S}_n$ operation for Γ_{rot} is the negative of its character in Γ_{trans}; in particular, the character of a reflection or an inversion for Γ_{rot} is found by multiplying its character in Γ_{trans} by -1.

We can now find the symmetry species of the rotations in H_2O. Taking $+1$ times the matrices (9.105) and -1 times the matrices (9.106), we obtain Γ_{rot}, which is seen to be reducible as follows:

$$\Gamma_{rot} = B_2 \oplus B_1 \oplus A_2 \tag{9.109}$$

Since Γ_{rot} is reducible to the direct sum of three one-dimensional representations, a rotation vector along any one of the three coordinate axes forms

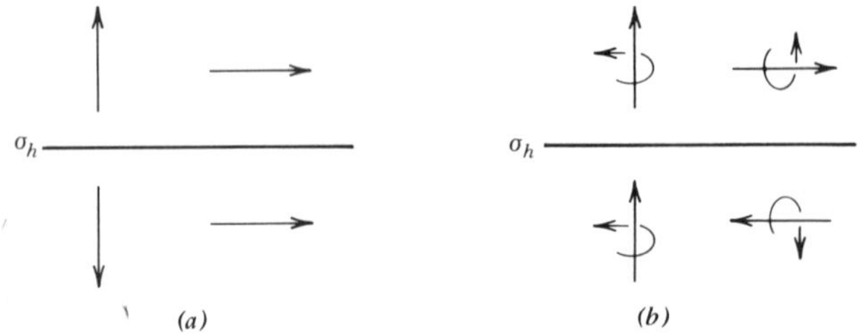

Fig. 9.8 The effect of reflection on: (a) polar vectors representing translations; (b) axial vectors representing rotations. In (b), the directions of the rotational-mode displacement vectors of the nuclei (symbolized by the curved lines) determine the directions of the axial vectors.

a basis for an irreducible representation; this is indicated in the $\mathcal{C}_{2v}$ character table by the symbols R_x, R_y, and R_z placed on the B_2, B_1, and A_2 lines, respectively.

Designating the representation generated by the $3N-6$ genuine vibrations as Γ_{vib}, we have

$$\Gamma_{3N}=\Gamma_{\text{vib}}\oplus\Gamma_{\text{rot}}\oplus\Gamma_{\text{trans}} \tag{9.110}$$

For H_2O, we find from (9.104), (9.107), and (9.109) that

$$\Gamma_{\text{vib}}=2A_1\oplus B_2 \tag{9.111}$$

There are two totally symmetric (a_1) normal modes and one b_2 normal mode. (The convention is to use lowercase letters for the symmetry species of the normal modes.) The symmetry species of the normal modes have been found without solving the vibrational secular equation. Moreover, since there is only one b_2 normal mode, the form of this vibration must be determined from symmetry considerations together with the requirement that the vibration have no translational or rotational energy associated with it. Thus (Fig 6.1), any bent XYX molecule has a b_2 normal mode with the X atoms vibrating along the X—Y bonds and the Y atom vibrating in the plane of the molecule and perpendicular to the symmetry axis. On the other hand, there are two a_1 symmetry coordinates and the two a_1 normal vibrations are linear combinations of the a_1 symmetry coordinates, where the coefficients are dependent on the nuclear masses and the force constants. Thus the angles between the displacement vectors of the X atoms and the X—Y bonds for the a_1 modes of a bent XYX molecule vary from molecule to molecule.

If the molecular plane of H_2O had been designated the xz plane rather than the yz plane, then the symmetry species of the ν_3 vibration would be b_1 rather than b_2. Of course, the labeling of the plane is arbitrary, but standard conventions have been agreed on for most molecules.[7]

Let us find general formulas for the characters $\chi_{3N}(\hat{R})$ of Γ_{3N}. As noted earlier, only displacement vectors on nuclei that are not shifted when the symmetry operation $\hat{R}$ is applied to the framework contribute to $\chi_{3N}(\hat{R})$. On each unshifted nucleus, the three displacement vectors are each transformed into linear combinations of the displacement vectors on the same nucleus. For a $\hat{C}_n$ operation with $n>1$, each unshifted nucleus lies on the C_n axis (which we shall call the z axis), and the matrix (9.12) tells how the three displacement vectors transform. Taking the trace of (9.12), we see that an unshifted nucleus gives a contribution of $1+2\cos(2\pi/n)$ to $\chi_{3N}(\hat{C}_n)$. If there are N_{C_n} unshifted nuclei for the $\hat{C}_n$ operation, then

$$\chi_{3N}(\hat{C}_n)=N_{C_n}[1+2\cos(2\pi/n)] \tag{9.112}$$

The trace of (9.13) gives

$$\chi_{3N}(\hat{C}_n^m)=N_{C_n^m}[1+2\cos(2\pi m/n)] \tag{9.113}$$

We know that

$$\chi_{3N}(\hat{E})=3N \tag{9.114}$$

which is a special case of (9.112) with $n=1$.

For an $\hat{S}_n$ operation with $n>1$, all unshifted nuclei lie on the S_n axis, and the matrix (9.14) tells how the displacement vectors on any unshifted nucleus transform. The trace of (9.14) gives

$$\chi_{3N}(\hat{S}_n)=N_{S_n}[-1+2\cos(2\pi/n)] \tag{9.115}$$

where N_{S_n} is the number of unshifted nuclei for the $\hat{S}_n$ operation. Also

$$\chi_{3N}(\hat{S}_n^m)=N_{S_n^m}[(-1)^m+2\cos(2\pi m/n)] \tag{9.116}$$

With $n=2$, Equation (9.115) reads

$$\chi_{3N}(\hat{i})=-3N_i \tag{9.117}$$

where N_i is 0 or 1. For a reflection, the unshifted nuclei lie in the reflection plane and the matrix describing the reflection is (9.16), whose trace is 1; we

[7] R. S. Mulliken, *J. Chem. Phys.*, **23**, 1997 (1955).

have

$$\chi_{3N}(\hat{\sigma}) = N_\sigma \tag{9.118}$$

where N_σ is the number of nuclei in the reflection plane.

The preceding equations can all be summarized in a single equation. Let $\hat{R}$ be a proper or improper rotation by an angle α; we have

$$\chi_{3N}(\hat{R}) = N_R(\pm 1 + 2\cos\alpha) \tag{9.119}$$

where N_R is the number of unmoved nuclei for $\hat{R}$, and where the plus and minus signs are used for proper and improper rotations, respectively.

[An objection that might be raised to the derivation of (9.119) is that we have taken all the proper and improper rotations as being performed about the z axis, whereas the symmetry operations of a molecule are, in general, performed about axes pointing in various directions. The answer to this objection is that although a different choice of coordinate axes will change the elements of a rotation matrix, it will not change the trace of that matrix: Each of the members of an arbitrary set of three unit displacement vectors can be expressed as a linear combination of unit displacement vectors along the x, y, and z axes; as shown in Section 9.6, taking linear combinations of a set of basis functions gives matrices that are related to the original matrices by a similarity transformation and thus have the same traces as the original matrices.]

Consider C_6H_6 as an example. The number of unshifted nuclei for the various symmetry operations are found to be

$\hat{R}$:	$\hat{E}$	$3\hat{C}_2$	$\hat{\sigma}_h$	$3\hat{\sigma}_v$	all others
N_R:	12	4	12	4	0

(9.120)

Application of (9.119) gives as the characters of Γ_{3N}

$\hat{E}$	$2\hat{C}_6$	$2\hat{C}_3$	$\hat{C}_2''$	$3\hat{C}_2$	$3\hat{C}_2'$	$\hat{\sigma}_h$	$3\hat{\sigma}_v$	$3\hat{\sigma}_d$	$2\hat{S}_6$	$2\hat{S}_3$	$\hat{i}$
36	0	0	0	−4	0	12	4	0	0	0	0

Equation (9.39) then gives

$$\Gamma_{3N} = 2A_{1g} \oplus 2A_{2g} \oplus 2A_{2u} \oplus 2B_{1u} \oplus 2B_{2g} \oplus 2B_{2u} \oplus 2E_{1g}$$

$$\oplus 4E_{1u} \oplus 4E_{2g} \oplus 2E_{2u} \tag{9.121}$$

The $\mathcal{D}_{6h}$ character table gives for the symmetry species of the nongenuine vibrations

$$\Gamma_{\text{trans}} = A_{2u} \oplus E_{1u}, \qquad \Gamma_{\text{rot}} = A_{2g} \oplus E_{1g} \tag{9.122}$$

Therefore,

$$\Gamma_{\text{vib}} = 2A_{1g} \oplus A_{2g} \oplus A_{2u} \oplus 2B_{1u} \oplus 2B_{2g} \oplus 2B_{2u} \oplus E_{1g} \oplus 3E_{1u} \oplus 4E_{2g} \oplus 2E_{2u} \tag{9.123}$$

Benzene thus has 10 nondegenerate and 20 doubly degenerate normal modes. The convention[8] is to number the vibrational frequencies according to the order their symmetry species are listed in the character table; modes of the same symmetry species are numbered in order of decreasing frequency. Thus the a_{2u} frequency is called ν_4, the frequency of the two degenerate e_{1g} modes is called ν_{11}, the lower of the two e_{2u} frequencies is called ν_{20}.

9.10 GROUP THEORY AND MOLECULAR ROTATION

We have seen that the molecular electronic and vibrational wave functions ψ_{el} and ψ_{vib} each transform according to the irreducible representations of the molecular point group. We now consider the rotational wave function ψ_{rot}.

The symmetry of a wave function is determined by the symmetry of the Hamiltonian of which it is an eigenfunction. The electronic and vibrational Hamiltonians each involve the specific configuration of the nuclei. The rotational Hamiltonian, however, does not involve the detailed nuclear configuration; rather, it involves the three principal moments of inertia and the rotational angular-momentum components along the three principal axes. The rotational Hamiltonian is unchanged by any proper or improper rotation of the molecule that sends the ellipsoid of inertia to a position physically indistinguishable from its original one. The symmetry of the rotational wave functions is therefore determined by the symmetry of the ellipsoid of inertia, whose symmetry is usually higher than that of the molecular point group.

The spherical-top Hamiltonian is $\hat{H}_{\text{rot}} = \hat{P}^2/2I$. The ellipsoid of inertia is a sphere. The spherical-top rotational wave functions can thus be classified according to the symmetry species of $\mathcal{K}_h$. (The angular-momentum operator $\hat{P}^2 = \hat{P}_a^2 + \hat{P}_b^2 + \hat{P}_c^2$ remains unchanged no matter how the *abc* axes are rotated.)

[8] R. S. Mulliken, *J. Chem. Phys.*, **23**, 1997 (1955).

The symmetric-top Hamiltonian is (5.61). The ellipsoid of inertia is an ellipsoid of revolution (symmetry $\mathcal{D}_{\infty h}$) and the symmetric-top ψ_{rot} can be classified according to the irreducible representations of $\mathcal{D}_{\infty h}$.

The asymmetric-top Hamiltonian is (5.74). The ellipsoid of inertia has three unequal principal axes, and its symmetry is $\mathcal{D}_{2h}$. The asymmetric-top rotational wave functions can be classified according to the symmetry species of $\mathcal{D}_{2h}$. The group $\mathcal{D}_2$ is a subgroup of $\mathcal{D}_{2h}$, and the asymmetric-top functions are usually classified according to $\mathcal{D}_2$, since this classification provides sufficient information about the symmetry of ψ_{rot}. The basis functions used in solving the asymmetric-top secular equation (5.78) are symmetric-top wave functions; the use of symmetry-adapted linear combinations (which transform according to irreducible representations of $\mathcal{D}_2$) of these basis functions reduces the secular equation to several secular equations of lower degree.[9]

Of course, the point group of a molecule is always a subgroup of the point group of its ellipsoid of inertia. For example, asymmetric tops can belong to the point groups

$$\mathcal{C}_1, \mathcal{C}_s, \mathcal{C}_i, \mathcal{C}_2, \mathcal{C}_{2h}, \mathcal{C}_{2v}, \mathcal{D}_2, \mathcal{D}_{2h}$$

each of which is a subgroup of $\mathcal{D}_{2h}$. We can, if desired, classify the rotational wave functions according to the symmetry species of the molecular point group.

The overall molecular wave function is the product of vibrational, rotational, and electronic wave functions and can be classified according to one of the symmetry species of the molecular point group.[10]

9.11 THE DIRECT-PRODUCT REPRESENTATION

Many quantum-mechanical applications of group theory involve the direct product of representations, which we now discuss.

Let $F_1, F_2, \ldots, F_m$ and $G_1, G_2, \ldots, G_n$ be two sets of functions forming bases for the representations Γ_F and Γ_G, respectively, of a certain group. We thus have [Eq. (9.41)]

$$\hat{O}_R F_j = \sum_{i=1}^{m} r_{ij,F} F_i, \qquad j=1,\ldots,m \tag{9.124}$$

$$\hat{O}_R G_l = \sum_{k=1}^{n} r_{kl,G} G_k, \qquad l=1,\ldots,n \tag{9.125}$$

[9] See *King*, pp. 305, 309–310.

[10] See J. T. Hougen, *J. Chem. Phys.*, **39**, 358 (1963); *Herzberg, Electronic Spectra of Polyatomic Molecules*, pp. 90–94, 99–101, 109–114.

where $r_{ij,F}$ and $r_{kl,G}$ are the elements of the matrices $\mathbf{R}_F$ and $\mathbf{R}_G$ that correspond to the symmetry operation $\hat{R}$ in the representations Γ_F and Γ_G.

Now consider the set of mn functions

$$F_j G_l \tag{9.126}$$

We shall order this set of functions by some convention and call the mn product functions J_q:

$$J_q = F_j G_l; \qquad q = 1,\ldots,mn; \qquad j = 1,\ldots,m; \qquad l = 1,\ldots,n \tag{9.127}$$

The ordering convention we use is (this is called *dictionary order*)

$$\begin{array}{cccc} J_1 = F_1 G_1 & J_2 = F_1 G_2 & \cdots & J_n = F_1 G_n \\ J_{n+1} = F_2 G_1 & J_{n+2} = F_2 G_2 & \cdots & J_{2n} = F_2 G_n \\ \cdot\;\cdot\;\cdot\;\cdot\;\cdot & \cdot\;\cdot\;\cdot\;\cdot\;\cdot & \cdots & \cdot\;\cdot\;\cdot\;\cdot \\ \cdot\;\cdot\;\cdot\;\cdot\;\cdot & \cdot\;\cdot\;\cdot\;\cdot\;\cdot & \cdots & J_{mn} = F_m G_n \end{array} \tag{9.128}$$

Let us examine how the functions (9.127) transform; we have

$$\hat{O}_R(F_j G_l) = (\hat{O}_R F_j)(\hat{O}_R G_l) = \sum_{i=1}^{m} \sum_{k=1}^{n} r_{ij,F} r_{kl,G} F_i G_k \tag{9.129}$$

Defining the constants $r_{pq,J}$ as

$$r_{pq,J} = r_{ij,F} r_{kl,G} \tag{9.130}$$

we write (9.129) as

$$\hat{O}_R J_q = \sum_{p=1}^{mn} r_{pq,J} J_p \qquad q = 1,\ldots,mn \tag{9.131}$$

The mn product functions (9.127) are thus transformed into linear combinations of one another by the symmetry operators of the group; they therefore form a basis for some mn-dimensional representation Γ_J of the group. The representation Γ_J is called the *direct product* (or *Kronecker product*) of the representations Γ_F and Γ_G; this is symbolized by

$$\Gamma_J = \Gamma_F \otimes \Gamma_G \tag{9.132}$$

The matrix of Γ_J that corresponds to the symmetry operation $\hat{R}$ has elements $r_{pq,J}$ and we shall call it $\mathbf{R}_J$; it is a square matrix of order mn and its elements are the $(mn)^2$ possible products of each of the m^2 elements of $\mathbf{R}_F$ with each of the n^2 elements of $\mathbf{R}_G$ [Equation (9.130)]. The matrix $\mathbf{R}_J$ is called the *direct product* of $\mathbf{R}_F$ and $\mathbf{R}_G$:

$$\mathbf{R}_J = \mathbf{R}_F \otimes \mathbf{R}_G \tag{9.133}$$

The direct product of two matrices is quite different from the ordinary matrix product.

How are the indices of the matrix elements of $\mathbf{R}_J$ related to the indices of the elements of $\mathbf{R}_F$ and $\mathbf{R}_G$? Comparison of (9.129) and (9.131) gives

$$J_p = F_i G_k, \qquad J_q = F_j G_l \tag{9.134}$$

Hence the subscript p in (9.130) is determined by the subscripts i and k, whereas q is determined by j and l. The dictionary-order convention (9.128) then gives

i:	1	1	$\cdots$	1	2	2	$\cdots$	2	$\cdots$	m
k:	1	2	$\cdots$	n	1	2	$\cdots$	n	$\cdots$	n
p:	1	2	$\cdots$	n	$n+1$	$n+2$	$\cdots$	$2n$	$\cdots$	mn

(9.135)

Table (9.135) applies with i, k, and p replaced by j, l, and q, respectively.

As an example, consider the direct product of a 2×2 matrix $\mathbf{A}$ and a 3×3 matrix $\mathbf{B}$. Table (9.135) gives for the indices

i	1	1	1	2	2	2
k	1	2	3	1	2	3
p	1	2	3	4	5	6

j	1	1	1	2	2	2
l	1	2	3	1	2	3
q	1	2	3	4	5	6

(9.136)

Equation (9.130) then gives

$$\begin{pmatrix} a_{11} & a_{12} \\ a_{21} & a_{22} \end{pmatrix} \otimes \begin{pmatrix} b_{11} & b_{12} & b_{13} \\ b_{21} & b_{22} & b_{23} \\ b_{31} & b_{32} & b_{33} \end{pmatrix}$$

$$= \begin{pmatrix} a_{11}b_{11} & a_{11}b_{12} & a_{11}b_{13} & a_{12}b_{11} & a_{12}b_{12} & a_{12}b_{13} \\ a_{11}b_{21} & a_{11}b_{22} & a_{11}b_{23} & a_{12}b_{21} & a_{12}b_{22} & a_{12}b_{23} \\ a_{11}b_{31} & a_{11}b_{32} & a_{11}b_{33} & a_{12}b_{31} & a_{12}b_{32} & a_{12}b_{33} \\ a_{21}b_{11} & a_{21}b_{12} & a_{21}b_{13} & a_{22}b_{11} & a_{22}b_{12} & a_{22}b_{13} \\ a_{21}b_{21} & a_{21}b_{22} & a_{21}b_{23} & a_{22}b_{21} & a_{22}b_{22} & a_{22}b_{23} \\ a_{21}b_{31} & a_{21}b_{32} & a_{21}b_{33} & a_{22}b_{31} & a_{22}b_{32} & a_{22}b_{33} \end{pmatrix}$$

This direct-product matrix $\mathbf{A}\otimes\mathbf{B}$ can be partitioned into four submatrices as follows:

$$\mathbf{A}\otimes\mathbf{B}=\left(\begin{array}{c|c} a_{11}\mathbf{B} & a_{12}\mathbf{B} \\ \hline a_{21}\mathbf{B} & a_{22}\mathbf{B} \end{array}\right) \tag{9.137}$$

This is an easy way to remember the direct product.

Now consider the character of $\hat{R}$ in the direct-product representation Γ_J:

$$\chi_J(\hat{R})=\mathrm{Tr}(\mathbf{R}_J)=\sum_{p=1}^{mn} r_{pp,J}$$

Since the index p is determined by the set of indices i and k, while q is determined by j and l, the equality $p=q$ (which defines the diagonal elements of $\mathbf{R}_J$) means that the set i and k must be the same as the set j and l: $i=j$, $k=l$. Hence (9.130) gives

$$\chi_J(\hat{R})=\sum_{j=1}^{m}\sum_{l=1}^{n} r_{jj,F}r_{ll,G}=\sum_{j=1}^{m} r_{jj,F}\sum_{l=1}^{n} r_{ll,G} \tag{9.138}$$

$$\chi_{F\otimes G}(\hat{R})=\chi_F(\hat{R})\chi_G(\hat{R}) \tag{9.139}$$

This is the desired result: The character of any symmetry operation in the direct-product representation $\Gamma_F\otimes\Gamma_G$ is the product of its characters in the representations Γ_F and Γ_G. (The direct product of matrices is not, in general, commutative; however, $\mathbf{A}\otimes\mathbf{B}$ and $\mathbf{B}\otimes\mathbf{A}$ have equal traces, and thus the corresponding direct-product representations are equivalent to each other.)

Let us look at some examples. For $\mathcal{C}_{2v}$, (9.139) and the character table give as the characters of the direct-product representation $A_2\otimes B_1$

$\hat{E}$	$\hat{C}_2(z)$	$\hat{\sigma}_v(xz)$	$\hat{\sigma}_v(yz)$
1	−1	−1	1

Hence

$$A_2\otimes B_1=B_2 \tag{9.140}$$

(For the direct product of two one-dimensional representations, the direct-product matrices are the same as the characters.)

For $\mathcal{C}_{3v}$, the characters of $E \otimes E$ are

$\hat{E}$	$2\hat{C}_3$	$3\hat{\sigma}_v$
4	1	0

Application of (9.39) gives

$$E \otimes E = A_1 \oplus A_2 \oplus E \tag{9.141}$$

The direct product of more than two representations can be considered; thus the direct product of $\Gamma_F \otimes \Gamma_G$ with some representation Γ_H is $(\Gamma_F \otimes \Gamma_G) \otimes \Gamma_H$. Applying (9.139) twice, we have

$$\chi_{(F\otimes G)\otimes H}(\hat{R}) = \chi_{F\otimes G}(\hat{R})\chi_H(\hat{R}) = \chi_F(\hat{R})\chi_G(\hat{R})\chi_H(\hat{R})$$

This result is independent of the order of Γ_F, Γ_G, and Γ_H in the direct product, so that for example, $(\Gamma_F \otimes \Gamma_G) \otimes \Gamma_H$ is equivalent to $(\Gamma_G \otimes \Gamma_H) \otimes \Gamma_F$; we denote all such equivalent direct products by $\Gamma_F \otimes \Gamma_G \otimes \Gamma_H$. Thus for $\mathcal{T}_d$, the characters of $A_2 \otimes T_1 \otimes T_2$ are

$\hat{E}$	$8\hat{C}_3$	$3\hat{C}_2$	$6\hat{\sigma}_d$	$6\hat{S}_4$
9	0	1	1	1

We find

$$A_2 \otimes T_1 \otimes T_2 = A_1 \oplus E \oplus T_1 \oplus T_2$$

An important theorem that follows from (9.139) is the following: *The direct product of two irreducible representations Γ_α and Γ_β contains the totally symmetric representation if and only if Γ_α is the complex-conjugate representation of Γ_β.* By the complex-conjugate representation, we mean the representation whose matrices are the complex conjugates of the original representation. (For representations whose matrices are real, the complex-conjugate representation is the same as the original representation.) The proof is as follows. Since the characters of the totally symmetric representation are all 1, Eq. (9.38) gives for the number of times this representation occurs in $\Gamma_{\alpha\otimes\beta}$:

$$\frac{1}{h}\sum_{\hat{R}} \chi_{\alpha\otimes\beta}(\hat{R}) = \frac{1}{h}\sum_{\hat{R}} \chi_\alpha(\hat{R})\chi_\beta(\hat{R}) = \frac{1}{h}\sum_{\hat{R}} \left[\chi_{\alpha^*}(\hat{R})\right]^* \chi_\beta(\hat{R}) \tag{9.142}$$

where Γ_{α^*} is the complex-conjugate representation of the irreducible representation Γ_α. We now use (9.31) to conclude that (9.142) vanishes unless

Γ_{α^*} and Γ_β are the same. Q.E.D. Equations (9.140) and (9.141) provide illustrations of this theorem.

Let us examine the direct product of a representation with itself. Let the two different sets of functions $F_1, F_2, \ldots, F_m$ and $\bar{F}_1, \bar{F}_2, \ldots, \bar{F}_m$ each form a basis for the same representation Γ_F of some point group. If we consider the symmetry properties of the m^2 functions

$$F_i\bar{F}_j; \qquad i=1,\ldots,m; \qquad j=1,\ldots,m \tag{9.143}$$

then (9.124)–(9.132) (with G replaced by $\bar{F}$ and n replaced by m) show that the set (9.143) forms a basis for the m^2-dimensional direct-product representation $\Gamma_F \otimes \Gamma_F$. (This case occurs when we consider two different m-fold degenerate energy levels, each of which belongs to the same irreducible representation Γ_F; we then have two different sets of wave functions that form bases for Γ_F.)

Suppose, however, we consider only the single set of functions $F_1, F_2, \ldots,$ F_m and form the products of these functions with each other:

$$K_q = F_j F_l, \qquad j \geqslant l \tag{9.144}$$

There are $\frac{1}{2}m(m-1)$ different functions K_q with $j \neq l$ (the factor $\frac{1}{2}$ avoids counting F_jF_l and F_lF_j as different functions) and m functions K_q with $j=l$; this gives $\frac{1}{2}m(m+1)$ functions of the form (9.144). How do they transform? Equations (9.129)–(9.131) with G replaced by F and n replaced by m give

$$\hat{O}_R(K_q) = \hat{O}_R(F_jF_l) = \sum_{i=1}^{m}\sum_{k=1}^{m} r_{ij,F}\, r_{kl,F}\, F_iF_k = \sum_{p=1}^{\frac{1}{2}m(m+1)} r_{pq,K}\, K_p \tag{9.145}$$

Since the functions (9.144) are transformed into linear combinations of one another by the symmetry operators of the group, they form a basis for a representation Γ_K of dimension $\frac{1}{2}m(m+1)$; this representation is called the *symmetric direct product* of Γ_F with itself. We write

$$\Gamma_K = [\Gamma_F \otimes \Gamma_F]^+ \tag{9.146}$$

where the plus sign indicates the symmetric direct product, rather than the ordinary direct product $\Gamma_F \otimes \Gamma_F$.

We shall not need the complete matrices of the symmetric direct product; only their traces are required. The character of $\hat{R}$ in the symmetric direct-product representation (9.146) can be shown to be

$$\chi_{[F\otimes F]^{+}}(\hat{R})=\tfrac{1}{2}\left\{\left[\chi_F(\hat{R})\right]^2+\chi_F(\hat{R}^2)\right\} \tag{9.147}$$

For the derivation of (9.147), see *Hamermesh*, Section 5-2.

As an example, consider the representation $[E\otimes E]^{+}$ for $\mathcal{C}_{3v}$. Equation (9.147) gives for the character of $\hat{C}_3$ in this representation

$$\chi(\hat{C}_3)=\tfrac{1}{2}\left\{\left[\chi_E(\hat{C}_3)\right]^2+\chi_E(\hat{C}_3^2)\right\}$$

$$\chi(\hat{C}_3)=\tfrac{1}{2}\left[(-1)^2+(-1)\right]=0 \tag{9.148}$$

The remaining characters are found in a similar manner, giving the characters for the $\mathcal{C}_{3v}$ representation $[E\otimes E]^{+}$ as

$\hat{E}$	$2\hat{C}_3$	$3\hat{\sigma}_v$
3	0	1

(9.149)

Equation (9.39) gives

$$[E\otimes E]^{+}=A_1\oplus E \tag{9.150}$$

which should be compared to the ordinary direct product, Eq. (9.141). The ordinary direct product is a four-dimensional representation with the basis functions

$$F_1\bar{F}_1,\qquad F_1\bar{F}_2,\qquad F_2\bar{F}_1,\qquad F_2\bar{F}_2$$

whereas the symmetric direct product is a three-dimensional representation with the basis functions

$$F_1^2,\qquad F_1F_2,\qquad F_2^2$$

The symmetric direct product of a one-dimensional representation with itself is clearly the same as the ordinary direct product of the representation with itself; if Γ_F is one-dimensional, then its matrices are the same as its characters, so that

$$\chi_F(\hat{R}^2)=\chi_F(\hat{R})\chi_F(\hat{R})$$

and (9.147) reduces to (9.139) with $F=G$.

We can take the symmetric direct product of a representation with itself any number of times. For example, if $F_1,\ldots,F_m$ form a basis for Γ_F, then the functions

$$F_j F_k F_l, \qquad j \geqslant k \geqslant l$$

form a basis for the symmetric direct product

$$[\Gamma_F \otimes \Gamma_F \otimes \Gamma_F]^+ \tag{9.151}$$

We shall abbreviate (9.151) as $[\Gamma_F^3]^+$ and shall use $[\Gamma_F^n]^+$ to indicate the symmetric direct product of Γ_F with itself n times. The basis functions for $[\Gamma_F^n]^+$ are

$$\underbrace{F_j F_k \cdots F_r}_{n \text{ factors}} \qquad j \geqslant k \geqslant \cdots \geqslant r \tag{9.152}$$

Each of the functions $F_j, F_k,\ldots,F_r$ in (9.152) is one of the basis functions $F_1, F_2,\ldots,F_m$ of Γ_F, so that the functions (9.152) can be written as

$$F_1^{a_1} F_2^{a_2} \cdots F_m^{a_m} \tag{9.153}$$

where $a_1, a_2,\ldots,a_m$ is any set of nonnegative integers that satisfy

$$a_1 + a_2 + \cdots + a_m = n \tag{9.154}$$

Let $\chi_{F^n}(\hat{R})$ be the character of $\hat{R}$ in the representation $[\Gamma_F^n]^+$; it can be shown that[11]

$$\chi_{F^n}(\hat{R}) = \tfrac{1}{2}\left[\chi_F(\hat{R})\chi_{F^{n-1}}(\hat{R}) + \chi_F(\hat{R}^n)\right] \tag{9.155}$$

Equation (9.155) is a recursion relation for the characters of the symmetric direct product of a representation with itself n times; note that it is consistent with (9.147).

The direct product enables one to find the symmetry of a wave function when the symmetries of its factors are known. For example, consider ψ_{vib}. In the harmonic-oscillator approximation, the vibrational wave function is the product of $3N-6$ harmonic-oscillator functions, one for each normal mode. To find the symmetry of ψ_{vib}, we first examine the symmetries of its factors. Let the distinct vibrational frequencies of the molecule be $\nu_1, \nu_2, \ldots, \nu_k, \ldots, \nu_n$, and let ν_k be d_k-fold degenerate; let the harmonic-oscillator

[11] V. Heine, *Group Theory in Quantum Mechanics*, Pergamon, Oxford, 1960, pp. 260–261; see also *Wilson, Decius, and Cross*, pp. 152–155.

factors in ψ_{vib} that correspond to frequency ν_k be

$$\psi_{1,\nu_k}, \psi_{2,\nu_k}, \ldots, \psi_{d_k,\nu_k} \tag{9.156}$$

We thus write [Equation (6.48)]

$$\psi_{\text{vib}} = \prod_{k=1}^{n} \left[\prod_{j=1}^{d_k} \psi_{j,\nu_k}(Q_{j,\nu_k}) \right] \tag{9.157}$$

where Q_{j,ν_k} is one of the normal coordinates corresponding to the frequency ν_k.

Consider the quantity

$$\prod_{j=1}^{d_k} \psi_{j,\nu_k}(Q_{j,\nu_k}) \tag{9.158}$$

The exponential factor in (9.158) is [(6.51) and (6.53)]

$$\exp\left(-\hbar^{-1} d_k \pi \nu_k \sum_{j=1}^{d_k} Q_{j,\nu_k}^2 \right) \tag{9.159}$$

It follows from (9.99) that the exponential (9.159) remains unchanged upon application of any symmetry operator. The symmetry of (9.158) is therefore determined by the transformation properties of the Hermite-polynomial factor

$$\prod_{j=1}^{d_k} H_{v(j,\nu_k)}(Q_{j,\nu_k}) \tag{9.160}$$

where $v(j,\nu_k)$ is the vibrational quantum number of ψ_{j,ν_k}. We are interested in those ψ_{vib} that belong to the same energy level, since these functions are transformed into linear combinations of one another by the symmetry operators. For such functions, in the harmonic-oscillator approximation the quantum numbers of the modes with frequency ν_k must satisfy the relation [see the vibrational-energy expression (6.98)]

$$v(1,\nu_k) + v(2,\nu_k) + \cdots + v(d_k,\nu_k) = v_k \tag{9.161}$$

where v_k is some constant. We therefore consider the transformation properties of all functions (9.158) whose quantum numbers satisfy (9.161). The polynomial $H_{v(j,\nu_k)}$ is of degree $v(j,\nu_k)$ in the normal coordinate Q_{j,ν_k}.

From (9.100), it follows that the effect of a symmetry operator on a term of the form

$$Q_{1,\nu_k}^{a} Q_{2,\nu_k}^{b} \cdots Q_{d_k,\nu_k}^{g} \tag{9.162}$$

is to convert it into a linear combination of terms each of degree $a + b + \cdots + g$. Therefore, the symmetry operators acting on the various products of Hermite polynomials (9.160) convert terms of the same degree into linear combinations of one another; the symmetry operators do not change the degree of any term they act on. Moreover, each Hermite polynomial is completely determined by its term of highest degree. Hence we need only examine the effect of the symmetry operators on the term of highest degree in (9.160) to determine their effect on the complete function (9.160). This highest-degree term is

$$(Q_{1,\nu_k})^{v(1,\nu_k)} (Q_{2,\nu_k})^{v(2,\nu_k)} \cdots (Q_{d_k,\nu_k})^{v(d_k,\nu_k)} \tag{9.163}$$

As shown in Section 9.9, the degenerate normal coordinates

$$Q_{1,\nu_k}, Q_{2,\nu_k}, \ldots, Q_{d_k,\nu_k}$$

form a basis for some irreducible representation Γ_k of the molecular point group. Comparison of (9.163) and (9.161) with (9.153) and (9.154) shows that the functions (9.163) that satisfy (9.161) transform according to the representation that is the symmetric direct product of Γ_k with itself v_k times; hence the function (9.158) belongs to the representation $[\Gamma_k^{v_k}]^+$, where v_k is the sum of the quantum numbers of the degenerate modes.

The complete vibrational wave function (9.157) is the product of functions (9.158) of different frequencies; since functions of different frequencies are transformed independently of one another, the function ψ_{vib} transforms according to the ordinary direct product of the representations of the functions (9.158). *Hence ψ_{vib} belongs to the representation*

$$[\Gamma_1^{v_1}]^+ \otimes [\Gamma_2^{v_2}]^+ \otimes \cdots \otimes [\Gamma_n^{v_n}]^+ \tag{9.164}$$

Consider CO_2 as an example. Equation (9.119) gives for the characters of Γ_{3N}

$\hat{E}$	$2\hat{C}^{\alpha}(z)$	$\infty\hat{\sigma}_v$	$\hat{i}$	$\hat{\sigma}_h$	$2\hat{S}^{\alpha}(z)$	$\infty\hat{C}_2$
9	$3+6\cos\alpha$	3	-3	1	$-1+2\cos\alpha$	-1

(9.165)

The $\mathcal{D}_{\infty h}$ character table gives

$$\Gamma_{\text{rot}}=\Pi_g, \qquad \Gamma_{\text{trans}}=\Sigma_u^+\oplus\Pi_u \tag{9.166}$$

(The representation Σ_g^- of R_z is omitted from Γ_{rot} because the linear molecule has no rotational angular momentum about the z axis.) Subtraction of the characters of Π_g, Σ_u^+, and Π_u from (9.165) gives as the characters of Γ_{vib}

$\hat{E}$	$2\hat{C}^\alpha(z)$	$\infty\hat{\sigma}_v$	$\hat{i}$	$\hat{\sigma}_h$	$2\hat{S}^\alpha(z)$	$\infty\hat{C}_2$
4	$2+2\cos\alpha$	2	-2	2	$2\cos\alpha$	0

(9.167)

Trial and error shows Γ_{vib} to reduce as follows:

$$\Gamma_{\text{vib}}=\Sigma_g^+\oplus\Pi_u\oplus\Sigma_u^+ \tag{9.168}$$

The symmetries of the modes shown in Fig. 6.2 are as in (9.168). The frequency of the two degenerate π_u modes is labeled ν_2, which is the one exception to the numbering rule given in Section 9.9.

Now consider the (123) vibrational level of CO_2 in the harmonic-oscillator approximation. According to (9.164), those ψ_{vib} that belong to the (123) harmonic-oscillator level form a basis for the representation

$$[\Sigma_g^+]\otimes[\Pi_u^2]^+\otimes\left[(\Sigma_u^+)^3\right]^+ \tag{9.169}$$

For a one-dimensional representation, the symmetric direct product is the same as the ordinary direct product of the representation with itself; hence

$$\left[(\Sigma_u^+)^3\right]^+=\Sigma_u^+\otimes\Sigma_u^+\otimes\Sigma_u^+ \tag{9.170}$$

Application of (9.139) gives as the characters of (9.170):

$\hat{E}$	$2\hat{C}^\alpha(z)$	$\infty\hat{\sigma}_v$	$\hat{i}$	$\hat{\sigma}_h$	$2\hat{S}^\alpha(z)$	$\infty\hat{C}_2$
1	1	1	-1	-1	-1	-1

(9.171)

Equation (9.147) gives for the characters of $[\Pi_u^2]^+$ [note that $(\hat{C}^\alpha)^2=\hat{C}^{2\alpha}$]:

$\hat{E}$	$2\hat{C}^\alpha(z)$	$\infty\hat{\sigma}_v$	$\hat{i}$	$\hat{\sigma}_h$	$2\hat{S}^\alpha(z)$	$\infty\hat{C}_2$
3	$1+2\cos 2\alpha$	1	3	3	$1+2\cos 2\alpha$	1

(9.172)

where we used $2\cos^2\alpha=1+\cos 2\alpha$. Multiplication of the characters (9.171) and (9.172) with those of Σ_g^+ gives for the characters of $\Gamma_{(123)}$:

$\hat{E}$	$2\hat{C}^\alpha(z)$	$\infty\hat{\sigma}_v$	$\hat{i}$	$\hat{\sigma}_h$	$2\hat{S}^\alpha(z)$	$\infty\hat{C}_2$
3	$1+2\cos 2\alpha$	1	-3	-3	$-1-2\cos 2\alpha$	-1

(9.173)

Trial and error shows that

$$\Gamma_{(123)} = \Sigma_u^+ \oplus \Delta_u \tag{9.174}$$

The wave functions of the harmonic-oscillator level (123) of CO_2 thus form a basis for a *reducible* representation of $\mathcal{D}_{\infty h}$. But we saw in Section 9.7 that (barring accidental degeneracy) the vibrational wave functions belonging to a degenerate level form a basis for an *irreducible* representation. Therefore, when the anharmonic terms in the vibrational potential energy are taken into account, we expect the (123) level to split into the two levels Σ_u^+ and Δ_u; these levels have vibrational angular-momentum components 0 and $2\hbar$ along the molecular axis and are thus designated (12^03) and (12^23). The group-theory results are consistent with the statements of Section 6.8, where a less general treatment was given.

For the (10^00) fundamental level of CO_2, (9.164) gives for the symmetry species of ψ_{vib}:

$$[\Sigma_g^+]^+ \otimes \left[(\Pi_u)^0\right]^+ \otimes \left[(\Sigma_u^+)^0\right]^+ = \Sigma_g^+ \tag{9.175}$$

Clearly, the symmetry species of any fundamental-level ψ_{vib} is the same as the symmetry of the excited vibrational mode. For CO_2, the symmetry species of these vibrational wave functions are

$$(10^00) : \Sigma_g^+ \qquad (01^10) : \Pi_u \qquad (00^01) : \Sigma_u^+ \tag{9.176}$$

The direct product is also useful for determining the symmetry of ψ_{el} from the symmetry of the occupied MOs. For two nonequivalent electrons outside closed subshells, the symmetry species of ψ_{el} is given by the direct product of the species of the orbitals of the two electrons. For example, the first excited electronic configuration of H_2O is

$$(1a_1)^2(2a_1)^2(1b_2)^2(3a_1)^2(1b_1)(4a_1)$$

The direct product of B_1 and A_1 for $\mathcal{C}_{2v}$ is B_1. Including spin, we see that the first excited configuration of water gives rise to $^3B_1$ and $^1B_1$ electronic terms.

As a second example, consider the π-electron configurations of C_6H_6. The ground electronic configuration was found to be $(a_{2u})^2(e_{1g})^4$, and the lowest excited configuration is $(a_{2u})^2(e_{1g})^3(e_{2u})$ (see Section 9.8). The $(a_{2u})^2$

closed-subshell part of the configuration can be ignored in finding the terms. Considering holes instead of electrons, we see that the terms of the lowest excited electron configuration are the same as those of an $(e_{1g})(e_{2u})$ configuration. For $\mathcal{D}_{6h}$, we find

$$E_{1g} \otimes E_{2u} = B_{1u} \oplus B_{2u} \oplus E_{1u}$$

The $(a_{2u})^2(e_{1g})^3(e_{2u})$ π-electron configuration of benzene thus gives rise to the six terms

$$^1B_{1u}, {}^3B_{1u}, {}^1B_{2u}, {}^3B_{2u}, {}^1E_{1u}, {}^3E_{1u} \tag{9.177}$$

For equivalent electrons (i.e., electrons in the same subshell), the Pauli principle must be taken into account; we omit consideration of this case.[12]

Before showing further applications of direct-product representations to quantum mechanics, we quote without proof a theorem we will need. Let $r_{ij,\alpha}$ and $r_{kl,\beta}$ be the elements of the matrices corresponding to the symmetry operation $\hat{R}$ in the two *different* nonequivalent irreducible representations Γ_α and Γ_β; it can be shown that

$$\sum_{\hat{R}} r^*_{ij,\alpha} r_{kl,\beta} = 0 \tag{9.178}$$

where i, j, k, and l have any fixed values. Actually, (9.178) is only valid for representations whose matrices are unitary. It can be shown that any given representation is equivalent to a representation all of whose matrices are unitary; moreover, the representation generated by a set of orthonormal basis functions can be shown to consist of unitary matrices.[13] Since we always use an orthonormal set of wave functions for each degenerate level, the requirement that the matrices be unitary is satisfied in quantum-mechanical applications. An illustration of (9.178) using the $\mathcal{C}_{3v}$ irreducible representations (9.28) and (9.29) as Γ_α and Γ_β and taking $i=1, j=2, k=l=1$ follows:

$$(0)(1)+\left(-\tfrac{1}{2}\sqrt{3}\right)(1)+\left(\tfrac{1}{2}\sqrt{3}\right)(1)+\left(\tfrac{1}{2}\sqrt{3}\right)(1)+\left(-\tfrac{1}{2}\sqrt{3}\right)(1)+(0)(1)=0$$

If Γ_α is taken to be the totally symmetric representation, then (9.178) becomes

$$\sum_{\hat{R}} r_{kl,\beta} = 0 \tag{9.179}$$

[12] See *Herzberg, Electronic Spectra of Polyatomic Molecules*, pp. 331–334.

[13] For the proof of these statements and of (9.178), see *Tinkham*, pp. 20–21, 23–24, 35–36; see also Problem 9.18.

where Γ_β is any irreducible representation other than the totally symmetric one.

Now we apply group theory to the evaluation of quantum-mechanical integrals. Let the set of linearly independent orthonormal functions $f_{1,\alpha}, f_{2,\alpha}, \ldots, f_{l,\alpha}$ form a basis for the irreducible representation Γ_α of a point group. We assert that the integral over all space of any one of these functions vanishes unless Γ_α is the totally symmetric representation. We begin the proof by noting that we can make any change of variables (i.e., any coordinate transformation) that we please in a definite integral without affecting its value; in particular, we can make a coordinate transformation that corresponds to one of the symmetry operations $\hat{R}$ of the point group. Hence

$$\int f_{j,\alpha}\, d\tau = \int \hat{O}_R f_{j,\alpha}\, d\tau \tag{9.180}$$

Use of (9.41) gives

$$\int f_{j,\alpha}\, d\tau = \sum_{i=1}^{l} \int f_{i,\alpha} r_{ij,\alpha}\, d\tau \tag{9.181}$$

We have an equation like (9.181) for each of the h symmetry operations of the group of order h; addition of these h equations gives

$$h\int f_{j,\alpha}\, d\tau = \sum_{i=1}^{l} \int f_{i,\alpha} \sum_{\hat{R}} r_{ij,\alpha}\, d\tau$$

Using (9.179) we conclude that

$$\int f_{j,\alpha}\, d\tau = 0 \tag{9.182}$$

unless Γ_α is the totally symmetric representation. Q.E.D.

A typical quantum-mechanical integral has the form

$$\int \psi_{i,\alpha}^* F \psi_{j,\beta}\, d\tau \tag{9.183}$$

where F is some function and $\psi_{i,\alpha}$ and $\psi_{j,\beta}$ are either exact or approximate wave functions. As shown in Section 9.7, the functions $\psi_{i,\alpha}$ and $\psi_{j,\beta}$ must each transform according to certain irreducible representations Γ_α and Γ_β of the molecular point group. In the applications we shall consider, the function F will be one of a set of functions that forms a basis for some irreducible representation Γ_F of the point group. The integrand therefore is one of a set of functions that forms a basis for the direct-product representation $\Gamma_{\alpha^*} \otimes \Gamma_F \otimes \Gamma_\beta$ of the molecular point group. This representation is reducible in some manner:

$$\Gamma_{\alpha^*}\otimes\Gamma_F\otimes\Gamma_\beta=\Gamma_1\oplus\Gamma_2\oplus\cdots \tag{9.184}$$

where $\Gamma_1, \Gamma_2, \ldots$ are irreducible representations of the molecular point group. As shown previously [see (9.66)], this means that we can take certain linear combinations of the basis functions of $\Gamma_{\alpha^*}\otimes\Gamma_F\otimes\Gamma_\beta$ to get sets of basis functions for the representations $\Gamma_1, \Gamma_2, \ldots$; conversely, the basis functions of $\Gamma_{\alpha^*}\otimes\Gamma_F\otimes\Gamma_\beta$ can each be expressed as certain linear combinations of basis functions for the irreducible representations $\Gamma_1, \Gamma_2, \ldots$:

$$\psi_{i,\alpha}^* F\psi_{j,\beta}=\sum_i c_i f_{i,1}+\sum_i d_i f_{i,2}+\cdots \tag{9.185}$$

where the $f_{i,1}$'s form a basis for Γ_1, and so on. Substitution of (9.185) into the integral (9.183) and use of the theorem (9.182) show that if none of the irreducible representations $\Gamma_1, \Gamma_2, \ldots$ is the totally symmetric representation, then the integral (9.183) vanishes.

This result is easily seen to be valid for any number of functions in the integrand: *The integral over all space of the product of the functions* $G_\alpha, G_\beta, \ldots, G_\epsilon$ *which transform according to the irreducible representations* $\Gamma_\alpha, \Gamma_\beta, \ldots, \Gamma_\epsilon$ *of the molecular point group vanishes if the direct-product representation* $\Gamma_\alpha\otimes\Gamma_\beta\otimes\cdots\otimes\Gamma_\epsilon$ *does not contain the totally symmetric representation*. (Note that we have not proved the converse; the integral might happen to vanish even if the direct-product representation contained the totally symmetric representation.)

An alternative statement of the results about the integral (9.183) can be given. The representation $\Gamma_{\alpha^*}\otimes\Gamma_F\otimes\Gamma_\beta$ is the direct product of Γ_{α^*} and $\Gamma_F\otimes\Gamma_\beta$. The representation $\Gamma_F\otimes\Gamma_\beta$ is decomposable in some manner to irreducible representations:

$$\Gamma_F\otimes\Gamma_\beta=\Gamma_\rho\oplus\Gamma_\sigma\oplus\cdots \tag{9.186}$$

Hence

$$\Gamma_{\alpha^*}\otimes\Gamma_F\otimes\Gamma_\beta=\Gamma_{\alpha^*}\otimes[\Gamma_\rho\oplus\Gamma_\sigma\oplus\cdots] \tag{9.187}$$

By the theorem preceding (9.142), the representation $\Gamma_{\alpha^*}\otimes\Gamma_F\otimes\Gamma_\beta$ contains the totally symmetric representation if and only if $\Gamma_F\otimes\Gamma_\beta$ contains the irreducible representation Γ_α. *The integral* (9.183) *vanishes if* $\Gamma_F\otimes\Gamma_\beta$ *does not contain* Γ_α.

An important case is where F in (9.183) is replaced by the Hamiltonian operator $\hat{H}$. The Hamiltonian is invariant under all symmetry operations, so that the function $\hat{H}\psi_{j,\beta}$ belongs to the irreducible representation Γ_β. For

$\Gamma_{\alpha^*} \otimes \Gamma_\beta$ to contain the totally symmetric representation, Γ_α and Γ_β must be the same irreducible representation. Hence *the integral*

$$\int \psi_i^* \hat{H} \psi_j \, d\tau \tag{9.188}$$

vanishes unless ψ_i *and* ψ_j *belong to the same irreducible representation.* We used this result in discussing MO theory in Section 9.8; the vanishing of matrix elements of $\hat{H}$ between symmetry-adapted orbitals belonging to different irreducible representations causes the secular determinant to be in block-diagonal form, thus reducing the secular equation to several lower-degree secular equations, one for each symmetry species.

The direct product of representations is basic to the determination of selection rules. We shall consider only transitions between vibrational levels of the same electronic state of a molecule, but the theory is applicable to electronic and rotational transitions as well.

According to (6.65), the IR vibrational selection rules are determined by the integral

$$\int \cdots \int [\psi'_{\text{vib}}(Q_1, \ldots, Q_{3N-6})]^* \mathbf{d}(Q_1, \ldots, Q_{3N-6}) \times \psi_{\text{vib}}(Q_1, \ldots, Q_{3N-6}) \, dQ_1 \cdots dQ_{3N-6} \tag{9.189}$$

where $\mathbf{d} = \mathbf{e}_a d_a + \mathbf{e}_b d_b + \mathbf{e}_c d_c$, with d_a, d_b, and d_c being the components of the dipole moment along the a, b, and c principal axes. The xyz molecule-fixed coordinate system that we have been using in this chapter coincides with the principal-axis system. (It would be inconvenient to use the principal-axis designations a, b, and c to discuss molecular symmetry, since which axis is which depends on the detailed molecular geometry, rather than on just the point group. For example, whether the C_2 axis of a bent ABA molecule is the a or b axis depends on the bond angle; hence it is simplest to call the C_2 axis of a $\mathcal{C}_{2v}$ molecule the z axis.) The quantities d_a, d_b, and d_c are thus the components of the dipole-moment vector $\mathbf{d}$ along the x, y, and z axes (though not necessarily in that order). The vector $\mathbf{d}$ transforms under the symmetry operations of the point group the same way a displacement vector $\mathbf{r}$ does (Sections 9.3 and 9.9); it therefore belongs to the representation Γ_{trans} and its components transform the same way that x, y, and z transform.

Let Γ_ψ and $\Gamma_{\psi'}$ be the irreducible representations to which the vibrational wave functions in (9.189) belong. According to the italicized statement following (9.187), *the integral* (9.189) *vanishes unless* $\Gamma_\psi \otimes \Gamma_{\text{trans}}$ *contains* $\Gamma_{\psi'}$. This is the basic IR selection rule. (Note that its deduction does not involve the harmonic-oscillator approximation or the approximation of neglecting higher derivatives of the dipole moment with respect to the normal coordinates, as in Section 6.5.)

Consider the infrared vibrational transitions of CO_2 as an example. We first ask which vibrational levels can combine with the ground (00^00) level. The ground vibrational level always belongs to the totally symmetric representation, which is Σ_g^+ for $\mathscr{D}_{\infty h}$. The character table yields

$$\Gamma_{\text{trans}} = \Sigma_u^+ \oplus \Pi_u \tag{9.190}$$

The direct product of Γ_ψ and Γ_{trans} is

$$(\Sigma_u^+ \oplus \Pi_u) \otimes \Sigma_g^+ = \Sigma_u^+ \oplus \Pi_u \tag{9.191}$$

since the characters of Σ_g^+ are all $+1$. Hence vibrational transitions are allowed from the ground level to Σ_u^+ and Π_u levels only. The fundamental levels have the symmetry given in (9.176); hence only the Π_u and Σ_u^+ fundamentals (ν_2 and ν_3) are IR active, as we concluded in Chapter 6.

Now consider IR transitions involving the CO_2 vibrational level (12^03), which has symmetry Σ_u^+ [Equation (9.174)]. The direct product $\Gamma_\psi \otimes \Gamma_{\text{trans}}$ is

$$\Sigma_u^+ \otimes (\Sigma_u^+ \oplus \Pi_u) \tag{9.192}$$

The characters of $\Sigma_u^+ \oplus \Pi_u$ are

$\hat{E}$	$2\hat{C}^\alpha(z)$	$\infty\hat{\sigma}_v$	$\hat{i}$	$\hat{\sigma}_h$	$2\hat{S}^\alpha(z)$	$\infty\hat{C}_2$
3	$1+2\cos\alpha$	1	-3	1	$-1+2\cos\alpha$	-1

(9.193)

Equation (9.139) gives for the characters of $\Gamma_\psi \otimes \Gamma_{\text{trans}}$

$\hat{E}$	$2\hat{C}^\alpha(z)$	$\infty\hat{\sigma}_v$	$\hat{i}$	$\hat{\sigma}_h$	$2\hat{S}^\alpha(z)$	$\infty\hat{C}_2$
3	$1+2\cos\alpha$	1	3	-1	$1-2\cos\alpha$	1

(9.194)

Trial and error yields

$$\Gamma_\psi \otimes \Gamma_{trans} = \Sigma_g^+ \oplus \Pi_g \tag{9.195}$$

Hence vibrational transitions are allowed between the (12^03) level and Σ_g^+ and Π_g levels. The only fundamental level that combines with (12^03) is (10^00). [Group theory gives only qualitative and not quantitative information. It tells us that an IR transition between the (12^03) and (10^00) levels is allowed, but it says nothing about the intensity of the transition. Since this transition involves a change of 2 in v_2 and a change of 3 in v_3, the considerations of Section 6.5 tell us that it will be extremely weak.]

The selection rules for Raman vibrational transitions are also readily derived from group theory. Here, the transition probability depends on integrals involving the components of the molecular polarizability matrix $\boldsymbol{\alpha}$. Since $\boldsymbol{\alpha}$ is symmetric, it has only six independent components: $\alpha_{xx}, \alpha_{xy}, \alpha_{xz}, \alpha_{yy}, \alpha_{yz}, \alpha_{zz}$. These six quantities can be shown[14] to transform the same way the six functions

$$x^2, \quad xy, \quad xz, \quad y^2, \quad yz, \quad z^2 \tag{9.196}$$

transform under symmetry operations. The set of functions (9.196) consists of the products of the functions x, y, and z with each other; since the set x,y,z transforms according to Γ_{trans}, the set (9.196) forms a basis for the symmetric direct-product representation $[\Gamma^2_{\text{trans}}]^+$. *Raman transitions are therefore allowed between* ψ'_{vib} *and* ψ_{vib} *if* $\Gamma_\psi \otimes [\Gamma^2_{\text{trans}}]^+$ *contains* $\Gamma_{\psi'}$. [To facilitate the finding of Raman selection rules, some character tables list the six functions (9.196) on the lines of the irreducible representations to which they belong.]

Recall the mutual exclusion rule stated in Section 6.7. The rule follows from the fact that the integral over all space of an odd (u) function is zero. The functions x, y, and z belong to u representations of molecules with a center of symmetry, since inversion converts each to its negative. Hence one of the functions ψ_{vib} and ψ'_{vib} must belong to a g representation and one to a u representation if the integrand of (9.189) is not to be odd. Thus only $g \leftrightarrow u$ IR transitions are allowed in molecules with a center of symmetry. In contrast, the functions (9.196) are all even (g), so that for centrosymmetric molecules only $g \leftrightarrow g$ and $u \leftrightarrow u$ Raman transitions are allowed. This proves the mutual exclusion rule.

Most IR and Raman absorption transitions originate from the ground vibrational state, which always belongs to the totally symmetric irreducible representation. Therefore, if ψ is the ground vibrational state, then $\Gamma_\psi \otimes \Gamma_{\text{trans}}$ equals Γ_{trans}. As noted above (and in Section 6.4), a fundamental vibrational level belongs to the same symmetry species as the excited normal mode. Hence it follows from the fundamental IR selection rule stated following (9.189) that a transition from the ground vibrational level to a fundamental level is allowed only if the normal mode being excited belongs to the same symmetry species as x or y or z. A normal mode of vibration is IR active if and only if its symmetry species is the same as the symmetry species of x or y or z. The same reasoning shows that a normal mode of vibration is Raman active if and only if its symmetry species is the same as one of the functions (9.196).

[14] See *Schonland*, pp. 266–267.

Molecular electronic selection rules are determined by whether or not the integral $\langle\psi'_{\text{el}}|\hat{\mathbf{d}}_{\text{el}}|\psi''_{\text{el}}\rangle$ in (7.6) vanishes. $\hat{\mathbf{d}}_{\text{el}}$ belongs to the representation Γ_{trans}, and comparison with (9.189) shows that the group-theory selection rules for electronic transitions are the same as for vibrational transitions, except that we must consider the symmetry species of the electronic wave functions, rather than the vibrational wave functions. One complication is that the molecular geometry may change on electronic excitation; in this case, we use the point group of lower symmetry to classify the wave functions and determine the selection rules.

McIver and Stanton used group theory to derive selection rules for the symmetry of transition states. For example, they proved that the exchange reaction $H_2+D_2 \rightarrow 2HD$ cannot have a tetrahedral transition state. See J. W. McIver and R. E. Stanton, *J. Am. Chem. Soc.*, **94**, 8618 (1972); J. W. McIver, *Accounts Chem. Res.*, **7**, 72 (1974).

9.12 CHARACTER TABLES OF SOME POINT GROUPS

Character tables for commonly occurring point groups follow.[15]

$\mathcal{C}_1$	$\hat{E}$	
A	1	x, y, z, R_x, R_y, R_z

$\mathcal{C}_s$	$\hat{E}$	$\hat{\sigma}(xy)$	
A'	1	1	x, y, R_z
A''	1	-1	z, R_x, R_y

$\mathcal{C}_i$	$\hat{E}$	$\hat{i}$	
A_g	1	1	R_x, R_y, R_z
A_u	1	-1	x, y, z

$\mathcal{C}_2$	$\hat{E}$	$\hat{C}_2(z)$	
A	1	1	z, R_z
B	1	-1	x, y, R_x, R_y

[15] For tables including more point groups, see *Schonland* or *Cotton* or R. M. Hochstrasser, *Molecular Aspects of Symmetry*, Benjamin, New York, 1966.

$\mathcal{C}_3$	$\hat{E}$	$\hat{C}_3(z)$	$\hat{C}_3^2$	$\omega = \exp(2\pi i/3)$
A	1	1	1	z, R_z
$E \begin{cases} \epsilon \end{cases}$	1	ω	ω^2	$(x,y); (R_x, R_y)$
$E \begin{cases} \epsilon^* \end{cases}$	1	ω^2	ω	

$\mathcal{C}_4$	$\hat{E}$	$\hat{C}_4(z)$	$\hat{C}_2$	$\hat{C}_4^3$	
A	1	1	1	1	z, R_z
B	1	-1	1	-1	
$E \begin{cases} \epsilon \end{cases}$	1	i	-1	$-i$	$(x,y); (R_x, R_y)$
$E \begin{cases} \epsilon^* \end{cases}$	1	$-i$	-1	i	

$\mathcal{C}_{2h}$	$\hat{E}$	$\hat{C}_2(z)$	$\hat{\sigma}_h$	$\hat{i}$	
A_g	1	1	1	1	R_z
A_u	1	1	-1	-1	z
B_g	1	-1	-1	1	R_x, R_y
B_u	1	-1	1	-1	x, y

$\mathcal{C}_{2v}$	$\hat{E}$	$\hat{C}_2(z)$	$\hat{\sigma}_v(xz)$	$\hat{\sigma}_v(yz)$	
A_1	1	1	1	1	z
A_2	1	1	-1	-1	R_z
B_1	1	-1	1	-1	x, R_y
B_2	1	-1	-1	1	y, R_x

$\mathcal{C}_{3v}$	$\hat{E}$	$2\hat{C}_3(z)$	$3\hat{\sigma}_v$	
A_1	1	1	1	z
A_2	1	1	-1	R_z
E	2	-1	0	$(x,y); (R_x, R_y)$

$\mathcal{S}_4$	$\hat{E}$	$\hat{S}_4(z)$	$\hat{C}_2$	$\hat{S}_4^3$	
A	1	1	1	1	R_z
B	1	-1	1	-1	z
$E \begin{cases} \epsilon \end{cases}$	1	i	-1	$-i$	$(x,y); (R_x, R_y)$
$E \begin{cases} \epsilon^* \end{cases}$	1	$-i$	-1	i	

$\mathcal{D}_2$	$\hat{E}$	$\hat{C}_2(z)$	$\hat{C}_2(y)$	$\hat{C}_2(x)$	
A	1	1	1	1	
B_1	1	1	-1	-1	z, R_z
B_2	1	-1	1	-1	y, R_y
B_3	1	-1	-1	1	x, R_x

$\mathcal{D}_{2d}$	$\hat{E}$	$2\hat{S}_4(z)$	$\hat{C}_2'(z)$	$2\hat{C}_2$	$2\hat{\sigma}_d$	
A_1	1	1	1	1	1	
A_2	1	1	1	-1	-1	R_z
B_1	1	-1	1	1	-1	
B_2	1	-1	1	-1	1	z
E	2	0	-2	0	0	(x,y); (R_x, R_y)

$\mathcal{D}_{3d}$	$\hat{E}$	$2\hat{S}_6(z)$	$2\hat{C}_3$	$\hat{i}$	$3\hat{C}_2$	$3\hat{\sigma}_d$	
A_{1g}	1	1	1	1	1	1	
A_{1u}	1	-1	1	-1	1	-1	
A_{2g}	1	1	1	1	-1	-1	R_z
A_{2u}	1	-1	1	-1	-1	1	z
E_g	2	-1	-1	2	0	0	(R_x, R_y)
E_u	2	1	-1	-2	0	0	(x,y)

$\mathcal{D}_{2h}$	$\hat{E}$	$\hat{\sigma}(xy)$	$\hat{\sigma}(xz)$	$\hat{\sigma}(yz)$	$\hat{i}$	$\hat{C}_2(z)$	$\hat{C}_2(y)$	$\hat{C}_2(x)$	
A_g	1	1	1	1	1	1	1	1	
A_u	1	-1	-1	-1	-1	1	1	1	
B_{1g}	1	1	-1	-1	1	1	-1	-1	R_z
B_{1u}	1	-1	1	1	-1	1	-1	-1	z
B_{2g}	1	-1	1	-1	1	-1	1	-1	R_y
B_{2u}	1	1	-1	1	-1	-1	1	-1	y
B_{3g}	1	-1	-1	1	1	-1	-1	1	R_x
B_{3u}	1	1	1	-1	-1	-1	-1	1	x

$\mathcal{D}_{3h}$	$\hat{E}$	$2\hat{C}_3(z)$	$3\hat{C}_2$	$\hat{\sigma}_h$	$2\hat{S}_3$	$3\hat{\sigma}_v$	
A_1'	1	1	1	1	1	1	
A_1''	1	1	1	-1	-1	-1	
A_2'	1	1	-1	1	1	-1	R_z
A_2''	1	1	-1	-1	-1	1	z
E'	2	-1	0	2	-1	0	(x,y)
E''	2	-1	0	-2	1	0	(R_x, R_y)

$\mathcal{D}_{6h}$	$\hat{E}$	$2\hat{C}_6(z)$	$2\hat{C}_3$	$\hat{C}_2''(z)$	$3\hat{C}_2$	$3\hat{C}_2'$	$\hat{\sigma}_h$	$3\hat{\sigma}_v$	$3\hat{\sigma}_d$	$2\hat{S}_6$	$2\hat{S}_3$	$\hat{i}$	
A_{1g}	1	1	1	1	1	1	1	1	1	1	1	1	
A_{1u}	1	1	1	1	1	1	−1	−1	−1	−1	−1	−1	
A_{2g}	1	1	1	1	−1	−1	1	−1	−1	1	1	1	R_z
A_{2u}	1	1	1	1	−1	−1	−1	1	1	−1	−1	−1	z
B_{1g}	1	−1	1	−1	1	−1	−1	−1	1	1	−1	1	
B_{1u}	1	−1	1	−1	1	−1	1	1	−1	−1	1	−1	
B_{2g}	1	−1	1	−1	−1	1	−1	1	−1	1	−1	1	
B_{2u}	1	−1	1	−1	−1	1	1	−1	1	−1	1	−1	
E_{1g}	2	1	−1	−2	0	0	−2	0	0	−1	1	2	(R_x, R_y)
E_{1u}	2	1	−1	−2	0	0	2	0	0	1	−1	−2	(x, y)
E_{2g}	2	−1	−1	2	0	0	2	0	0	−1	−1	2	
E_{2u}	2	−1	−1	2	0	0	−2	0	0	1	1	−2	

$\mathcal{T}_d$	$\hat{E}$	$8\hat{C}_3$	$3\hat{C}_2$	$6\hat{\sigma}_d$	$6\hat{S}_4$	
A_1	1	1	1	1	1	
A_2	1	1	1	−1	−1	
E	2	−1	2	0	0	
T_1	3	0	−1	−1	1	(R_x, R_y, R_z)
T_2	3	0	−1	1	−1	(x, y, z)

$\mathcal{O}_h$	$\hat{E}$	$8\hat{C}_3$	$6\hat{C}_2$	$6\hat{C}_4$	$3\hat{C}_2''\ (=\hat{C}_4^2)$	$3\hat{\sigma}_h$	$6\hat{\sigma}_d$	$\hat{i}$	$8\hat{S}_6$	$6\hat{S}_4$	
A_{1g}	1	1	1	1	1	1	1	1	1	1	
A_{1u}	1	1	1	1	1	−1	−1	−1	−1	−1	
A_{2g}	1	1	−1	−1	1	1	−1	1	1	−1	
A_{2u}	1	1	−1	−1	1	−1	1	−1	−1	1	
E_g	2	−1	0	0	2	2	0	2	−1	0	
E_u	2	−1	0	0	2	−2	0	−2	1	0	
T_{1g}	3	0	−1	1	−1	−1	−1	3	0	1	(R_x, R_y, R_z)
T_{1u}	3	0	−1	1	−1	1	1	−3	0	−1	(x, y, z)
T_{2g}	3	0	1	−1	−1	−1	1	3	0	−1	
T_{2u}	3	0	1	−1	−1	1	−1	−3	0	1	

$\mathcal{C}_{\infty v}$	$\hat{E}$	$2\hat{C}^{\alpha}(z)$	$\cdots$	$\infty\hat{\sigma}_v$	
$\Sigma^+ \equiv A_1$	1	1	$\cdots$	1	z
$\Sigma^- \equiv A_2$	1	1	$\cdots$	-1	R_z
$\Pi \equiv E_1$	2	$2\cos\alpha$	$\cdots$	0	(x,y); (R_x, R_y)
$\Delta \equiv E_2$	2	$2\cos 2\alpha$	$\cdots$	0	
$\Phi \equiv E_3$	2	$2\cos 3\alpha$	$\cdots$	0	
.	.	.	$\cdots$	.	
.	.	.	$\cdots$	.	
.	.	.	$\cdots$	.	

$\mathcal{D}_{\infty h}$	$\hat{E}$	$2\hat{C}^{\alpha}(z)$	$\cdots$	$\infty\hat{\sigma}_v$	$\hat{i}$	$\hat{\sigma}_h$	$2\hat{S}^{\alpha}(z)$	$\cdots$	$\infty\hat{C}_2$	
$\Sigma_g^+ \equiv A_{1g}$	1	1	$\cdots$	1	1	1	1	$\cdots$	1	
$\Sigma_u^+ \equiv A_{1u}$	1	1	$\cdots$	1	-1	-1	-1	$\cdots$	-1	z
$\Sigma_g^- \equiv A_{2g}$	1	1	$\cdots$	-1	1	1	1	$\cdots$	-1	R_z
$\Sigma_u^- \equiv A_{2u}$	1	1	$\cdots$	-1	-1	-1	-1	$\cdots$	1	
$\Pi_g \equiv E_{1g}$	2	$2\cos\alpha$	$\cdots$	0	2	-2	$-2\cos\alpha$	$\cdots$	0	(R_x, R_y)
$\Pi_u \equiv E_{1u}$	2	$2\cos\alpha$	$\cdots$	0	-2	2	$2\cos\alpha$	$\cdots$	0	(x,y)
$\Delta_g \equiv E_{2g}$	2	$2\cos 2\alpha$	$\cdots$	0	2	2	$2\cos 2\alpha$	$\cdots$	0	
$\Delta_u \equiv E_{2u}$	2	$2\cos 2\alpha$	$\cdots$	0	-2	-2	$-2\cos 2\alpha$	$\cdots$	0	
.	.	.	$\cdots$	.	.	.	.	$\cdots$	.	
.	.	.	$\cdots$	.	.	.	.	$\cdots$	.	
.	.	.	$\cdots$	.	.	.	.	$\cdots$	.	

PROBLEMS

9.1 Which of these arithmetical operations are associative: addition, subtraction, multiplication, division.

9.2 Which of the following are groups? (a) All rational numbers except zero, under multiplication. (b) $1, 0, -1$, under addition. (c) $1, 0, -1$, under multiplication. (d) $1, -1, i, -i$, under multiplication.

9.3 Let A and B be elements of a group. Show that $(AB)^{-1} = B^{-1}A^{-1}$.

9.4 Give the order of each of the following groups: (a) $\mathcal{C}_1$; (b) $\mathcal{C}_{40}$; (c) $\mathcal{D}_{40}$; (d) $\mathcal{S}_{40}$.

9.5 Which standard point group is each of the following groups the same as? (a) $\mathcal{C}_{1h}$; (b) $\mathcal{C}_{1v}$; (c) $\mathcal{D}_1$; (d) $\mathcal{D}_{1h}$; (e) $\mathcal{D}_{1d}$.

9.6 Give the multiplication table for each of the following groups: (a) $\mathcal{C}_1$; (b) $\mathcal{C}_{2h}$; (c) $\mathcal{D}_3$. Which are Abelian?

9.7 Consider the following multiplication table:

	I	A	B	C	D
I	I	A	B	C	D
A	A	I	D	B	C
B	B	C	I	D	A
C	C	D	A	I	B
D	D	B	C	A	I

Each row and each column contains each element exactly once (as must be true for the multiplication table of a group). Show that the above table is *not* the multiplication table of a group.

9.8 Which point groups are subgroups of (a) $\mathcal{D}_2$; (b) $\mathcal{C}_3$; (c) $\mathcal{S}_4$.

9.9 Are the point groups $\mathcal{C}_{2v}$ and $\mathcal{D}_2$ isomorphic? Justify your answer.

9.10 Prove that if P and R are group elements each conjugate to the element Q, then P and R are conjugate to each other.

9.11 Show that (9.2) is the only possible multiplication table for a group of order 3.

9.12 Derive the possible multiplication tables for a group of order 4. (If you seem to find more than two forms, recall that the elements of a group are not ordered in any particular way; hence the labels A, B, C are arbitrary and may be interchanged.)

9.13 (a) Prove that $\hat{O}_R$ in (9.24) is linear. (b) Prove that $\hat{O}_R\hat{O}_S = \hat{O}_{RS}$. You will find it useful to define $g(\mathbf{w}) \equiv f(\mathbf{S}^{-1}\mathbf{w})$.

9.14 (a) For the group $\mathcal{D}_2$, write down the 3×3 matrices that describe the effects of the symmetry operations on the point (x,y,z). Reduce this representation to the direct sum of irreducible representations. Use the determinants of the 3×3 matrices to get a fourth irreducible representation. Verify these representations by using the multiplication table of $\mathcal{D}_2$. (b) For the group $\mathcal{C}_4$, write down the 3×3 matrices corresponding to the symmetry operations. Express this representation as the direct sum of one- and two-dimensional representations. Use Theorems 3 and 5 of Section 9.5 to see whether the two-dimensional representation is irreducible.

9.15 Find the dimensions of the irreducible representations for groups with the following order and number of classes: (a) order 12, 4 classes; (b) order 24, 5 classes; (c) order 32, 11 classes (two possible answers).

9.16 (a) For the groups of Problem 9.15, verify that the dimensions of the irreducible representations are divisors of the order of the group. (This is a general theorem.) (b) Why must a group whose order is a prime number be Abelian?

9.17 Does the set of all nonsingular square matrices of order 13 form a group with matrix multiplication as the rule of combination? If so, what is the order of this group?

9.18 Let the functions $F_1, \ldots, F_n$ form a basis for a representation Γ_F of some point group. Since a symmetry operation $\hat{R}$ amounts to a rotation (and possibly a reflection) of coordinates, it cannot change the value of a definite integral over all space; we have

$$\int (\hat{O}_R F_j)^* (\hat{O}_R F_k)\, dv = \int F_j^* F_k\, dv$$

Prove that if the set of functions F_i is orthonormal, then the representation Γ_F consists of unitary matrices.

9.19 The following are the characters of certain reducible representations. Use (9.39) to write each representation as the direct sum of irreducible representations.

(a) Group $\mathcal{C}_{2v}$:

$\hat{E}$	$\hat{C}_2(z)$	$\hat{\sigma}_v(xz)$	$\hat{\sigma}_v(yz)$
5	-3	3	-1

(b) Group $\mathcal{C}_{3v}$:

$\hat{E}$	$2\hat{C}_3$	$3\hat{\sigma}_v$
292	-119	8

(c) Group $\mathcal{C}_3$:

$\hat{E}$	$\hat{C}_3$	$\hat{C}_3^2$
4	1	1

9.20 In Problem 9.14, a two-dimensional reducible representation was found for the group $\mathcal{C}_4$. What irreducible representations is this reducible representation the sum of?

9.21 For the group $\mathcal{C}_{3v}$, find $\chi(\hat{C}_3^2)$ for the representation $13A_1 \oplus 46A_2 \oplus 287E$.

9.22 Prove that if the irreducible representation Γ_j is one dimensional, then the function g of (9.67) is a symmetry-adapted function that transforms according to Γ_j. Start by writing the basis function f as some linear combination of symmetry-adapted functions g_i; apply $\hat{O}_R$ to this equation; then multiply by $\chi_j^*(\hat{R})$ and sum over $\hat{R}$.

9.23 For each of the following molecules, find the characters of Γ_{AO} for a minimal-basis-set SCF MO calculation. Then reduce Γ_{AO} to find the symmetry species of the lowest MOs. (Choose the axes as indicated.) (a)

Ethylene (x axis perpendicular to molecular plane, z axis through carbons); (b) NH_3; (c) CH_4 (x, y, and z axes each coincident with an S_4 axis); (d) naphthalene (x axis perpendicular to molecular plane, z axis through carbons 9 and 10); (e) C_2H_6; (f) cubane, C_8H_8 (x, y, and z axes each coincident with a C_4 axis); (g) acetylene.

9.24 Construct symmetry orbitals for each of the molecules in Problem 9.23.

9.25 For each of the molecules of Problem 9.23, find the characters of Γ_{3N}. Then find the symmetry species of the $3N-6$ normal modes of vibration. State the number of distinct vibrational frequencies for each molecule.

9.26 Show that the characters of Γ_{vib} are

$$\chi_{\text{vib}}(\hat{R}) = (N_R - 2)(1 + 2\cos\alpha)$$

for proper rotations, and

$$\chi_{\text{vib}}(\hat{R}) = N_R(-1 + 2\cos\alpha)$$

for improper rotations.

9.27 Write each of the following direct products as the sum of irreducible representations. Verify that the results satisfy the theorem stated preceding (9.142). (a) $A_2 \otimes E$ for $\mathcal{C}_{3v}$; (b) $T_1 \otimes E$ for $\mathcal{T}_d$; (c) $E \otimes E \otimes E$ for $\mathcal{C}_{3v}$; (d) $E' \otimes E'$ for $\mathcal{D}_{3h}$; (e) $[E' \otimes E']^+$ for $\mathcal{D}_{3h}$; (f) $\epsilon \otimes \epsilon^*$ for $\mathcal{C}_3$.

9.28 Find $\mathbf{A} \otimes \mathbf{B}$, where

$$\mathbf{A} = \begin{pmatrix} 1 & -1 \\ 3 & 2 \end{pmatrix} \qquad \mathbf{B} = \begin{pmatrix} 5 & -3 \\ 4 & -2 \end{pmatrix}$$

9.29 Find the symmetry species of the CO_2 vibrational wave functions that belong to the harmonic-oscillator level (241). When anharmonicity is included, how many vibrational levels do we get?

9.30 Find the electronic terms that arise from the following configurations: (a) the $\ldots(3a_1)(1b_1)^2(4a_1)$ configuration of water; (b) the $\ldots(3a_1)^2(1b_1)(2b_2)$ configuration of water; (c) the lowest configuration of H_2O^+; (d) the lowest excited configuration of ethylene; (e) the lowest excited configuration of *s-trans*-1,3-butadiene.

9.31 For the point groups of the molecules in Problem 9.23, find the symmetry species to which the functions (9.196) belong.

9.32 (a) Give the number of fundamental IR bands for each of the molecules in Problem (9.23). (b) Give the number of fundamental vibrational Raman bands for each of the molecules in Problem 9.23. (See Problem 9.31.)

9.33 Explain why Fermi resonance (Section 6.8) cannot occur between vibrational states belonging to different symmetry species.

9.34 Find what IR transitions can occur between pairs of the H_2O vibrational levels of Problem 6.9.

9.35 For CH_4, we have (Problem 9.25)

$$\Gamma_{\text{vib}} = A_1 \oplus E \oplus 2T_2$$

A harmonic-oscillator level with a degenerate vibration excited consists of several state of varying symmetry. (Anharmonicity splits the degenerate harmonic-oscillator levels into closely spaced levels of different symmetry.) If two harmonic-oscillator levels have states whose symmetries are such that an IR transition is allowed between them, then a transition is allowed between the harmonic-oscillator levels. Which of the following methane vibrational transitions are allowed? (a) (0000)→(0300); (b) (0000)→(0020); (c) (1010)→(0112).

APPENDIX

Table A.1 Physical Constants[a]

Constant	Symbol	Value	Error Limit[b]
Speed of light in vacuum	c	$2.99792458 \times 10^{10}$ cm/sec	4
Proton charge	e	4.80324×10^{-10} statcoul	4
Avogadro constant	N_0	6.02204×10^{23} mole^{-1}	9
Electron rest mass	m	9.10953×10^{-28} g	14
Proton rest mass	m_p	1.672648×10^{-24} g	26
Neutron rest mass	m_n	1.674954×10^{-24} g	26
Planck's constant	h	6.62618×10^{-27} erg sec	11
	$\hbar = h/2\pi$	1.054589×10^{-27} erg sec	17
Faraday	$F = N_0 e$	2.892534×10^{14} statcoul/mole	25
	F	9.64846×10^{4} C/mole	8
Proton–electron mass ratio	m_p/m	1836.1515	21
Fine-structure constant	$\alpha = e^2/\hbar c$	7.297351×10^{-3}	18
	$1/\alpha$	137.03604	33
Bohr radius	$a_0 = \hbar^2/me^2$	0.5291771×10^{-8} cm	13
Bohr magneton	$\beta_e = e\hbar/2mc$	9.27408×10^{-21} erg/G	11
Nuclear magneton	$\beta_p = e\hbar/2m_p c$	5.05082×10^{-24} erg/G	6
Electron g value	g_e	2.002319313	11
Proton g value	g_p	5.585691	7
Gas constant	R	8.3144×10^{7} erg/°K mole	8
Boltzmann constant	$k = R/N_0$	1.38066×10^{-16} erg/°K	13
Gravitational constant	G	6.672×10^{-8} dyn cm^2/g^2	12
Rydberg constant for infinite nuclear mass	$R_\infty = e^2/2a_0 hc$	109,737.32 cm^{-1}	2

[a] E. R. Cohen and B. N. Taylor, *J. Phys. Chem. Ref. Data*, **2**, 663 (1973).
[b] The listed error limit is three standard deviations and applies to the final digits of the quoted value.

Table A.2 Definitions of Standards and Equivalents

Quantity	Definition
meter	1,650,763.73 times the vacuum wavelength of the orange-red line of Kr^{86}
kilogram	mass of the international kilogram at Sèvres, France
second	9,192,631,770 times the period of the hyperfine transition between the sublevels $F=4$, $M_F=0$ and $F=3$, $M_F=0$ of the $^2S_{1/2}$ ground level of Cs^{133}
degree Kelvin	defined in the thermodynamic scale by assigning 273.16°K to the triple point of water
Avogadro's number	the number of atoms in exactly 12 g of C^{12}
thermochemical calorie	4.1840×10^7 ergs

Table A.3 Energy Conversion Factors

		erg	eV	cm^{-1}
1 erg	=	1	6.2415×10^{11}	5.0340×10^{15}
1 eV	=	1.6022×10^{-12}	1	8065.5
1 cm^{-1}	=	1.9865×10^{-16}	1.23985×10^{-4}	1
1 kcal/mole	=	6.9478×10^{-14}	4.3364×10^{-2}	349.75
1 hartree	=	4.3598×10^{-11}	27.212	219,474.6

		kcal/mole	hartree
1 erg	=	1.4393×10^{13}	2.2937×10^{10}
1 eV	=	23.060	3.6749×10^{-2}
1 cm^{-1}	=	2.8591×10^{-3}	4.55634×10^{-6}
1 kcal/mole	=	1	1.5936×10^{-3}
1 hartree	=	627.51	1

Table A.5 allows one to convert an equation from SI units to Gaussian units; the quantities in the SI column are replaced by those in the Gaussian column. For example, Coulomb's law in SI units is

$$F=\frac{q_1q_2}{4\pi\epsilon_0 r^2}$$

Replacement of ϵ_0 by $1/4\pi$ gives for Coulomb's law in Gaussian units

$$F=\frac{q_1q_2}{r^2}$$

Table A.4 Conversion Table for Units[a]

Quantity	SI System	Gaussian System
mass (m)	1 kg	$=10^3$ g
length (l)	1 meter	$=10^2$ cm
time (t)	1 sec	$=1$ sec
force (**F**)	1 newton	$=10^5$ dynes
energy (E)	1 joule	$=10^7$ ergs
power (P)	1 watt	$=10^7$ ergs/sec
charge (q)	1 coulomb	$=$ "3" $\times 10^9$ statcoulombs
current (I)	1 ampere	$=$ "3" $\times 10^9$ statamp
electric field (**E**)	1 volt/meter	$=(1/$"3"$)\times 10^{-4}$ statvolt/cm
electric potential (ϕ or V)	1 volt	$=(1/$"3"$)\times 10^{-2}$ statvolt
electric displacement (**D**)	1 coul/m^2	$=$ "3" $\times 4\pi \times 10^5$ statcoul/cm^2
magnetic induction (**B**)	1 tesla	$=10^4$ gauss
magnetic field strength (**H**)	1 amp/m	$=4\pi \times 10^{-3}$ oersted

[a] The symbol "3" stands for 2.9979..., the digits in the speed of light.

Table A.5 Conversion Table for Symbols[a]

Quantity	SI	Gaussian
permittivity constant	ϵ_0	$1/4\pi$
permeability constant	μ_0	$4\pi/c^2$
electric field	$\mathbf{E}$	$\mathbf{E}$
electric displacement	$\mathbf{D}$	$\mathbf{D}/4\pi$
magnetic induction	$\mathbf{B}$	$\mathbf{B}/c$
magnetic field strength	$\mathbf{H}$	$c\mathbf{H}/4\pi$
electric potential	ϕ	ϕ
magnetic vector potential	$\mathbf{A}$	$\mathbf{A}/c$
Poynting vector	$\mathbf{S}$	$\mathbf{S}$
magnetization	$\mathbf{M}$	$c\mathbf{M}$
magnetic dipole moment	$\boldsymbol{\mu}$	$c\boldsymbol{\mu}$

[a] c is the speed of light.

Table A.6 Properties of Some Isotopes[a]

Isotope	% Abundance	I	g_N	Atomic Mass
neutron		$\frac{1}{2}$	−3.8263	1.008665
H^1	99.985[b]	$\frac{1}{2}$	5.5856	1.007825
H^2	0.015[b]	1	0.8574	2.014102
Li^7	92.58	$\frac{3}{2}$	2.17	7.016004
B^{10}	19	3	0.6002	10.012939
B^{11}	81	$\frac{3}{2}$	1.792	11.009305
C^{12}	98.892	0	—	12.000...
C^{13}	1.108	$\frac{1}{2}$	1.4048	13.003354
N^{14}	99.635	1	0.4036	14.003074
N^{15}	0.365	$\frac{1}{2}$	−0.566	15.0001
O^{16}	99.76	0	—	15.994915
O^{17}	0.037	$\frac{5}{2}$	−0.7575	16.99913
O^{18}	0.20	0	—	17.99916
F^{19}	100	$\frac{1}{2}$	5.257	18.99840
Na^{23}	100	$\frac{3}{2}$	1.478	22.98977
P^{31}	100	$\frac{1}{2}$	2.263	30.97376
S^{32}	95.0	0	—	31.97207
Cl^{35}	75.5	$\frac{3}{2}$	0.5479	34.96885
Cl^{37}	24.5	$\frac{3}{2}$	0.4560	36.96590
Br^{79}	50.6	$\frac{3}{2}$	1.404	78.91833
Br^{81}	49.4	$\frac{3}{2}$	1.513	80.91629
I^{127}	100	$\frac{5}{2}$	1.124	126.90447

[a] Data from E. U. Condon and H. Odishaw, eds., *Handbook of Physics*, Second Edition, McGraw-Hill, New York, 1967, p. **9**–65; C. M. Lederer, J. M. Hollander, and I. Perlman, *Table of Isotopes*, Sixth Edition, Wiley, New York, 1967. Abundances are for the earth's crust; atomic masses are for the neutral atoms on the C^{12} scale; g_N and I are defined in Section 8.2.

[b] Lake Michigan water.

BIBLIOGRAPHY

Quantum Chemistry

Atkins, P. W., *Molecular Quantum Mechanics*, Oxford University Press, London, 1970.

Bethe, H. A., and E. E. Salpeter, *Quantum Mechanics of One- and Two-Electron Atoms*, Academic Press, New York, 1957.

Eyring, H., J. Walter, and G. E. Kimball, *Quantum Chemistry*, Wiley, New York, 1944.

Gatz, C. R., *Introduction to Quantum Chemistry*, Merrill, Columbus, Ohio, 1971.

Kauzmann, W., *Quantum Chemistry*, Academic Press, New York, 1957.

Levine, I. N., *Quantum Chemistry*, Second Edition, Allyn and Bacon, Boston, 1974.

McWeeny, R., and B. T. Sutcliffe, *Methods of Molecular Quantum Mechanics*, Academic Press, London, 1969.

Murrell, J. N., and A. J. Harget, *Semi-empirical Self-Consistent-Field Molecular Orbital Theory of Molecules*, Wiley-Interscience, London, 1971.

Offenhartz, P. O'D., *Atomic and Molecular Orbital Theory*, McGraw-Hill, New York, 1970.

Parr, R. G., *Quantum Theory of Molecular Electronic Structure*, Benjamin, New York, 1963.

Pauling, L., and E. B. Wilson, Jr., *Introduction to Quantum Mechanics*, McGraw-Hill, New York, 1935.

Pilar, F. L., *Elementary Quantum Chemistry*, McGraw-Hill, New York, 1968.

Pople, J. A., and D. L. Beveridge, *Approximate Molecular Orbital Theory*, McGraw-Hill, New York, 1970.

Rapp, D., *Quantum Mechanics*, Holt, Rinehart and Winston, New York, 1971.

Salem, L., *The Molecular Orbital Theory of Conjugated Systems*, Benjamin, New York, 1966.

Schaefer, H. F., *The Electronic Structure of Atoms and Molecules*, Addison-Wesley, Reading, 1972.

Quantum Mechanics

Bates, D. R., ed., *Quantum Theory*, three volumes, Academic Press, New York, 1961.

Bethe, H. A., and R. W. Jackiw, *Intermediate Quantum Mechanics*, Second Edition, Benjamin, New York, 1968.

Fong, P., *Elementary Quantum Mechanics*, Addison-Wesley, Reading, 1962.

Hameka, H. F., *Introduction to Quantum Theory*, Harper and Row, New York, 1967.

Landau, L. D., and E. M. Lifshitz, *Quantum Mechanics; Non-Relativistic Theory*, Second Edition, Addison-Wesley, Reading, 1965.

Merzbacher, E., *Quantum Mechanics*, Second Edition, Wiley, New York, 1970.

Spectroscopy

Barrow, G., *Introduction to Molecular Spectroscopy*, McGraw-Hill, New York, 1962. An introductory text for undergraduates.

Becker, E. D., *High Resolution NMR*, Academic Press, New York, 1969. A useful intermediate-level text.

Bersohn, M., and J. C. Baird, *An Introduction to Electron Paramagnetic Resonance*, Benjamin, New York, 1966. An elementary text with extensive discussion of applications of ESR to chemistry and biology.

Bourcier, S., *Spectroscopic Data Relative to Diatomic Molecules*, vol. 17 of Tables of Constants and Numerical Data, Pergamon, New York, 1970. A comprehensive collection of experimental data.

Carrington, A., and A. D. McLachlan, *Introduction to Magnetic Resonance*, Harper and Row, New York, 1967. An excellent graduate-level treatment of ESR and NMR; the emphasis is on principles rather than applications.

Chang, R., *Basic Principles of Spectroscopy*, McGraw-Hill, New York, 1971. A concise, informative treatment of many branches of spectroscopy.

Emsley, J. W., J. Feeney, and L. H. Sutcliffe, *High Resolution Nuclear Magnetic Resonance Spectroscopy*, vols. I and II, Pergamon, London, 1965, 1966. An encyclopedic treatise. Volume I presents the theory of high-resolution NMR; volume II discusses applications and contains many tables of data and many literature references.

Harmony, M. D., *Introduction to Molecular Energies and Spectra*, Holt, Rinehart and Winston, New York, 1972. Covers microwave, IR, NMR, ESR, and electronic spectroscopy.

Herzberg, G., *Atomic Spectra and Atomic Structure*, Dover, New York, 1944. A short nonmathematical, introductory book.

Herzberg, G., *Molecular Spectra and Molecular Structure*; vol. I: *Spectra of Diatomic Molecules*, Second Edition, 1950; vol. II: *Infrared and Raman Spectra of Polyatomic Molecules*, 1945; vol. III: *Electronic Spectra and Electronic Structure of Polyatomic Molecules*, 1966; Van Nostrand, Princeton, New Jersey. Very comprehensive treatments with many tabulations of experimental data. (The data in vols. I and II have been largely superseded by more recent work.)

King, G. W., *Spectroscopy and Molecular Structure*, Holt, Rinehart and Winston, New York, 1964. A good treatment of electronic, vibrational, and rotational spectroscopy.

Mathieson, D. W., ed., *Nuclear Magnetic Resonance for Organic Chemists*, Academic Press, New York, 1967. A good practical guide to the analysis of NMR spectra.

Murrell, J. N., *The Theory of the Electronic Spectra of Organic Molecules*, Wiley, New York, 1963. Good discussions of the applications of HMO, FE MO, and Pariser–Parr–Pople theory to organic compounds.

Orchin, M., and H. H. Jaffé, *Symmetry, Orbitals, and Spectra*, Wiley, New York, 1971. Applies MO theory to electronic spectra.

Pople, J. A., W. G. Schneider, and H. J. Bernstein, *High Resolution Nuclear Magnetic Resonance*, McGraw-Hill, New York, 1959. A comprehensive treatment.

Robinson, J. W., ed., *CRC Handbook of Spectroscopy*, Chemical Rubber Co., Cleveland, 1974. Covers the major fields of spectroscopy in two volumes.

Sugden, T. M., and C. N. Kenney, *Microwave Spectroscopy of Gases*, Van Nostrand, Princeton, New Jersey, 1965. A good introductory book with a valuable bibliography of molecules studied by microwave spectroscopy.

Suzuki, H., *Electronic Absorption Spectra and Geometry of Organic Molecules*, Academic Press, New York, 1967. The development is in terms of MO theory.

Townes, C. H., and A. L. Schawlow, *Microwave Spectroscopy*, McGraw-Hill, New York, 1955. Comprehensive discussions of theory and experimental techniques; useful appendices for calculating asymmetric-top energy levels. For internal rotation, however, consult *Sugden and Kenney* or *Wollrab*.

Wertz, J., and J. R. Bolton, *Electron Spin Resonance*, McGraw-Hill, New York, 1972. A good treatment of ESR theory and applications.

Wilson, E. B., Jr., J. C. Decius, and P. C. Cross, *Molecular Vibrations*, McGraw-Hill, New York, 1955. Excellent presentation of the theory of infrared and Raman spectra.

Wollrab, J. E., *Rotational Spectra and Molecular Structure*, Academic Press, New York, 1967. More advanced than *Sugden and Kenney*; has an extensive bibliography of research papers including titles.

Group Theory

Cotton, F. A., *Chemical Applications of Group Theory*, Second Edition, Wiley, New York, 1970. A popular, nonrigorous introduction with applications to MO theory and molecular vibrations.

Hamermesh, M., *Group Theory*, Addison-Wesley, Reading 1962. A nicely done, rather advanced treatment; the applications given are mainly of interest to physicists.

Jaffé, H. H., and M. Orchin, *Symmetry in Chemistry*, Wiley, New York, 1965. An elementary, nonmathematical treatment with applications to MO theory, molecular vibrations, and crystal symmetry.

Schonland, D. S., *Molecular Symmetry*, Van Nostrand, Princeton, New Jersey, 1965. An outstanding introductory text for chemists; careful discussions of difficult points; applications to MO theory, electronic spectra, and molecular vibrations.

Tinkham, M., *Group Theory and Quantum Mechanics*, McGraw-Hill, New York, 1964. The applications given are of interest to both chemists and physicists, but the treatment is a bit too advanced for many chemists.

Miscellaneous

Annual Review of Physical Chemistry, Annual Reviews, Palo Alto, California.

Davidson, N., *Statistical Mechanics*, McGraw-Hill, New York, 1962.

Feynman, R. P., R. B. Leighton, and M. Sands, *The Feynman Lectures on Physics*, vols. I, II, and III, Addison-Wesley, Reading, 1963–1965.

Halliday, D., and R. Resnick, *Physics*, parts I and II, Wiley, New York, 1966.

Margenau, H., and G. M. Murphy, *The Mathematics of Physics and Chemistry*, Second Edition, Van Nostrand, Princeton, New Jersey, 1956.

Sokolnikoff, I. S., and R. M. Redheffer, *Mathematics of Physics and Modern Engineering*, Second Edition, McGraw-Hill, New York, 1966.

Taylor, A. E., and W. R. Mann, *Advanced Calculus*, Second Edition, Xerox, Lexington, 1972.

ANSWERS TO SELECTED PROBLEMS

1.2 $\frac{1}{2}\hbar$. **1.7** $E+(2b\hbar^2/m)(4bx^6-3x^2)$, where E is the energy of the state.

2.1 Row vectors of **AB**: 3, 15, 12; −9, −20, 5; 1, 5, 8. Row vectors of **BA**: −5, −14, 2; −2, −11, 5; −1, 3, 7. All products are defined. **2.8** (a) **C**; (b) **A**, **C**; (c) **C**. **2.9** (a) **A**; (b) **A**, **C**. **2.15** (a) **A** eigenvalues 5, −2; one eigenvector has elements 3/5, 4/5; the second has elements $2^{-1/2}$, $-2^{-1/2}$. **2.19** (b) eigenvalues $+\hbar$, 0, $-\hbar$. **2.20** (a). **2.28** (a) unitary; (c) not unitary.

3.1 (a) 0.029 statvolt/cm; (b) 0.029 gauss; (c) 3.0×10^{17}. **3.2** No. **3.4** One quantum number changes by an odd integer, while the other two remain unchanged. **3.5** One quantum number changes by ±1, while the other two remain unchanged. **3.6** 0, ±2. **3.13** $\Delta J=\pm1$, $\Delta M=0, \pm1$. **3.17** 0.0021 eV.

4.1 (c) $0<R<2R_e$. **4.2** 1.4×10^{-8}. **4.7** 8.9 cm^{-1}. **4.9** (a) 51.58 cm^{-1}; (b) 2988.7 cm^{-1}. **4.13** $\frac{1}{2}(\Delta^{1/2}-1)/R_e^2$. **4.15** (c) Yes. (d) No. **4.17** (a) 57,635.97 MHz, 0.184 MHz; (b) 525.0 MHz; (c) 57,898.5 MHz; (e) 490.7 MHz, 0.168 MHz; (f) 55,101.02 MHz, 55,346.4 MHz; (g) $C^{12}O^{16}$: $R_e=1.12832$, $R_0=1.13089$; $C^{13}O^{16}$: $R_e=1.12832$, $R_0=1.13083$. **4.19** (a) 0.017_5 cm^{-1}; (b) 1.930_2 cm^{-1}. **4.20** 90. **4.22** H^2, N^{14}, O^{17}, B^{10}, C^{13}. **4.26** (a) 1, 9, 5; (b) 6, 9, 30; (c) 6, 18, 30; (d) 1, 0, 5.

5.1 (a) spherical; (b) symmetric; (c) asymmetric; (d) symmetric; (e) spherical; (k) asymmetric; (p) symmetric. **5.2** $\mathcal{T}_d$, $\mathcal{O}_h$, $\mathcal{I}_h$, $\mathcal{T}$, $\mathcal{O}$, $\mathcal{I}$, $\mathcal{T}_h$.

5.3 $\mathcal{C}_n$ for $n \geqslant 3$, $\mathcal{C}_{nv}$ for $n \geqslant 3$, $\mathcal{C}_{nh}$ for $n \geqslant 3$, $\mathfrak{D}_n$ for $n \geqslant 3$, $\mathfrak{D}_{nd}$ for $n \geqslant 2$, $\mathfrak{D}_{nh}$ for $n \geqslant 3$, $\mathcal{S}_{2n}$ for $n \geqslant 2$. **5.4** (b) 823 GHz, 438 GHz, and 286 GHz; (d) only 1109 GHz; (e) 537 GHz. **5.7** 1.06 Å and 1.16 Å. **5.9** $E/h = A+B+4C,\ A+4B+C,\ 4A+B+C,\ 2(A+B+C)\pm 2(A^2+B^2+C^2-AB-BC-AC)^{1/2}$. **5.10** 41 kcal/mole. **5.11** only (a). **5.12** $\langle P_a^2 \rangle = 0, \hbar^2, \hbar^2$ in order of increasing τ. **5.17** $J=4$. **5.21** (a) Yes. (b) No.

6.1 (a) 1; (b) 4; (c) 3; (d) 4; (e) 15. **6.4** (a), (c), (e), (f). **6.6** (a) a_u; (b) a_g; (c) b_{1u}; (d) b_{2u}. **6.7** (b) Inadequacy of the valence-force approximation. **6.9** (a) B_2; (b) A_1; (c) B_2. **6.10** B_{2g}. **6.11** (c), (d), and (f) are IR active. **6.13** 2600 cm^{-1}. **6.18** ellipse. **6.23** (a) 28, 36; (b) 78, 66. **6.24** (a) 1; (b) 1; (c) 1; (d) n; (e) n; (k) 12; (1) 24. **6.25** (a) 1.54 Å. (If you got 1.09 Å, reread Section 6.10.) **6.26** $R_0 = 1.551$ Å. **6.28** (b) *s-trans*. **6.29** (b) $\nu_2+\nu_3-\nu_2$. **6.30** The last two bands are $2\nu_1+\nu_2$ and $2\nu_1+\nu_3$.

7.6 (b) 0.32, 0.62, 0.91, 1.19. **7.7** Use the equivalence of holes and electrons. **7.9** (a) one; (b) 15 eV. **7.10** (a) two. **7.12** (a) one, 280 eV; (b) one near 500 eV and two peaks of equal intensity near 280 eV.

8.2 (a) 2.44×10^{-23} erg/G. **8.4** (a) 4.575 MHz. **8.5** 23,500 G and 51,700 G. **8.6** 1.000016. **8.12** (b) 60 Hz, 100 Hz. **8.15** (a) A_4; (b) A_3X; (c) A_3; (e) ABC; (j) AA′BB′; (k) ABCD; (1) AA′BB′. **8.18** (a). **8.23** No. **8.24** (a) $E = \pm\frac{1}{2}B_0(g_e\beta_e - g_p\beta_p) + \frac{1}{4}ha,\ -\frac{1}{4}ha \pm \frac{1}{2}[a^2 + B_0^2(g_e\beta_e + g_p\beta_p)^2]^{1/2}$. **8.26** (a) 25; (b) 75; (e) 35; (h) 18. **8.30** only (b). **8.38** (a) 6; (b) 12; (c) 6.

9.1 addition and multiplication. **9.2** (a), (d). **9.4** (a) 1; (b) 40. **9.5** (a) $\mathcal{C}_s$; (b) $\mathcal{C}_s$; (c) $\mathcal{C}_2$; (d) $\mathcal{C}_{2v}$; (e) $\mathcal{C}_{2h}$. **9.8** (a) $\mathcal{C}_1$, $\mathcal{C}_2$, $\mathfrak{D}_2$. **9.15** (a) 3, 1, 1, 1; (c) 4, 2, 2, and eight one-dimensional; or seven two-dimensional and four one-dimensional. **9.19** (b) $13A_1 \oplus 5A_2 \oplus 137E$; (c) $2A_1 \oplus \epsilon \oplus \epsilon^*$. **9.23** (a) $4A_g \oplus 4B_{1u} \oplus B_{2g} \oplus 2B_{2u} \oplus 2B_{3g} \oplus B_{3u} \oplus B_{2g}$; (c) $3A_1 + 2T_2$. **9.25** (a) $3A_g \oplus A_u \oplus 2B_{1u} \oplus B_{2g} \oplus 2B_{2u} \oplus 2B_{3g} \oplus B_{3u}$. Twelve distinct frequencies. (Many workers use a nonstandard choice of axes, and hence give different labels to the normal modes.) (c) $A_1 \oplus E \oplus 2T_2$. Four distinct frequencies. **9.27** (a) E; (d) $A_1' \oplus A_2' \oplus E'$; (e) $A_1' \oplus E'$. (For tables of direct products, see *Herzberg, Electronic Spectra of Polyatomic Molecules*.) **9.30** (a) ${}^1A_1$, ${}^3A_1$; (b) ${}^1A_2$, ${}^3A_2$. **9.31** For $\mathfrak{D}_{2h}$: x^2, y^2, z^2—A_g; xy—B_{1g}; xz—B_{2g}; yz—B_{3g}. **9.32** (a) C_2H_4—5; CH_4—2. (c) C_2H_4—6; CH_4—4. **9.34** All 30 transitions are allowed.

INDEX

AB case in NMR, 339-346
Abel, N., 387
Abelian group, 389, 392, 393, 406-407
Ab initio methods, 70. *See also* Hartree-Fock method
Absolute value, 13
Absorption of radiation, 118
Accidental degeneracy, 253, 415
Acetaldehyde, NMR spectrum of, 352, 353
Acetylene:
  degenerate normal modes, 275, 276
  nuclear statistical weights, 288
Adiabatic ionization potential, 316
Adjoint of a matrix, 85
Adjoint of an operator, 107-108
a levels, 181, 287
Allerhand, A., 364
Alternant hydrocarbons, 73, 380
Ammonia:
  in interstellar space, 231
  inversion in, 228-229, 282
  in planetary atmospheres, 269
  vibrational potential for, 244-245
A_mX_n case in NMR, 349-352
A_n case in NMR, 348-349, 352-354
Angstrom, 23, 132
Angular momentum:
  addition of, 49
  in diatomics, 58-59
  ladder operators for, 30-32
  orbital, 27-29, 49
  rotational, 203, 204-208, 209
  rotational, operators for, 204-208
  spin, 42-45, 58, 59
  total, 50
  vibrational, 271, 273-277, 451
Anharmonicity:
  in diatomics, 156, 159
  in polyatomics, 254-255, 274-275
Anharmonic oscillator, 150-156
Anthracene ions, 380
Anti-Stokes lines, 186
Antisymmetric function, 45
Arnold, J. T., 331
Associativity, 12, 387
Astrophysics, 231, 269
Asymmetric top, 199, 200-202
  correlation with symmetric-top levels, 216
  definition of, 199
  degeneracy for, 215
  energies of, 213-216
  microwave spectrum of, 218-219
  selection rules for, 218-219, 258, 259-260
  wave functions of, 213-215, 440
Atomic spectroscopy, 130
Atomic units, 23
Atoms:
  many-electron, 47-51
  one-electron, 41-42, 124-126
  selection rules for, 127-128
a-type transition, 219
Auroras, 129
Average value, 2
AX case in NMR, 346-347
Axes of inertia, 198-202
Axial vector, 434
Axis of symmetry, 53

Band head, 172, 303-304
Band origin, 168
Band spectra, 297
Band structure of electronic transitions, 303, 306-307
Barrier to internal rotation, 226-228, 364-365

Barrier to inversion, 228–229
Bartholomé, E., 268
Basis functions, 64, 65, 66, 96
  in group theory, 409–412
Basis vectors, 97
Benzene:
  character table of, 406
  Fermi resonance in, 279
  Hartree–Fock calculation for, 317
  HMO treatment of, 72–73
  and Jahn–Teller theorem, 314
  NMR spectrum of, 348
  normal modes of, 438–439
  ordering of orbitals in, 317
  photoelectron spectrum of, 317
  pi MOs of, 427
  symmetry number of, 289
  symmetry orbitals for, 426–427
  symmetry species of normal modes of, 438–439
  terms of excited configuration of, 451–452
Benzene negative ion, 314, 367, 374, 375, 378
Beveridge, D. L., 380
Binomial coefficients, 351, 374
Biochemical applications:
  of ESR, 380–381
  of IR spectroscopy, 268
  of NMR, 357, 363–364
  of Raman spectroscopy, 270–271
Birge–Sponer extrapolation, 304–305
Blackbody radiation, 121–122
Bloch, F., 328
Block-diagonal form, 15
Bohr (unit), 23
Bohr magneton, 51, 368
Bohr radius, 42
Bolometer, 260
Boltzmann distribution law, 121, 135, 361
Bond lengths, 222–223
Born, M., 94
Born–Oppenheimer approximation, 56–57, 300–301
Boron trifluoride normal modes, 281, 290
Boson, 46
Bracket notation, 5
Branches, 171–173, 188, 218, 303
Brillouin's theorem, 311
Broadening of spectral lines, 133–135
Broad-line NMR, 359
b-type transition, 219
Butadiene excited states, 312–313

Canonical orbitals, 69, 310
Carbon dioxide:
  Fermi resonance in, 278–279
  IR bands, table of, 264
  normal modes, 242, 248, 262, 265
  nuclear statistical weights for, 287–288
  vibrational constants of, 275
  vibrational levels of, 253
  vibrations of, and group theory, 449–451, 456
Carbon monoxide:
  Cameron system of, 298
  fundamental IR band of, table of, 174
  rotational transitions of, table of, 168
Carbon-13 NMR, 356–357
Cartesian displacement coordinates, 236
Case (a) coupling, 188–189, 212
Case (b) coupling, 190–191, 212
Cayley, A., 78, 387
Center of symmetry, 53
Central-force problem, 38
Centrifugal distortion:
  in diatomics, 158, 166–167
  in polyatomics, 213, 216, 218
Chain rule, 20–21
Characters:
  complex, 416–417, 429–430
  definition of, 403
  of direct product, 443
  of elements in same class, 403
  of symmetric direct product, 445–446
  theorems on, 403–404
Character tables, 402–409
  for some point groups, 458–462
Chemical analysis, 130, 231, 319–320
Chemical lasers, 137–138
Chemically equivalent nuclei, 348
Chemical shifts:
  chart of, 338
  in ESCA, 319–320
  in NMR, 331–337
Chromophores, 310
Classically forbidden region, 27
Class of a group, 393, 406, 407
Cleeton, C. E., 219
Closure, 387, 388

CNDO method, 71, 75, 76
and electronic spectra, 313
and force constants, 245
and ionization potentials, 318
and quadrupole coupling constants, 231
Cofactor, 14
Coherent radiation, 139
Collision broadening, 134
Collision of the first kind, 129
Collision of the second kind, 129
Column matrix, 79
Column vectors, 80, 82
orthogonal, 86
orthonormal, 86
Combination level, 252
Commutator, 8, 13
Commuting operators, 8
Complete set, 4
expansion in terms of, 6-8
Complex conjugate, 13-14
Complex numbers, 13-14
Computer of averaged transients, 330
Computers, 76
and asymmetric-top secular equation, 216
and NMR spectra, 355
and vibrational secular equation, 244
Configuration, 48, 58, 66
Configuration interaction, 69-70, 311, 312-313
Conformational isomers, 229
Conjugated molecules, 71-75
Conjugate group elements, 392-393
Continuum states, 6
Conversion factors for energy, table of, 468
Coordinate transformation, 22, 87, 394-398
Core Hamiltonian, 64
Correlation diagram for asymmetric top, 216, 217
Correlation energy, 69
Costain, C. C., 223
Coulomb operator, 64
Coupling constants, ESR, 372, 374, 376-380
calculation of, 380
Coupling constants, NMR, 339, 345-346, 352-354
absolute signs of, 360
calculation of, 360
F^{19}-F^{19}, 357
in liquids and gases, 359-360
in olefins, 346
proton-C^{13}, 356
proton-F^{19}, 345, 357
proton-proton, 345-346
in solids, 357-359
Cramer's rule, 16
Cross product, 12
c-type transition, 219
CW laser, 137
Cyclobutadiene, 269

Davis, R. E., 320
Degeneracy, 6
accidental, 253, 415
for asymmetric top, 215
of atomic levels, 50
and group theory, 412-415, 416-417
molecular electronic, 58-62, 313-314, 417-418
orbital, 61, 62
for particle in a cubic box, 25
for spherical top, 210
for symmetric top, 211
for two-particle rigid rotor, 41
of vibrational levels, 252-253, 271-277, 451
Degenerate perturbation theory, 36-38
Degenerate vibrational modes, 241, 246-247, 271-276, 430
Degrees of freedom, 235
Del, 12
Delta chemical-shift scale, 335-336, 356
Delta function, 18, 371
Density matrix, 108
Determinants, 14-15, 78, 84
Dewar, M. J. S., 75
Diagonal form, 15
Diagonalization of matrices, 92, 101, 102
Diagonal matrix, 83
Diamagnetic contribution to shielding constant, 332
Diatomic molecules:
dissociation of, 156-157
electronic spectra of, 298-306
electronic structure of, 57-59
energy of, 148, 155
normal mode of, 242-244
nuclear motion in, 142-159
and nuclear spin, 178-185

parity of wave functions, 175–178
rotational energies of, 188–191
rotational spectra of, 165–167
table of constants for, 160
vibration–rotation spectra of, 168–174
Dictionary order, 441
Difference operator, 170
Differential equations, 18–19
Dimension of representation, 399, 403, 404
Dipole moment, electric, 62–63, 69
of asymmetric top, 218–219
of HD, 190
of hydrocarbons, 225
induced, 187
measurement of, 225
of methane, 225
of symmetric top, 217
and transitions, 164
Dipole-moment function, 164–165
Dipole-moment operator, 62
Dirac delta function, 18, 371
Direction cosines, 21–22, 257, 394–397
Direct product, 440–458
characters of, 443
definition of, 441
of matrices, 441–443
symmetric, 445–447, 449
of three representations, 444
and totally symmetric representation, 444–445
Direct sum of representations, 402
Dissociation energy, 304–306
Dobosh, P. A., 380
Doppler broadening, 134, 135
Dot product, 11
Double resonance, 354, 356
DPPH, 369
Dummy variable, 81
Dunham coefficients, 158, 159
Dye lasers, 138

Eigenfunctions, 2, 98–102
Eigenvalues, 2, 5, 98–102
continuous, 7
of a matrix, 88–92, 99–102
Eigenvectors, 88–92, 99–102
Einstein coefficients, 121–122
Electric dipole moment, *see* Dipole moment, electric
Electric-dipole transition, 123
Electric-field-gradient tensor, 230
Electric potential, 230
Electric quadrupole moment, 229–231, 365–366
Electric-quadrupole transition, 123, 127
Electromagnetic radiation, 114–117. *See also* Radiation, electromagnetic
Electromagnetic spectrum, 115
Electronic energy, 57, 64, 148
Electronic spectra, 130, 296–314
of diatomics, 298–306
and molecular structure, 311
of polyatomics, 71–72, 73, 75, 306–314
selection rules for, 297–301, 306–307
Electronic structure of molecules, 56–76
Electron spectroscopy for chemical analysis (ESCA), 319–320
Electron spin resonance (ESR), 130, 366–381
biochemical applications of, 380–381
coupling constants in, 376–380
electron–C^{13} interaction in, 375
electron–N^{14} interaction in, 375–376
electron spin–nuclear spin interactions in, 369–380
energy levels for, 368, 373
experimental procedures in, 368–369
hyperfine structure in, 369–380
of organic radicals in solution, 367–381
of paramagnetic ions, 367
of radicals in solids, 367
and rate processes, 380
selection rules for, 373
of triplet-state molecules, 367
Electron volt, 131
Ellipsoid of inertia, 198–202, 439–440
Emission of radiation:
spontaneous, 121–122
stimulated, 118, 120, 122, 135–139
Energy conversion factors, 468
Energy-localized orbitals, 69, 103–104
Equilibrium frequencies, 147, 262
Equivalent representations, 400
Ethane:
deuterated, dipole moment of, 225
rotational barrier in, 226–228
Ethanol NMR spectrum, 331, 333–334, 337, 352, 353, 365
Ethylene:
normal modes of, 291

nuclear statistical weights of, 288
structure of, 265
Ethyl radical, 375
Eulerian angles, 203–204, 257
Even function, 32, 33
Exchange operator, 64
Excited-state wave functions, 311–313
Extended basis set, 66
Extended Hückel (EH) method, 71, 73–74, 231

FE MO method, 71–72
Fermi contact term, 371–372
Fermion, 46
Fermi resonance, 278–279
Fine-structure constant, 45
Fingerprint region, 266
First-order NMR approximation, 349–352
Flame spectroscopy, 296
Fluorescence, 128
Forbidden transitions, 123
Force constant(s), 25
calculation of, 245
for diatomics, 146
for polyatomics, 237, 244–245
Formaldehyde, ellipsoid of inertia, 198, 200
Formaldoxime, 223–224, 225
Formamide, 225
Formic acid, 264
Fourier transform IR, 331
Fourier transform NMR, 330–331, 356
Franck–Condon factor, 303
Franck–Condon principle, 302–303, 307
Free-electron molecular-orbital (FE MO) method, 71–72
Free particle, 25
Free radicals, 269, 296. *See also* Electron spin resonance
Fundamental band, 169, 262
Fundamental frequencies, 262
Fundamental level, 252

Galois, E., 387
Gauss (unit), 52
Gaussian line shape, 135, 369
Gaussian-type orbitals, 66
Gaussian units, 23
g factor:
electron, 45, 376
nuclear, 324
Golay cell, 260
Group:
Abelian, 389, 392, 393, 406–407
classes of, 393, 406, 407
definition of, 387–388
elements of, 387
and isomorphism, 392
multiplication table of, 389–392
order of, 388
representations of, 399–409
rule of combination for, 387
structure of, 392
subgroup of, 392
of symmetry operations, 53–56, 388–389
Group frequencies, 266–268, 270
Group theory, 387–462
and degeneracy, 412–415, 416–417
and electronic structure, 417–427
and evaluation of integrals, 453–455
and molecular rotation, 439–440
and molecular vibration, 427–439, 447–451, 455–457
and normal modes, 430–439
and quantum mechanics, 412–417
and selection rules, 455–458
Gwinn, W. D., 244

Half-width, 133
Hamiltonian operator, 3, 4–5, 47, 56–57
Harmonic oscillator, 25–27, 124, 251
matrix elements for, 152–155
Hartree (unit), 23
Hartree–Fock method, 52, 63–70, 102
for benzene, 317
for excited states, 311–312
and force constants, 245
and inversion barriers, 229
and ionization potentials, 318
and oscillator strengths, 313
and quadrupole coupling constants, 231
restricted, 312
and rotational barriers, 229
unrestricted, 311–312
Hartree–Fock operator, 63–64
Hartree–Fock–Roothaan method, 64–68
Heisenberg, W., 93–94
Helium, 47, 128
Helium–neon laser, 137
Helmholz, L., 73
Hemoglobin, 131

Hermite polynomials, 25–26
Hermitian conjugate of a matrix, 85
Hermitian matrix, 84–85, 89–90, 91–92, 96
Hermitian operators, 4, 5–8
Herschel, W., 260
Hertz (unit), 132
High-resolution NMR, 331
HMO method, 71, 72–73, 313, 379
Hoffmann, R., 73
Homogeneous linear equations, 16
Hot bands, 170, 262
Hückel molecular-orbital (HMO) method, 71, 72–73, 313, 379
Hughes, R. H., 220
Hund, F., 188
Hund's coupling cases, 188–191
Hydrocarbons:
  conjugated, 71–75
  dipole moments of, 225
  radicals of, 367, 376–380
Hydrogen atom, 41–42, 124–126, 374
Hydrogen bonding, 264, 334, 366
Hydrogen chloride IR bands, table of, 171
Hydrogen deuteride, 190
Hydrogen molecule, 148, 150, 159, 184–185
  chemical shift in, 335, 337
  molecular orbitals of, 68
  ortho and para forms of, 185
  rotation in liquid and solid phases, 263
  symmetry orbitals for, 420–423
Hyperfine coupling constant, 372
Hyperfine structure, 52, 129, 374
  in ESR, 369–380

Identity element, 387–388
Identity operation, 54, 395
Improper axis of symmetry, 53
Improper rotation, 396
Index of refraction, 132
INDO method, 71, 75–76
  and ESR coupling constants, 380
  and force constants, 245
  and ionization potentials, 318
  and NMR coupling constants, 360
Induced dipole moment, 187
Inertial defect, 224–225
Inertia tensor, 201
Infrared spectroscopy, 130, 260–269
  and astrophysics, 269
  biochemical applications of, 268
  experimental procedures in, 260–261
  Fourier transform method in, 331
  and group frequencies, 266–268, 270
  and lasers, 261–262
  and matrix isolation method, 269
  and molecular structure, 265
  rotational fine structure in, 171–174, 259–260, 265
  see also Vibration
Inhomogeneous linear equations, 16
Inner product, 98
Integrals and group theory, 453–455
Intensity:
  of radiation, 115, 119
  of a transition, 122
Intensity distribution in IR band, 173–174
Intersteller space, molecules in, 231
Inverse:
  of a group element, 388
  of a matrix, 84
  of an operator, 108
Inversion of molecules, 228–229, 282
Inversion operation, 393, 395, 405
Ionization potentials, 317–320
Irreducible representations, 402, 403–409, 415–416
Isobutane, dipole moment of, 225
Isomorphic groups, 392
Isotopes, table of properties of, 470
Isotopic substitution:
  in ESR, 379
  in IR spectroscopy, 244
  in microwave spectroscopy, 221–223
  in NMR, 354–355

Jacobi method, 101
Jahn-Teller theorem, 313–314
Javan, A., 137
j-j coupling, 51
Jolly, W. L., 320
Jordan, P., 94

Karplus, M., 360
Ketene, 226
Ket notation, 5
Kinetic energy, 3
Klystron, 219–220, 368
Koopmans, T. C., 318
Koopmans' theorem, 318

Kraitchman, J., 223
Krebiozen, 268
Kronecker delta, 6
Kronecker product, 441

Ladder operators, 30-32, 44, 327
Lambda-type doubling, 59, 277
Laplacian, 3
Laporte rule, 126
Lasers, 135-140
  chemical, 137-138
  CW, 137
  dye, 138
  in electronic spectroscopy, 296
  helium-neon, 137
  in IR spectroscopy, 261-262
  pulse, 137, 139
  in Raman spectroscopy, 270, 271
  ruby, 135-137
  semiconductor, 138
  tunable, 138
LCAO MO, 65
Left form, 282
Levels:
  atomic, 50
  diatomic, 59
Lifetime, 122
Light, *see* Radiation electromagnetic
Light, speed of, 140
Linear independence, 16
Linear molecules:
  nuclear statistical weights for, 284-288
  parity of wave functions of, 280
  rotational energy of, 211-212
Linear momentum, 27
Linear operator, 3
Linear variation method, 34-35
Line shape, 133-134, 135
Line spectra, 297
Line width, 133-134, 139
  in NMR, 362-363
Localized orbitals, 69, 103-104, 309-310
Lorentzian shape, 133-134, 135, 369
Lowering operator, 30
l-type doubling, 277

McConnell, H. M., 379
McConnell equation, 379-380
McIver, J. W., 360, 458
Magnetically equivalent nuclei, 348
Magnetic dipole moment, 29, 45. *See also* Magnetic moment
Magnetic-dipole transition, 123, 127, 221, 326
Magnetic field strength, 51-52
Magnetic flux density, 51-52
Magnetic induction, 51
Magnetic moment:
  electron spin, 45, 368
  nuclear, 323-325
Magnetic resonance, 129-130, 323-381. *See also* Nuclear magnetic resonance; Electron spin resonance
Magnetogyric ratio, 325
Maiman, T. H., 137
Maser, 137
Mass-weighted Cartesian displacement coordinates, 236, 239
Matrix (-ces), 78-104, 187
  addition of, 79
  adjoint of, 85
  associate, 85
  column, 79
  complex conjugate of, 85
  and coordinate transformations, 394-398
  determinant of, 84
  diagonal, 83
  diagonalization of, 92, 101, 102
  direct product of, 441-443
  direct sum of, 402
  eigenvalues of, 88-92, 99-102
  eigenvectors of, 88-92, 99-102
  elements of, 78
  Hermitian, 84-85, 89-90, 91-92, 96
  Hermitian conjugate of, 85
  inverse of, 84
  modal, 106
  multiplication of, 80-82, 152
  nonsingular, 84
  null, 83
  order of, 82
  orthogonal, 85-86, 87
  partitioned, 88
  principal diagonal of, 82
  and quantum mechanics, 93-102
  real, 84
  row, 79
  similar, 101-102
  square, 82
  symmetric, 84

trace of, 82–83
transpose of, 85, 236
unit, 83
unitary, 85–86, 87
Matrix elements, 5, 94–96
Matrix isolation method, 269
Matrix mechanics, 93–102
Matrix representatives:
of functions, 96–97
of operators, 94–96
Maxwell, J. C., 114
Measurement, 9–10, 113–114
Metastable states, 128–129
Methane:
dipole moment of, 225
meta, ortho, and para, 289
in planetary atmospheres, 269
symmetry number of, 289
Methyl radical, 367, 379
Meyer, W., 317
Micrometer, 132
Micron, 132
Microwave spectrometer, 219–221
Microwave spectroscopy, 130, 219–231
compilations of results of, 231
dipole-moment measurements in, 225
experimental procedures in, 219–221
frequency measurements in, 220
and molecular structure, 221–225
and rotational barriers, 226–228
and vibrational frequencies, 225–226
Mid infrared, 261
MINDO method, 71, 76
and force constants, 245
and ionization potentials, 318–319
Minimal basis set, 65
Minor, 14
Modal matrix, 106
Molecular orbitals:
for diatomics, 58
and group theory, 418–427
for polyatomics, 66
see also Hartree–Fock method, Orbitals
Molecule-fixed axes, 175, 178, 196
Momental ellipsoid, 198–202, 439–444
Moment of inertia, 40, 148, 149, 197, 200
Morse function, 161, 162
Mössbauer spectroscopy, 130–131
Mulliken, R. S., 405
Multiconfiguration SCF method, 70
Multiplication table of a group, 389–392
Multiplicity, 50, 58
Mutual exclusion rule, 270, 457

Nanometer, 132
Naphthalene negative ion, 367, 379, 380
Natural line width, 134
Near-degenerate vibrational levels, 278–279
Near infrared, 261
Negative levels, 177, 280
Negative spin density, 378
Neutron, 178, 324
Nitrogen molecule:
calculation of ionization potentials of, 318
photoelectron spectrum of, 316–317
Nongenuine vibrations, 432
Nonsingular matrix, 84
Nonradiative transitions, 129, 183–185
n orbital, 310
Normal coordinates, 239–240, 246–247, 272
Normalization, 6
Normal modes, 240–244
of benzene, 438–439
of boron trifluoride, 281, 290
of carbon dioxide, 242, 248, 262, 265
of ethylene, 291
and group frequencies, 266–268
IR active, 457
Raman active, 457
and symmetry, 246–249, 430–439
of water, 431–437
Normal operator, 108
Nuclear *g* factor, 324
Nuclear magnetic moments, 323–325
Nuclear magnetic resonance, 129–130, 323–366
AB case in, 339–346
A_mX_n case in, 349–352
A_n case in, 348–349, 352–354
AX case in, 346–347
for B^{11} compounds, 357
biochemical applications of, 357, 363–364
broad-line, 359
for C^{13} compounds, 356–357
chemical shifts in, 331–337
classification of spectra in, 346–348
coupling constants for, 339, 345–346, 352–354, 360

delta scale for, 356
deuterium substitution in, 354–355
double resonance in, 354, 356
energy levels for, 325, 327, 333, 342–343
experimental procedures in, 328–331, 354
for F^{19} compounds, 357
first-order approximation in, 349–352
Fourier-transform, 330–331, 356
four-spin cases, 348
frequencies for nuclei, 327–328
high-resolution, 331
intensities in, 337, 343–344, 351, 352
and rate processes, 364–365
and rotational barriers, 364–365
secular equation for, 355
selection rules for, 326–327, 333, 343
of solids, 357–359
spin relaxation in, 361–364
spin–spin coupling in, 337–356, 357–360
three-spin cases, 347–348
of water in tissues, 363–364
Nuclear magneton, 324
Nuclear Overhauser effect, 354
Nuclear quadrupole moment, 229–231, 365–366
Nuclear quadrupole resonance (NQR) spectroscopy, 366
Nuclear Schrödinger equation, 57, 142–149, 195–197
Nuclear spin, 178–185, 229, 283–289, 323–327
Nuclear statistical weights, 182, 284–289
Null matrix, 83
Null operator, 12

Oblate symmetric top, 199, 211
O branch, 188
Odd function, 32, 33
Olefins:
electronic spectra of, 310
NMR coupling constants of, 346
One-electron methods, 71
Operator(s), 1–11, 12–13
adjoint of, 107–108
Coulomb, 64
difference, 170
eigenfunctions of, 2, 98–102
eigenvalues of, 2, 5, 98–102
for electric dipole moment, 62
exchange, 64
Hamiltonian, 3, 4–5, 47, 56–57
Hartree–Fock, 63–64
Hermitian, 4, 5–8
inverse of, 108
linear, 3
and matrices, 94–96
matrix elements of, 94–96
matrix representatives of, 94–96
normal, 108
null, 12
parity, 32–33, 56
projection, 425, 426
self-adjoint, 108
symmetry, 55–56, 59–62, 398
unit, 12
unitary, 108
Optical pumping, 136
Optimized valence configuration method, 70, 245
Orbital angular momentum, 27–29, 49
Orbital degeneracy, 61, 62
Orbital energy, 64
Orbital exponent, 65
Orbitals, 48
canonical, 69, 310
energy-localized, 69, 103–104
and group theory, 418–427
Hartree–Fock, *see* Hartree–Fock method
localized, 69, 103–104, 309–310
n, 310
pi, 68, 310, 317
pi-star, 309, 310
Rydberg, 308–309
SCF, 65
sigma, 68
Slater-type, 65
symmetry, 67, 419–427
valence, 309
virtual, 68, 69
Order of a group, 388
Order of a square matrix, 82
Ortho deuterium, 185
Ortho forms of homonuclear diatomics, 183–185
Orthogonality, 5–6
Orthogonal matrix, 85–86, 87
Orthogonal transformation, 93
Orthogonal vectors, 86
Ortho hydrogen, 185
Ortho methane, 289

Orthonormality, 6
Orthonormal vectors, 86
Ortho water, 288
Oscillator strength, 307–308, 313
OVC CI method, 70, 245
Overtone band, 169
Overtone level, 252
Oxygen molecule:
  localized MOs for, 104
  microwave spectrum of, 221
  rotational levels of, 191

Para forms of homonuclear diatomics, 183–185
Para hydrogen, 185
Parallel band, 259, 265
Paramagnetic contribution to shielding constant, 332
Para water, 288
Pariser–Parr–Pople (PPP) method, 71, 74–75, 313
Parity, 32–33, 126–127, 128
  in atoms, 52
  in diatomics, 59, 175–178
  in polyatomics, 276, 279–282
  and selection rules, 126–127, 128, 178, 299
Parity operator, 32–33, 56
Particle in a box, 23–25, 71, 124
Partitioned matrix, 88
Pascal's triangle, 351
Pauli principle, 45–47, 178–182, 284–287
Pauli spin matrices, 96
P branch, 171–173, 218, 303
"Peanuts," 320
Perpendicular band, 259, 265
Perturbation, spectroscopic, 283
Perturbation theory, 35–38, 102
  degenerate, 36–38
  for nuclear motion, 149–159
  time-dependent, 110–114
Phase, 13
Phenol, 225
Phosphorescence, 128
Phosphorous trichloride, structure of, 222, 223
Photoelectron spectroscopy, 314–320
  UV, 315–318
  x-ray, 319–320
Photosynthesis, 381
Physical constants, table of, 467
Pi-electron approximation, 71
Pigpen, 320
Pi orbitals, 68, 310, 317
Pi-star orbital, 309, 310
Planarity, 224–225
Planck, M., 122
Plane of symmetry, 53
Plane-polarized wave, 114
Plane polar normal coordinates, 272
Point groups, 53–56, 388–389
  character tables of, 458–462
  see also Group theory
Polarizability, 187, 188, 202, 269, 270, 457
Polarizability ellipsoid, 202, 270
Polar vector, 434
Polyenes, conjugated, 71–72
Polymethine ions, 72
Polywater, 320
Pople, J. A., 75, 337, 360, 380
Population inversion, 135
Positive levels, 177, 280
Potential energy, 3
Potential-energy functions:
  for diatomics, 144–145, 160–162
  for polyatomics, 244–245
Poynting vector, 115
Precession, 52
Predissociation, 306
Pressure broadening, 134
Principal axes, 198–202
Principal diagonal, 15
Principal moments of inertia, 198–202
Probability, 9
Probability density, 9
Probability of a transition, 119–122
Products of inertia, 200
Progression, 304
Projection operators, 425, 426
Project Ozma, 374*n*.
Prolate symmetric top, 199, 211
Propane, dipole moment of, 225
Proper axis of symmetry, 53
Proper rotation, 395–396
Proton, 178
Pseudovector, 434
Pulse laser, 137, 139
Purcell, E. M., 328, 360
Purely electronic energy, 57
Pure-rotation spectra, 165

of diatomics, 165–167
of polyatomics, 216–231

Q branch, 173, 218, 265, 303
Quadratic form, 92–93, 200–201, 238
Quadratic integrability, 1
Quadrupole coupling constants, 231, 366
Quadrupole moments, 229–231, 365–366
Quantum Chemistry Program Exchange, 76, 244, 355
Quantum electrodynamics, 116
Quantum field theory, 116
Quantum mechanics, 1–11
matrix formulation of, 93–102
Quenching of radiation, 129

Rabi, I. I., 328
Radiation, electromagnetic, 114–117
absorption of, 118
coherent, 139
density of, 119
electric and magnetic fields of, 114–116
intensity of, 115, 119
polarized, 114
scattering of, 186, 188
spontaneous emission of, 121–122
stimulated emission of, 118, 120, 122, 135–139
Radiationless transitions, 129, 183–185
Raising operator, 30
Raman, C. V., 186
Raman spectroscopy:
of diatomics, 186–188
lasers in, 270, 271
of polyatomics, 269–271
selection rules for, 187–188, 270, 457
Ramsey, N. F., 332, 360
Rank of a tensor, 187
Rate processes:
and ESR, 380
and NMR, 364–365
Rayleigh scattering, 186
Ray's asymmetry parameter, 217
R branch, 171–172, 218, 303
R-centroid, 301
Real matrix, 84
Real quadratic form, 92–93, 200–201, 238
Reduced mass, 39
Reduction of representations, 407–409
Reduction of the state function, 9–10, 114
Reflection, 397
Relaxation, spin, 361–364
Representation(s), 399–409
basis functions for, 409–412
definition of, 399
dimension of, 399, 403, 404
direct product of, 441 (*see also* Direct product)
direct sum of, 402
equivalent, 400
irreducible, 402, 403–409, 415–416
nomenclature of, 405–406
reducible, 401–402
reduction of, 407–409
symmetric direct product of, 445–447, 449
theorems for, 403–404
totally symmetric, 402
Representatives, 94–97
Restricted Hartree-Fock (RHF) method, 312
Right form, 282
Rigid rotor:
classical treatment of, 197–203
quantum treatment of, 203–216
two-particle, 40–41
RKR method, 161
Roothaan, C. C. J., 64
Rotation:
classical treatment of, 197–203
of diatomics, 144, 148, 149, 188–191
and group theory, 439–440
of linear molecules, 211–212
of polyatomics, 195–231
Rotational angular momentum operators, 204–208
Rotational barriers, 226–228, 364–365
Rotational constants:
of diatomics, 148, 157, 158, 165
of polyatomics, 210, 213, 222
Rotational spectra, *see* Microwave spectroscopy; Pure-rotation spectra
Rotational structure of electronic bands, 303–304, 306
Rotational structure of IR bands:
for diatomics, 171–174
for polyatomics, 259–260, 265
Rotational subgroup, 289*n*, 392
Rotation of coordinates, 22, 87, 394–398
Rotations, symmetry species of, 434–436

Rousseau, D. L., 320
Rovibronic wave function, 283
Row matrix, 79
Row vector, 80
Ruby laser, 135–137
Rule of mutual exclusion, 270, 457
Running number, 173
Russell–Saunders coupling, 51
Rydberg, R., 161
Rydberg constant, 126
Rydberg defect, 308
Rydberg orbital, 308–309
Rydberg transition, 308

Saturation broadening, 134, 363
S branch, 188
Scalar, 187
Scalar product, 98
Scattering of radiation, 186, 188
SCF orbitals, 65. *See also* Hartree–Fock method
Schawlow, A. L., 137
Schmidt orthogonalization, 6
Schrödinger equation, 4, 102
  electronic, 56–57
  nuclear, 57, 142–149, 195–197
  time-dependent, 3, 110, 112
Screening constant, 332, 334
Secular determinant, 37–38
Secular equation, 34, 37, 91, 99
  for asymmetric top, 213–216, 218–219
  in ESR, 373, 374
  and group theory, 419
  in NMR, 341–342, 355
  vibrational, 238–239, 243, 244
Selection rules, 123–129
  for atoms, 127–128
  for electronic transitions of molecules, 297–301, 306–307, 458
  for ESR, 373
  and group theory, 455–458
  for harmonic oscillator, 124, 126–127
  for hydrogen atom, 124–126, 127
  for NMR, 326–327, 333, 343
  and parity, 126–127, 128, 178, 299
  for particle in a box, 124, 127
  for positive and negative levels, 178
  for Raman spectra, 187–188, 270, 457
  rotational, for diatomics, 163–164
  rotational, for polyatomics, 216–219, 257–258
  for *s* and *a* levels, 184
  for spin, 125, 127–128, 298
  for transition states, 458
  vibrational, 164–165, 455–457
  vibration–rotation, for diatomics, 162–165
  vibration–rotation, for polyatomics, 255–260
Self-adjoint operator, 108
Semiclassical treatment of radiation absorption, 116
Semiempirical methods, 70–76
Separation of variables, 19–20
Shielding constant, 332, 334
  calculation of, 337
Shielding tensor, 332
Sigma orbitals, 68
Similarity transformation, 87, 92, 101, 412
Similar matrices, 87, 101–102
Simultaneous linear equations, 15–16, 87
Singlet, 58, 310–311
Singly excited function, 311
Singularity, 17
SI units, 23
Skeletal vibrations, 266, 268
Slater determinant, 48
Slater-type orbitals, 65
s levels, 181, 287
Snow, 328*n*
Space-fixed axes, 143, 175, 178, 196, 203
Spectrometer:
  ESR, 368–369
  IR, 260–261
  microwave, 219–221
  NMR, 328–330, 354
Spectroscopic perturbations, 283
Spectroscopy, 110, 129–133. *See also under the various kinds of spectroscopy*
Spherical harmonics, 29, 38
Spherical polar coordinates, 21
Spherical top:
  definition of, 199
  degeneracy for, 210
  degenerate vibrational modes in, 276
  energies of, 209–210
  microwave spectrum of, 217, 225
  polarizability, 202
  wave functions of, 209, 211

Spin, 42–45
electron, 231
nuclear, 178–185, 229, 283–289, 323–327
selection rule for, 125, 127–128, 298
Spin density, 378
Spin labeling, 381
Spin-lattice relaxation, 361–362
Spin magnetic moment, *see* Magnetic moment
Spin matrices, 96
Spin-orbital, 48
Spin-orbit interaction, 50, 59
Spin relaxation, 361–364
Spin-spin coupling, NMR, 337–356, 357–360, 366
Spin-spin coupling constants, *see* Coupling constants, NMR
Spin-spin relaxation, 362–363
Spontaneous emission, 121–122
Standard deviation, 10
Stanton, R. E., 458
Stark modulation, 220
State, 1
State function, 1
States, atomic, 51
State vector, 97
Stationary state, 5
Statistical weights, 284–289
Steele, D., 161
Stimulated emission, 118, 120, 122, 135–139
Stokes lines, 186
Stothers, J. B., 356
Structure of molecules, determination of:
by electronic spectroscopy, 311
by IR spectroscopy, 265
by microwave spectroscopy, 221–225
by NMR, 359
by x-ray diffraction, 130
Subgroup, 392
Submatrix, 88
Sub-Rydberg orbital, 309
Sulfur tetrafluoride, 357, 364
Sum rule, 10–11
Sutherland parameter, 161
Sweat, 320
Symmetric direct product, 445–447, 449
Symmetric function, 45
Symmetric matrix, 84
Symmetric top, 199–200
degeneracy for, 211
energies of, 210–211
microwave spectrum of, 217–218, 221–222
nuclear statistical weights for, 288–289
oblate, 199
polarizability of, 202
prolate, 199
selection rules for, 217, 258, 259, 265
wave functions of, 211
Symmetry:
and momental ellipsoid, 198–200
and normal modes, 246–249
in polyatomics, 59–62
see also Group theory
Symmetry-adapted orbitals, 419–427
Symmetry coordinates, 431
Symmetry element, 53
Symmetry number, 289
Symmetry operation, 53–54, 55
Symmetry operator, 55–56, 59–62, 398
Symmetry orbitals, 67
for benzene, 426–427
generation of, 420–427
for hydrogen molecule, 420–423
for water, 423–426
Symmetry point groups, 53–56, 388–389. *See also* Group theory
Symmetry species, 60–62, 415–416
of electronic wave functions, 451–452
of MOs, 66–68
of normal modes, 247–248, 430–439
of rotations, 434–436
of translations, 433–434
of vibrational wave functions, 253–254, 447–451

Tau chemical-shift scale, 336
Taylor series, 17–18, 237
Teller, E., 268
Tensor, 187
electric-field-gradient, 230
inertia, 201
polarizability, 187, 202
shielding, 332
Terenin, A. N., 315
Terms:
for atoms, 49–50
for diatomic molecules, 58–59
for polyatomic molecules, 59–62

Tesla, 52
Tetramethylsilane (TMS), 336, 356
Tiers, G. V. D., 336
Time averaging, 329–331
Time-dependent perturbation theory, 110–114
Time-dependent Schrödinger equation, 3, 110, 112
Time-independent Schrödinger equation, 4, 102
Torque constant, 245
Totally symmetric representation, 402
Totally symmetric symmetry species, 60–61
Totally symmetric vibrations, 247
Townes, C., 137
Trace, 82–83
Transition, 114
Transition intensity, 122
Transition moment, 120, 297–298, 299–301
Transition probability, 119–122
Transition rate, 120
Transition states, 458
Translations, symmetry species of, 433–434
Transpose of a matrix, 85, 236
Triphenylmethyl radical, 367, 378
Triplet terms, 58, 310–311
Trivial solution, 16
Tunable lasers, 138
Turner, D. W., 315
Two-electron methods, 71
Two-particle problem, 39
Two-photon transition, 139

Uncertainty, 10
Uncertainty broadening, 134
Uncertainty principle, 10
  for energy and time, 10, 133, 134
Unitary matrix, 85–86, 87
Unitary operator, 108
Unit matrix, 83
Unit operator, 12
Units, 23
Unity, roots of, 14
Unrestricted Hartree–Fock (UHF) method, 311–312, 378
Urey–Bradley potential function, 245
U.S. Navy, 219

Vacuum wavelength, 132
Valence-electron approximation, 71
Valence-force approximation, 244–245
Valence orbitals, 309
Valence state, 74
van der Waals forces, 156
Variation method, 33–35
Variation–perturbation method, 36
Varshni function, 162
Vectors, 11–12
Vector space, 97
Vertical ionization potential, 316
Vibration:
  of diatomics, 147–150, 156–158, 168–174
  and group theory, 427–439, 447–451, 455–457
  of polyatomics, 195–197
  of polyatomics, classical treatment of, 235–245
  of polyatomics, quantum treatment of, 249–255
  and selection rules, 162–165, 255–260, 455–457
Vibrational angular momentum, 271, 273–277
Vibrational degrees of freedom, 235
Vibrational modes:
  degenerate, 241, 271–276, 430
  IR active, 262, 457
  IR inactive, 262, 457
Vibrational structure of electronic transitions, 301–304, 306–307
Vibration frequencies, 147, 158–159
  dependence on phase, 263–264
  equilibrium, 147, 262
  fundamental, 262
  group, 266–268, 270
  from microwave spectroscopy, 225–226
  numbering of, 248, 439
  of polyatomics, 241, 244, 251
  of polyatomics, tabulation of, 252*n*
  zero-order, 262
Vibration–rotation coupling constant, 157–158, 213, 216
Vibronic wave function, 283
Vinyl radical, 104, 377–378
Virtual orbitals, 68, 69

Water:
  calculation of, force constants of, 245

ionization potentials of, 317–318
CI calculations on, 317–318
excited-state calculations on, 313, 317–318
Fermi resonance in, 279
in interstellar space, 231
IR bands of, table of, 263
on Mars, 269
MOs of, 66–69
normal modes of, 242, 247–248, 262, 431–437
nuclear statistical weights for, 288
proton shielding constant in, 337
Rydberg orbitals of, 308–309
symmetry orbitals for, 423–426
symmetry species for, 60–62, 66
symmetry of vibrational levels of, 254
in tissues, NMR spectrum of, 363–364
vibrational constants of, 255
Wave functions, 5. *See also* Asymmetric top; Harmonic oscillator, *etc.*
Waveguide, 220
Wavelength, 114, 131, 132
Wave number, 131, 132–133
Weisskopf, V. F., 133
Weisskopf–Wigner line shape, 133–134
Well-behaved functions, 1
Width of spectral lines, 133–134, 139
Wigner, E., 133
Wilson, E. B., 220
Wolfsberg, M., 73

X-ray photoelectron spectroscopy, 319–320
X-ray spectra, 130

Zavoisky, E., 366
Zeeman effect, 51
Zero differential overlap (ZDO) approximation, 75
Zero-order frequencies, 262
Zero-point energy:
of diatomics, 158, 159
of polyatomics, 251
Zero-point vibrations, 222